Caterina Consani
Matilde Marcolli (Eds.)

Noncommutative Geometry and Number Theory

Aspects of Mathematics

Edited by Klas Diederich

*A Publication of the Max-Planck-Institute for Mathematics, Bonn

Caterina Consani
Matilde Marcolli
(Eds.)

Noncommutative Geometry and Number Theory

**Where Arithmetic meets
Geometry and Physics**

A Publication of the Max-Planck-Institute
for Mathematics, Bonn

vieweg

Bibliografische information published by Die Deutsche Bibliothek
Die Deutsche Bibliothek lists this publication in the Deutschen Nationalbibliografie;
detailed bibliographic data is available in the Internet at <http://dnb.ddb.de>.

Prof. Dr. Caterina Consani
Department of Mathematics
The Johns Hopkins University
3400 North Charles Street
Baltimore, MD 21218, USA

kc@jhu.edu

Prof. Dr. Matilde Marcolli
Max-Planck-Institut für Mathematik
Vivatsgasse 7
D-53111 Bonn

marcolli@mpim-bonn.mpg.de

Prof. Dr. Klas Diederich (Series Editor)
Fachbereich Mathematik
Bergische Universität Wuppertal
Gaußstraße 20
D-42119 Wuppertal

klas@math.uni-wuppertal.de

Mathematics Subject Classification
Primary: 58B34, 11X
Secondary: 11F23, 11S30, 11F37, 11F41, 11F70, 11F80, 11J71, 11B57, 11K36, 11F32, 11F75, 11G18, 14A22, 14F42, 14G05, 14G40, 14K10, 14G35, 18E30, 19D55, 20G05, 22E50, 32G05, 46L55, 58B34, 58E20, 70S15, 81T40.

First edition, March 2006

Softcover re-print of the Hardcover 1st edition 2006

Editorial Office: Ulrike Schmickler-Hirzebruch / Petra Rußkamp

Vieweg is a company in the specialist publishing group Springer Science+Business Media.
www.vieweg.de

Cover design: Ulrike Weigel, www.CorporateDesignGroup.de
Printing and binding: MercedesDruck, Berlin
Printed on acid-free paper

ISBN 3-8348-2673-1   ISSN 0179-2156

Preface

In recent years, number theory and arithmetic geometry have been enriched by new techniques from noncommutative geometry, operator algebras, dynamical systems, and K-Theory. Research across these fields has now reached an important turning point, as shows the increasing interest with which the mathematical community approaches these topics.

This volume collects and presents up-to-date research topics in arithmetic and noncommutative geometry and ideas from physics that point to possible new connections between the fields of number theory, algebraic geometry and noncommutative geometry.

The contributions to this volume partly reflect the two workshops "Noncommutative Geometry and Number Theory" that took place at the Max–Planck–Institut für Mathematik in Bonn, in August 2003 and June 2004. The two workshops were the first activity entirely dedicated to the interplay between these two fields of mathematics. An important part of the activities, which is also reflected in this volume, came from the hindsight of physics which often provides new perspectives on number theoretic problems that make it possible to employ the tools of noncommutative geometry, well designed to describe the quantum world.

Some contributions to the volume (Aubert–Baum–Plymen, Meyer, Nistor) center on the theory of reductive p-adic groups and their Hecke algebras, a promising direction where noncommutative geometry provides valuable tools for the study of objects of number theoretic and arithmetic interest. A generalization of the classical Burnside theorem using noncommutative geometry is discussed in the paper by Fel'shtyn and Troitsky. The contribution of Laca and van Frankenhuijsen represents another direction in which substantial progress was recently made in applying tools of noncommutative geometry to number theory: the construction of quantum statistical mechanical systems associated to number fields and the relation of their KMS equilibrium states to abelian class field theory. The theory of Shimura varieties is considered from the number theoretic side in the contribution of Blasius, on the Weight-Monodromy conjecture for Shimura varieties associated to quaternion algebras over totally real fields and the Ramanujan conjecture for Hilbert modular forms. An approach via noncommutative geometry to the boundaries of Shimura varieties is discussed in Paugam's paper. Modular forms can be studied using techniques from noncommutative geometry, via the Hopf algebra symmetries of the modular Hecke algebras, as discussed in the paper of Connes and Moscovici. The general underlying theory of Hopf cyclic cohomology in noncommutative geometry is presented in the paper of Khalkhali and Rangipour. Further results in arithmetic geometry include a reinterpretation of the archimedean cohomology of the fibers at archimedean primes of arithmetic varieties (Consani–Marcolli), and Kim's paper on a noncommutative method of Chabauty. Arithmetic aspects of noncommutative tori are also discussed (Boca–Zaharescu and Polishchuk). The input from physics and its interactions with number theory and noncommutative geometry is represented by the contributions of Kreimer, Landi, Marcolli–Mathai, and Ponge.

The workshops were generously funded by the Humboldt Foundation and the ZIP Program of the German Federal Government, through the Sofja Kovalevskaya Prize awarded to Marcolli in 2001.

We are very grateful to Yuri Manin and to Alain Connes, who helped us with the organization of the workshops, and whose ideas and results contributed crucially to bring number theory in touch with noncommutative geometry.

Caterina Consani and Matilde Marcolli

Contents

CONTENTS

The Hecke algebra of a reductive p-adic group: a geometric conjecture

Anne-Marie Aubert, Paul Baum, and Roger Plymen

ABSTRACT. Let $\mathcal{H}(G)$ be the Hecke algebra of a reductive p-adic group G. We formulate a conjecture for the ideals in the Bernstein decomposition of $\mathcal{H}(G)$. The conjecture says that each ideal is *geometrically equivalent* to an algebraic variety. Our conjecture is closely related to Lusztig's conjecture on the asymptotic Hecke algebra. We prove our conjecture for SL(2) and GL(n). We also prove part (1) of the conjecture for the Iwahori ideals of the groups PGL(n) and SO(5). The conjecture, if true, leads to a parametrization of the smooth dual of G by the points in a complex affine locally algebraic variety.

1. Introduction

The reciprocity laws in number theory have a long development, starting from conjectures of Euler, and including contributions of Legendre, Gauss, Dirichlet, Jacobi, Eisenstein, Takagi and Artin. For the details of this development, see [**Le**]. The local reciprocity law for a local field F, which concerns the finite Galois extensions E/F such that $\mathrm{Gal}(E/F)$ is *commutative*, is stated and proved in [**N**, p. 320]. This local reciprocity law was dramatically generalized by Langlands. The local Langlands correspondence for GL(n) is a noncommutative generalization of the reciprocity law of local class field theory. The local Langlands conjectures, and the global Langlands conjectures, all involve, inter alia, the representations of reductive p-adic groups, see [**BG**].

To each reductive p-adic group G there is associated the Hecke algebra $\mathcal{H}(G)$, which we now define. Let K be a compact open subgroup of G, and define $\mathcal{H}(G//K)$ as the convolution algebra of all complex-valued, compactly-supported functions on G such that $f(k_1 x k_2) = f(x)$ for all k_1, k_2 in K. The Hecke algebra $\mathcal{H}(G)$ is then defined as

$$\mathcal{H}(G) := \bigcup_K \mathcal{H}(G//K).$$

The smooth representations of G on a complex vector space V correspond bijectively to the nondegenerate representations of $\mathcal{H}(G)$ on V, see [**B**, p.2].

In this article, we consider $\mathcal{H}(G)$ from the point of view of noncommutative (algebraic) geometry.

We recall that the coordinate rings of affine algebraic varieties are precisely the commutative, unital, finitely generated, reduced $\mathbb{C}$-algebras, see [**EH**, II.1.1].

The Hecke algebra $\mathcal{H}(G)$ is a non-commutative, non-unital, non-finitely-generated, non-reduced $\mathbb{C}$-algebra, and so cannot be the coordinate ring of an affine algebraic variety.

The Hecke algebra $\mathcal{H}(G)$ is non-unital, but it admits *local units*, see [**B**, p.2].

The algebra $\mathcal{H}(G)$ admits a canonical decomposition into ideals, the Bernstein decomposition [**B**]:

$$\mathcal{H}(G) = \bigoplus_{\mathfrak{s} \in \mathfrak{B}(G)} \mathcal{H}^{\mathfrak{s}}(G).$$

Each ideal $\mathcal{H}^{\mathfrak{s}}(G)$ is a non-commutative, non-unital, non-finitely-generated, non-reduced $\mathbb{C}$-algebra, and so cannot be the coordinate ring of an affine algebraic variety.

In section 2, we define the *extended centre* $\widetilde{\mathfrak{Z}}(G)$ of G. At a crucial point in the construction of the centre $\mathfrak{Z}(G)$ of the category of smooth representations of G, certain quotients are made: we replace each ordinary quotient by the *extended quotient* to create the *extended centre*.

In section 3 we define *morita contexts*, following [**CDN**].

In section 4 we prove that each ideal $\mathcal{H}^{\mathfrak{s}}(G)$ is Morita equivalent to a unital k-algebra of finite type, where k is the coordinate ring of a complex affine algebraic variety. We think of the ideal $\mathcal{H}^{\mathfrak{s}}(G)$ as a noncommutative algebraic variety, and $\mathcal{H}(G)$ as a noncommutative scheme.

In section 5 we formulate our conjecture. We conjecture that each ideal $\mathcal{H}^{\mathfrak{s}}(G)$ is *geometrically equivalent* (in a sense which we make precise) to the coordinate ring of a complex affine algebraic variety $X^{\mathfrak{s}}$:

$$\mathcal{H}^{\mathfrak{s}}(G) \asymp \mathcal{O}(X^{\mathfrak{s}}) = \widetilde{\mathfrak{Z}^{\mathfrak{s}}}(G).$$

The ring $\widetilde{\mathfrak{Z}^{\mathfrak{s}}}(G)$ is the $\mathfrak{s}$-factor in the *extended centre* of G. The ideals $\mathcal{H}^{\mathfrak{s}}(G)$ therefore qualify as noncommutative algebraic varieties.

We have stripped away the homology and cohomology which play such a dominant role in [**BN**], [**BP**], leaving behind three crucial moves: *Morita equivalence, morphisms which are spectrum-preserving with respect to filtrations, and deformation of central character.* These three moves generate the notion of geometric equivalence.

In section 6 we prove the conjecture for all *generic* points $\mathfrak{s} \in \mathfrak{B}(G)$.

In section 7 we prove our conjecture for SL(2).

In section 8 we discuss some general features used in proving the conjecture for certain examples.

In section 9 we review the asymptotic Hecke algebra of Lusztig.

The asymptotic Hecke algebra J plays a vital role in our conjecture, as we now proceed to explain. One of the Bernstein ideals in $\mathcal{H}(G)$ corresponds to the point $\mathfrak{i} \in \mathfrak{B}(G)$, where $\mathfrak{i}$ is the quotient variety $\Psi(T)/W_f$. Here, T is a maximal torus in G, $\Psi(T)$ is the complex torus of unramified quasicharacters of T, and W_f is the finite Weyl group of G. Let I denote an Iwahori subgroup of G, and define e as follows:

$$e(x) = \begin{cases} \mathrm{vol}(I)^{-1} & \text{if } x \in I, \\ 0 & \text{otherwise.} \end{cases}$$

Then the *Iwahori ideal* is the two-sided ideal generated by e:

$$\mathcal{H}^{\mathfrak{i}}(G) := \mathcal{H}(G)e\mathcal{H}(G).$$

There is a Morita equivalence $\mathcal{H}(G)e\mathcal{H}(G) \sim e\mathcal{H}(G)e$ and in fact we have

$$\mathcal{H}^{\mathfrak{i}}(G) := \mathcal{H}(G)e\mathcal{H}(G) \asymp e\mathcal{H}(G)e \cong \mathcal{H}(G//I) \cong \mathcal{H}(W, q_F) \rtimes J$$

where $\mathcal{H}(W, q_F)$ is an (extended) affine Hecke algebra based on the (extended) Coxeter group W, J is the asymptotic Hecke algebra, and $\asymp$ denotes geometric equivalence. Now J admits a decomposition into finitely many two-sided ideals

$$J = \bigoplus_{\mathbf{c}} J_{\mathbf{c}}$$

labelled by the two-sided cells $\mathbf{c}$ in W. We therefore have

$$\mathcal{H}^{\mathfrak{i}}(G) \asymp \oplus J_{\mathbf{c}}.$$

This canonical decomposition of J is well-adapted to our conjecture.

In section 10 we prove that

$$J_{\mathbf{c}_0} \asymp \mathfrak{Z}^{\mathfrak{i}}(G)$$

where $\mathbf{c}_0$ is the lowest two-sided cell, for any connected F-split adjoint simple p-adic group G. We note that $\mathfrak{Z}^{\mathfrak{i}}(G)$ is the ring of regular functions on the *ordinary quotient* $\Psi(T)/W_f$.

In section 11 we prove the conjecture for $\mathrm{GL}(n)$. We establish that for each point $\mathfrak{s} \in \mathfrak{B}(G)$, we have

$$\mathcal{H}^{\mathfrak{s}}(\mathrm{GL}(n)) \asymp \widetilde{\mathfrak{Z}^{\mathfrak{s}}}(\mathrm{GL}(n)).$$

In section 12 we prove part (1) of the conjecture for the Iwahori ideal in $\mathcal{H}(\mathrm{PGL}(n))$.

In section 13 we prove part (1) of the conjecture for the Iwahori ideal in $\mathcal{H}(\mathrm{SO}(5))$. Our proofs depend crucially on Xi's affirmation, in certain special cases, of Lusztig's conjecture on the asymptotic Hecke algebra J (see [**L4**, §10]).

In section 14 we discuss some consequences of the conjecture.

We thank the referee for his many detailed and constructive comments, which forced us thoroughly to revise the article. We also thank Nigel Higson, Ralf Meyer and Victor Nistor for valuable discussions, and Gene Abrams for an exchange of emails concerning rings with local units.

2. The extended centre

Let G be the set of rational points of a reductive group defined over a local nonarchimedean field F, and let $\mathfrak{R}(G)$ denote the category of smooth G-modules. Let (L, σ) denote a *cuspidal pair*: L is a Levi subgroup of G and σ is an irreducible supercuspidal representation of L. The group $\Psi(L)$ of unramified quasicharacters of L has the structure of a complex torus.

We write $[L, \sigma]_G$ for the equivalence class of (L, σ) and $\mathfrak{B}(G)$ for the set of equivalence classes, where the equivalence relation is defined by $(L, \sigma) \sim (L', \sigma')$ if $gLg^{-1} = L'$ and ${}^g\sigma \simeq \nu'\sigma'$, for some $g \in G$ and some $\nu' \in \Psi(L')$. For $\mathfrak{s} = [L, \sigma]_G$, let $\mathfrak{R}^{\mathfrak{s}}(G)$ denote the full subcategory of $\mathfrak{R}(G)$ whose objects are the representations Π such that each irreducible subquotient of Π is a subquotient of a parabolically induced representation $\iota_P^G(\nu\sigma)$ where P is a parabolic subgroup of G with Levi subgroup L and $\nu \in \Psi(L)$. The action (by conjugation) of $N_G(L)$ on L induces an action of $W(L) = N_G(L)/L$ on $\mathfrak{B}(L)$. Let $W_{\mathfrak{t}}$ denote the stabilizer of $\mathfrak{t} = [L, \sigma]_L$ in $W(L)$. Thus $W_{\mathfrak{t}} = N_{\mathfrak{t}}/L$ where

$$N_{\mathfrak{t}} = \{n \in N_G(L) : {}^n\sigma \simeq \nu\sigma, \text{ for some } \nu \in \Psi(L)\}.$$

It acts (via conjugation) on $\mathrm{Irr}^{\mathfrak{t}} L$, the set of isomorphism classes of irreducible objects in $\mathfrak{R}^{\mathfrak{t}}(L)$.

Let $\Omega(G)$ denote the set of G-conjugacy classes of cuspidal pairs (L, σ). The groups $\Psi(L)$ create orbits in $\Omega(G)$. Each orbit is of the form $D_\sigma/W_{\mathfrak{t}}$ where $D_\sigma = \mathrm{Irr}^{\mathfrak{t}} L$ is a complex torus.

We have

$$\Omega(G) = \bigsqcup D_\sigma/W_{\mathfrak{t}}.$$

Let $\mathfrak{Z}(G)$ denote the *centre* of the category $\mathfrak{R}(G)$. The centre of an abelian category (with a small skeleton) is the endomorphism ring of the identity functor. An element z of the centre assigns to each object A in $\mathfrak{R}(G)$ a morphism $z(A)$ such that

$$f \cdot z(A) = z(B) \cdot f$$

for each morphism $f \in \mathrm{Hom}(A, B)$.

According to Bernstein's theorem [**B**] we have the explicit decomposition of $\mathfrak{R}(G)$:

$$\mathfrak{R}(G) = \prod_{\mathfrak{s} \in \mathfrak{B}(G)} \mathfrak{R}^{\mathfrak{s}}(G).$$

We also have

$$\mathfrak{Z}(G) \cong \prod \mathfrak{Z}^{\mathfrak{s}}$$

where

$$\mathfrak{Z}^{\mathfrak{s}}(G) = \mathcal{O}(D_\sigma/W_{\mathfrak{t}})$$

is the centre of the category $\mathfrak{R}^{\mathfrak{s}}(G)$.

Let the finite group Γ act on the space X. We define, as in [**BC**],

$$\widetilde{X} := \{(\gamma, x) : \gamma x = x\} \subset \Gamma \times X$$

and define the Γ-action on $\widetilde{X}$ as follows:

$$\gamma_1(\gamma, x) := (\gamma_1 \gamma \gamma_1^{-1}, \gamma_1 x).$$

The *extended quotient* of X by Γ is defined to be the ordinary quotient $\widetilde{X}/\Gamma$. If Γ acts freely, then we have

$$\widetilde{X} = \{(1, x) : x \in X\} \cong X$$

and, in this case, $\widetilde{X}/\Gamma = X/\Gamma$.

We will write

$$\widetilde{\mathfrak{Z}}^{\mathfrak{s}}(G) := \mathcal{O}(\widetilde{D}_\sigma/W_{\mathfrak{t}}).$$

We now form the *extended centre*

$$\widetilde{\mathfrak{Z}}(G) := \prod_{\mathfrak{s} \in \mathfrak{B}(G)} \mathcal{O}(\widetilde{D}_\sigma/W_{\mathfrak{t}}).$$

We will write

$$k := \mathcal{O}(D_\sigma/W_{\mathfrak{t}}) = \mathfrak{Z}^{\mathfrak{s}}(G).$$

3. Morita contexts and the central character

A fundamental theorem of Morita says that the categories of modules over two rings with identity R and S are equivalent if and only if there exists a strict Morita context connecting R and S.

A Morita context connecting R and S is a datum

$$(R, S, {}_R M_{S}, {}_S N_R, \phi, \psi)$$

where M is an R-S-bimodule, N is an S-R-bimodule, $\phi : M \otimes_S N \to R$ is a morphism of R-R-bimodules and $\psi : N \otimes_R M \to S$ is a morphism of S-S-bimodules such that

$$\phi(m \otimes n)m' = m\psi(n \otimes m')$$

$$\psi(n \otimes m)n' = n\psi(m \otimes n')$$

for any $m, m' \in M, n, n' \in N$. A Morita context is *strict* if both maps ϕ, ψ are isomorphisms, see [**CDN**].

Let R be a ring with local units, i.e. for any finite subset X of R, there exists an idempotent element $e \in R$ such that $ex = xe = x$ for any $x \in X$. An R-module M is unital if $RM = M$. Unital modules are called non-degenerate in [**B**].

Let R and S be two rings with local units, and $R - MOD$ and $S - MOD$ be the associated categories of unital modules. A Morita context for R and S is a datum $(R, S, {}_R M_{S}, {}_S N_R, \phi, \psi)$ with the condition that M and N are unital modules to the left and to the right. If the Morita context is strict, then we obtain by [**CDN, Theorem 4.3**] an equivalence of categories $R - MOD$ and $S - MOD$.

If R and S are rings with local units which are connected by a strict Morita context, then we will say that R and S are *Morita equivalent* and write

$$R \sim_{morita} S.$$

We will use repeatedly the following elementary lemmas.

LEMMA 1. *Let R be a ring with local units. Let $M_n(R)$ denote $n \times n$ matrices over R. Then we have*

$$M_n(R) \sim_{morita} R.$$

PROOF. Let $M_{i \times j}(R)$ denote $i \times j$ matrices over R. Then we have a strict Morita context

$$(R, M_{n \times n}(R), M_{1 \times n}(R), M_{n \times 1}(R), \phi, \psi)$$

where ϕ, ψ denote matrix multiplication. □

LEMMA 2. *Let R be a ring with local units. Let e be an idempotent in R. Then we have*

$$ReR \sim_{morita} eRe.$$

PROOF. Given $r \in R$, there is an idempotent $f \in R$ such that $r = fr \in R^2$ so that $R \subset R^2 \subset R$. A ring R with local units is *idempotent*: $R = R^2$. This creates a strict Morita context

$$(eRe, ReR, eR, Re, \phi, \psi)$$

where ϕ, ψ are the obvious multiplication maps in R. We now check identities such as $(eR)(Re) = eRe, (Re)(eR) = ReR$. □

Let X be a complex affine algebraic variety, let $\mathcal{O}(X)$ denote the algebra of regular functions on X. Let
$$k := \mathcal{O}(X).$$
A k-algebra A is a $\mathbb{C}$-algebra which is also a k-module such that
$$1_k \cdot a = a, \quad \omega(a_1 a_2) = (\omega a_1)a_2 = a_1(\omega a_2), \quad \omega(\lambda a) = \lambda(\omega a) = (\lambda \omega)a$$
for all $a, a_1, a_2 \in A, \omega \in k, \lambda \in \mathbb{C}$.

If A has a unit then a k-algebra is a $\mathbb{C}$-algebra with a given unital homomorphism of $\mathbb{C}$-algebras from k to the centre of A. If A does not have a unit, then a k-algebra is a $\mathbb{C}$-algebra with a given unital homomorphism of $\mathbb{C}$-algebras from k to the centre of the multiplier algebra $\mathcal{M}(A)$ of A, see [**B**].

A k-algebra A is of *finite type* if, as a k-module, A is finitely generated.

If A, B are k-algebras, then a morphism of k-algebras is a morphism of $\mathbb{C}$-algebras which is also a morphism of k-modules.

If A is a k-algebra and Y is a unital A-module then the action of A on Y extends uniquely to an action of $\mathcal{M}(A)$ on Y such that Y is a unital $\mathcal{M}(A)$-module [**B**]. In this way the category $A - MOD$ of unital A-modules is equivalent to the category $\mathcal{M}(A) - MOD$ of unital $\mathcal{M}(A)$-modules. Since the k-algebra structure for A can be viewed as a unital homomorphism of $\mathbb{C}$-algebras from k to the centre of $\mathcal{M}(A)$, it now follows that given any unital A-module Y, Y is canonically a k-module with the following compatibility between the A-action and the k-action:
$$1_k \cdot y = y, \quad \omega(ay) = (\omega a)y = a(\omega y), \quad \omega(\lambda y) = \lambda(\omega y) = (\lambda \omega)y$$
for all $a \in A, y \in Y, \omega \in k, \lambda \in \mathbb{C}$.

If A, B are k-algebras then a strict Morita context (in the sense of k-algebras) connecting A and B is a 6-tuple
$$(A, B, M, N, \phi, \psi)$$
which is a strict Morita context connecting the rings A, B. We require, in addition, that
$$\omega y = y\omega$$
for all $y \in M, \omega \in k$. Similarly for N. When this is satisfied, we will say that the k-algebras A and B are *Morita equivalent* and write
$$A \sim_{morita} B.$$

By *central character* or *infinitesimal character* we mean the following. Let M be a simple A-module. Schur's lemma implies that each $\theta \in k$ acts on M via a complex number $\lambda(\theta)$. Then $\theta \mapsto \lambda(\theta)$ is a morphism of $\mathbb{C}$-algebras $k \to \mathbb{C}$ and is therefore given by evaluation at a $\mathbb{C}$-point of X. The map $\mathrm{Irr}(A) \to X$ so obtained is the central character (or infinitesimal character). The notation for the infinitesimal character is as follows:
$$inf.\,ch. : \mathrm{Irr}(A) \longrightarrow X.$$

If A and B are k-algebras connected by a strict Morita context then we have a commutative diagram in which the top horizontal arrow is bijective:

$$
\begin{array}{ccc}
\mathrm{Irr}(A) & \longrightarrow & \mathrm{Irr}(B) \\
{\scriptstyle inf.ch.}\downarrow & & \downarrow{\scriptstyle inf.ch.} \\
X & \xrightarrow{\;id\;} & X
\end{array}
$$

The algebras that occur in this paper have the property that

$$\mathrm{Prim}(A) = \mathrm{Irr}(A).$$

In particular, each Bernstein ideal $\mathcal{H}^{\mathfrak{s}}(G)$ has this property and any k-algebra of finite type has this property, see [**BN**]. Here, $\mathrm{Irr}(A)$ is the set of equivalence classes of simple unital (i.e. non-degenerate) A-modules and $\mathrm{Prim}(A)$ is the set of primitive ideals in A.

4. A Morita equivalence

Let $\mathcal{H} = \mathcal{H}(G)$ denote the Hecke algebra of G. Note that $\mathcal{H}(G)$ admits a set E of local units. For let K be a compact open subgroup of G and define

$$e_K(x) = \begin{cases} \mathrm{vol}(K)^{-1} & \text{if } x \in K, \\ 0 & \text{otherwise.} \end{cases}$$

Given a finite set $X \subset \mathcal{H}(G)$, we choose K sufficiently small. Then we have $e_K x = x = x e_K$ for all $x \in X$. It follows that $\mathcal{H}$ is an idempotent algebra: $\mathcal{H} = \mathcal{H}^2$.

We have the Bernstein decomposition

$$\mathcal{H}(G) = \bigoplus_{\mathfrak{s} \in \mathfrak{B}(G)} \mathcal{H}^{\mathfrak{s}}(G)$$

of the Hecke algebra $\mathcal{H}(G)$ into two-sided ideals.

LEMMA 3. *Each Bernstein ideal $\mathcal{H}^{\mathfrak{s}}(G)$ admits a set of local units.*

PROOF. Define

$$E^{\mathfrak{s}} := E \cap \mathcal{H}^{\mathfrak{s}}(G).$$

Then $E^{\mathfrak{s}}$ is a set of local units for $\mathcal{H}^{\mathfrak{s}}(G)$. $\qquad \square$

We recall the notation from section 2:

$$\mathfrak{s} \in \mathfrak{B}(G), \quad \mathfrak{s} = [L, \sigma]_G, \quad D_\sigma = \Psi(L)/\mathcal{G}.$$

THEOREM 1. *Let $\mathfrak{s} \in \mathfrak{B}(G), k = \mathcal{O}(D_\sigma)$. The ideal $\mathcal{H}^{\mathfrak{s}}(G)$ is a k-algebra Morita equivalent to a unital k-algebra of finite type. If $W_{\mathfrak{t}} = \{1\}$ then $\mathcal{H}^{\mathfrak{s}}(G)$ is Morita equivalent to k.*

PROOF. Let $\mathfrak{s} = [L, \sigma]_G$. We will write

$$\mathcal{H}(G) = \mathcal{H} = \mathcal{H}^{\mathfrak{s}} \oplus \mathcal{H}'$$

where $\mathcal{H}'$ denotes the sum of all $\mathcal{H}^{\mathfrak{t}}$ with $\mathfrak{t} \in \mathfrak{B}(G)$ and $\mathfrak{t} \neq \mathfrak{s}$. It follows from [**B**, (3.7)] that there is a compact open subgroup K of G with the property that $V^K \neq 0$ for every irreducible representation (π, V) with $\mathfrak{I}(\pi) = \mathfrak{s}$. We will write $e_K = e + e'$ with $e \in \mathcal{H}^{\mathfrak{s}}, e' \in \mathcal{H}'$. Both e and e' are idempotent and we have $ee' = 0 = e'e$.

Given $h \in \mathcal{H}$, we will write $h = h^{\mathfrak{s}} + h'$ with $h^{\mathfrak{s}} \in \mathcal{H}^{\mathfrak{s}}, h' \in \mathcal{H}'$. By [**BK2**, 3.1, 3.4 – 3.6], we have $\mathcal{H}^{\mathfrak{s}}(G) \cong \mathcal{H}e\mathcal{H}$. We note that this follows from the general considerations in [**BK2**, §3], and does *not* use the existence or construction of types. The ideal $\mathcal{H}^{\mathfrak{s}}$ is the idempotent two-sided ideal generated by e. By Lemma 2 we have

$$\mathcal{H}^{\mathfrak{s}} = \mathcal{H}e\mathcal{H} \sim_{morita} e\mathcal{H}e.$$

Since $eh' = he' = 0, e'h^{\mathfrak{s}} = h^{\mathfrak{s}}e' = 0$ we also have

$$e\mathcal{H}e = e\mathcal{H}^{\mathfrak{s}}e = e_K \mathcal{H}^{\mathfrak{s}} e_K.$$

It follows that

$$\mathcal{H}^s \sim_{morita} e_K \mathcal{H}^s e_K.$$

Let B be the unital algebra defined as follows:

$$B := e_K \mathcal{H}^s(G) e_K.$$

We will use the notation in [**BR**, p.80]. Let $\Omega = D_\sigma / W_t$ and let $\Pi := \Pi(\Omega)^K, \Lambda := \operatorname{End} \Pi$. There is a strict Morita context

$$(B, \Lambda, \Pi, \Pi^*, \phi, \psi).$$

We therefore have

$$\mathcal{H}^s(G) \sim_{morita} B \sim_{morita} \Lambda.$$

The intertwining operators A_w generate Λ over $\Lambda(D_\sigma)$, see [**BR**, p.81]. We also have [**BR**, p. 73]

$$\Lambda(D_\sigma) = \mathcal{O}(\Psi(L)) \rtimes \mathcal{G}$$

where the finite abelian group $\mathcal{G}$ is defined as follows:

$$\mathcal{G} := \{\psi \in \Psi(G) : \psi\sigma \cong \sigma\}.$$

Note that $\mathcal{G}$ acts freely on the complex torus $\Psi(L)$.

The k-algebras $\mathcal{O}(\Psi(L)) \rtimes \mathcal{G}$ and $\mathcal{O}(\Psi(L)/\mathcal{G})$ are connected by the following strict Morita context:

$$(\mathcal{O}(\Psi(L)) \rtimes \mathcal{G}, \mathcal{O}(\Psi(L)/\mathcal{G}), \mathcal{O}(\Psi(L)), \mathcal{O}(\Psi(L)), \phi, \psi).$$

We conclude that

$$\Lambda(D_\sigma) = \mathcal{O}(\Psi(L)) \rtimes \mathcal{G} \sim_{morita} \mathcal{O}(\Psi(L)/\mathcal{G}) = \mathcal{O}(D_\sigma) = k.$$

Therefore, Λ is finitely generated as a k-module. Note that

$$Centre\, \Lambda(D_\sigma) \cong k.$$

If $W_t = \{1\}$ then there are no intertwining operators, so that $\Lambda = \Lambda(D_\sigma)$. In that case we have

$$\mathcal{H}^s(G) \sim_{morita} k.$$

$\square$

5. The conjecture

Let k be the coordinate ring of a complex affine algebraic variety X, $k = \mathcal{O}(X)$. Let A be an associative $\mathbb{C}$-algebra which is also a k-algebra. We work with the collection of all k-algebras A which are countably generated. As a $\mathbb{C}$-vector space, A admits a finite or countable basis.

We will define an equivalence relation, called *geometric equivalence*, on the collection of such algebras A. This equivalence relation will be denoted $\asymp$.

(1) MORITA EQUIVALENCE OF k-ALGEBRAS WITH LOCAL UNITS. Let A and B be k-algebras, each with a countable set of local units. If A and B are connected by a strict Morita context, then $A \asymp B$. Periodic cyclic homology is preserved, see [**C**, Theorem 1].

(2) SPECTRUM PRESERVING MORPHISMS WITH RESPECT TO FILTRATIONS OF k-ALGEBRAS OF FINITE TYPE, as in [**BN**]. Such filtrations are automatically finite, since k is Noetherian.

A morphism $\phi \colon A \to B$ of k-algebras of finite type is called

- *spectrum preserving* if, for each primitive ideal $\mathfrak{q}$ of B, there exists a unique primitive ideal $\mathfrak{p}$ of A containing $\phi^{-1}(\mathfrak{q})$, and the resulting map $\mathfrak{q} \mapsto \mathfrak{p}$ is a bijection from $\mathrm{Prim}(B)$ onto $\mathrm{Prim}(A)$;
- *spectrum preserving with respect to filtrations* if there exist increasing filtrations by ideals

$$(0) = I_0 \subset I_1 \subset I_2 \subset \cdots \subset I_{r-1} \subset I_r \subset A$$

$$(0) = J_0 \subset J_1 \subset J_2 \subset \cdots \subset J_{r-1} \subset J_r \subset B,$$

such that, for all j, we have $\phi(I_j) \subset J_j$ and the induced morphism

$$\phi_* : I_j/I_{j-1} \to J_j/J_{j-1}$$

is spectrum preserving. If A, B are k-algebras of finite type such that there exists a morphism $\phi : A \to B$ of k-algebras which is spectrum preserving with respect to filtrations then $A \asymp B$.

(3) DEFORMATION OF CENTRAL CHARACTER. Let A be a unital algebra over the complex numbers. Form the algebra $A[t, t^{-1}]$ of Laurent polynomials with coefficients in A. If q is a non-zero complex number, then we have the evaluation-at-q map of algebras

$$A[t, t^{-1}] \longrightarrow A$$

which sends a Laurent polynomial $P(t)$ to $P(q)$. Suppose now that $A[t, t^{-1}]$ has been given the structure of a k-algebra i.e. we are given a unital map of algebras over the complex numbers from k to the centre of $A[t, t^{-1}]$. We assume that for any non-zero complex number q the composed map

$$k \longrightarrow A[t, t^{-1}] \longrightarrow A$$

where the second arrow is the above evaluation-at-q map makes A into a k-algebra of finite type. For q a non-zero complex number, denote the finite type k-algebra so obtained by $A(q)$. Then we decree that if q_1 and q_2 are any two non-zero complex numbers, $A(q_1)$ is equivalent to $A(q_2)$.

We fix k. The first two moves preserve the central character. This third move allows us to algebraically deform the central character.

Let $\asymp$ be the equivalence relation generated by (1), (2), (3); we say that A and B are *geometrically equivalent* if $A \asymp B$.

Since each move induces an isomorphism in periodic cyclic homology [**BN**] [**C**], we have

$$A \asymp B \Longrightarrow HP_*(A) \cong HP_*(B).$$

In order to formulate our conjecture, we need to review certain results and definitions.

The primitive ideal space of $\widetilde{\mathfrak{Z}^{\mathfrak{s}}}(G)$ is the set of $\mathbb{C}$-points of the variety $\widetilde{D_\sigma}/W_{\mathfrak{t}}$ in the Zariski topology.

We have an isomorphism

$$HP_*(\mathcal{O}(\widetilde{D_\sigma}/W_{\mathfrak{t}})) \cong H^*(\widetilde{D_\sigma}/W_{\mathfrak{t}}; \mathbb{C}).$$

This is a special case of the Feigin-Tsygan theorem; for a proof of this theorem which proceeds by reduction to the case of smooth varieties, see [**KNS**].

Let E_σ be the maximal compact subgroup of the complex torus D_σ, so that E_σ is a compact torus.

Let $\mathrm{Prim}^t \, \mathcal{H}^{\mathfrak{s}}(G)$ denote the set of primitive ideals attached to tempered, simple $\mathcal{H}^{\mathfrak{s}}(G)$-modules.

CONJECTURE 1. *Let* $\mathfrak{s} \in \mathfrak{B}(G)$. *Then*

(1) $\mathcal{H}^{\mathfrak{s}}(G)$ *is geometrically equivalent to the commutative algebra* $\widetilde{\mathfrak{Z}}^{\mathfrak{s}}(G)$:

$$\mathcal{H}^{\mathfrak{s}}(G) \asymp \widetilde{\mathfrak{Z}}^{\mathfrak{s}}(G).$$

(2) *The resulting bijection of primitive ideal spaces*

$$\mathrm{Prim}\,\mathcal{H}^{\mathfrak{s}}(G) \longleftrightarrow \widetilde{D_\sigma}/W_{\mathfrak{t}}$$

restricts to give a bijection

$$\mathrm{Prim}^{\mathfrak{t}}\,\mathcal{H}^{\mathfrak{s}}(G) \longleftrightarrow \widetilde{E_\sigma}/W_{\mathfrak{t}}.$$

REMARK 1. The geometric equivalence of part (1) induces an isomorphism

$$HP_*(\mathcal{H}^{\mathfrak{s}}(G)) \cong H^*(\widetilde{D_\sigma}/W_{\mathfrak{t}}; \mathbb{C}).$$

REMARK 2. The referee has posed a very interesting question: if A, B are geometrically equivalent, what does this imply about the categories $A - MOD$ and $B - MOD$ of unital modules?

It seems likely that for each ideal $\mathcal{H}^{\mathfrak{s}}$ the category $\mathcal{H}^{\mathfrak{s}} - MOD$ will have some resemblance to $\widetilde{\mathfrak{Z}}^{\mathfrak{s}} - MOD$. If the conjecture is true, then $\mathrm{Prim}\,\mathcal{H}^{\mathfrak{s}}$ is in bijection with $\widetilde{D_\sigma}/W_{\mathfrak{t}}$. However, as the referee has indicated, there may be further resemblances between $\mathcal{H}^{\mathfrak{s}} - MOD$ and $\widetilde{\mathfrak{Z}}^{\mathfrak{s}} - MOD$.

6. Generic points in the Bernstein spectrum

We begin with a definition.

DEFINITION 1. The point $\mathfrak{s} \in \mathfrak{B}(G)$ is *generic* if $W_{\mathfrak{t}} = \{1\}$.

For example, let $\mathfrak{s} = [G, \sigma]_G$ with σ an irreducible supercuspidal representation of G. Then $\mathfrak{s}$ is a generic point in $\mathfrak{B}(G)$. For a second example, let $\mathfrak{s} = [\mathrm{GL}(2) \times \mathrm{GL}(2), \sigma_1 \otimes \sigma_2]_{\mathrm{GL}(4)}$ with σ_1 not equivalent to σ_2 (after unramified twist). Then $\mathfrak{s}$ is a generic point in $\mathfrak{B}(\mathrm{GL}(4))$.

THEOREM 2. *The conjecture is true if $\mathfrak{s}$ is a generic point in* $\mathfrak{B}(G)$.

PROOF. Part (1). This is immediate from Theorem 1. We conclude that

$$\mathcal{H}^{\mathfrak{s}}(G) \asymp \mathcal{O}(D_\sigma) = \mathfrak{Z}^{\mathfrak{s}}(G) = \widetilde{\mathfrak{Z}}^{\mathfrak{s}}(G).$$

Part (2). Let $\mathcal{C}(G)$ denote the Harish-Chandra Schwartz algebra of G, see [**W**]. We choose e, e_K as in the proof of Theorem 1. Let $\mathfrak{s} \in \mathfrak{B}(G)$ and let $\mathcal{C}^{\mathfrak{s}}(G)$ be the corresponding Bernstein ideal in $\mathcal{C}(G)$. As in the proof of Theorem 1, $\mathcal{C}^{\mathfrak{s}}(G)$ is the two-sided ideal generated by e. We have $\mathcal{C}^{\mathfrak{s}}(G) = \mathcal{C}(G)e\mathcal{C}(G)$. By Lemma 2, we have

$$\mathcal{C}^{\mathfrak{s}}(G) = \mathcal{C}(G)e\mathcal{C}(G) \sim_{morita} e\mathcal{C}(G)e = e_K\mathcal{C}(G)e_K.$$

According to Mischenko's theorem [**Mi**], the Fourier transform induces an isomorphism of unital Fréchet algebras:

$$e_K\mathcal{C}(G)e_K \cong C^\infty(E_\sigma, \mathrm{End}\,E^K).$$

We also have

$$C^\infty(E_\sigma, \mathrm{End}\,E^K) \cong M_n(C^\infty(E_\sigma))$$

with $n = \dim_{\mathbb{C}}(E^K)$. By Lemma 1 we have

$$\mathcal{C}^{\mathfrak{s}}(G) \sim_{morita} C^\infty(E_\sigma).$$

We now exploit the liminality of the reductive group G, and take the primitive ideal space of each side. We conclude that there is a bijection

$$\operatorname{Prim}^t \mathcal{H}^s(G) \longleftrightarrow E_\sigma.$$

$\square$

7. The Hecke algebra of SL(2)

Let $G = \mathrm{SL}(2) = \mathrm{SL}(2, F)$, a group not of adjoint type. Let W be the Coxeter group with 2 generators:

$$W = \langle s_1, s_2 \rangle = \mathbb{Z} \rtimes W_f$$

where $W_f = \mathbb{Z}/2\mathbb{Z}$. Then W is the infinite dihedral group. It has the property (see section 8) that

$$\mathcal{H}(W, q_F) = \mathcal{H}(\mathrm{SL}(2)//I).$$

There is a unique isomorphism of $\mathbb{C}$-algebras between $\mathcal{H}(W, q_F)$ and $\mathbb{C}[W]$ such that

$$T_{s_1} \mapsto \frac{q_F + 1}{2} \cdot s_1 + \frac{q_F - 1}{2}, \quad T_{s_2} \mapsto \frac{q_F + 1}{2} \cdot s_2 + \frac{q_F - 1}{2},$$

where T_{s_i} is the element of $\mathcal{H}(W, q_F)$ corresponding to s_i. We note also that

$$\mathbb{C}[\mathbb{Z} \rtimes \mathbb{Z}/2\mathbb{Z}] \cong \mathbb{M} \rtimes \mathbb{Z}/2\mathbb{Z}$$

where

$$\mathbb{M} := \mathbb{C}[t, t^{-1}]$$

denotes the $\mathbb{Z}/2\mathbb{Z}$-graded algebra of Laurent polynomials in one indeterminate t. Let α denote the generator of $\mathbb{Z}/2\mathbb{Z}$. The group $\mathbb{Z}/2\mathbb{Z}$ acts as automorphism of $\mathbb{M}$, with $\alpha(t) = t^{-1}$. We define

$$\mathbb{L} := \{P \in \mathbb{M} : \alpha(P) = P\}$$

as the algebra of balanced Laurent polynomials. We will write

$$\mathbb{L}^* := \{P \in \mathbb{M} : \alpha(P) = -P\}.$$

Then $\mathbb{L}^*$ is a free of rank 1 module over $\mathbb{L}$, with generator $t - t^{-1}$. We will refer to the elements of $\mathbb{L}^*$ as *anti-balanced* Laurent polynomials.

LEMMA 4. *We have*

$$\mathbb{M} \rtimes \mathbb{Z}/2\mathbb{Z} \asymp \mathbb{C}^2 \oplus \mathbb{L}.$$

PROOF. We will realize the crossed product as follows:

$$\mathbb{M} \rtimes \mathbb{Z}/2\mathbb{Z} = \{f \in \mathcal{O}(\mathbb{C}^\times, M_2(\mathbb{C})) : f(z^{-1}) = \mathfrak{a} \cdot f(z) \cdot \mathfrak{a}^{-1}\}$$

where

$$\mathfrak{a} = \begin{pmatrix} 1 & 0 \\ 0 & -1 \end{pmatrix}$$

We then have

$$\mathbb{M} \rtimes \mathbb{Z}/2\mathbb{Z} = \begin{pmatrix} \mathbb{L} & \mathbb{L}^* \\ \mathbb{L}^* & \mathbb{L} \end{pmatrix}$$

There is an algebra map

$$\begin{pmatrix} \mathbb{L} & \mathbb{L}^* \\ \mathbb{L}^* & \mathbb{L} \end{pmatrix} \to \begin{pmatrix} \mathbb{L} & \mathbb{L} \\ \mathbb{L} & \mathbb{L} \end{pmatrix}$$

as follows:

$$x_{11} \mapsto x_{11}, \quad x_{22} \mapsto x_{22}, \quad x_{12} \mapsto x_{12}(t - t^{-1}), \quad x_{21} \mapsto x_{21}(t - t^{-1})^{-1}.$$

This map, combined with evaluation of x_{22} at $t = 1$, $t = -1$, creates an algebra map

$$\mathbb{M} \rtimes \mathbb{Z}/2\mathbb{Z} \to M_2(\mathbb{L}) \oplus \mathbb{C} \oplus \mathbb{C}.$$

This map is spectrum-preserving with respect to the following filtrations:

$$0 \subset M_2(\mathbb{L}) \subset M_2(\mathbb{L}) \oplus \mathbb{C} \oplus \mathbb{C}$$

$$0 \subset I \subset \mathbb{M} \rtimes \mathbb{Z}/2\mathbb{Z}$$

where I is the ideal of $\mathbb{M} \rtimes \mathbb{Z}/2\mathbb{Z}$ defined by the conditions

$$x_{22}(1) = 0 = x_{22}(-1).$$

We therefore have

$$\mathbb{M} \rtimes \mathbb{Z}/2\mathbb{Z} \asymp M_2(\mathbb{L}) \oplus \mathbb{C}^2 \asymp \mathbb{L} \oplus \mathbb{C}^2.$$

It is worth noting that $\mathbb{M} \rtimes \mathbb{Z}/2\mathbb{Z}$ is *not* Morita equivalent to $\mathbb{L} \oplus \mathbb{C}^2$. For the primitive ideal space $\mathrm{Prim}(\mathbb{M} \rtimes \mathbb{Z}/2\mathbb{Z})$ is connected, whereas the primitive ideal space $\mathrm{Prim}(\mathbb{L} \oplus \mathbb{C}^2)$ is disconnected (it has 3 connected components). $\square$

THEOREM 3. *The conjecture is true for* $\mathrm{SL}(2, F)$.

PROOF. For the Iwahori ideal $\mathcal{H}^i(\mathrm{SL}(2))$ we have

$$\mathcal{H}^i(G) \asymp \mathcal{H}(W, q_F) \asymp \mathbb{C}[W].$$

Let χ be a unitary character of $F^\times$ of exact order two. Let $G = \mathrm{SL}(2, F)$ and let $\mathfrak{j} = \mathfrak{j}(\lambda) = [T, \lambda]_G \in \mathfrak{B}(G)$ with λ defined as follows:

$$\lambda \colon \begin{pmatrix} x & 0 \\ 0 & x^{-1} \end{pmatrix} \mapsto \chi(x).$$

Let c be the least integer $n \geq 1$ such that $1 + \mathfrak{p}_F^n \subset \ker(\chi)$. We set

$$J_\chi := \begin{pmatrix} \mathfrak{o}_F^\times & \mathfrak{p}_F^{[c/2]} \\ \mathfrak{p}_F^{[(c+1)/2]} & \mathfrak{o}_F^\times \end{pmatrix} \cap \mathrm{SL}(2, F).$$

Let τ_χ denote the restriction of λ to the compact torus

$$\left\{ \begin{pmatrix} x & 0 \\ 0 & x^{-1} \end{pmatrix} \colon x \in \mathfrak{o}_F^\times \right\}.$$

Then (J_χ, τ_χ) is an $\mathfrak{s}$-type in $\mathrm{SL}(2, F)$ and (for instance, as a special case of [**GR**, Theorem 11.1]) the Hecke algebra $\mathcal{H}(G, \tau_\chi)$ is isomorphic to $\mathcal{H}(W, q_F)$. We have

$$\mathcal{H}^{\mathfrak{j}}(G) \asymp \mathcal{H}(G, \tau_\chi) \cong \mathcal{H}(W, q_F) \asymp \mathbb{C}[W].$$

To summarize: if $\mathfrak{s} = \mathfrak{i}$ or $\mathfrak{j}(\lambda)$ then we have

$$\mathcal{H}^{\mathfrak{s}}(G) \asymp \mathbb{M} \rtimes \mathbb{Z}/2\mathbb{Z}.$$

We have $W_{\mathfrak{t}} = W_f = \mathbb{Z}/2\mathbb{Z}$. Let Ω be the variety which corresponds to $\mathfrak{j} \in \mathfrak{B}(G)$, so that $\Omega = D/W_f$ with D a complex torus of dimension 1. Each unramified quasicharacter of the maximal torus $T \subset \mathrm{SL}(2)$ is given by

$$\begin{pmatrix} x & 0 \\ 0 & x^{-1} \end{pmatrix} \mapsto s^{\mathrm{val}(x)}$$

with $z \in \mathbb{C}^\times$. Let $\mathbb{V}(zw-1)$ denote the algebraic curve in $\mathbb{C}^2$ defined by the equation $zw - 1 = 0$. This algebraic curve is a hyperbola. The map $\mathbb{C}^\times \to \mathbb{V}(zw - 1)$, $z \mapsto (z, z^{-1})$ defines the structure of algebraic curve on $\mathbb{C}^\times$. The generator of W_f sends a point z on this curve to z^{-1} and there are two fixed points, 1 and -1. The coordinate algebra of the quotient curve is given by

$$\mathbb{C}[D/W_f] = \mathbb{L}$$

and the coordinate algebra of the extended quotient is given by

$$\mathbb{C}[\tilde{D}/W_f] = \mathbb{L} \oplus \mathbb{C} \oplus \mathbb{C}.$$

Then it follows from Lemma 4 that

$$\mathcal{H}^{\mathfrak{s}}(\mathrm{SL}(2)) \asymp \tilde{3}^{\mathfrak{s}}(\mathrm{SL}(2))$$

if $\mathfrak{s} = \mathfrak{i}$ or $\mathfrak{j}(\lambda)$.

If $\mathfrak{s} \neq \mathfrak{i}, \mathfrak{j}(\lambda)$ then $\mathfrak{s}$ is a generic point in $\mathfrak{B}(\mathrm{SL}(2))$ and we apply Theorem 1. Part (1) of the conjecture is proved.

Recall that a representation of G is called *elliptic* if its character is not identically zero on the elliptic set of G.

It follows from [**Go**, Theorem 3.4] (see also [**SM**, Prop. 1]) that the (normalized) induced representation $\mathrm{Ind}_{TU}^{G}(\lambda \otimes 1)$ has an elliptic constituent (and then the other constituent is also elliptic) if and only if the character χ is of order 2.

We then have

$$\mathrm{Ind}_{TU}^{G}(\lambda \otimes 1) = \pi^{+} \oplus \pi^{-},$$

and we have the identity $\theta^{+} = \theta^{-} = 0$, between the characters θ^{+}, θ^{-} of π^{+}, π^{-}.

Let $s \in \mathbb{C}$, $|s| = 1$. The corresponding (normalized) induced representation will be denoted $\pi(s)$:

$$\pi(s) := \mathrm{Ind}_{TU}^{G}(\chi_{s}\lambda \otimes 1).$$

The representations $\pi(1)$, $\pi(-1)$ are reducible, and split into irreducible components:

$$\pi(1) = \pi^{+}(1) \oplus \pi^{-}(1)$$

$$\pi(-1) = \pi^{+}(-1) \oplus \pi^{-}(-1)$$

These are *elliptic* representations: their characters

$$\theta^{+}(1), \theta^{-}(1), \theta^{+}(-1), \theta^{-}(-1)$$

are not identically zero on the elliptic set, although we do have the identities

$$\theta^{+}(1) + \theta^{-}(1) = 0$$

$$\theta^{+}(-1) + \theta^{-}(-1) = 0.$$

Concerning infinitesimal characters, we have

$$inf.ch.\,\pi^{+}(1) = inf.ch.\,\pi^{-}(1) = \lambda$$

$$inf.ch.\,\pi^{+}(-1) = inf.ch.\,\pi^{-}(-1) = (-1)^{\mathrm{val}_F} \otimes \lambda.$$

Note that $E_{\sigma} = \mathbb{T}$, and recall that $\pi(s) \cong \pi(s^{-1})$. The induced bijection

$$\mathrm{Prim}^{t}\mathcal{H}^{\mathfrak{s}}(G) \to \widetilde{E_{\sigma}}/W_{t}$$

is as follows. With $s \in \mathbb{C}, |s| = 1$:

$$\pi(s) \mapsto \{s, s^{-1}\}, \qquad s^{2} \neq 1$$

$$\pi^{+}(1) \mapsto 1, \quad \pi^{-}(1) \mapsto a$$

$$\pi^{+}(-1) \mapsto -1, \quad \pi^{-}(-1) \mapsto b$$

where a, b are the two isolated points in the compact extended quotient of $\mathbb{T}$ by $\mathbb{Z}/2\mathbb{Z}$. We note that the pair $\pi^+(1), \pi^-(1)$ are *L-indistinguishable*. The corresponding Langlands parameter fails to distinguish them from each other. The pair $\pi^+(-1), \pi^-(-1)$ are also L-indistinguishable.

The unramified unitary principal series is defined as follows:

$$\omega(s) := \mathrm{Ind}_{TU}^G (\chi_s \otimes 1).$$

Recall that $\omega(s) = \omega(s^{-1})$. The representation $\omega(-1)$ is reducible:

$$\omega(-1) = \omega^+(-1) \oplus \omega^-(-1).$$

For that part of the tempered spectrum which admits non-zero Iwahori-fixed vectors, the induced bijection

$$\mathrm{Prim}^t \mathcal{H}^{\mathfrak{s}}(G) \to \widetilde{E_\sigma}/W_{\mathfrak{t}}$$

is as follows. With $s \in \mathbb{C}, |s| = 1$:

$$\omega(s) \mapsto \{s, s^{-1}\}, \qquad s \neq -1$$
$$\omega^+(-1) \mapsto -1, \quad \omega^-(-1) \mapsto c$$
$$\mathrm{St}(2) \mapsto d$$

where c, d are the two isolated points in the compact extended quotient of $\mathbb{T}$ by $\mathbb{Z}/2\mathbb{Z}$, and $\mathrm{St}(2)$ denotes the Steinberg representation of $\mathrm{SL}(2)$.

These maps are induced by the geometric equivalences and are therefore not quite canonical, because the geometric equivalences are not canonical.

If $\mathfrak{s} \neq \mathrm{i}, \mathrm{j}(\lambda)$ then $\mathfrak{s}$ is a generic point in $\mathfrak{B}(\mathrm{SL}(2))$ and the induced bijection is as follows:

$$\mathrm{Prim}^t \mathcal{H}^{\mathfrak{s}}(G) \to \widetilde{E_\sigma}/W_{\mathfrak{t}}$$

takes the form of the identity map $\mathbb{T} \to \mathbb{T}$ or the identity map $pt \to pt$. $\square$

There are 2 non-generic points in $\mathfrak{B}(\mathrm{SL}(2, \mathbb{Q}_p))$ with $p > 3$; there are 4 non-generic points in $\mathfrak{B}(\mathrm{SL}(2, \mathbb{Q}_2))$.

8. Iwahori-Hecke algebras

The proof of Theorem 1 shows that $\mathcal{H}^{\mathfrak{s}}(G)$ is Morita equivalent to a unital k-algebra which we will denote by $\mathcal{A}_{\mathfrak{s}}$. The next step in proving the conjecture will be to relate this algebra $\mathcal{A}_{\mathfrak{s}}$ to a *generalized Iwahori-Hecke algebra*, as defined below.

Let W' be a Coxeter group with generators $(s)_{s \in S}$ and relations

$$(ss')^{m_{s,s'}} = 1, \quad \text{for any } s, s' \in S \text{ such that } m_{s,s'} < +\infty,$$

and let L be a weight function on W', that is, a map $L: W' \to \mathbb{Z}$ such that $L(ww') = L(w) + L(w')$ for any w, w' in W' such that $\ell(ww') = \ell(w) + \ell(w')$, where ℓ is the usual length function on W'. Clearly, the function ℓ is itself a weight function.

Let Ω be a finite group acting on the Coxeter system (W', S). The group $W := W' \rtimes \Omega$ will be called an *extended Coxeter group*. We extend L to W by setting $L(w\omega) := L(w)$, for $w \in W', \omega \in \Omega$.

Let $A := \mathbb{Z}[v, v^{-1}]$ where v is an indeterminate. We set $u := v^2$ and $v_s := v^{L(s)}$ for any $s \in S$. Let $\bar{\ }: A \to A$ be the ring involution which takes v^n to v^{-n} for any $n \in \mathbb{Z}$.

Let $\mathcal{H}(W, u) = \mathcal{H}(W, L, u)$ denote the A-algebra defined by the generators $(T_s)_{s \in S}$ and the relations

$$(T_s - v_s)(T_s + v_s^{-1}) = 0 \quad \text{for } s \in S,$$

$$\underbrace{T_s T_{s'} T_s \cdots}_{m_{s,s'} \text{ factors}} = \underbrace{T_{s'} T_s T_{s'} \cdots}_{m_{s,s'} \text{ factors}}, \quad \text{for any } s \neq s' \text{ in } S \text{ such that } m_{s,s'} < +\infty.$$

For $w \in W$, we define $T_w \in \mathcal{H}(W, u)$ by $T_w = T_{s_1} T_{s_2} \cdots T_{s_m}$, where $w = s_1 s_2 \cdots s_m$ is a reduced expression in W. We have $T_1 = 1$, the unit element of $\mathcal{H}(W, u)$, and $(T_w)_{w \in W}$ is an A-basis of $\mathcal{H}(W, u)$. The v_s are called the parameters of $\mathcal{H}(W, u)$.

Let $\mathcal{H}(W', u)$ be the A-subspace of $\mathcal{H}(W, u)$ spanned by all T_w with $w \in W'$. For each $q \in \mathbb{C}^\times$, we set $\mathcal{H}(W, q) := \mathcal{H}(W, u) \otimes_A \mathbb{C}$, where $\mathbb{C}$ is regarded as an A-algebra with u acting as scalar multiplication by q. The algebras of the form $\mathcal{H}(W, q)$ where W is an extended Coxeter group and $q \in \mathbb{C}$ will be called *extended Iwahori-Hecke algebras*. In the case when the Coxeter group W' is an affine Weyl group, we will say that $\mathcal{H}(W, q)$ is an *extended affine Iwahori-Hecke algebra*.

We now observe that $W_{\mathfrak{t}}$ is a (finite) extended Coxeter group. Indeed, there exists a root system $\Phi_{\mathfrak{t}}$ with associate Weyl group denoted $W_{\mathfrak{t}}'$ and a subset $\Phi_{\mathfrak{t}}^+$ of positive roots in $\Phi_{\mathfrak{t}}$, such that, setting

$$C_{\mathfrak{t}} := \left\{ w \in W_{\mathfrak{t}} : w(\Phi_{\mathfrak{t}}^+) \subset \Phi_{\mathfrak{t}}^+ \right\},$$

we have

$$W_{\mathfrak{t}} = W_{\mathfrak{t}}' \rtimes C_{\mathfrak{t}}.$$

This follows from [**He**, Prop. 4.2] and [**Ho**, Lem. 2].

It is expected, and proved, using the theory of types of [**BK2**], for level-zero representations in [**M1**], [**M2**], for principal series representations of split groups in [**R2**], for the group $\mathrm{GL}(n, F)$ in [**BK1**], [**BK3**], for the group $\mathrm{SL}(n, F)$ in [**GR**], for the group $\mathrm{Sp}(4)$ in [**BB**], and for a large class of representations of classical groups in [**Ki1**], [**Ki2**], that there exists always an extended affine Iwahori-Hecke algebra $\mathcal{H}_{\mathfrak{s}}'$ such that the following holds:

(1) there exists a (finite) Iwahori-Hecke algebra $H_{\mathfrak{s}}'$ with corresponding Coxeter group $W_{\mathfrak{t}}'$ and a Laurent polynomial algebra $\mathcal{B}_{\mathfrak{t}}$ satisfying $\mathcal{H}_{\mathfrak{s}}' = H_{\mathfrak{s}}' \otimes_{\mathbb{C}} \mathcal{B}_{\mathfrak{t}}$;

(2) there exists a two-cocycle $\mu \colon C_{\mathfrak{t}} \times C_{\mathfrak{t}} \to \mathbb{C}^\times$ and an injective homomorphism of groups $\iota \colon C_{\mathfrak{t}} \to \mathrm{Aut}_{\mathbb{C}-\mathrm{alg}} \mathcal{H}_{\mathfrak{s}}'$ such that $\mathcal{A}_{\mathfrak{s}}$ is Morita equivalent to the twisted tensor product algebra $\mathcal{H}_{\mathfrak{s}}' \tilde{\otimes}_\iota \mathbb{C}[C_{\mathfrak{t}}]_\mu$.

In the case of $\mathrm{GL}(n, F)$ (see [**BK1**]), and in the case of principal series representations of split groups with connected centre (see [**R2**]), we always have $C_{\mathfrak{t}} = \{1\}$. The references quoted above give examples in which $C_{\mathfrak{t}} \neq \{1\}$. The results in [**GR**] also show that the algebra $\mathcal{H}_{\mathfrak{s}}' \tilde{\otimes}_\iota \mathbb{C}[C_{\mathfrak{t}}]_\mu$ is not always isomorphic to an extended Iwahori-Hecke algebra.

There are no known example in which the cocycle μ is non-trivial. In the case of unipotent level zero representations [**L6**], [**L1**], of principal series representations [**R2**], and of the group Sp_4 [**BB**], it has been proved that μ is trivial.

From now we restrict attention to the case where $C_{\mathfrak{t}} = \{1\}$, so that $\mathcal{A}_{\mathfrak{s}}$ is expected to be Morita equivalent to a generalized affine Iwahori-Hecke algebra $\mathcal{H}_{\mathfrak{s}}'$.

In particular, if L is a torus and $\mathfrak{s} = [T,1]_G$, then $\mathcal{A}_\mathfrak{s}$ is isomorphic to the commuting algebra $\mathcal{H}(G//I)$ in G of the induced representation from the trivial representation of an Iwahori subgroup I of G. We have

$$\mathcal{H}(G//I) \simeq \mathcal{H}(W, q_F),$$

where q_F is the order of the residue field of F and W is defined as follows (see for instance [**Ca**, §3.2 and 3.5]). Here the weight function is taken to be equal to the length function. In particular, we are in the equal parameters case.

Let T be a maximal split torus in G, and let $X^*(T)$, $X_*(T)$ denote its groups of characters and cocharacters, respectively. Let $\Phi(G,T) \subset X^*(T)$, $\Phi^\vee(G,T) \subset X_*(T)$ be the corresponding root and coroot systems, and W_f the associated (finite) Weyl group. Then

$$W = X_*(T) \rtimes W_f,$$

Now let $X'_*(T)$ denote the subgroup of $X_*(T)$ generated by $\Phi^\vee(G,T)$. Then $W' := X'_*(T) \rtimes W_f$ is a Coxeter group (an affine Weyl group) and $W = W' \rtimes \Omega$, where Ω is the group of elements in W of length zero.

Let $^L G^0$ be the Langlands dual group of G, and let $^L T^0$ denote the Langlands dual of T, a maximal torus of $^L G^0$. By Langlands duality, we have

$$W = X_*(T) \rtimes W_f = X^*(^L T^0) \rtimes W_f.$$

The isomorphism

$$^L T^0 \cong \Psi(T), \quad t \mapsto \chi_t$$

is fixed by the relation

$$\chi_t(\phi(\varpi_F)) = \phi(t)$$

for $t \in {}^L T^0, \phi \in X_*(T) = X^*(^L T^0)$, and ϖ_F a uniformizer in F. This isomorphism commutes with the W_f-action, see [**GS**, Section I.2.3].

The group W_f acts on $^L T^0$, and we form the quotient variety $^L T^0 / W_f$.

Let $\mathfrak{i} \in \mathfrak{B}(G)$ be determined by the cuspidal pair $(T,1)$. We have

$$\mathfrak{Z}^\mathfrak{i} = \mathbb{C}[^L T^0 / W_f],$$

$$\widetilde{\mathfrak{Z}}^\mathfrak{i} = \mathbb{C}[\widetilde{^L T^0} / W_f].$$

9. The asymptotic Hecke algebra

There is a unique algebra involution $h \mapsto h^\dagger$ of $\mathcal{H}(W', u)$ such that $T_s^\dagger = -T_s^{-1}$ for any $s \in S$, and a unique endomorphism $h \mapsto \bar{h}$ of $\mathcal{H}(W', u)$ which is A-semilinear with respect to $\bar{\ }: A \to A$ and satisfies $\bar{T}_s = T_s^{-1}$ for any $s \in S$. Let

$$A_{\leq 0} := \bigoplus_{m \leq 0} \mathbb{Z} v^m = \mathbb{Z}[v^{-1}], \quad A_{<0} := \bigoplus_{m < 0} \mathbb{Z} v^m,$$

$$\mathcal{H}(W', u)_{\leq 0} := \bigoplus_{w \in W'} A_{\leq 0} T_w, \quad \mathcal{H}(W', u)_{<0} := \bigoplus_{w \in W'} A_{<0} T_w.$$

Let $z \in W'$. There is a unique $c_z \in \mathcal{H}(W', u)_{\leq 0}$ such that $\bar{c}_z = c_z$ and $c_z = T_z$ mod $\mathcal{H}(W', u)_{<0}$, [**L7**, Theorem 5.2 (a)]. We write $c_z = \sum_{y \in W'} p_{y,z} T_y$, where $p_{y,z} \in A_{\leq 0}$. For $y \in W'$, $\omega, \omega' \in \Omega$, we define $p_{y\omega, z\omega'}$ as $p_{y,z}$ if $\omega = \omega'$ and as 0 otherwise. For $w \in W$, we set $c_w := \sum_{y \in W} p_{y,w} T_y$. Then it follows from [**L7**, Theorem 5.2 (b)] that $(c_w)_{w \in W}$ is an A-basis of $\mathcal{H}(W, u)$.

For x, y, z in W, we define $f_{x,y,z} \in A$ by

$$T_x T_y = \sum_{z \in W} f_{x,y,z} T_z.$$

From now on, we assume that W' is a bounded weighted Coxeter group, that is, that there exists an integer $N \in \mathbb{N}$ such that $v^{-N} f_{x,y,z} \in A_{\leq 0}$ for all x, y, z in W.

For x, y, z in W, we define $h_{x,y,z} \in A$ by

$$c_x \cdot c_y = \sum_{z \in W} h_{x,y,z} c_z.$$

It follows from [**L7**, §13.6] that, for any $z \in W$, there exists an integer $\mathbf{a}(z) \in [0, N]$ such that

$$h_{x,y,z} \in v^{\mathbf{a}(z)} \mathbb{Z}[v^{-1}] \quad \text{for all } x, y \in W,$$

$$h_{x,y,z} \notin v^{\mathbf{a}(z)-1} \mathbb{Z}[v^{-1}] \quad \text{for some } x, y \in W.$$

Let $\gamma_{x,y,z^{-1}}$ be the coefficient of $v^{\mathbf{a}(z)}$ in $h_{x,y,z}$.

Let $\Delta(z) \geq 0$ be the integer defined by

$$p_{1,z} = n_z v^{-\Delta(z)} + \text{strictly smaller powers of } v, \quad n_z \in \mathbb{Z} - \{0\},$$

and let $\mathcal{D}$ denote the following (finite) subset of W:

$$\mathcal{D} := \{z \in W : \mathbf{a}(z) = \Delta(z)\}.$$

In [**L7**, chap. 14.2] Lusztig stated a list of 15 conjectures $P_1, \ldots, P_{15}$ and proved them in several cases [**L7**, chap. 15, 16, 17]. Assuming the validity of the conjectures, Lusztig was able to define in [**L7**] partitions of W into left cells, right cells and two-sided cells, which extend the theory of Kazhdan-Lusztig from the case of equal parameters (that is, $v_s = v_{s'}$ for any $(s, s') \in S^2$) to the general case. In the case of equal parameters the conjectures mentioned above are known to be true.

From now on we shall assume the validity of these conjectures. Let us recall some of them below:

P1. For any $z \in W$ we have $\mathbf{a}(z) \leq \Delta(z)$.

P2. If $d \in \mathcal{D}$ and $x, y \in W$ satisfy $\gamma_{x,y,d} \neq 0$, then $x = y^{-1}$.

P3. If $y \in W$, there exists a unique $d \in \mathcal{D}$ such that $\gamma_{y,y^{-1},d} \neq 0$.

P4. If $z' \leq_{\mathcal{LR}} z$ then $\mathbf{a}(z') \geq \mathbf{a}(z)$. Hence, if $z' \sim_{\mathcal{LR}} z$, the $\mathbf{a}(z') = \mathbf{a}(z)$.

P5. If $d \in \mathcal{D}$, $y \in W$, $\gamma_{y^{-1},y,d} \neq 0$, then $\gamma_{y^{-1},y,d} = n_d = \pm 1$.

P6. If $d \in \mathcal{D}$, then $d^2 = 1$.

P7. For any x, y, z in W we have $\gamma_{x,y,z} = \gamma_{y,z,x}$.

P8. Let x, y, z in W be such that $\gamma_{x,y,z} \neq 0$. Then $x \sim_{\mathcal{L}} y^{-1}$, $y \sim_{\mathcal{L}} z^{-1}$, $z \sim_{\mathcal{L}} x^{-1}$.

For any $z \in W$, we set $\hat{n}_z := n_d$ where d is the unique element of $\mathcal{D}$ such that $d \sim_{\mathcal{L}} z^{-1}$.

Let J denote the free Abelian group with basis $(t_w)_{w \in W}$. We set

$$t_x \cdot t_y := \sum_{z \in W} \gamma_{x,y,z^{-1}} t_z.$$

(This is a finite sum.) This defines an associative ring structure on J. The ring J is called the *based ring* of W. It has a unit element $\sum_{d \in \mathcal{D}} t_d$ (see [**L7**, §18.3]).

The $\mathbb{C}$-algebra $J(W) := J \otimes_{\mathbb{Z}} \mathbb{C}$ is called the *asymptotic Hecke algebra* of W.

According to property (P8), for each two-sided cell $\mathbf{c}$ in W, the subspace $J_{\mathbf{c}}$ spanned by the t_w, $w \in \mathbf{c}$, is a two-sided ideal of J. The ideal $J_{\mathbf{c}}$ is in fact an associative ring with unit $\sum_{d \in \mathcal{D} \cap \mathbf{c}} T_d$, which is called the based ring of the two-sided cell $\mathbf{c}$, [**L7**, §18.3], and

$$(1) \qquad\qquad J = \bigoplus_{\mathbf{c}} J_{\mathbf{c}}$$

is a direct sum decomposition of J as a ring.

Let $J(W, u) := A \otimes_{\mathbb{Z}} J$. We recall from [**L7**, Theorem 18.9] (which extends [**L2**, 2.4]) that the A-linear map $\phi \colon \mathcal{H}(W, u) \to J(W, u)$ given by

$$\phi(c_w^\dagger) := \sum_{\substack{z \in W, d \in \mathcal{D} \\ \mathbf{a}(d) = \mathbf{a}(z)}} h_{x,d,z}\, \hat{n}_z\, t_z \qquad (x \in W)$$

is a homomorphism of A-algebra with unit (note that the conjecture (P_{15}) is used here).

Let

$$(2) \qquad\qquad \phi_q \colon \mathcal{H}(W, q) \to J \otimes_{\mathbb{Z}} \mathbb{C}$$

be the $\mathbb{C}$-algebra homomorphism induced by ϕ.

Let $\mathcal{H}(W, q)^{\geq i}$ be the $\mathbb{C}$-subspace of $\mathcal{H}(W, q)$ spanned by all the $c_w^\dagger$ with $w \in W$ and $\mathbf{a}(w) \geq i$. This a two-sided ideal of $\mathcal{H}(W, q)$, because of [**L7**, §13.1] and (P7). Let

$$\mathcal{H}(W, q)^i := \mathcal{H}(W, q)^{\geq i} / \mathcal{H}(W, q)^{\geq i+1};$$

this is an $\mathcal{H}(W, q)$-bimodule. It has as $\mathbb{C}$-basis the images $[c_w^\dagger]$ of the $c_w^\dagger \in \mathcal{H}(W, q)^{\geq i}$ such that $\mathbf{a}(w) = i$.

We may regard $\mathcal{H}(W, q)^i$ as a J-bimodule with multiplication defined by the rule:

$$t_x * [c_w^\dagger] = \sum_{\substack{z \in W \\ \mathbf{a}(z) = i}} \gamma_{x,w,z^{-1}}\, \hat{n}_w\, \hat{n}_z\, c_z^\dagger,$$

$$(3) \qquad [c_w^\dagger] * t_x = \sum_{\substack{z \in W \\ \mathbf{a}(z) = i}} \gamma_{w,x,z^{-1}}\, \hat{n}_w\, \hat{n}_z\, c_z^\dagger, \qquad (w, x \in W, \mathbf{a}(w) = i).$$

We have (see [**L7**, 18.10]):

$$(4) \qquad\qquad hf = \phi_q(h) * f, \quad \text{for all } f \in \mathcal{H}(W, q)^i,\ h \in \mathcal{H}(W, q).$$

On the other side:

$$(5) \qquad (j * f)h = j * (fh), \quad \text{for all } f \in \mathcal{H}(W, q)^i,\ h \in \mathcal{H}(W, q),\ j \in J.$$

Let

$$f_i := \sum_{\substack{d \in \mathcal{D} \\ \mathbf{a}(d) = i}} [c_d^\dagger] \in \mathcal{H}(W, q)^i.$$

LEMMA 5. *We have*

$$t_x * f_i = f_i * t_x = \hat{n}_x\, [c_x^\dagger].$$

PROOF. By definition of $*$, we have

$$t_x * f_i = \sum_{\substack{d \in \mathcal{D} \\ \mathbf{a}(d)=i}} \sum_{\substack{z \in W \\ \mathbf{a}(z)=i}} \gamma_{x,d,z^{-1}} \, \hat{n}_d \, \hat{n}_z \, c_z^\dagger.$$

Since (because of P7) $\gamma_{x,d,z^{-1}} = \gamma_{d,z^{-1},x} = \gamma_{z^{-1},x,d}$, it follows from (P2) that if $\gamma_{x,d,z^{-1}} \neq 0$ then $z = x$. Then, using (P3) and (P5), we obtain $t_x * f_i = \hat{n}_x \, c_x^\dagger$. The proof for $f_i * t_x$ is similar. $\square$

Let k denote the centre of $\mathcal{H}(W,q)$. Lusztig proved the following result in [**L3**, Proposition 1.6 (i)] in the case of equal parameters. Our proof will follow the same lines.

PROPOSITION 1. *The centre of $J \otimes_{\mathbb{Z}} \mathbb{C}$ contains $\phi_q(k)$.*

PROOF. It is enough to show that $\phi_q(z) \cdot t_x = t_x \cdot \phi_q(z)$ for any $z \in k$, $x \in W$. Assume that $\mathbf{a}(x) = i$. Let $z \in k$. Using Lemma 5, we obtain

$$(\phi_q(z)t_x) * f_i = \phi_q(z) * t_x * f_i = \hat{n}_x \phi_q(z) * [c_x^\dagger] = \hat{n}_x z[c_x^\dagger].$$

On the other side, using equation (4), we get

$$(t_x \phi_q(z)) * f_i = t_x * (\phi_q(z) * f_i) = t_x * (z f_i).$$

Since $z f_i = f_i z$, it gives

$$(t_x \phi_q(z)) * f_i = t_x * (f_i z),$$

and then, using equation (5) and again Lemma 5, we obtain

$$(t_x \phi_q(z)) * f_i = (t_x * f_i) z = \hat{n}_x [c_x^\dagger] z.$$

Now, since $z \in k$, we have $z[c_x^\dagger] = [c_x^\dagger] z$. Hence

(6) $$(\phi_q(z)t_x) * f_i = (t_x \phi_q(z)) * f_i.$$

It follows from the combination of (P4) and (P8) that $\gamma_{x,y,z} \neq 0$ implies $\mathbf{a}(x) = \mathbf{a}(y) = \mathbf{a}(z)$. Hence we have

$$\phi_q(z)t_x = \sum_{\substack{x' \in W \\ \mathbf{a}(x')=i}} \alpha_{x'} t_{x'},$$

$$t_x \phi_q(z) = \sum_{\substack{x' \in W \\ \mathbf{a}(x')=i}} \beta_{x'} t_{x'},$$

with $\alpha_{x'}$, $\beta_{x'}$ in $\mathbb{C}$. Then (6) implies that

$$\sum_{\substack{x' \in W \\ \mathbf{a}(x')=i}} \alpha_{x'} [c_{x'}^\dagger] = \sum_{\substack{x' \in W \\ \mathbf{a}(x')=i}} \beta_{x'} [c_{x'}^\dagger].$$

Hence $\alpha_{x'} = \beta_{x'}$ for all $x' \in W$ such that $\mathbf{a}(x') = i$. It gives $\phi_q(z)t_x = t_x \phi_q(z)$, as required. $\square$

REMARK 3. The above proposition provides $J \otimes_{\mathbb{Z}} \mathbb{C}$ (and also each $J_{\mathbf{c}}$) with a structure of k-algebra. This k-algebra structure is not canonical: it depends on q. Our move (3) precisely allows us to pass from one k-algebra structure, depending on q_1, to another k-algebra structure, depending on q_2.

From now on we will assume that the weight function is equal to the length function ℓ. We will assume that $q = 1$, in which case $H(W, q)$ is the group algebra of W; or q is not a root of unity, in which case we can take for q the order q_F of the residue field of F.

Let E be a simple $\mathcal{H}(W, q)$-module (resp. $J \otimes_\mathbb{Z} \mathbb{C}$-module). We attach to E an integer $\mathbf{a}_E$ by the following two requirements:

(1) $c_w E = 0$ (resp. $t_w E = 0$) for any w with $\mathbf{a}(w) > \mathbf{a}_E$;
(2) $c_w E \neq 0$ (resp. $t_w E \neq 0$) for some w such $\mathbf{a}(w) = \mathbf{a}_E$.

Then Lusztig proved in [**L3**, Cor. 3.6] (see also [**L5**, Th. 8.1]) that there is a unique bijection $E \mapsto E'$ between the set of isomorphism classes of simple $\mathcal{H}(W, q)$-modules and the set of isomorphism classes of simple $J \otimes_\mathbb{Z} \mathbb{C}$-modules such that $\mathbf{a}_{E'} = \mathbf{a}_E$ and such that the restriction of E' to $\mathcal{H}(W, q)$ via ϕ_q is an $\mathcal{H}(W, q)$-module with exactly one composition factor isomorphic to E and all other composition factors of the form $\bar{E}$ with $\mathbf{a}_{\bar{E}} < \mathbf{a}_E$.

As shown in [**BN**, Th. 9], it follows that ϕ_q is spectrum preserving with respect to filtrations. Hence

$$(7) \qquad \qquad \mathcal{H}(W, q) \asymp J \otimes_\mathbb{Z} \mathbb{C}.$$

Let G be a connected F-split adjoint simple p-adic group. By Langlands duality we have

$$(8) \qquad \qquad W := X_*(T) \rtimes W_f = X^*({}^L T^0(\mathbb{C})) \rtimes W_f.$$

Lusztig proved in [**L4**, Theorem 4.8] that the unipotent conjugacy classes in ${}^L G^0$ are in bijection with the two-sided cells in W.

Let J be the based ring attached to W. We fix a two-sided cell $\mathbf{c}$ in W. Let $\mathcal{O}_\mathbf{c}$ be the unipotent conjugacy class in ${}^L G^0$ corresponding to $\mathbf{c}$. Let $\varphi \colon \mathrm{SL}(2)(\mathbb{C}) \to {}^L G^0$ be a homomorphism of algebraic groups such that $u = \varphi \begin{pmatrix} 1 & 1 \\ 0 & 1 \end{pmatrix}$ belongs to $\mathcal{O}_\mathbf{c}$ and let $F_\mathbf{c}$ be a maximal reductive algebraic subgroup of the centralizer $C_{L G^0}(u)$. The reductive group $F_\mathbf{c}$ may be disconnected: the identity component of $F_\mathbf{c}$ will be denoted $F_\mathbf{c}^0$.

Let Y be a finite $F_\mathbf{c}$-set (that is, a set with an algebraic action of $F_\mathbf{c}$; thus, $F_\mathbf{c}^0$ acts trivially). An $F_\mathbf{c}$-vector bundle on Y is a collection of finite dimensional $\mathbb{C}$-vector spaces V_y ($y \in Y$) with a given algebraic representation of $F_\mathbf{c}$ on $\bigoplus_{y \in Y} V_y$ such that $G \cdot V_y = V_{gy}$ for all $g \in F_\mathbf{c}$, $y \in Y$. We now consider the finite $F_\mathbf{c}$-set $Y \times Y$ with diagonal action of $F_\mathbf{c}$ and denote by $K_{F_\mathbf{c}}(Y \times Y)$ the Grothendieck group of the category of $F_\mathbf{c}$-vector bundles on $Y \times Y$. One can define an associative ring structure on $K_{F_\mathbf{c}}(Y \times Y)$ (see [**L4**, §10.2]).

Then the conjecture of Lusztig in [**L4**, §10.5] states in particular that there should exist a finite $F_\mathbf{c}$-set Y and a bijection π from $\mathbf{c}$ onto the set of irreducible $F_\mathbf{c}$-vector bundles on $Y \times Y$ (up to isomorphism) such the $\mathbb{C}$-linear map $J_\mathbf{c} \to K_{F_\mathbf{c}}(Y \times Y) \otimes \mathbb{C}$ sending t_w to $\pi(w)$ is an algebra isomorphism (preserving the unit element).

Let $|Y|$ denote the cardinality of Y. This number is expected to be the number of left cells contained in $\mathbf{c}$. When $F_\mathbf{c}$ is connected, $K_{F_\mathbf{c}}(Y \times Y)$ is isomorphic to the $|Y| \times |Y|$ matrix algebra $\mathrm{M}_{|Y|}(R_\mathbb{C}(F_\mathbf{c}))$ over the (complexified) rational representation ring $R_\mathbb{C}(F_\mathbf{c})$ of $F_\mathbf{c}$. It is important to note: when $F_\mathbf{c}$ is connected, the Lusztig conjecture asserts that $J_\mathbf{c}$ is Morita equivalent to a *commutative* algebra.

The Lusztig conjecture has been proved by Xi for any two-sided cell $\mathbf{c}$ when G is one of the following groups $GL(n)$, $PGL(n)$, $SL(2)$, $SO(5)$ and G_2, and for the lowest two-sided cell $\mathbf{c}_0$ (see next section) when G is any connected F-split adjoint simple p-adic group.

10. The ideal $J_{\mathbf{c}_0}$ in J

As above we assume that G is a connected F-split adjoint simple p-adic group. Let J be the based ring attached to W, with W as in (8). The centralizer of 1 is of course $^L G^0$. Under the bijection cited above, the unipotent class 1 corresponds to the *lowest two-sided cell* $\mathbf{c}_0$, that is the subset of all the elements w in W such that $\mathbf{a}(w)$ equals the number of positive roots in the root system of W_f.

Xi proved the Lusztig conjecture for this ideal $J_{\mathbf{c}_0}$ in [**X3**, Theorem 1.10]. According to his result, we have a ring isomorphism

$$J_{\mathbf{c}_0} \cong \mathrm{M}_{|W_f|}(R_{\mathbb{C}}(^L G^0)).$$

The character map Ch creates an isomorphism

$$R_{\mathbb{C}}(^L G^0) \cong (R_{\mathbb{C}}(^L T^0))^{W_f}.$$

The W_f-invariant subring of the (complexified) representation ring of $^L T^0$ is precisely the coordinate ring of the quotient torus $^L T^0 / W_f$.

Since

$$\Psi(T) = {}^L T^0$$

we have a Morita equivalence

$$J_{\mathbf{c}_0} \sim 3^{\mathrm{i}}(G)$$

where i is the quotient variety $\Psi(T)/W_f$. Therefore, we obtain the following result.

THEOREM 4. *Let G be a connected F-split adjoint simple p-adic group. There is a Morita equivalence between $J_{\mathbf{c}_0}$ and the coordinate ring of the Bernstein variety* $\Psi(T)/W_f$.

According to our conjecture, *the other ideals $J_{\mathbf{c}}$ account (up to geometric equivalence) for the rest of the extended quotient of $\Psi(T)$ by W_f.*

The classical Satake isomorphism is an isomorphism between the spherical Hecke algebra $\mathcal{H}(G//K)$ and the ring $R_{\mathbb{C}}(^L G^0)$. Further, a theorem of Bernstein (see e.g. [**L0**, Proposition 8.6]) asserts that the centre $Z(\mathcal{H}(G//I))$ of the Iwahori-Hecke algebra $\mathcal{H}(G//I)$ is also isomorphic to $R_{\mathbb{C}}(^L G^0)$.

At this point, we need the map $\phi_{q_F, \mathbf{c}_0}$ defined in section 1.7 of Xi's paper [**X3**]. This map is the composition of ϕ_{q_F} and of the projection of J onto $J_{\mathbf{c}_0}$.

Xi has proved in [**X3**, Theorem 3.6] that the image $\phi_{q_F, \mathbf{c}_0}(Z(H(G//I))$ is the centre $Z(J_{\mathbf{c}_0})$ of the algebra $J_{\mathbf{c}_0}$. This creates the following diagram:

$$
\begin{array}{ccc}
\mathcal{H}(G//K) & \longrightarrow & R_{\mathbb{C}}(^L G^0) \\
\downarrow & & \downarrow \\
\mathcal{H}(G//I) & \xrightarrow{\phi_{q_F, \mathbf{c}_0}} & J_{\mathbf{c}_0}
\end{array}
$$

in which the top horizontal map is the Satake isomorphism, the left vertical map is induced by the inclusion $K \subset I$, the right vertical map sends $R_{\mathbb{C}}(^L G^0)$ onto the centre of $J_{\mathbf{c}_0}$ and the bottom horizontal map is Xi's map $\phi_{q_F, \mathbf{c}_0}$. The vertical maps are injective. We expect that this diagram is commutative.

11. The Hecke algebra of $\mathrm{GL}(n)$

THEOREM 5. *The conjecture is true for* $\mathrm{GL}(n)$.

PROOF. In this proof, we follow [**BP**] rather closely; we have refined the proof at certain points. The occurrence of an extended quotient in the smooth dual of $\mathrm{GL}(n)$ was first recorded in [**HP**], in the context of Deligne-Langlands parameters.

Let $G := \mathrm{GL}(n)$, $\mathfrak{s} = [L, \sigma]_G \in \mathfrak{B}(G)$ and $\mathfrak{t} = [L, \sigma]_L \in \mathfrak{B}(L)$. We can think of $\mathfrak{t}$ as a vector of irreducible supercuspidal representations of smaller general linear groups. If the vector is

$$(\sigma_1, \ldots, \sigma_1, \ldots, \sigma_t, \ldots, \sigma_t)$$

with σ_i repeated e_i times, $1 \le i \le t$, and $\sigma_1, \ldots, \sigma_t$ pairwise distinct (after unramified twist) then we say that $\mathfrak{t}$ has *exponents* $e_1, \ldots, e_t$.

Each representation σ_i of $G_i := \mathrm{GL}(m_i)$ has a *torsion number*: the order of the cyclic group of all those unramified characters η for which $\sigma_i \otimes \eta \cong \sigma_i$. The torsion number of σ_i will be denoted r_i.

Hence

$$L \simeq \prod_{i=1}^{t} G_i^{e_i} \quad \text{and} \quad \sigma \simeq \bigotimes_{i=1}^{t} \sigma_i^{\otimes e_i},$$

Each σ_i contains a maximal simple type (K_i, λ_i) in G_i [**BK1**]. Let

$$K_L := \prod_{i=1}^{t} K_i^{e_i} \quad \text{and} \quad \tau_L := \bigotimes_{i=1}^{t} \lambda_i^{\otimes e_i}.$$

Then (K_L, τ_L) is a $\mathfrak{t}$-type in L. We have

$$W_{\mathfrak{t}} \simeq \prod_{i=1}^{t} S_{e_i}.$$

Let W_{e_i} denote the extended affine Weyl group associated to $\mathrm{GL}(e_i, \mathbb{C})$.

Let (K, τ) be a semisimple $\mathfrak{s}$-type, see [**BK1, BK2, BK3**]. It is worth pointing out that we do not need the type explicitly. Instead, we need certain items attached to the type: the idempotent e_τ and the endomorphism-valued Hecke algebra $\mathcal{H}(G, \tau)$. Let e_τ be the idempotent attached to the type (K, τ) as in [**BK2**, Definition 2.9]:

$$e_\tau(x) = \begin{cases} (\mathrm{vol}\, K)^{-1} (\dim \tau) \, tr(\tau(x^{-1})) & \text{if } x \in K, \\ 0 & \text{if } x \in G, x \notin K. \end{cases}$$

The idempotent e_τ is then a *special* idempotent in the Hecke algebra $\mathcal{H}(G)$ according to [**BK2**, Definition 3.11]. Let $\mathcal{H} = \mathcal{H}(G)$. It follows from [**BK2**, §3] that

$$\mathcal{H}^{\mathfrak{s}}(G) = \mathcal{H} * e_\tau * \mathcal{H}.$$

We then have a Morita equivalence

$$\mathcal{H} * e_\tau * \mathcal{H} \sim_{morita} e_\tau * \mathcal{H} * e_\tau.$$

Now let $\mathcal{H}(K, \tau)$ be the endomorphism-valued Hecke algebra attached to the semisimple type (K, τ). By [**BK2**, 2.12] we have a canonical isomorphism of unital $\mathbb{C}$-algebras :

$$\mathcal{H}(G, \tau) \otimes_{\mathbb{C}} \mathrm{End}_{\mathbb{C}} W \cong e_\tau * \mathcal{H}(G) * e_\tau$$

so that the algebra $e_\tau * \mathcal{H}(G) * e_\tau$ is Morita equivalent to the algebra $\mathcal{H}(G, \tau)$. Now we quote the main theorem for semisimple types in $\mathrm{GL}(n)$ [**BK3**, 1.5]: there is an isomorphism of unital $\mathbb{C}$-algebras

$$\mathcal{H}(G, \tau) \cong \bigotimes_{i=1}^{t} \mathcal{H}(W_{e_i}, q_F^{r_i}).$$

The factors $\mathcal{H}(W_{e_i}, q_F^{r_i})$ are (extended) affine Hecke algebras whose structure is given explicitly in [**BK1**, 5.4.6, 5.6.6].

We conclude that

$$\mathcal{H}^{\mathfrak{s}}(\dot{G}) \asymp \bigotimes_{i=1}^{t} \mathcal{H}(W_{e_i}, q_F^{r_i}).$$

On the other hand, from (7), we have

$$\bigotimes_{i=1}^{t} \mathcal{H}(W_{e_i}, q_F^{r_i}) \asymp \bigotimes_{i=1}^{t} J(W_{e_i}).$$

Finally we will prove that that

$$\bigotimes_{i=1}^{t} J(W_{e_i}) \asymp \widetilde{3^{\mathfrak{s}}}.$$

Let $^L T^0$ be the maximal standard torus of $^L G^0 = \mathrm{GL}(n, \mathbb{C})$ and let W be the extended affine Weyl group associated to $\mathrm{GL}(n, \mathbb{C})$. We have $W := X^*(^L T^0) \rtimes S_n = W_n$. For each two-sided cell $\mathbf{c}$ of W we have a corresponding partition λ of n. Let μ be the dual partition of λ. Let u be a unipotent element in $\mathrm{GL}(n, \mathbb{C})$ whose Jordan blocks are determined by the partition μ. Let the distinct parts of the dual partition μ be $\mu_1, \ldots, \mu_p$ with μ_r repeated n_r times, $1 \leq r \leq p$.

Let $C_G(u)$ be the centralizer of u in $G = \mathrm{GL}(n, \mathbb{C})$. Then the maximal reductive subgroup $F_{\mathbf{c}}$ of $C_G(u)$ is isomorphic to $\mathrm{GL}(n_1, \mathbb{C}) \times \mathrm{GL}(n_2, \mathbb{C}) \times \cdots \times \mathrm{GL}(n_p, \mathbb{C})$. For the non-trivial combinatorics which underlies this statement, see [**G**, §2.6].

Let J be the based ring of W. For each two-sided cell in W, let $|Y|$ be the number of left cells contained in $\mathbf{c}$. The Lusztig conjecture says that there is a ring isomorphism

$$J_{\mathbf{c}} \simeq \mathrm{M}_{|Y|}(R_{F_{\mathbf{c}}}), \quad t_w \mapsto \pi(w)$$

where $R_{F_{\mathbf{c}}}$ is the rational representation ring of $F_{\mathbf{c}}$. This conjecture for $\mathrm{GL}(n, \mathbb{C})$ has been proved by Xi [**X1**, 1.5, 4.1, 8.2].

Since $F_{\mathbf{c}}$ is isomorphic to a direct product of the general linear groups $\mathrm{GL}(n_i, \mathbb{C})$ $(1 \leq i \leq p)$ we see that $R_{F_{\mathbf{c}}}$ is isomorphic to the tensor product over $\mathbb{Z}$ of the representation rings $R_{\mathrm{GL}(n_i, \mathbb{C})}, 1 \leq i \leq p$. For the ring $R(\mathrm{GL}(n, \mathbb{C}))$ we have

$$R(\mathrm{GL}(n, \mathbb{C})) = \mathbb{Z}[X^*(T(\mathbb{C}))]^{S_n}$$

where $T(\mathbb{C})$ is the standard maximal torus in $\mathrm{GL}(n, \mathbb{C})$, and $X^*(T(\mathbb{C}))$ is the set of rational characters of $T(\mathbb{C})$, by [**Bo**, Chapter VIII]. Therefore we have

$$R_{F_{\mathbf{c}}} \otimes_{\mathbb{Z}} \mathbb{C} \simeq \mathbb{C}[\mathrm{Sym}^{n_1} \mathbb{C}^\times \times \cdots \times \mathrm{Sym}^{n_p} \mathbb{C}^\times].$$

Let $\gamma \in S_n$ have cycle type μ, let $X = (\mathbb{C}^\times)^n$. Then

$$
\begin{aligned}
X^\gamma &\simeq (\mathbb{C}^\times)^{n_1} \times \cdots \times (\mathbb{C}^\times)^{n_p} \\
Z(\gamma) &\simeq (\mathbb{Z}/\mu_1\mathbb{Z}) \wr S_{n_1} \times \cdots \times (\mathbb{Z}/\mu_p\mathbb{Z}) \wr S_{n_p} \\
X^\gamma/Z(\gamma) &\simeq \mathrm{Sym}^{n_1}\mathbb{C}^\times \times \cdots \times \mathrm{Sym}^{n_p}\mathbb{C}^\times
\end{aligned}
$$

and so

$$
R_{F_\mathbf{c}} \otimes_\mathbb{Z} \mathbb{C} \simeq \mathbb{C}[X^\gamma/Z(\gamma)].
$$

Then, using (1), we obtain

$$
J \otimes_\mathbb{Z} \mathbb{C} = \bigoplus_\mathbf{c} (J_\mathbf{c} \otimes_\mathbb{Z} \mathbb{C}) \sim \bigoplus_\mathbf{c} (R_{F_\mathbf{c}} \otimes_\mathbb{Z} \mathbb{C}) \simeq \mathbb{C}[\widetilde{X}/S_n].
$$

The algebra $J \otimes_\mathbb{Z} \mathbb{C}$ is Morita equivalent to a reduced, finitely generated, commutative unital $\mathbb{C}$-algebra, namely the coordinate ring of the extended quotient $\widetilde{X}/S_n$. This finishes the proof of part (1) of the conjecture.

Part (2) of the conjecture for $\mathrm{GL}(n)$ is a consequence of [**P**, Theorem Theorem 5.1]. $\square$

12. The Iwahori ideal in $\mathcal{H}(\mathrm{PGL}(n))$

Let $G = \mathrm{PGL}(n)$, let T be its standard maximal torus. Let $W := X_*(T) \rtimes W_f$. Then ${}^L G^0 = \mathrm{SL}(n, \mathbb{C})$ is the Langlands dual group. Its maximal torus will be denoted ${}^L T^0$.

The discrete group W is an *extended Coxeter group*:

$$
W = \langle s_1, s_2, \ldots, s_n \rangle \rtimes \mathbb{Z}/n\mathbb{Z}
$$

where $\mathbb{Z}/n\mathbb{Z}$ permutes cyclically the generators $s_1, \ldots, s_n$. We have

$$
\mathcal{H}(W, q_F) = \mathcal{H}(G//I).
$$

The symmetric group $W_f = S_n$ acts on ${}^L T^0$ by permuting coordinates, and we form the quotient variety ${}^L T^0/S_n$.

Let $\mathfrak{i} \in \mathfrak{B}(G)$ be determined by the cuspidal pair $(T, 1)$. We have

$$
\mathfrak{Z}^\mathfrak{i} = \mathbb{C}[{}^L T^0/S_n],
$$

$$
\widetilde{\mathfrak{Z}}^\mathfrak{i} = \mathbb{C}[\widetilde{{}^L T^0}/S_n].
$$

THEOREM 6. *Let $\mathcal{H}^\mathfrak{i}(G)$ denote the Iwahori ideal in $\mathcal{H}(G)$. Then $\mathcal{H}^\mathfrak{i}(G)$ is geometrically equivalent to the extended quotient of ${}^L T^0$ by the symmetric group S_n:*

$$
\mathcal{H}^\mathfrak{i}(G) \asymp \mathbb{C}[\widetilde{{}^L T^0}/S_n].
$$

PROOF. The non-unital algebra $\mathcal{H}^\mathfrak{i}(G)$ is Morita equivalent to the unital affine Hecke algebra $\mathcal{H}(W, q_F)$:

$$
\mathcal{H}^\mathfrak{i}(G) = \mathcal{H}e\mathcal{H} \sim_{morita} e\mathcal{H}e \cong \mathcal{H}(W, q_F).
$$

From (7), we have

$$
\mathcal{H}(W, q_F) \asymp J \otimes_\mathbb{Z} \mathbb{C}.
$$

The Langlands dual of $\mathrm{PGL}(n, F)$ is $\mathrm{SL}(n, \mathbb{C})$. For each two-sided cell $\mathbf{c}$ of W we have a corresponding partition λ of n. Let μ be the dual partition of λ. Let u be a unipotent element in $\mathrm{SL}(n, \mathbb{C})$ whose Jordan blocks are determined by the partition μ. Let the distinct parts of the dual partition μ be $\mu_1 < \cdots < \mu_p$ with μ_r repeated n_r times, $1 \le r \le p$.

Let $C_G(u)$ be the centralizer of u in $G = \mathrm{SL}(n, \mathbb{C})$. Then the maximal reductive subgroup F'_λ of $C_G(u)$ is isomorphic to $(\mathrm{GL}(n_1, \mathbb{C}) \times \mathrm{GL}(n_2, \mathbb{C}) \times \cdots \times \mathrm{GL}(n_p, \mathbb{C})) \cap \mathrm{SL}(n, \mathbb{C})$. For details of the injective map

$$(\mathrm{GL}(n_1, \mathbb{C}) \times \mathrm{GL}(n_2, \mathbb{C}) \times \cdots \times \mathrm{GL}(n_p, \mathbb{C})) \cap \mathrm{SL}(n, \mathbb{C}) \longrightarrow \mathrm{SL}(n, \mathbb{C})$$

see [**G**].

As a special case, let the two-sided cell **c** correspond to the partition $\lambda = (1, 1, 1, \ldots, 1)$ of n. Then the dual partition $\mu = (n)$. The unipotent matrix u has one Jordan block, and its centralizer $C_G(u) = Z$ the centre of $\mathrm{SL}(n, \mathbb{C})$. The maximal reductive subgroup F'_λ of $C_G(u)$ is the finite group Z. This is the case $p = 1$, $\mu_1 = n$, $n_1 = 1$.

By the theorem of Xi [**X1**, 8.4] we have

$$J_{\mathbf{c}} \otimes_{\mathbb{Z}} \mathbb{C} \sim_{morita} R_{\mathbb{C}}(F'_\lambda) = R_{\mathbb{C}}(Z) = \mathbb{C}^n.$$

Let γ have cycle type (n). Then the fixed set $({}^L T^0)^\gamma$ comprises the n fixed points

$$\mathrm{diag}(\omega^j, \ldots, \omega^j) \in {}^L T^0$$

where $\omega = \exp(2\pi i/n)$ and $0 \le j \le n - 1$. These n fixed points correspond to the n generators in the commutative ring $\mathbb{C}^n$.

We expect that the corresponding points in $\mathrm{Irr}(\mathrm{PGL}(N))$ arise as follows. The unramified unitary twist

$$z^{\mathrm{valodet}} \otimes \mathrm{St}(n)$$

of the Steinberg representation of $\mathrm{GL}(n)$ has trivial central character if and only if z is an nth root of unity. For these values $1, \omega, \omega^2, \ldots, \omega^{n-1}$ of z, we obtain n irreducible smooth representations of $\mathrm{PGL}(n)$.

From now on, we will assume that $\lambda \ne (1, 1, 1, \ldots, 1)$. Then F'_λ is a *connected* Lie group.

We will write $T_\lambda(\mathbb{C})$ for the standard maximal torus of F'_λ. The Weyl group is then

$$W(\lambda) = S_{n_1} \times \cdots \times S_{n_p}.$$

According to Bourbaki [**Bo**, Chapter 8], the map Ch, sending each (virtual) representation to its (virtual) character, creates an isomorphism:

$$Ch : R(F'_\lambda) \cong \mathbb{Z}[X^*(T_\lambda(\mathbb{C})]^{W(\lambda)}.$$

Note that a complex linear combination of rational characters of $T_\lambda(\mathbb{C})$ is precisely a regular function on $T_\lambda(\mathbb{C})$.

For each two-sided cell **c** of W the $\mathbb{Z}$-submodule $J_{\mathbf{c}}$ of J, spanned by all t_w, $w \in \mathbf{c}$, is a two-sided ideal of J. The ring $J_{\mathbf{c}}$ is the based ring of the two-sided cell **c**. Now apply the theorem of Xi [**X1**, 8.4]. We get

$$J_{\mathbf{c}} \otimes_{\mathbb{Z}} \mathbb{C} \sim_{morita} R_{\mathbb{C}}(F'_\lambda) \cong \mathbb{C}[T_\lambda(\mathbb{C})]^{W(\lambda)} \cong \mathbb{C}[T_\lambda(\mathbb{C})/W(\lambda)].$$

Let $\gamma \in S_n$ have cycle type μ. Then the γ-centralizer is a direct product of wreath products:

$$Z(\gamma) \simeq (\mathbb{Z}/\mu_1 \mathbb{Z}) \wr S_{n_1} \times \cdots \times (\mathbb{Z}/\mu_p \mathbb{Z}) \wr S_{n_p}.$$

The image of $T_\lambda(\mathbb{C})$ in the inclusion $T_\lambda(\mathbb{C}) \to {}^L T^0$ is precisely the subtorus of ${}^L T^0$ fixed by $\mathbb{Z}/\mu_1 \mathbb{Z} \times \cdots \times \mathbb{Z}/\mu_p \mathbb{Z}$. We therefore have

$$({}^L T^0)^\gamma / Z(\gamma) \simeq T_\lambda(\mathbb{C})/W(\lambda).$$

We conclude that

$$R(F'_\lambda) \otimes_{\mathbb{Z}} \mathbb{C} \simeq \mathbb{C}[(^L T^0)^\gamma / Z(\gamma)]$$

Then

$$J \otimes_{\mathbb{Z}} \mathbb{C} = \oplus_{\mathbf{c}} (J_{\mathbf{c}} \otimes_{\mathbb{Z}} \mathbb{C}) \sim \oplus_{\mathbf{c}} (R(F'_\lambda) \otimes_{\mathbb{Z}} \mathbb{C}) \simeq \mathbb{C}[\widetilde{^L T^0} / S_n]$$

The algebra $J \otimes_{\mathbb{Z}} \mathbb{C}$ is Morita equivalent to a reduced, finitely generated, commutative unital $\mathbb{C}$-algebra, namely the coordinate ring of the extended quotient $\widetilde{^L T^0} / S_n$. □

13. The Iwahori ideal in $\mathcal{H}(\mathrm{SO}(5))$

Let G denote the special orthogonal group $\mathrm{SO}(5, F)$. We view it as the group of elements of determinant 1 which stabilise the symmetric bilinear form

$$\begin{pmatrix} 0 & 0 & 0 & 0 & 1 \\ 0 & 0 & 0 & 1 & 0 \\ 0 & 0 & 1 & 0 & 0 \\ 0 & 1 & 0 & 0 & 0 \\ 1 & 0 & 0 & 0 & 0 \end{pmatrix}.$$

Let T be the group of diagonal matrices

$$\begin{pmatrix} \lambda_1 & 0 & 0 & 0 & 0 \\ 0 & \lambda_2 & 0 & 0 & 0 \\ 0 & 0 & 1 & 0 & 0 \\ 0 & 0 & 0 & \lambda_2^{-1} & 0 \\ 0 & 0 & 0 & 0 & \lambda_1^{-1} \end{pmatrix}, \quad \lambda_1, \lambda_2 \in F^\times.$$

The extended affine Weyl group $W = X_*(T) \rtimes W_f$ is of type $\tilde{B}_2$, with $W_f \simeq S_2 \ltimes (\mathbb{Z}/2)^2$ a finite Weyl group of type B_2.

Here $^L G^0 = \mathrm{Sp}(4, \mathbb{C})$, and $^L T^0$ is the group of diagonal matrices

$$d(t_1, t_2) := \begin{pmatrix} t_1 & 0 & 0 & 0 \\ 0 & t_2 & 0 & 0 \\ 0 & 0 & t_2^{-1} & 0 \\ 0 & 0 & 0 & t_1^{-1} \end{pmatrix}, \quad t_1, t_2 \in \mathbb{C}^\times.$$

We have $\mathrm{N}_{^L G^0}(^L T^0)/^L T^0 \simeq W_f$. The group $^L G^0 = \mathrm{Sp}(4, \mathbb{C})$ is simply connected. In [**X2**, §11.1], Xi has proved Lusztig's conjecture for the group W.

The extended Coxeter group $W = W' \rtimes \Omega$ has four two-sided cells c_e, c_1, c_2 and c_0 (see [**X2**, §11.1]):

$$c_e = \{w \in W : \mathbf{a}(w) = 0\} = \{e, \omega\} = \Omega,$$

$$c_1 = \{w \in W : \mathbf{a}(w) = 1\},$$

$$c_2 = \{w \in W : \mathbf{a}(w) = 2\},$$

$$c_0 = \{w \in W : \mathbf{a}(w) = 4\} \quad \text{(the lowest two-sided cell)}.$$

We have

$$J = J_{\mathbf{c}_e} \oplus J_{\mathbf{c}_1} \oplus J_{\mathbf{c}_2} \oplus J_{\mathbf{c}_0}.$$

Let $\mathfrak{i} \in \mathfrak{B}(G)$ be determined by the cuspidal pair $(T, 1)$. We have

$$\mathfrak{Z}^{\mathfrak{i}} = \mathbb{C}[^L T^0 / W_f], \quad \widetilde{\mathfrak{Z}}^{\mathfrak{i}} = \mathbb{C}[\widetilde{^L T^0} / W_f].$$

LEMMA 6. *Let*

$$\mathbb{L} := \mathbb{C}[t, t^{-1}]^{\mathbb{Z}/2\mathbb{Z}}$$

denote the balanced Laurent polynomials in one indeterminate t, where the generator α of $\mathbb{Z}/2\mathbb{Z}$ acts as follows: $\alpha(t) = t^{-1}$. Then the coordinate algebra of the extended quotient of $^L T^0$ by W_f is the $\mathbb{C}$-algebra

$$\mathbb{C}^5 \oplus \mathbb{L}^3 \oplus \mathbb{C}[^L T^0 / W_f].$$

PROOF. Let $X := {}^L T^0$. The 8 elements $\gamma_1, \ldots, \gamma_8$ of W_f can be described as follows:

$$\gamma_1(\mathrm{d}(t_1, t_2)) = (\mathrm{d}(t_1, t_2)) \qquad \gamma_2(\mathrm{d}(t_1, t_2)) = \mathrm{d}(t_2, t_1)$$
$$\gamma_3(\mathrm{d}(t_1, t_2)) = \mathrm{d}(t_1^{-1}, t_2) \qquad \gamma_4(\mathrm{d}(t_1, t_2)) = \mathrm{d}(t_1, t_2^{-1})$$
$$\gamma_5(\mathrm{d}(t_1, t_2)) = \mathrm{d}(t_2, t_1^{-1}) \qquad \gamma_6(\mathrm{d}(t_1, t_2)) = \mathrm{d}(t_1^{-1}, t_2^{-1})$$
$$\gamma_7(\mathrm{d}(t_1, t_2)) = \mathrm{d}(t_2^{-1}, t_1) \qquad \gamma_8(\mathrm{d}(t_1, t_2)) = \mathrm{d}(t_2^{-1}, t_1^{-1})$$

We have $\gamma_5 = \gamma_2 \gamma_3 = \gamma_4 \gamma_2$, $\gamma_6 = \gamma_5^2 = \gamma_3 \gamma_4 = \gamma_4 \gamma_3$, $\gamma_7 = \gamma_5^3 = \gamma_2 \gamma_4 = \gamma_3 \gamma_2$, $\gamma_8 = \gamma_4 \gamma_2 \gamma_4 = \gamma_2 \gamma_4 \gamma_3$. The elements γ_2, γ_3, γ_4, γ_6 and γ_8 are of order 2, the elements γ_5 and γ_7 are of order 4. We obtain

$$X^{\gamma_1} = X, \quad X^{\gamma_2} = \left\{ \mathrm{d}(t, t) : t \in \mathbb{C}^\times \right\},$$

$$X^{\gamma_3} = X^{\gamma_4} = \left\{ \mathrm{d}(1, t_2), \mathrm{d}(-1, t_2) : t_2 \in \mathbb{C}^\times \right\},$$

$$X^{\gamma_5} = X^{\gamma_7} = \left\{ \mathrm{d}(1, 1), \mathrm{d}(-1, -1) \right\},$$

$$X^{\gamma_6} = \left\{ \mathrm{d}(1, 1), \mathrm{d}(1, -1), \mathrm{d}(-1, 1), \mathrm{d}(-1, -1) \right\}.$$

The elements γ_1, γ_6 are central, and we have

$$Z(\gamma_2) = \left\{ \gamma_1, \gamma_2, \gamma_6, \gamma_8 \right\},$$

$$Z(\gamma_3) = \left\{ \gamma_1, \gamma_3, \gamma_4, \gamma_6 \right\}, \quad Z(\gamma_7) = \left\{ \gamma_1, \gamma_5, \gamma_6, \gamma_7 \right\}.$$

There are five W_f-conjugacy classes:

$$\{\gamma_1\}, \{\gamma_6\}, \{\gamma_2, \gamma_8\}, \{\gamma_3, \gamma_4\}, \{\gamma_5, \gamma_7\}.$$

As representatives, we will take $\gamma_1, \gamma_2, \gamma_3, \gamma_5, \gamma_6$.

- We have

$$\mathbb{C}[X^{\gamma_6}/Z(\gamma_6)] = \mathbb{C}[X^{\gamma_6}/W_f] = \mathbb{C} \oplus \mathbb{C} \oplus \mathbb{C}$$

since there are three W_f-orbits in X^{γ_6}, namely

$$\{\mathrm{d}(1, 1)\}, \{\mathrm{d}(1, -1), \mathrm{d}(-1, 1)\}, \{\mathrm{d}(-1, -1)\}$$

- We have

$$\mathbb{C}[X^{\gamma_5}/Z(\gamma_5)] = \mathbb{C}[X^{\gamma_5}] = \mathbb{C} \oplus \mathbb{C} = R_\mathbb{C}(Z) = \mathbb{C}[\Omega] = J_{\mathbf{c}_e}$$

since $Z(\gamma_5)$ acts trivially on X^{γ_5}.

- We have

$$X^{\gamma_3} = \{\mathrm{d}(1, t); t \in \mathbb{C}^\times\} \sqcup \{\mathrm{d}(-1, t) : t \in \mathbb{C}^\times\}$$

The $Z(\gamma_3)$-orbit of $\mathrm{d}(1, t)$ is the unordered pair $\{\mathrm{d}(1, t), \mathrm{d}(1, t^{-1})\}$ and the $Z(\gamma_3)$-orbit of $\mathrm{d}(-1, t)$ is the unordered pair $\{\mathrm{d}(-1, t), \mathrm{d}(-1, t^{-1})\}$. Therefore we have

$$\mathbb{C}[X^{\gamma_3}/Z(\gamma_3)] = \mathbb{L} \oplus \mathbb{L}.$$

- We have $X^{\gamma_2} = \{d(t,t) : t \in \mathbb{C}^\times\} \cong \{t : t \in \mathbb{C}^\times\}$. The $Z(\gamma_2)$-orbit of the point t is the unordered pair $\{t, t^{-1}\}$. So we have

$$X^{\gamma_2}/Z(\gamma_2) \cong \mathbb{C}^\times/\mathbb{Z}/2$$

and $\mathbb{C}[X^{\gamma_2}/Z(\gamma_2)] = \mathbb{L}$.
- We have $\mathbb{C}[X^{\gamma_1}/Z(\gamma_1)] = \mathbb{C}[^L T^0/W_f]$.

$\square$

The reductive group $F_{\mathbf{c}_e}$ is the center of $^L G^0$, $F_{\mathbf{c}_1} = (\mathbb{Z}/2\mathbb{Z}) \ltimes \mathbb{C}^\times$ where $\mathbb{Z}/2\mathbb{Z}$ acts on $\mathbb{C}^\times$ by $z \mapsto z^{-1}$, and $F_{\mathbf{c}_2} = (\mathbb{Z}/2\mathbb{Z}) \times \mathrm{SL}(2,\mathbb{C})$ and there is a ring isomorphism $J_{\mathbf{c}_2} \simeq \mathrm{M}_4(R_{F_{\mathbf{c}_2}})$, where $R_{F_{\mathbf{c}_2}}$ is the rational representation ring of $F_{\mathbf{c}_2}$, [**X2**, Theorem 11.2].

We have $F_{\mathbf{c}_1} = \langle \alpha \rangle \ltimes \mathbb{C}^\times$ where α generates $\mathbb{Z}/2\mathbb{Z}$. Note the crucial relation

$$\alpha z = z^{-1}\alpha$$

with $z \in \mathbb{C}^\times$. The (semisimple) conjugacy classes in $F_{\mathbf{c}_1}$ are:

$$\{1\}, \quad \{-1\}, \quad \{\{z, z^{-1}\} : z \in \mathbb{C}^\times, z^2 \neq 1\}, \quad \alpha \cdot \mathbb{C}^\times$$

LEMMA 7. *Let* $\mathbb{M} := \mathbb{C}[t, t^{-1}]$ *denote the Laurent polynomials in one indeterminate* t. *We have*

$$J_{\mathbf{c}_1} \asymp \mathbb{C} \oplus (\mathbb{M} \rtimes \mathbb{Z}/2\mathbb{Z})$$

where the generator α *of* $\mathbb{Z}/2\mathbb{Z}$ *acts as follows:* $\alpha(t) = t^{-1}$.

PROOF. Let $F = F_{\mathbf{c}_1}$, $F = \langle \alpha \rangle \ltimes \mathbb{C}^\times$. We have to construct the simple $J_{\mathbf{c}_1}$-modules explicitly, following Xi [**X2**, p. 51, 107]. We will use Xi's explicit proof of the Lusztig conjecture for B_2. Let $Y = \{1, 2, 3, 4\}$ be the F-set such that as F-sets we have $\{1\} \cong \{2\} \cong F/F$ and $\{3, 4\} \cong F/F^0$. The simple $J_{\mathbf{c}_1}$-modules are given by

$$E_{s,\rho} := \mathrm{Hom}_{A(s)}(\rho, H_*(Y^s))$$

where $A(s)$ denotes the component group $C_F(s)/C_F(s)^0$ and ρ is a simple $A(s)$-module which appears in the homology group $H_*(Y^s)$. The set Y^s denotes the s-fixed set: $Y^s = \{y \in Y : sy = y\}$. The pair (s, ρ) is chosen up to F-conjugacy.

(A). $s = 1$, $\rho = 1$. Then $Y^s = \{1, 2, 3, 4\}$. Also $H_*(Y^s)$ is the free $\mathbb{C}$-vector space $V := \mathbb{C}^4$ on $\{1, 2, 3, 4\}$: we will denote its basis by $\{e_1, e_2, e_3, e_4\}$. $A(s) = \mathbb{Z}/2\mathbb{Z}$. The generator of $A(s)$ permutes e_3, e_4. Let V_1 denote the span of $e_1, e_2, e_3 + e_4$, let V_2 denote the span of $e_3 - e_4$. Then

$$E_{1,1} := \mathrm{Hom}_{A(s)}(\rho, V) = V_1 = \mathbb{C}^3.$$

(B). $s = 1$, $\rho = \epsilon$ where ϵ is the sign representation of $\mathbb{Z}/2\mathbb{Z}$. We have

$$E_{1,\epsilon} := \mathrm{Hom}_{A(s)}(\rho, V) = V_2 = \mathbb{C}.$$

Note that we have

$$E_{1,1} \oplus E_{1,\epsilon} = V = \mathbb{C}^4$$

as $A(s)$-modules.

(C). $s = -1$, $\rho = 1$. We have $Y^s = Y$, $A(s) = \mathbb{Z}/2\mathbb{Z}$ and

$$E_{-1,1} := \mathrm{Hom}_{A(s)}(\rho, V) = V_1 = \mathbb{C}^3$$

(D). $s = -1$, $\rho = \epsilon$. We have $Y^s = Y$, $A(s) = \mathbb{Z}/2\mathbb{Z}$ and

$$E_{-1,\epsilon} := \mathrm{Hom}_{A(s)}(\rho, V) = V_2 = \mathbb{C}$$

Note that we have

$$E_{-1,1} \oplus E_{-1,\epsilon} = V = \mathbb{C}^4$$

as $A(s)$-modules.

(E) $s = z$, $\rho = 1$ where $z \in \mathbb{C}^\times$, $z^2 \neq 1$. We have $Y^s = Y$, $A(s) = \{1\}$ and

$$E_{z,1} := \operatorname{Hom}(\mathbb{C}, V) = V = \mathbb{C}^4$$

(F). $s = \alpha$, $\rho = 1$. We have $Y^s = \{1, 2\}$, $H_*(Y^s) = \mathbb{C}^2$, $A(s) = \{1, -1, \alpha, -\alpha\} = \mathbb{Z}/2\mathbb{Z} \times \mathbb{Z}/2\mathbb{Z}$. We have

$$E_{\alpha,1} := \operatorname{Hom}_{A(s)}(\mathbb{C}, \mathbb{C}^2) = \mathbb{C}^2.$$

This concludes the list of simple $J_{\mathbf{c}_1}$-modules. We now turn to the Lusztig-Xi isomorphism of unital algebras:

$$J_{\mathbf{c}_1} \cong K_F(Y \times Y) \otimes_{\mathbb{Z}} \mathbb{C}.$$

The equivariant K-theory $K_F(Y \times Y) \otimes_{\mathbb{Z}} \mathbb{C}$ is equipped with the convolution product. The action of F on the set Y leads to the following description. We identify $K_F(Y \times Y) \otimes_{\mathbb{Z}} \mathbb{C}$ with the $\mathbb{C}$-algebra of 4×4 matrices (a_{ij}) where $a_{11}, a_{12}, a_{21}, a_{22} \in R(F) \otimes_{\mathbb{Z}} \mathbb{C}$ and all other entries are in $R(\mathbb{C}^\times) \otimes_{\mathbb{Z}} \mathbb{C}$, subject to the following conditions:

$$a_{14} = \overline{a_{13}}, \ a_{24} = \overline{a_{23}}, \ a_{41} = \overline{a_{31}}, \ a_{42} = \overline{a_{32}}, \ a_{44} = \overline{a_{33}}, \ a_{43} = \overline{a_{34}}$$

where, for all $z \in \mathbb{C}^\times$,

$$\overline{a_{ij}}(z) = a_{ij}(z^{-1}).$$

Let

$$\mathbb{M} := \mathbb{C}[t, t^{-1}]$$

denote the Laurent polynomials in one indeterminate t. We have an injective homomorphism of unital $\mathbb{C}$-algebras:

$$\psi : \mathbb{C} \oplus (\mathbb{M} \rtimes \mathbb{Z}/2\mathbb{Z}) \longrightarrow K_F(Y \times Y) \otimes_{\mathbb{Z}} \mathbb{C}$$

$$(\lambda, p + \alpha \cdot q) \mapsto \begin{pmatrix} \lambda & 0 & 0 & 0 \\ 0 & \lambda & 0 & 0 \\ 0 & 0 & p & \overline{q} \\ 0 & 0 & q & \overline{p} \end{pmatrix}$$

We claim that this map is a spectrum-preserving morphism of unital finite-type k-algebras. We view each entry in $K_F(Y \times Y)$ as a virtual character of F or $\mathbb{C}^\times$. We can then *evaluate* each matrix entry at $z \in \mathbb{C}^\times$. Evaluation at $z \in \mathbb{C}^\times$ determines an algebra homomorphism

$$K_F(Y \times Y) \otimes_{\mathbb{Z}} \mathbb{C} \longrightarrow M_4(\mathbb{C}).$$

This homomorphism gives $\mathbb{C}^4$ the structure of $J_{\mathbf{c}_1}$-module.

- When $z^2 \neq 1$, this homomorphism gives $\mathbb{C}^4$ the structure of a simple $J_{\mathbf{c}_1}$-module, namely $E_{z,1}$.
- When $z = 1$, the module $\mathbb{C}^4$ splits into simple $J_{\mathbf{c}_1}$-modules of dimensions 3 and 1, namely $E_{1,1}$ and $E_{1,\epsilon}$.
- When $z = -1$, the module $\mathbb{C}^4$ splits into simple $J_{\mathbf{c}_1}$-modules of dimensions 3 and 1, namely $E_{-1,1}$ and $E_{-1,\epsilon}$.

When restricted to the lower right 2×2 block of the 4×4 matrix algebra (a_{ij}), each simple 4-dimensional module splits into the direct sum of a 2-dimensional 0-module and a 2-dimensional simple module over the crossed product $\mathbb{M} \rtimes \mathbb{Z}/2\mathbb{Z}$. At 1 and -1, the module $\mathbb{C}^4$ restricts to a 2-dimensional 0-module and two 1-dimensional simple modules. It now follows that $\mathrm{Prim}(J_{\mathbf{c}_1})$ with one deleted point is in bijection (via our morphism of algebras) with $\mathrm{Prim}(\mathbb{M} \rtimes \mathbb{Z}/2\mathbb{Z})$. There is one remaining point in $\mathrm{Prim}(J_{\mathbf{c}_1})$, namely $E_{\alpha,1}$. This point in $\mathrm{Prim}(J_{\mathbf{c}_1})$ maps, via our morphism of algebras, to the one remaining primitive ideal $0 \oplus (\mathbb{M} \rtimes \mathbb{Z}/2\mathbb{Z})$.

We conclude from this that

$$J_{\mathbf{c}_1} \asymp \mathbb{C} \oplus (\mathbb{M} \rtimes \mathbb{Z}/2\mathbb{Z}).$$

$\square$

THEOREM 7. *Let* $\mathcal{H}^{\mathrm{i}}(\mathrm{SO}(5))$ *denote the Iwahori ideal in* $\mathrm{SO}(5)$. *We have*

$$\mathcal{H}^{\mathrm{i}}(\mathrm{SO}(5)) \asymp \widetilde{\mathfrak{Z}}^{\mathrm{i}}(\mathrm{SO}(5)).$$

PROOF. We note that

- $J_{\mathbf{c}_e} = \mathbb{C}[\Omega] = \mathbb{C}^2$
- $J_{\mathbf{c}_2} \asymp R_{\mathbb{C}}(F_{\mathbf{c}_2}) = \mathbb{L}^2$ since

$$R_{\mathbb{C}}(F_{\mathbf{c}_2}) = R_{\mathbb{C}}(\mathbb{Z}/2\mathbb{Z} \times \mathrm{SL}(2,\mathbb{C})) = R_{\mathbb{C}}(\mathbb{Z}/2\mathbb{Z}) \otimes R_{\mathbb{C}}(\mathrm{SL}(2,\mathbb{C})) = \mathbb{L} \oplus \mathbb{L}.$$

By Lemmas 4 and 7, we have

$$J = J_{\mathbf{c}_e} \oplus J_{\mathbf{c}_1} \oplus J_{\mathbf{c}_2} \oplus J_{\mathbf{c}_0} \asymp \mathbb{C}^2 \oplus (\mathbb{C}^3 \oplus \mathbb{L}) \oplus \mathbb{L}^2 \oplus J_{\mathbf{c}_0}.$$

By Lemma 6, we have

$$\mathbb{C}[\widetilde{{}^L T^0}/W_f] = \mathbb{C}^5 \oplus \mathbb{L}^3 \oplus \mathbb{C}[{}^L T^0/W_f].$$

We conclude, by Theorem 4, that

$$\mathcal{H}^{\mathrm{i}}(\mathrm{SO}(5)) \asymp J \asymp \mathbb{C}[\widetilde{X}/W_f] = \widetilde{\mathfrak{Z}}^{\mathrm{i}}(\mathrm{SO}(5))$$

as required. This confirms part (1) of our conjecture for the Iwahori ideal of $\mathcal{H}(\mathrm{SO}(5))$.

$\square$

It is worth noting that, in this example, there are 4 two sided cells and 5 W_f-conjugacy classes.

14. Consequences of the conjecture

PARAMETRIZATION OF THE SMOOTH DUAL. In this section we will suppose that the conjecture is true for the Iwahori ideal $\mathcal{H}^{\mathrm{i}}(G)$. We then have

$$\mathcal{H}^{\mathrm{i}}(G) \asymp \mathcal{O}(\widetilde{{}^L T^0}/W_f).$$

We now take the primitive ideal space of each side. We obtain a bijection

$$\mathrm{Prim}\,\mathcal{H}^{\mathrm{i}}(G) \longleftrightarrow \widetilde{{}^L T^0}/W_f.$$

Now $\mathrm{Prim}\,\mathcal{H}^{\mathrm{i}}(G)$ may be identified with the subset $\mathrm{Irr}_I(G)$ of the smooth dual $\mathrm{Irr}(G)$ which admits nonzero Iwahori-fixed vectors. This leads to a parametrization of $\mathrm{Irr}_I(G)$:

$$\mathrm{Irr}_I(G) \longleftrightarrow \widetilde{{}^L T^0}/W_f.$$

This is a *parametrization of* $\mathrm{Irr}_I(G)$ *by the* $\mathbb{C}$-*points in a complex affine algebraic variety (with several components)*. This parametrization is not quite canonical: it depends on the specific finite sequence of elementary steps (and filtrations) connecting $\mathcal{H}^i(G)$ to $\mathcal{O}(\widetilde{{}^LT^0}/W_f)$.

This parametrization will assign to each $\omega \in \mathrm{Irr}_I(G)$ a pair

$$(s, \gamma)$$

with $s \in {}^L T^0$, γ a W_f-conjugacy class.

More generally, the conjecture leads to a parametrization of $\mathrm{Irr}(G)$ by the $\mathbb{C}$-points in a complex affine locally algebraic variety (with countably many components). The dimensions of the components are less than or equal to the rank of G.

LANGLANDS PARAMETERS. We wish to make a comparison with Langlands parameters. We first recall some background material.

Let $\mathcal{W}_F$ denote the Weil group of F, and let $\mathcal{I}_F \subset \mathcal{W}_F$ be the inertia subgroup so that $\mathcal{W}_F / \mathcal{I}_F \simeq \mathbb{Z}$. Denote the Frobenius generator by Frob in order that $\mathcal{W}_F = \mathcal{I}_F \rtimes \langle \mathrm{Frob} \rangle$.

By *Langlands parameter* we mean a continuous homomorphism

$$\varphi \colon W_F \times \mathrm{SL}(2, \mathbb{C}) \to {}^L G^0$$

which is rational on $\mathrm{SL}(2, \mathbb{C})$ and such that $\varphi(\mathrm{Frob})$ is semisimple.

Call φ *unramified* if it has trivial restriction to $\mathcal{I}_F$. The unramified Langlands parameters are parameterized by pairs (s, u) with $s \in {}^L G^0$ semisimple, $u \in {}^L G^0$ unipotent with $sus^{-1} = u^{q_F}$.

To each Langlands parameter φ should correspond a finite packet Π_φ of irreducible smooth representations of G. According to the Langlands philosophy, the unramified Langlands parameters should be those for which the representations in Π_φ are unipotent in the sense of [L6]. This has been proved by Lusztig [L6] when G is the group of F-points of a split adjoint simple algebraic group.

More precisely, let Z be the centre of ${}^L G^0$, let $C(g)$ denote the centralizer in ${}^L G^0$ of $g \in {}^L G^0$, and $C(s, u) := C(s) \cap C(u)$. We denote by $C(s, u)^0$ the connected component of $C(s, u)$ and set $A(s, u) := C(s, u)/Z \cdot C(s, u)^0$. It is proved in [L6] that the isomorphism classes of unipotent representations of G are naturally in one-to-one correspondence with the set of triples (s, u, ρ) (modulo the natural action of ${}^L G^0$) where s, u as above and ρ an irreducible representation (up to isomorphism) of $A(s, u)$.

In the case when we restrict ourselves to representations in $\mathrm{Irr}_I(G)$ the above classification specializes to the Kazhdan-Lusztig classification, which provides a proof of a refined form of a conjecture of Deligne and Langlands: if G is the group of F-points of any split reductive group over F, the set $\mathrm{Irr}_I(G)$ is naturally in bijection with the set of triples (s, u, ρ) as above (modulo the natural action of ${}^L G^0$) such that ρ appears in the homology of the variety $\mathcal{B}_u^s$ of the Borel subgroups of ${}^L G^0$ containing both s and u (see [KL], [Re]).

For $\mathrm{GL}(n)$, the simultaneous centralizer $C(s, u)$ is connected, and so each Deligne-Langlands parameter is a pair (s, u). There is a bijection between unipotent classes in $\mathrm{GL}(n, \mathbb{C})$ and partitions of n (via Jordan canonical form). There is a second bijection between partitions of n and conjugacy classes in $W_f = S_n$ the symmetric group. In this way we obtain a perfect match

$$(s, u) \longleftrightarrow (s, \gamma).$$

The Deligne-Langlands parameters can be arranged to form an extended quotient of the complex torus $\mathbb{C}^\times$ by the symmetric group S_n. The details of this correspondence were recorded in [**HP**].

IRREDUCIBILITY OF INDUCED REPRESENTATIONS. If $W_\mathfrak{t}$ acts freely then $\widetilde{D_\sigma}/W_\mathfrak{t} \cong D_\sigma/W_\mathfrak{t}$ and the conjecture in this case predicts irreducibility of the induced representations $\iota_P^G(\sigma \otimes \chi)$. This situation is discussed in [**R1**], with any maximal Levi subgroup of G, and also with the following (non maximal) Levi subgroup

$$L = (\mathrm{GL}(2) \times \mathrm{GL}(2) \times \mathrm{GL}(4)) \cap \mathrm{SL}(8)$$

of $\mathrm{SL}(8)$ and σ defined as in [**R1**, p.127].

Acknowledgment. The second author was partially supported by an NSF grant.

References

[BC] P. Baum, A. Connes, *The Chern character for discrete groups, A Fête of topology*, Academic Press, New York, 1988, 163–232.

[BN] P. Baum, V. Nistor, *Periodic cyclic homology of Iwahori-Hecke algebras*, K-Theory **27** (2002), 329–357.

[BR] J. Bernstein, *Representations of p-adic groups*, Notes by K.E. Rumelhart, Harvard University 1992.

[B] J. Bernstein, P. Deligne, D. Kazhdan, M.-F. Vigneras, *Représentations des groupes réductifs sur un corps local*, Travaux en cours, Hermann, 1984.

[BG] J. Bernstein and S. Gelbart, *An introduction to the Langlands program*. Birkhauser, Basel, 2003.

[BB] L. Blasco, C. Blondel, *Algèbres de Hecke et séries principales généralisées de* $\mathrm{Sp}_4(F)$, Proc. London Math. Soc. **85** (2002), 659–685.

[Bo] N. Bourbaki, *Lie groups and Lie algebras*, Chapters 7–9, Springer 2005.

[BP] J. Brodzki, R.J. Plymen, *Complex structure on the smooth dual of* $\mathrm{GL}(n)$, Documenta Math. **7** (2002), 91–112.

[BK1] C.J. Bushnell, P.C. Kutzko, *The admissible dual of* $\mathrm{GL}(n)$ *via compact open subgroups*, Ann. Math. Study **129**, Princeton Univ. Press 1993.

[BK2] C.J. Bushnell, P.C. Kutzko, *Smooth representations of reductive p-adic groups: Structure theory via types*, Proc. London Math. Soc. **77** (1998), 582–634.

[BK3] C.J. Bushnell, P.C. Kutzko, *Semisimple types in* GL_n, Comp. Math. **119** (1999), 53–97.

[Ca] P. Cartier, *Representations of p-adic groups: a survey*, Proc. Symp. Pure Math. **33** (1979), part 1, 111–155.

[CDN] N. Chifan, S. Dascalescu and C. Nastasescu, *Wide Morita contexts, relative injectivity and equivalence results*, J. Algebra **284** (2005), 705–736.

[C] J. Cuntz, *Morita invariance in cyclic homology for nonunital algebras*, K-Theory **15** (1998), 301 - 305.

[CST] J. Cuntz, G. Skandalis, B. Tsygan, *Cyclic homology in noncommutative geometry*, EMS **121**, 2004.

[EH] D. Eisenbud, J. Harris, *The geometry of schemes*, Springer Graduate Text **197**, 2001.

[G] M. Geck, *An introduction to algebraic geometry and algebraic groups*, Oxford Graduate Texts in Mathematics **10**, Oxford University Press, 2003.

[GS] S. Gelbart, F. Shahidi, *Analytic properties of automorphic L-functions*, Perspectives in Math. **6** (1988).

[Go] D. Goldberg, *R-groups and elliptic representations for* SL_n, Pacific J. of Math. **165** (1994), 77–92.

[GR] D. Goldberg, A. Roche, *Hecke algebras and* SL_n-*types*, Proc. London Math. Soc. **90** (2005), 87–131.

[He] V. Heiermann, *Décomposition spectrale et représentations spéciales d'un groupe réductif p–adique*, Journal de l'Institut Math. Jussieu **3** (2004), 327–395.

[HP] J.E. Hodgins and R.J. Plymen, *The representation theory of p-adic* GL(n) *and Deligne-Langlands parameters*, Texts and Readings in Mathematics **10** (1996), 54–72.

[Ho] R. B. Howlett, *Normalizers of parabolic subgroups of reflection groups*, J. London Math. Soc. **21** (1980), 62–80.

[KL] D. Kazhdan, G. Lusztig, *Proof of the Deligne-Langlands conjecture for Hecke algebras*, Invent. Math. **87** (1987) 153–215.

[KNS] D. Kazhdan, V. Nistor, P. Schneider, *Hochschild and cyclic homology of finite type algebras*, Sel. Math., New ser. **4** (1998), 321–359.

[Ki1] J.-L. Kim, *Hecke algebras of classical groups over p-adic fields and supercuspidal representations*, Amer. J. Math. **121** (1999), 967–1029.

[Ki2] J.-L. Kim, *Hecke algebras of classical groups over p-adic fields II*, Compositio Math. **127** (2001), 117–167.

[Le] F. Lemmermeyer, *Reciprocity laws from Euler to Eisenstein*, Springer Monograph in Mathematics, Berlin, 2000.

[JL] J.-L. Loday, *Cyclic homology*, Springer-Verlag, Berlin, 1992.

[L0] G. Lusztig, *Singularities, character formulas and a q-analog of weight multiplicities*, Astérisque **101 −102** (1983), 208–229.

[L1] G. Lusztig, *Classification of unipotent representations of simple p-adic groups, II*, Represent. Theory **6** (2002), 243–289.

[L2] G. Lusztig, *Cells in affine Weyl groups, II*, J. Algebra **109** (1987), 536–548.

[L3] G. Lusztig, *Cells in affine Weyl groups, III*, J. Fac. Sci. Univ. Tokyo, Sect. IA, Math, **34** (1987), 223–243.

[L4] G. Lusztig, *Cells in affine Weyl groups, IV*, J. Fac. Sci. Univ. Tokyo, Sect. IA, Math, **36** (1989), 297–328.

[L5] G. Lusztig, *Representations of affine Hecke algebras*, Astérisque **171-172** (1989), 73-84.

[L6] G. Lusztig, *Classification of unipotent representations of simple p-adic groups*, Internat. Math. Res. Notices **11** (1995), 517–589.

[L7] G. Lusztig, *Hecke algebras with unequal parameters*, CRM Monograph Series **18**, Amer. Math. Soc. 2003.

[Mi] P. Mischenko, *Invariant tempered distributions on the reductive p-adic group* $GL_n(F_p)$, C. R. Math. Rep. Acad. Sci. Canada **4** (1982),123–127.

[M1] L. Morris, *Tamely ramified intertwining algebras*, Invent. Math. **114** (1993), 1–54.

[M2] L. Morris, *Level zero G-types*, Compositio Math. **118** (1999), 135–157.

[N] J. Neukirch, *Algebraic number theory*, Springer-Verlag, 1999.

[P] R.J. Plymen, *Reduced C^*-algebra of the p-adic group* GL(n) *II*, J. Functional Analysis **196** (2002) 119–134.

[Re] M. Reeder, *Isogenies of Hecke algebras and a Langlands correspondence for ramified principal series representations*, Represent. Theory **6** (2002), 101–126.

[R1] A. Roche, *Parabolic induction and the Bernstein decomposition*, Compositio Math. **134** (2002), 113–133.

[R2] A. Roche, *Types and Hecke algebras for principal series representations of split reductive p-adic groups*, Ann. scient. Éc. Norm. Sup. **31** (1998), 361–413.

[SM] G. Sanje Mpacko, *Types for elliptic non-discrete series representations of* $SL_N(F)$, *N prime and F a p-adic field*, J. Lie Theory **5** (1995), 101–127.

[W] J.-L. Waldspurger, *La formule de Plancherel d'après Harish-Chandra*. Journal de l'Institut Math. Jussieu 2 (2003) 235–333.

[X1] N. Xi, *The based ring of two-sided cells of affine Weyl groups of type* $\widetilde{A}_{n-1}$, Amer. Math. Soc. Memoir **749**, 2002.

[X2] N. Xi, *Representations of affine Hecke algebras*, Lecture Notes in Math. **1587**, Springer, 1994.

[X3] N. Xi, *The based ring of the lowest two-sided cell of an affine Weyl group*, J. of Algebra **134** (1990) 356–368.

INSTITUT DE MATHÉMATIQUES DE JUSSIEU, U.M.R. 7586 DU C.N.R.S., PARIS, FRANCE
E-mail address: aubert@math.jussieu.fr

MATHEMATICS DEPARTMENT, PENNSYLVANIA STATE UNIVERSITY, UNIVERSITY PARK, PA 16802, USA
E-mail address: baum@math.psu.edu

SCHOOL OF MATHEMATICS, MANCHESTER UNIVERSITY, MANCHESTER M13 9PL, ENGLAND
E-mail address: plymen@manchester.ac.uk

Hilbert modular forms and the Ramanujan conjecture

Don Blasius

ABSTRACT. This paper completes the proof, at all finite places, of the Ramanujan Conjecture for motivic holomorphic Hilbert modular forms which belong to the discrete series at the infinite places. In addition, the Weight-Monodromy Conjecture of Deligne is proven for the Shimura varieties attached to $GL(2)$ and its inner forms, and the conjecture of Langlands, often today called the *local-global compatibility*, is established at all places for these varieties. This latter conjecture gives, for a finite place v of the field of definition F, an automorphic description of the action of decomposition group Γ_v of the Galois group $Gal(\overline{F}/F)$ on the l-adic cohomology of the the variety, at least if l is distinct from the residue characteristic of v. In particular, the Hasse-Weil zeta functions of these varieties are computed at all places.

Let F be a totally real field. In this paper we study the Ramanujan Conjecture for Hilbert modular forms and the Weight-Monodromy Conjecture for the Shimura varieties attached to quaternion algebras over F. As a consequence, we deduce, at all finite places of the field of definition, the full automorphic description conjectured by Langlands of the zeta functions of these varieties. Concerning the first problem, our main result is the following:

THEOREM 1. *The Ramanujan conjecture holds at all finite places for any cuspidal holomorphic automorphic representation π of $GL(2, \mathbf{A}_F)$ having weights all congruent modulo 2 and at least 2 at each infinite place of F.*

See below (2.2) for a more precise statement. For background, we note that the above result has been known for any such π at all but finitely many places, and without the congruence restriction, since 1984 ([**BrLa**]), as a consequence of the direct local computation of the trace of Frobenius on the intersection cohomology of a Hilbert modular variety. Additionally, the local method of [**Ca**] is easily seen to yield the result at *all* finite places, for the forms π which satisfy the restrictive hypothesis that either $[F : \mathbf{Q}]$ is odd or the local component π_v is discrete series at some finite place v. Hence, the novel cases in Theorem 1 are essentially those of the forms π attached to F of even degree, and which belong to the principal series at all finite v.

To prove Theorem 1, we here proceed globally, using the fact ([**Ca**], [**Oh**], [**T1**], [**W**]) that there exist two dimensional irreducible ([**BR1**], [**T2**]) l-adic representations $\rho_l^T(\pi)$ of the Galois group of $\overline{F}$ over F attached to such forms π. Crucial to us is the fact that these representations satisfy the Global Langlands Correspondence, i.e. that at every finite place v whose residue characteristic is different from l, the representations of the Weil-Deligne group defined by π_v and $\rho_l^T(\pi)$ ([**Ca**],[**T1**], [**BR1**], [**T2**], [**W**]) are isomorphic. Thus we get information about π_v from that about the local Galois representation $\rho_l^T(\pi)|D_v$ whenever we realize $\rho_l^T(\pi)$, or a closely related representation $\rho_l'(\pi)$, in some l-adic cohomology. Many such realizations are provided by the Shimura varieties attached to inner forms of $GL(2)/F$, and to the unitary groups $GU(2)/K$ and $GU(3)/K$ where K is a totally real solvable extension of F. Actually, to go beyond the case of lowest discrete series at ∞, in order to obtain cohomological realizations of these Galois representations $\rho_l^T(\pi)$ it is necessary to consider fiber systems of abelian varieties over these unitary Shimura varieties. However, we need no explicit treatment of them here since the result is contained in [**BR1**]. The fact that these are realizations of ρ_l^T follows from suitable local Hasse-Weil zeta function computations at all but finitely many *good places*; it is important to note that in this paper no new such computations at bad places are done.

To actually get the results, there are several overlapping methods:

A. If one of the weights is greater than 2, or if either (a) $[F : \mathbf{Q}]$ is odd or (b) there is a finite place at which π is discrete series, the result follows easily from a basic theorem of De Jong ([**DJ**]), the Local Langlands Correspondence, and the classification of unitary representations of $GL(2)$ over a local field. In all these cases there is a direct realization of $\rho_l^T(\pi)$ as a subquotient of an l-adic cohomology group of a variety.

B. If all the weights are 2, we proceed, using a known case of Langlands functoriality, by finding a geometric realization of a Galois representation $\rho_l'(\pi)$, made using $\rho_l^T(\pi)$, and from which we can deduce crucial constraints on the Frobenius eigenvalues of $\rho_l^T(\pi)$ at an unramified place under study. While several approaches are possible, we here use one for which the L-function of $\rho_l'(\pi)$ is, after a formal base change to a field L, a Rankin product L- function defined by $\pi|_L$ and a Galois twist $^\tau\pi|_L$. Unlike case (A) above, to conclude Ramanujan by an extension of that method we use a stronger, global Ramanujan estimate ([**Sha**]) for $GL(2)$ which the local analytic theory cannot provide. Although several alternative constructions of $\rho_l'(\pi)$ are possible, the present method has the merit that, further developed, it enables progress on the p-adic analogue of the Langlands correspondence for these forms. Nevertheless, in order not to obscure the simple formal structure of the paper, we defer p-adic questions to a sequel.

C. If all the weights are 2, we can give (See 4.2) prove Theorem 1 by a geometric argument (found after that of B.) using the fact that the Weight-Monodromy Conjecture is a theorem for surfaces. We give both arguments since, the method of B., although a little longer, has a chance to be applicable to other cases, such as regular algbraic forms on $GL(N)$ where $N > 2$.

In this paper, we have restricted our study to the case of forms having weights all congruent modulo 2. However, the method may extend to all holomorphic forms whose weights are all at least 2 at the infinite places. A key fact, already present in [**BR1**], is that a suitable twist $\pi' = \pi_K \otimes \chi$ of a CM quadratic base change π_K of π defines motivic forms on appropriate unitary groups $GU(2)$ and $GU(3)$. Once the Global Langlands Correspondence (See below, Section 2.3)is known for these forms, the Ramanujan Conjecture will follow by the methods of this paper. One natural approach is to generalize, in the setting of those $GU(2)$ which define curves, the results of Carayol ([**Ca**]), and then to extend by congruences ([**T1**]), to the general case.

The second main goal of the paper is to provide new examples, of arbitrary dimension, and with N (See the text for definitions) of many different, often highly decomposable, types, of the Weight-Monodromy Conjecture (WMC) ([**D1**]).

THEOREM 2. *Let Sh_B be the Shimura variety attached to a quaternion algebra B over a totally real field F. Then WMC holds for the l-adic cohomology of Sh_B at all finite places v whose residue characteristic is different from l.*

Remarks.

1. Sh_B is a projective limit of varieties $Sh_{B,W}$, where W is an open compact subgroup of the finite adele group of the reductive $\mathbf{Q}$-group $G = G_B = Res_{F/\mathbf{Q}}(B*)$ associated to the multiplicative group of B. Each $Sh_{B,W}$ is defined over the canonically defined number field F', named by Shimura the *reflex field*; the definition is recalled below in Section 3. We say that WMC holds for Sh_B if it holds for each smooth variety $Sh_{B,W}$.

2. The Shimura variety is not proper exactly when $B = M_2(F)$, in which case the connected components of the $Sh_{B,W}$ are the classical Hilbert modular varieties. In this case, the theorem is understood to refer to the l-adic intersection cohomologies of the Baily-Borel compactification of $Sh_{B,W}$.

3. Several authors have recently made significant progress on cases of WMC involving Shimura varieties. In [**It2**], instances of WMC are shown for certain Shimura varieties Sh associated to unitary groups. In fact, WMC is shown at places v at which Sh admit p-adic uniformization. In [**DS**], the p-adic extension of WMC is shown for a similar class of varieties: here v divides p. As already noted, this is a case not treated at all in this paper. Finally, in [**TY**] Taylor and Yoshida establish WMC, by careful study of the Rapoport-Zink spectral sequence, for all Shimura varieties associated to the unitary groups defined by division algebras over a CM field which are definite at all but one infinite place. This is the key class studied in [**HT**], and is a vast generalization of the Shimura curves studied in [**Ca**]. As a consequence, WMC is true for the l-adic representations attached to the class of essentially self-dual regular automorphic cusp forms on $GL(N, \mathbf{A}_F)$. This result implies Theorem 1 for π which are discrete series at some finite place, in which case the result is due to Carayol ([**Ca**]).

As a corollary of the above result, we achieve easily the third main goal of the paper: the proof of Langlands' conjecture ([**L1**]) which describes, in automorphic terms, the Frobenius semisimplification of the action of a decomposition group D_v for v on the l-adic Galois cohomology of the quaternionic Shimura varieties. Here v is any finite place of the reflex field, and l is a prime different from the residue characteristic of v. . This result completes the zeta function computations of Langlands ([**L1**]), Brylinksi-Labesse ([**BrLa**]) and Reimann ([**Re2**])).

THEOREM 3. *Let B be a quaternion algebra over a totally real field F having $B_v \cong M_2(\mathbf{R})$ for $r > 0$ infinite places v of F. Let F' be the canonical field of definition of the r-fold Sh_B attached to B. Let π' be a cuspidal holomorphic representation of $G) = (B \otimes \mathbf{A}_F)^*$ such that*

    (1) *π'_v has weight 2 at each split infinite place,*

    (2) *π'_v is one-dimensional at each ramified infinite place,*

(3) *the central character ω of π' has the form $\omega = |\cdot|^{-1}\Psi$, with a character of finite order Ψ.*

Let l be a rational prime. Then for each finite place v of F' whose residue characteristic is different from l, the isomorphism class of the Frobenius semisimple parameter $(\rho^*_{W,v}, N_{W,v})$ of the Weil-Deligne group WD_v of F'_v defined by the restriction to a decomposition group for v of the action of $Gal(\overline{\mathbf{Q}}/F')$ on

$$H^r(Sh_{B,W}, \overline{\mathbf{Q}_l})(\pi'_{f,W})$$

coincides with the class of

$$m(\pi'_f, W) r_B(\sigma(JL(\pi')_p)|_{WD_{F'_v}}),$$

where $m(\pi'_f, W)$ is defined by

$$dim(H^r(Sh_{B,W}, \overline{\mathbf{Q}_l})(\pi'_{f,W})) = 2^r m(\pi'_f, W).$$

Here, for p the place of $\mathbf{Q}$ lying under v,

(1) $JL(\pi')_p$ is the p-component of the cuspidal representation of $GL(2, \mathbf{A}_F)$, obtained from π' via the Jacquet-Langlands correspondence JL.
(2) $\sigma(JL(\pi')_p)$ is the homomorphism of WD_p into the L-group $^L G$ which is, as usual, identified with the L-group of the $\mathbf{Q}$-group $R_{F/\mathbf{Q}}(GL(2))$.
(3) r_B is the complex representation of dimension 2^r defined by Langlands.
(4) Let $\mathcal{H}_W$ be the level W Hecke algebra of G which consists of the convolution algebra of left and right W invariant compactly supported functions on $G(\mathbf{A}_f)$. Then $\pi'_{f,W}$ is the representation of $\mathcal{H}_W$ on the subspace $(\pi'_f)^W$ of π'_f consisting of the vectors fixed by all of W.

For an exposition of (2), see [**BR2**], 3.5, and [**Ku**]. For an exposition of (3), defined by Langlands ([**L1**]), see [**BR2**] esp. 5.1, 7.2. Definitions are briefly recalled as needed in the paper. Note that we are computing the L-functions as Euler products over the primes of F', not as Euler products over primes of $\mathbf{Q}$.

This result may be the first verification, for some Shimura varieties of dimension greater than one, at all places and levels, of Langlands' general conjecture. Nevertheless, for the last two theorems, the proofs are rather formal and do not involve new direct local verifications of difficult facts. On the contrary, one key principle is that the semisimplification of the *global*, i.e. $Gal(\overline{\mathbf{Q}}/F')$, Galois action on the l-adic cohomology of any variety in Theorem 2 is computable in simple ways from globally *irreducible* l-adic representations which satisfy WMC and the Global Langlands correspondence at each place. Of course this type of fact does not hold locally: the WMC concerns, for each place v of F', the nature of the associated indecomposable, and in general non-irreducible, Frobenius semisimple representations of the Weil-Deligne group.

1. Background

1.1. Weil Numbers. Let q be power of a rational prime p. An *integral q-Weil number of weight $j \in \mathbf{Z}$* is an algebraic integer α having the property that, for each automorphism σ of $\overline{\mathbf{Q}}$, we have

$$|\sigma(\alpha)| = q^{j/2},$$

with a fixed j independent of σ. We omit reference to q or the weight j when convenient. An algebraic number of the form $\beta = \alpha q^n$, for some $n \in \mathbf{Z}$ and an integral q-Weil Number α is called a q-*Weil number*, or simply a Weil number, if the q is clear from context. Obvious facts about Weil numbers include: (i) the q-Weil numbers form a group under multiplication;(ii) if $q = q_0^f$, then α is a q-Weil number of weight j if and only if $\sqrt[f]{\alpha}$ is a q_0-Weil number of weight j; (iii) all roots of unity are q-Weil numbers of weight 0 for all q.

1.2. l-adic Representations. Let K be a field and let $\Gamma_K = Gal(\overline{K}/K)$ be the group of K-linear automorphisms of its algebraic closure $\overline{K}$, endowed with the usual topology. For a prime l, let V be a finite dimensional vector space over $\overline{\mathbf{Q}_l}$, and let $\rho : \Gamma_K \to GL(V)$ be a homomorphism. We say that ρ is an l-*adic representation* if there exists a finite extension T of $\mathbf{Q}_l$, a T vector space V_0, and a continuous homomorphism $\rho_0 : \Gamma_K \to GL(V_0)$ which becomes isomorphic to ρ after extension of scalars on V_0 from T to $\overline{\mathbf{Q}_l}$. We use the notation ρ, V, and (V, ρ) at will to denote such a representation. An l-adic representation (V, ρ) is called *motivic* if there is a smooth projective variety X over K such that (V, ρ) is isomorphic to a Tate twist of a subquotient of the Γ_K-module $H^*(\overline{X}, \overline{\mathbf{Q}_l})$ where $\overline{X}$ is the scalar extension of X to the algebraic closure $\overline{K}$ of K. Here, for a Γ_K-module (V, ρ), and $m \in \mathbf{Z}$, the Tate twisted module is the pair $(V(m), \rho(m))$ where $V(m) = V$, $\rho(m) = \rho \otimes \chi_l^m$, and χ_l is the usual l-adic cyclotomic character.

1.3. Local Weil group. For the rest of this paper, K will denote a local field of characteristic 0 and residue characteristic p. We denote by q the number of the residue field. Of course, q is a power of p. We let l be any rational prime different from p. We recall some basics about the Weil group $W_K \subset \Gamma_K$ of K. Let I be the inertia subgroup of W_K. Then W_K/I is isomorphic to the subgroup $q^{\mathbf{Z}}$ of $\mathbf{Q}^*$; the isomorphism is that induced by the homomorphism that sends an element w of W_K to the power $|w|$ of q to which it raises the prime-to-p roots of unity in the maximal unramified extension of K. Any element Φ of W_K for which $|\Phi| = q^{-1}$ is called a *Frobenius*. Let I_w be the subgroup of wild inertia, i.e. the maximal pro-p subgroup of I. Let I_t denote the quotient I/I_w and let $W_{K,t}$ denote W_K/I_w. We call these groups the *tame inertia* group and the *tame Weil* group, respectively. Then I_t is non-canonically isomorphic to the product

$$\prod_{l \neq p} \mathbf{Z}_l,$$

and $W_{K,t}$ is isomorphic to the semidirect product of $\mathbf{Z}$ and I_t ; the action of W_K on I_t is given by

$$wxw^{-1} = |w|x$$

for all $x \in I_t$ and all $w \in W_K$. Choose, once and for all, an isomorphism $t = (t_l)_{l \neq p}$ of I_t with

$$\prod_{l \neq p} \mathbf{Z}_l.$$

Let (V, ρ) be an l-adic representation of Γ_K. We extend, replacing Γ_K by W_K, the definition of an l-adic representation to W_K, and thus each l-adic representation ρ of Γ_K gives rise, by restriction, to an l-adic representation of W_K which we also denote by (V, ρ).

1.4. Grothendieck's Theorem. According to a basic result of Grothendieck ([**ST**], Appendix), there is a subgroup J of finite index in I such that, for $\sigma \in J$,

$$\rho(\sigma) = exp(t_l(\sigma)N)$$

where $N \in End(V)$ is a uniquely determined nilpotent endomorphism.

If we can take $J = I$ in this theorem, (V, ρ) is said to be *semistable*. It is well-known that there exists a finite extension L of K such that $(\rho|_L, V)$ is semistable.

1.5. Weil-Deligne parametrization of l-adic representations.
Fix a choice Φ of a Frobenius in W_K. Define, for this Φ, and any σ in W_K, an automorphism

$$\rho_{WD}(\sigma) = \rho(\sigma)exp(-t_l(\Phi^{-log_q(|\sigma|)}\sigma)N)$$

of V. Then $\sigma \to \rho_{WD}(\sigma)$ is a continuous representation of W_K whose restriction to I has finite image. The triple (V, ρ_{WD}, N) depends on the choice of t_l and Φ. Such a triple (V', ρ'_{WD}, N') arises from an l-adic representation on V of Γ_K if and only if the relation

$$\rho'_{WD}(\sigma)N'\rho'_{WD}(\sigma)^{-1} = |\sigma|N',$$

holds for all $\sigma \in W_K$. Note that (V, ρ) is semistable if and only if ρ_{WD} is unramified, i.e. trivial on I.

1.6. Frobenius semisimplification. Following Deligne ([**D3**], 8.5), let

$$\rho_{WD}(\Phi) = \rho_{WD}(\Phi)^{ss}u$$

be the Jordan decomposition of $\rho_{WD}(\Phi)$ as the product of a diagonalizable matrix $\rho_{WD}(\Phi)^{ss}$ and a unipotent matrix u. Define, for $\sigma \in W_K$,

$$\rho_{WD}^{ss}(\sigma) = \rho_{WD}(\sigma)u^{log_q(|\sigma|)}.$$

Then ρ_{WD}^{ss} is a semisimple representation of W_K and, for all σ, $\rho_{WD}^{ss}(\sigma)$ is semisimple. The representation (V, ρ_{WD}^{ss}) is called the Φ-semisimplification of (V, ρ_{WD}) and the triple (V, ρ_{WD}^{ss}, N) is called the Φ-semisimplification of (V, ρ_{WD}, N).

Let now ι_l be an isomorphism of $\overline{\mathbf{Q}_l}$ with the complex numbers $\mathbf{C}$. This will be fixed in any discussion, and, to avoid a cumbersome notation, we will identify $\overline{\mathbf{Q}_l}$ with $\mathbf{C}$, suppressing explicit reference to ι_l. We will use ι_l to define complex representations of the Weil-Deligne group (c.f. [**D3**], 8.3, [**Ta**], 4.1, or [**Roh**]), via the triples (V, ρ_{WD}^{ss}, N).

1.7. WD_K. The *Weil-Deligne* group WD_K of K is the semidirect product of W_K with $\mathbf{C}$ defined by the relation

$$\sigma z \sigma^{-1} = |\sigma|z$$

for all $\sigma \in W_K$ and $z \in \mathbf{C}$. Using ι_l we regard V as a finite dimensional complex vector space (i.e. if $z \in \mathbf{C}$ and $v \in V$, we put $zv = \iota_l^{-1}(z)v$). Then ρ_{WD}^{ss} is a continuous representation of W_K on V, and N is a nilpotent endomorphism of V. The complex triple (V, ρ_{WD}^{ss}, N) defines, in view of (1.5), a representation ρ^* of WD_K by the rule

$$\rho^*((z, \sigma)) = exp(zN)\rho_{WD}^{ss}(\sigma)$$

for all $(z, \sigma) \in WD_K$. Then ρ^* satisfies

(i) the restriction to W_K is semisimple, and

(ii) the restriction to $\mathbf{C} = \mathbf{G}_a(\mathbf{C})$ is algebraic.

We denote the family of all complex representations satisfying (i) and (ii) by $Rep^s(WD_K)$ and denote members by pairs (V, ρ'); a triple giving rise to (V, ρ') by the construction above is given by

$$(V, \rho'|_{W_K}, N_{\rho'})$$

where log of a unipotent matrix M is the standard polynomial in $M - 1$ inverting exponentiation on nilpotents and

$$N_{\rho'} = log(\rho'((1,1))).$$

Henceforth an element of $Rep^s(WD_K)$ is identified with the triple it defines. Note that a member of $Rep^s(WD_K)$ is actually a semisimple representation of WD_K if and only if it factors through the quotient W_K. A member of $Rep^s(WD_K)$ is called *semistable* if it is trivial on I. We denote by $Rep^{ss}(WD_K)$ the subfamily of $Rep^s(WD_K)$ consisting of semistable representations. Of course, if (V, ρ) is a semistable l-adic representation if and only if the associated element (V, ρ') of $Rep^s(WD_K)$ belongs to $Rep^{ss}(WD_K)$.

As shown in [**D3**], the isomorphism class of the $(V, \rho') \in Rep^s(WD_K)$ gotten from an l-adic representation of Γ_K is independent of the choices of Φ and t_l. The class of (V, ρ') does depend on the choice of ι_l, but, since any ι_l' has the form $\iota_l' = \eta\iota_l$ for an automorphism η of $\mathbf{C}$, we see that, after such a change, ρ' is just replaced by the conjugate $\eta\rho'$.

1.8. Structure of semistable modules.
Recall that a WD_K-module is *indecomposable* if it cannot be written as the direct sum two proper submodules. We have the following basic structure results ([Roh]) for the members of $Rep^{ss}(WD_K)$:

(i) Any member of $Rep^{ss}(WD_K)$ is isomorphic to a direct sum of indecomposable modules, hence of $V_{\alpha,t}$'s. As such the decomposition is unique up to re-ordering the factors, and replacing factors by isomorphic factors.(ii) Any indecomposable member of $Rep^{ss}(WD_K)$ is isomorphic to exactly one of the form $V_{\alpha,t} = (\mathbf{C}^{t+1}, \rho_{\alpha,t}, N_t)$, where α is a non-zero complex number, t is a non-negative integer, and $\rho_{\alpha,t}$ is the unramified representation of W_K defined by the rule:

$$\rho_{\alpha,t}(\Phi) = Diag(\alpha, q^{-1}\alpha, ..., q^{-t}\alpha),$$

where $Diag$ denote diagonal matrix, and $N = (n_{ij})$, where $n_{ij} = 0$ unless $i = j+1$, in which case $n_{ij} = 1$.

1.9. Structure of Frobenius semisimple modules.
We have:

(i) Any member of $Rep^s(WD_K)$ is a direct sum of indecomposable submodules. As such the decomposition is unique up to re-ordering the factors, and replacing factors by isomorphic factors.

(ii) Any indecomposable representation is isomorphic to one of the form $V_{\Lambda,t} \overset{def}{=} \Lambda \otimes V_{q^{t/2},t}$ where Λ (and hence t) is a uniquely determined irreducible representation of W_K, and any such representation is indecomposable. Such a representation is irreducible iff $t = 0$.

(iii) if Λ is an irreducible representation of W_K and Φ is any Frobenius element in W_K, and α is an eigenvalue of Φ in Λ, then $|\alpha|$ is independent of α.

To see the last claim, note that we can find a Galois extension L of K such that the restriction to $WD_L \subseteq WD_K$ of Λ is unramified, hence a direct sum of unramified characters χ_k. Since Λ is irreducible, the χ_k are permuted transitively by the natural action of $\Gamma(L/K)$. Regarding them, via local class field theory, as characters of L^*, and letting τ be an element of $\Gamma(L/K)$, the action is just that sending χ_k to $\chi_k \circ \tau = \chi_k$. Hence all χ_k are the same character χ. Now let χ_0 be an unramified character of W_K such that $\chi_0 \circ N_{L/K} = \chi$, and consider the irreducible representation $\Lambda_0 = \Lambda \otimes \chi_0^{-1}$. Then the restriction to L of Λ_0 is trivial, and hence Λ_0 has finite image. In particular, $\Lambda(\Phi) = \Lambda_0(\Phi)\chi_0(\Phi)$, and so each eigenvalue α of Φ in Λ is of the form $\alpha = \zeta\chi_0(\Phi)$ with a root of unity ζ. This proves (iii).

Let Λ be an irreducible representation of WD_K. We call the real number $w(\Lambda) = 2log_q(|\alpha|)$, where α is any eigenvalue of any Φ, the *weight* of Λ. It is independent of the choices.

1.10. Pure modules. Fix an integer j. An indecomposable module $V_{\Lambda,t}$ for K as above is *q-pure of weight j*, or simply *pure*, if

(i) the eigenvalues of Φ in $V_{\Lambda,t}$ are q-Weil numbers, and

(ii) $w(\Lambda) = t + j$.

By the argument at the end of the previous subsection, changing Φ will change the eigenvalues of $\Lambda(\Phi)$ only by roots of unity, and hence both conditions are independent of the choice of Φ. Also, an indecomposable $V_{\Lambda,t}$ is q_K-pure of weight j if and only if, for each finite extension L of K, the restriction $V_{\Lambda,t}|_L$ of $V_{\Lambda,t}$ to $WD_L \subseteq WD_K$ is q_L-pure of (the same) weight j. To see this, note since the condition is obviously stable under passage from K to L, it is enough to show the descent statement from an L, as above, such that $\Lambda|L$ is unramified. In this case, if $f = f(L/K)$ is the degree of the residue field extension, then Φ^f is a Frobenius element for W_L, and, in the above notation, $\chi(\Phi^f) = \chi_0(\Phi)^f$. Hence $\alpha = (\chi(\Phi^f))^{1/f}\zeta$, for some f-th root of $\chi(\Phi^f)$. Suppose now that $\chi(\Phi^f)$ is a q_L-Weil number of weight j. Then, since $q_L = q_K^f$, α is a q_K-Weil number of weight j also. This shows (i) holds over K if it holds over L. To see (ii), just note that $w(\Lambda)$ is unchanged when $q_L = q_K^f$ is replaced by q_K and $|\chi(\Phi^f)|$ is replaced by $|(\chi(\Phi^f))^{1/f}|$. This proves the claim.

We say that a general member V of $Rep^s(WD_K)$ is pure of weight w ($w \in \mathbf{R}$) if each indecomposable constituent is pure of weight w. Of course, if the module V is pure of weight w, then w is uniquely determined. Furthermore, if V is pure of weight w, then any conjugate ηV, for $\eta \in Aut(\mathbf{C})$ is also pure of weight w.

Finally, we say that an l-adic representation V of Γ_K is

(i) q-pure of weight w if one, and hence any, associated member of $Rep^s(WD_K)$ is q-pure of weight w, and

(ii) pure if it is q-pure of weight w for some w.

Here are summarized some basic facts about elements of $Rep^s(WD_K)$:

Proposition 1

Let $V_1, ..., V_n$ be l-adic representations of Γ_K.

(i)Let V be the direct sum of the V_i. Then V is q-pure of weight j if and only if each V_i is q-pure of weight j.

(ii) V is q-pure of weight j if and only if its contragredient V^* is q-pure of weight $-j$.

(iii) if V is q-pure of weight j, then the Tate twisted module $V(m) = V \otimes \chi_l^m$, where χ_l is the usual l-adic cyclotomic character, and $m \in \mathbf{Z}$, is q-pure of weight $j - 2m$.

(iv) If V and W are q-pure of weights k and l, their tensor product $V \otimes W$ is q-pure of weight $k + l$.

1.11. Weight-Monodromy Conjecture. This is the following statement ([II]):

> Let X be a projective smooth variety defined over the local field K. Let, as usual, for a rational prime l which is different from the residue characteristic of K, $H^j(\overline{X}, \mathbf{Q}_l)$ be the l-adic étale cohomology of $\overline{X}$, regarded as a Γ_K-module. Then the Γ_K-module $H^j(\overline{X}, \mathbf{Q}_l)$ is q-pure of weight j, where q is the cardinality of the residue field of K.

Remark. Since X is projective, its cohomology is polarizable, and so Φ acts on $det(H^j(\overline{X}, \mathbf{Q}_l))$ as $q^{jb_j/2}$ where b_j is the dimension of $H^j(\overline{X}, \mathbf{Q}_l)$. On the other hand, if $H^j(\overline{X}, \mathbf{Q}_l)$ is q-pure of some weight k, we have also that Φ acts on this space as $q^{kb_j/2}$. Hence, we say simply that $H^j(\overline{X}, \mathbf{Q}_l)$ satisfies WMC if it is q-pure.

If, contrary to the convention of this paper, K and its residue field both have characteristic p, then WMC is a theorem of Deligne ([**D4**], Theorem 1.8.4). In mixed characteristic, WMC is known for curves and abelian varieties ([**SGA7-I**]), for surfaces ([**RZ**], Theorem 2.13, [**DJ**], and see below, (4.3)), certain threefolds ([**It1**]) and, as mentioned in the introduction, a class of Shimura varieties associated to division algebras over CM fields ([**It2**], [**TY**]).

As De Jong remarks in the Introduction to [**DJ**], it follows from his theory of alterations that condition (i) of the definition of purity is always satisfied for l-adic representations that are subquotients of the l-adic étale cohomology of a quasiprojective-variety X over a non-archimedian local field K. We sketch this result, for the case that X is smooth and projective, since it is basic.

Proposition 2.(De Jong). Let X be a smooth projective variety defined over the local field K. Let $\Phi \in \Gamma_K$ be a Frobenius. Then the eigenvalues of Φ on $H^j(\overline{X}, \overline{\mathbf{Q}_l})$ are integral Weil numbers.

PROOF. Let L be a finite extension of K over which there exists an L-alteration $a : X' \to X_L$ such that X' is the generic fiber of a strictly semistable scheme $\mathcal{X}'$ defined over the ring of integers $\mathcal{O}_L$ of L. Since an alteration is surjective and generically finite, we may regard $H^j(\overline{X}, \overline{\mathbf{Q}_l})$ as a submodule of $H^j(\overline{X'}, \overline{\mathbf{Q}_l})$ via a^*. Let $\underline{\mathcal{X}}'$ be the geometric special fiber of $\mathcal{X}'$. Since $\mathcal{X}'$ is strictly semistable, its cohomology is computable via the Γ_L-equivariant weight spectral sequence of Rapoport and Zink (c.f.[**RZ**], Section 2, and [**II**],3.8). In the notation of [**II**], we have

$$_W E_1^{ij} = H^{i+j}(\underline{\mathcal{X}}', gr_{-i}^W R\Psi(\overline{\mathbf{Q}_l})) \implies H^{i+j}(\overline{X'}, \overline{\mathbf{Q}_l}).$$

Thus, it suffices to show that the eigenvalues of Φ on each $_WE_1^{ij}$ are integral Weil numbers. But each $_WE_1^{i,j}$ is a direct sum of cohomology groups of the form

$$H^{j+i-2l}(\underline{\mathcal{X}}'^{(2l-i+1)}, \overline{\mathbf{Q}_l})(i-l)$$

where $l \geq max(0,i)$, and $\underline{\mathcal{X}}'^{(2l-i+1)}$ is the disjoint union of smooth proper subvarieties of $\underline{\mathcal{X}}'$ defined by taking $(2l-i+1)$- fold intersections of the irreducible divisors provided in the definition of the strict semistability of $\mathcal{X}'$. The result now follows from the Weil conjectures.

$$\square$$

1.12. Weight-Monodromy: Background Facts. Let W be a finite set of q-Weil numbers and $m : W \to \mathbf{Z}_{\geq 0}$ be a function. For us, $m(\alpha)$ is the multiplicity of α in the spectrum of a Frobenius in a semistable module. The pair (W,m) is said to be *wm-q-pure of weight j* if ,

(i) whenever $|\alpha| > q^{j/2}$, $m(q^{-1}\alpha) \geq m(\alpha)$, and,

(ii) for all α, $m(\alpha) = m(q^{-s_\alpha}\alpha)$, where $s_\alpha = 2log_q(|\alpha|q^{-j/2})$.

Let, for $\alpha \in W$, $|\alpha| \geq q^{j/2}$,

$$\delta(\alpha) = m(\alpha) - m(q\alpha).$$

Let W^+ be the subset of W of all α such that $|\alpha| \geq q^{j/2}$.

If (V,ρ) is a q-pure of weight j semistable representation of WD_K, let $b(V)$ denote the number of indecomposable factors in any representation of V as a direct sum of such so that $b(V) = dim(ker(N_\rho))$. More generally, for any nilpotent endomorphism N of V, let $b(N) = dim(ker(N))$ denote the number of indecomposable Jordan blocks in the representation of N as a direct sum of such. Evidently, if V is a q-pure of weight j semistable representation of WD_K, the associated pair (W_V, m_V) is wm-q-pure of weight j. Conversely, if (W,m) is wm-q-pure of weight j, then

$$\bigoplus_{\alpha \in W^+} V_{\alpha,s_\alpha}^{\delta(\alpha)}$$

belongs to the unique isomorphism class of q-pure of weight j semistable representations $(V_{(W,m)}, \rho_{(W,m)})$ of WD_K that give rise to (W,m).
In this case let

$$b(W,m) = b(V_W) = \Sigma_{\alpha \in W^+}\delta(\alpha)$$

.

The following elementary result is key to our work in this paper.

Proposition 3. Let K be a local field and let V be a finite dimensional representation in $Rep^s(WD_K)$. Let $F^\cdot V$ be a filtration of V by WD_K-stable submodules. Suppose that the graded Galois module $Gr_F(V)$ is q-pure of weight j. Then V is q-pure of weight j.

PROOF. Restricting from K to a suitable extension L, we can assume that V is semistable.

Let E_V be an endomorphism of a vector space V, and suppose that we have an E_V stable short exact sequence

$$0 \to S \to V \to Q \to 0$$

with induced endomorphisms E_S and E_Q on S and Q. Let K_S, K_V, and K_Q be the kernels of these operators. Then

$$dim(K_S) + dim(K_Q) \geq dim(K_V).$$

This is evident since we have a short exact sequence

$$0 \to K_V \cap S \to K_V \to \frac{K_V + S}{S} \to 0$$

and $K_V \cap S = K_S$ and $\frac{K_V+S}{S}$ is a subspace of K_Q.

Hence, by induction, if E_V is a filtered endomorphism of $F^\cdot V$, inducing $Gr_F(E_V)$ on $Gr_F(V)$, then

$$dim(ker(Gr_F(E_V))) \geq dim(ker(E_V)).$$

We apply this to the case $E_V = N$.

Lemma. Let (W, m) be wm-q-pure of weight j. Let (V, ρ) be a semistable representation of W_K such that $(W_V, m_V) = (W, m)$. Then

$$b(V) \geq b(W, m).$$

Further,

$$b(V) = b(W, m)$$

if and only if (V, ρ) is q-pure of weight j.

PROOF. Obvious. □

To conclude the proof of the Proposition, we note that $Gr_F(V)$ defines the same pair (W, m) as V. Since we assume $Gr_F(V)$ is pure, we have $b(W, m) = b(Gr_F(V))$. By the remarks just above, we always have $b(Gr_F(V)) \geq b(V)$ and $b(V) \geq b(W, m)$. Hence $b(V) = b(W, m)$ and so V is q-pure of weight j.

1.13. A problem on abelian varieties. Proposition 4.

Let A be an abelian variety defined over a number field J. Let M be an irreducible motive, defined over J in the category of motives for absolute Hodge cycles generated by $A([\mathbf{DM}])$. Then for each prime l and each finite place v of J, the l-adic cohomology M_l of M satisfies the WMC.

PROOF. This is, of course, trivial: any irreducible M is of then form $M_0(n)$ where M_0 is a submotive of the motive $\otimes^k H^1(A, \overline{Q})$, k is a non-negative integer, and n is the n-fold Tate twist. The l-adic cohomology $M_{0,l}$ of M_0 is a $Gal(\overline{J}/J)$ direct summand of $\otimes^k H^1(A, \overline{Q}_l)$ and hence everywhere locally satisfies WMC since $H^1(A, \overline{Q}_l)$ does. □

Problem: Is the conjugacy class of N_l(in $GL(M_l)$) independent of l? Evidently, this amounts to asking whether the Frobenius eigenvalues on the semisimplification of M_l is independent of l. Of course, these statements are consequences of the standard l-independence conjecture of Serre and Tate which asserts that, for any motive M over J, and any non-archimedean completion $J_v = K$, the isomorphism classes of the elements of $Rep^s(WD_K)$ gotten from the l-adic étale cohomology groups of M are all the same.

2. Automorphic Forms.

2.1. Basic Conventions. Let F be a number field with adele ring $\mathbf{A}_F$. Let $A_0(F, n)$ be the set of irreducible cuspidal unitary summands of the space

$$L^2(GL(n, \mathbf{A}_F)/GL(n, F)).$$

Each constituent π of such a space is isomorphic to a restricted tensor product $\pi = \otimes_v \pi_v$ where π_v is an infinite dimensional irreducible unitary representation. If v is finite, each π_v is classified up to isomorphism by an associated isomorphism class $\sigma(\pi_v)$ of n- dimensional members of $Rep^s(WD_v)$, where we denote by WD_v the Weil- Deligne group of F_v ([**HT**],[**Ku**]). As is customary, we denote also by $\sigma(\pi_v)$ any member of its class. Let W_v be the Weil group of F_v. Then $\sigma(\pi_v)$ is isomorphic, as in 1.9, to a direct sum of indecomposable modules of the form $V_{\Lambda_i, t} = \Lambda_i \otimes V_{q^{t/2}, t}$, $1 \le i \le n_v$, with irreducible representations Λ_i of W_v.

2.2. Ramanujan Conjecture. This is the assertion:

> Let $\pi \in A_0(F, n)$. Let v be a finite place of F and define, as above, the set of representations Λ_i of W_v for π_v. Then the image of each Λ_i is bounded.

Remark: This form of the conjecture is equivalent to the more elementary statement, independent of the Local Langlands Correspondence, which asserts that each π_v is tempered. However, we work exclusively with the formulation via Weil-Deligne groups in this paper.

Suppose now that $n = 2$. Then, at non-archimedian v, the local components π_v of a cuspidal π are classified into several types:

(i) π_v is supercuspidal,

(ii) π_v is a twist of the Steinberg representation: $\pi_v = St_v \otimes \psi(det)$, so that $\sigma(\pi_v) = \psi \otimes V_{q^{1/2}, 1}$ with a character ψ of $F_v^* = W_v^{ab}$.

(iii) π_v is principal series.

In cases (i) and (ii), $\sigma(\pi_v)$ is indecomposable and if π is unitary, $\Lambda = \Lambda_1$ is bounded. (In case (i), Λ is irreducible, $t = 0$, and $det(\Lambda)$ is the unitary central character of π, so Λ is bounded; in case (ii), $t = 1$, so $\Lambda = \psi$ is one-dimensional, and Λ^2 is the unitary central character of π, so Λ has bounded image.

For case (iii), $\sigma(\pi_v)$ is a direct sum of 2 quasicharacters ψ_1 and ψ_2 of F_v^* whose product is the central character of π, hence unitary. The classification of unitary representations shows that either (a) $|\psi_1| = |\psi_2| = 1$ or (b) there are quasicharacters $\mu = |\cdot|^t$ where $0 < t < 1/2$ and ψ of F_v^* such that $\sigma(\pi_v)$ is the sum of quasicharacters $\mu\psi$ and $\mu^{-1}\psi$. Hence, the Ramanujan Conjecture amounts to the assertion that representations of this type (*complementary series*) don't occur as local components of cusp forms. Note that at such a place, the local central character $\omega_{\pi, v}$ of π is ψ^2. For the forms of interest in this paper, F is totally real, and the infinity type π_∞ of π is discrete series and has the property that the idele class character ω_π takes the form $\omega_\pi = \nu_F^j \otimes \phi$, where ν_F is the norm, j is an integer, and ϕ is a character of finite order. Hence, $\psi^2 = \phi_v$ and so ψ has finite order. Invoking the Gruenwald-Hasse-Wang theorem, we see that there is an idele class character of finite order η such that the local identity $\eta_v = \psi$ holds. Thus, replacing π by a form of the same infinity type $\pi' = \pi \otimes \eta^{-1}$, we see that to establish the conjecture for all local components of all cusp forms π' of the given discrete series infinity type, it is enough to prove it for all π of the given type at all v that are unramified for π. Although this easy argument is special to $GL(2)$, it may be worth noting that solvable base change for $GL(n)$ should provide a reduction of Ramanujan to the case of semistable representations (i.e. to those whose local components $\sigma(\pi_v)$ are semistable.)

2.3. Global Langlands Correspondence. Let F be a number field and let (V_l, ρ_l) be an irreducible n-dimensional l-adic representation of Γ_F. Fix, for the rest of the paper, an isomorphism $\iota_l : \overline{\mathbf{Q}_l} \to \mathbf{C}$. For each finite place v of F, whose residue characteristic is different from l, choose t_l and Φ_l, as before. Let $\rho_{l,v}^*$ be the associated member of $Rep^s(WD_v)$ so defined.

Global Langlands Correspondence(GLC). This is the assertion:

> Suppose that the irreducible l-adic representation (V_l, ρ_l) is motivic. Then there are cuspidal representations $\pi \in A_0(F, n)$ and $\chi \in A_0(F, 1)$ such that, for all v whose residue characteristic is different from l, $\sigma(\pi_v) \otimes \chi_v$ is the class of $\rho_{l,v}^*$.

Remark 1. Since the statement of the GLC presupposes the existence of the Local Langlands Correspondence and an l-adic representation of the absolute Galois group of a global field, the GLC is often called the problem of *Local-Global Compatibility.*

Remark 2. If the residue characteristic of v is l the classes $\sigma(\pi_v) \otimes \chi_v$ can be predicted using methods of p-adic cohomology ([Fo]). This done, the above conjecture is extended to all finite places.

Remark 3. It is usual, especially to treat compatibility questions as l and ι_l vary, to formulate the conjecture in terms of a motive M and its Galois representations. However, as we do not treat compatibility questions in any essential way in this paper, there is no benefit to this viewpoint.

Remark 4. There is a converse conjecture: if π_∞ is algebraic ([C1]) then there should exist a (V_l, ρ_l) corresponding to π as above.

2.4. GLC and WMC. Proposition 5
Suppose that the GLC holds for the motivic l-adic representation (V_l, ρ_l) over F. Then WMC holds for (V_l, ρ_l).

PROOF. The conjecture is invariant under Tate twist, so we may assume that (V_l, ρ_l) is isomorphic to a subquotient of $H^i(X, \overline{\mathbf{Q}}_l)$ for some smooth projective X over F. For almost all places v, π_v is unramified. At such a place, the parameter $\sigma(\pi_v)$ consists of $n = dim(V_l)$ unramified quasicharacters of F_v^*, whose values on a prime element of F_v determine the unordered n-tuple $\{\alpha_j | j = 1, ..., n\}$. Each α_j is a Weil number, and, if we further restrict v to be a place of good reduction of X, then we have $|\alpha_j| = q^{i/2}$ for all j, for some i which is independent of j. Consider the cuspidal representation $\pi' = \pi \otimes |\cdot|^{i/2}$. Then π' is unitary because its central character is unitary; this holds at all unramified places v and hence everywhere. Let v_0 be finite place which we wish to study. The classification ([**Tad**], see [**Ku**])of unitary representations of $GL(n, F_{v_0})$ shows that $\sigma(\pi'_{v_0})$ is a direct sum of indecomposables $\Lambda \otimes V_{q^{t/2},t}$ where

$$-1/2 < w(\Lambda) < 1/2.$$

Hence

$$(i-1)/2 < w(\Lambda \otimes |\cdot|^{-i/2}) < (i+1)/2.$$

Since (V_l, ρ_l) is motivic, Proposition 1 shows that $w(\Lambda \otimes |\cdot|^{-i/2})$ is an integer in this interval. Hence $w(\Lambda \otimes |\cdot|^{-i/2}) = i$, which means $w(\Lambda) = 0$, as was to be shown. □

Remark. The proof of Proposition 5 uses only the fact that the weight of Λ is an integer, not the fact that the eigenvalues of Frobenius are algebraic numbers.

3. Zeta functions of quaternionic Shimura varieties.

Assume henceforth that F is totally real and let G be an inner form of $GL(2)/F$, so that $G(F) = B^*$ with a quaternion algebra B over F. Let $J_{F,nc} = \{\tau_1, ..., \tau_r\}$ be the set of real embeddings (= infinite places) of F where B is indefinite; assume that $J_{F,nc}$ is non-empty and contains $\tau_1 = 1_F$. To B is attached a Shimura variety Sh_B defined over F', the smallest extension of $\mathbf{Q}$ containing all elements $\tau_1(f)+...+\tau_r(f)$ for all $f \in F$. See ([**Shi**], [**D2**]) for constructions of Sh_B. It is the projective limit of quasi-projective r-folds $Sh_{B,W}$, where W is an open compact subgroup of $G(\mathbf{A}_{F,f}) = (B \otimes \mathbf{A}_{F,f})^*$. Each $Sh_{B,W}$ is defined over F', and is a finite disjoint union of connected r- folds. These components are proper if G is not $GL(2)/F$ and smooth if W is small enough. Any such Shimura variety is called a *quaternionic Shimura variety*. The Hasse-Weil zeta function, at almost all places, of the l-adic étale cohomology of such Shimura varieties has been computed by Reimann (See [**Re1**], Theorem 11.6), in the case $B \neq GL(2)/F$, and Brylinski and Labesse in the case $G = GL(2)/F$. In the latter case, it is the zeta function of an intersection cohomology which has been computed, and it is this cohomology that we consider in the following, using the same notation as the other cases. The zeta functions of the l-adic cohomology groups of $Sh_{B,W}$ have, at almost all places of F', the form conjectured by Langlands ([**L1**])and proved by him in the case $r = [F : \mathbf{Q}]$ ([**L1**]). See ([**BR2**], Sections 3.5, 5.1, and 7.2) for an expository treatment of the result but not the proof.

For our purposes, it is sufficient to give a global description of the result over a Galois extension L of $\mathbf{Q}$ which contains F. Thus, for each $j \in \{1, ..., r\}$, let $\overline{\tau_j}$ be an extension of τ_j to $\overline{L}$. Let π be a cuspidal automorphic representation of weight $(2, ..., 2)$ of $GL(2, \mathbf{A}_F)$ which is discrete series at any finite place of F at which B is ramified. Choose π so that its central character ω_π satisfies $\omega_\pi = \Psi |\cdot|^{-1}$ with a character Ψ of finite order. Let T be the number field generated by the traces $tr(\sigma(\pi_v)(\Phi))$ for all v which are unramified for π. As shown by Taylor ([**T1**], [**T2**]), there is an irreducible two-dimensional l-adic representation ρ_l^T, depending only on π and ι_l, which satisfies GLC relative to π. Let $\rho_{l,L}^T$ be the restriction to Γ_L of ρ_l^T. Let $^{[\overline{\tau_j}]}\rho_{l,L}^T$ be the representation defined, for $\eta \in \Gamma_L$, by

$$^{[\overline{\tau_j}]}\rho_{l,L}^T(\eta) = \rho_{l,L}^T(\overline{\tau_j}\eta\overline{\tau_j}^{-1}).$$

Let

$$R_l(\pi) = R_{l,J_{F,nc}}(\pi) = \otimes_{j=1,...,r} {}^{[\overline{\tau_j}]}\rho_{l,L}^T.$$

Then $R_l(\pi)$ is a semisimple l-adic representation of Γ_L of dimension 2^r.

3.1. Semisimple cohomology of Sh_B. Let, if it exists, π' be an automorphic representation of $G(\mathbf{A}_F)$ such that π'_v is isomorphic to π_v at all places v of F which are unramified for B. Thus $\pi = JL(\pi')$, where JL denotes the Jacquet-Langlands correspondence. Choose an open compact subgroup W as above so that $Sh_{B,W}$ is smooth and π'_f has a non-zero space of W-invariants. Let $\pi'_{f,W}$ denote the representation of the level W Hecke algebra $\mathcal{H}_W$ gotten from π'_f. Let

$$H^r(Sh_{B,W}, \overline{\mathbf{Q}_l})(\pi'_{f,W})$$

be the $\pi'_{f,W}$-isotypic component of $H^r(Sh_{B,W}, \overline{\mathbf{Q}_l})$.

Proposition 6

The irreducible subquotients of the action of Γ_L on $H^r(Sh_{B,W}, \overline{\mathbf{Q}_l})(\pi'_{f,W})|L$ are exactly the irreducible subquotients of $R_l(JL(\pi'))$.

PROOF. By the l-adic Cebotarev Theorem ([**Se**]), it suffices to show that the semisimplification of the Galois action on $H^r(Sh_{B,W}, \overline{\mathbf{Q}_l})(\pi'_{f,W})|L$ is a multiple of $R_l(JL(\pi'))$. But, up to notation and the base change to L, this is given by Theorem 11.6 of [**Re1**], and by the main theorem of [**BrLa**] in the non-compact case. See Section 5.3 for some explicit review of the zeta function.

4. Ramanujan and Weight-Monodromy for Hilbert modular forms

Let F be a totally real field and let $\pi = \pi_\infty \otimes \pi_f$ be a holomorphic cuspidal automorphic representation of $GL(2, \mathbf{A}_F)$. Up to twist, the isomorphism class of π at the infinite places of F is specified, as usual, by a tuple of positive integral weights $k = (k(\tau))$, where the variable τ runs over the real embeddings of F. We normalize π by assuming that its central character ω_π satisfies $|\omega_\pi| = |\cdot|^{1-k}$ where k is the maximum of the $k(\tau)$'s. It is natural to classify the holomorphic cuspidal π's into several types, depending on π_∞, i.e. on the classical weights at each infinite place:

(i) type G: all the weights are 1.

(ii) type MC: all the weights are at least 2 and they are all congruent modulo 2;

(iii) type NMC: all the weights are at least 2 and they are not all congruent modulo 2;

(iv) type NC: at least one, but not all , of the weights are 1.

Types G and MC are well-studied. RC is known at all places for type G ([**W**], [**RT**]); there is a 2-dimensional Artin representation ρ of Γ_F that satisfies the GLC. In this paper we prove RC at all places for the class of forms MC. As mentioned in the Introduction, the method of this paper should apply to type NMC but we do not consider this case in the paper. Type NC, except for the case of CM forms of this type, is completely open. Even in the case where the weights are all congruent mod 2, we do not know any motivic realization of associated Galois representations ([**J**]).

4.1. Proof of Theorem 1. We must show that the Ramanujan Conjecture (RC) holds for all Hilbert modular forms of type MC.

Let π be a representation of type MC of classical weight $k = (k(\tau))$. Let T be the field generated by almost all Hecke eigenvalues of π. Let ρ_l^T be one of the $[T : \mathbf{Q}]$ two dimensional l-adic representations attached to π which satisfy GLC. As shown in [**BR1**], these representations are motivic except possibly in the case where $[F : \mathbf{Q}]$ is even, $k(\tau) = 2$ for all τ, and π_v belongs to the principal series for all finite v. Hence, except in this case, RC follows from Proposition 5.

Let v be a finite place at which we will prove that π_v satisfies RC. Changing l, if necessary, we assume that v does not lie above l. Replacing π by a twist $\pi \otimes \Psi$, we may assume that π is unramified at all finite places of F which lie above the rational prime p under v. Let τ_1 be the tautological infinite place of F and let τ_2 be another infinite place. Let B be the quaternion algebra over F which is unramified at τ_1, τ_2, and at all finite places, and which is ramified at the remaining infinite places. Let G be the inner form of $GL(2)$ over F such that $G(F) = B^*$. Let L be a Galois extension of $\mathbf{Q}$ which contains F. By Proposition 6, the 4-dimensional l-adic representation $R_l(\pi)$ of Γ_L, made using $\rho_l^T|_L$ with $J_{F,nc} = \{\tau_1, \tau_2\}$, is isomorphic to the sum of irreducible subquotients of the cohomology $H^2(Sh_{B,W}, \overline{\mathbf{Q}_l})$ of the quaterionic Shimura surface $Sh_{B,W}$, for small enough open compact subgroup W of the finite adele group $(B \otimes \mathbf{A}_{F,f})^*$.

Now we need to make explicit the action on $R_l(\pi)$ of a decomposition group of a place w of L dividing v. Choose a decomposition group $D_{\overline{w}} \subset \Gamma_F$ for w, and denote by $R_{l,w}(\pi)$ the restriction of $\rho_l^T|_L$ to $D_{\overline{w}} \cap \Gamma_L$. Let $\tau_2 v$ be the place of F lying below $\overline{\tau_2}w$. Let f_1 and f_2 be degrees of the residue field extensions associated to L_w/F_v, and $L_{\overline{\tau_2}w}/F_{\tau_2 v}$, respectively. For each place v' of F above p, the restriction of $\rho_l^T|_L$ to $D_{v'}$ is unramified. Denote the eigenvalues of Φ_v by α_v and β_v. Denote the eigenvalues of $\Phi_{\tau_2 v}$ by $\gamma_{\tau_2 v}$ and $\delta_{\tau_2 v}$. Over L, Φ_w acts via $\rho_l^T|_L$ with eigenvalues $\alpha_v^{f_1} = \alpha_w$ and $\beta_v^{f_1} = \beta_w$. Likewise, $\Phi_{\overline{\tau_2}w}$ acts with eigenvalues $\gamma_{\tau_2 v}^{f_2} = \gamma_{\overline{\tau_2}w}$ and $\delta_{\overline{\tau_2}w}^{f_2} = \delta_{\overline{\tau_2}w}$. Hence Φ_w acts via $^{[\tau_2]}\rho_l^T|_L$ with eigenvalues $\gamma_{\overline{\tau_2}w}$ and $\delta_{\overline{\tau_2}w}$. Note the product relation $\gamma_{\overline{\tau_2}w}\delta_{\overline{\tau_2}w} = \zeta q_{\overline{\tau_2}w} = \zeta q_w$, where ζ is a root of unity, which is obvious since, by the global Langlands correspondence, $det(\rho_l^T(\Phi_v)) = \mu q_v$ with a root of unity μ, where q_v, q_w, and $q_{\overline{\tau_2}w}$ are the numbers of elements of the residue fields associated to v, w, and $\overline{\tau_2}w$.

By definition of $R_l(\pi)$, Φ_w acts via $R_l(\pi)$ with eigenvalues $a = \alpha_w \gamma_{\overline{\tau_2}w}$, $b = \alpha_w \delta_{\overline{\tau_2}w}$, $c = \beta_w \gamma_{\overline{\tau_2}w}$, and $d = \beta_w \delta_{\overline{\tau_2}w}$.

By Proposition 2, a, b, c and d are q_w-Weil numbers. Hence $ab = \alpha_w^2 \zeta q_w$, and so α_w^2 is a q_w-Weil number. Since $\alpha_w^2 = (\alpha_v^2)^{f_1}$, we see that α_v^2 is a q_v-Weil number. Likewise, β_v^2 is a q_v-Weil number. If we let $|\alpha_v^2| = q_v^{l/2}$ and $|\beta_v^2| = q_v^{m/2}$ then $|\alpha_v| = q_v^{l/4}$ and $|\beta_v| = q_v^{m/4}$, for integers l and m.

Recall now the following ([**Sha**]) Ramanujan estimate:

$$q_v^{-1/5} < |\alpha_v|, |\beta_v| < q_v^{1/5},$$

which applies to all unitary cusp forms π' for $GL(2)$ over any number field, at an unramified place v for π'. Since in our case $\pi \otimes |\cdot|^{1/2}$ is unitary, we see that the exponents $q_v^{\frac{l-2}{4}}$ and $q_v^{\frac{m-2}{4}}$ must be compatible with this estimate. Evidently this happens if and only if $l = m = 2$, which is precisely what we needed.

Of course, this result about π implies something about ρ_l^T:

Corollary 7. Let π be a Hilbert modular form of type MC. Then any l- adic representation ρ_l^T which satisfies the GLC satsifies WMC at all places v prime to l.

Remark. An easy extension of the above method shows RC at all places for all F, at least under the congruence condition on the weights. To prove RC at the place v, it is enough to choose any totally real quadratic extension K of F. Then, defining B over K as above, one proves RC, by the method here, at each place of K for the base change π_K of π from F to K. But it is easy to see that RC holds for π_K at a place w of K iff it holds for π at the place of F under w. Thus, RC may be proved for all Hilbert modular forms which satisfy the congruence condition at infinity by a uniform method which reduces the problem to the calculation of [**Re1**] and Shahidi's estimate.

Proof of the Corollary. Indeed, it only remained, in view of the work of Carayol ([**Ca**]), to establish the result at the unramified places of π, and this is precisely the RC.

4.2. Geometric Proof of Theorem 1. There exists a finite extension $\overline{L_{\overline{u}}}$ of L_w over which the generic fiber $\overline{Sh_{B,W}}$ of a semistable alteration of $Sh_{B,W}$ is defined. Then $H^2(Sh_{B,W}, \overline{\mathbf{Q}_l})$ is direct summand, as $D_{\overline{u}}$-module of $H^2(\overline{Sh_{B,W}}, \overline{\mathbf{Q}_l})$. This latter group satisfies WMC by [**RZ**]. Now $R_l(\pi)$ is, after some base change, a tensor product, and its associated Weil-Deligne parameter is thus a tensor product as well. We now note the following simple result whose proof is left to the reader:

Lemma. Let V_1 and V_2 be 2-dimensional representations of a Weil-Deligne group WD. Suppose that $V_1 \otimes V_2$ satisfies WMC, and suppose that the modules $\Lambda^2(V_i)$ are pure of weight 2. Then each V_i is pure of weight 1.

Applying this with $V_1 = \rho_l^T$, we conclude that $\rho_l^T|_{D_{\overline{u}}}$ is pure of weight 1 at u, and hence ρ_l^T is pure of weight 1 at v. Hence, since Local-Global Compatibility is known ([**T1**]), π_v satisfies RC.

5. Weight-Monodromy Conjecture for Quaternionic Shimura Varieties.

5.1. Proposition 8. Let F be a number field and let V be a variety defined over F. Let l be a rational prime with $(v, l) = 1$. Let v be a finite place of F. Then if the WMC holds at the place v for the semisimplification of $H^j(X, \overline{\mathbf{Q}_l})$ as a Γ_F-module, then the WMC holds for $H^j(X, \overline{\mathbf{Q}_l})$ at v.

PROOF. This is just a geometric restatement of a special case of Proposition 3.
□

5.2. Proof of Theorem 2:WMC for Quaternionic Shimura varieties.
Let, with notations as above, $Sh_{B,W}$ be a quaternionic Shimura variety of dimension r. The Hecke algebra at level W acts semisimply, and $H^r(Sh_{B,W}, \overline{\mathbf{Q}_l})$ is thus a direct sum of isotypic components $H^r(Sh_{B,W}, \overline{\mathbf{Q}_l})(\pi'_{f,W})$. It is sufficient, in view of Proposition 1, to show that WMC holds for each of these components. If π_f is one-dimensional, the result of Reimann ([**Re1**], Theorem 11.6) shows that after a finite base change to a number field L, the character of the Galois action on $H^*(Sh_{B,W}, \overline{\mathbf{Q}_l})(\pi_{f,W})$ is a sum of powers of the cyclotomic character at almost all, and hence, by Cebotarev, all, finite places. Since $H^j(Sh_{B,W}, \overline{\mathbf{Q}_l})(\pi_{f,W})$ is pure of

weight j at almost all finite places v, the Galois action on it over L is a multiple of χ_l^{-j} where χ_l is the l-adic cyclotomic character. Hence $H^j(Sh_{B,W}, \overline{\mathbf{Q}_l})(\pi_{f,W})$ is pure of weight j at all places. (We note that this fact is much less deep: it is not hard to see that all of $H^j(Sh_{B,W}, \overline{\mathbf{Q}_l})(\pi_{f,W})$ for $j \neq r$ is algebraic, generated on a single geometrically connected component by the j-fold products of the r Chern classes of the r line bundles defined by the factors of automorphy attached to the non-compact archimedian places.) Thus, to prove Theorem 3, we need only consider $H^j(Sh_{B,W}, \overline{\mathbf{Q}_l})(\pi_{f,W})$ where π_f is infinite dimensional. In this case, π_∞ is discrete series and $H^j(Sh_{B,W}, \overline{\mathbf{Q}_l})(\pi_{f,W}) \neq 0$ iff $j = r$. Let C be an irreducible subquotient of $H^r(Sh_{B,W}, \overline{\mathbf{Q}_l})(\pi_{f,W})$. Then, by Proposition 6, C is a direct summand of $R_l(JL(\pi'))$ for some π'. Note that each $^{[\overline{\tau_j}]}\rho_{l,L}^T$ is semisimple and satisfies the WMC at each finite place, since ρ_l^T has these properties. Since, for L as before, $R_l(JL(\pi')|_L$ is a tensor product of such representations, it also satisfies WMC at each finite place. Hence the summand C satisfies WMC and consequently the semisimplification of $H^r(Sh_{B,W}, \overline{\mathbf{Q}_l})$ as Γ_L-module also satisfies WMC at each finite place. By Proposition 3, this means $H^r(Sh_{B,W}, \overline{\mathbf{Q}_l})$ itself satisfies WMC.

5.3. Proof of Theorem 3: Langlands' Conjecture. We recall the statement. Let π' be a cuspidal holomorphic automorphic representation of $G(\mathbf{A}_F)$ having weight 2 at each unramified infinite place and with central character $|\cdot|^{-1}\overline{\Psi}$ where Ψ is a character of finite order. Let F' be the canonical field of definition of $Sh_{B,W}$. Then, for each finite place v of F', the element

$$(\rho_{W,v}^*, N_{W,v})$$

of $Rep^s(WD_{F_v})$ defined by the restriction of the $\Gamma_{F'}$ action on $H^r(Sh_{B,W}, \overline{\mathbf{Q}_l})(\pi'_f)$ to a decomposition group for v, coincides with the class

$$m(\pi'_f, W) r_B(\sigma(JL(\pi')_p)|_{WD_{F'_v}}),$$

where

(1) the non-negative integer $m(\pi'_f, W)$ is defined by

$$dim(H^r(Sh_{B,W}, \overline{\mathbf{Q}_l})(\pi'_{f,W})) = 2^r m(\pi'_f, W),$$

(2) p is the rational prime under v and $JL(\pi')$ is regarded as an automorphic representation of the $\mathbf{Q}$-group $Res_{F/\mathbf{Q}}(GL(2))$ whose Langlands parameter at p is

$$\sigma(JL(\pi'))_p : WD_p \to^L Res_{F/\mathbf{Q}}(GL(2))$$

([**BR2**], 3.5), and

(3)

$$r :^L Res_{F/\mathbf{Q}}(GL(2))|_{F'} \to GL(2^r, \mathbf{C})$$

is the representation defined by Langlands (c.f. [**BR2**], 5.1, 7.2).

PROOF. As before, although the statement is local, for each v, the proof proceeds via the global Galois representations. In order to see clearly what is being claimed, we review the key definitions.

Let

$$R_l^T = r_B(Ind_{\mathbf{Q}}^F(\rho_l^T(JL(\pi')))|_{\Gamma_{F'}}).$$

Here, for any two dimensional l-adic representation ρ of Γ_F,

$$Ind_{\mathbf{Q}}^F(\rho) : \Gamma_{\mathbf{Q}} \to^L Res_{F/\mathbf{Q}}(GL(2))_l$$

is a representation of $\Gamma_{\mathbf{Q}}$ into the l-adic L-group

$$^L Res_{F/\mathbf{Q}}(GL(2))_l.$$

This latter group is defined in general as in [**BR2**], 3.5 using groups $\hat{G}_l = GL(2, \overline{\mathbf{Q}_l})$ in lieu of the complex groups $\hat{G} = GL(2, \mathbf{C})$. Thus, in this case,

$$^L Res_{F/\mathbf{Q}}(GL(2))_l = GL(2, \overline{\mathbf{Q}_l})^{Hom(F,\mathbf{R})} \times \Gamma_{\mathbf{Q}}$$

is the semidirect product defined via the action: if $g = (g_\tau)_{\tau \in Hom(F,\mathbf{R})}$ and $\eta \in \Gamma_{\mathbf{Q}}$, then $\eta(g)_\tau = g_{\eta^{-1}\tau}$. The homomorphism

$$I = Ind_{\mathbf{Q}}^F(\rho_l^T(R))$$

is defined as

$$I(\eta) = ((\rho_l^T(\eta_{\overline{\tau}})_{\tau \in Hom(F,\mathbf{R})}), \eta)$$

where the $\overline{\tau}$ are a set of representatives in $\Gamma_{\mathbf{Q}}$ for the τ, and $\eta_{\overline{\tau}}$ is defined by the identity

$$\overline{\eta\eta^{-1}\tau} = \overline{\tau}\eta_{\overline{\tau}},$$

for all η and all τ.

Of course, if η fixes the Galois closure of F, then

$$\eta_{\overline{\tau}} = \overline{\tau}^{-1}\eta\overline{\tau},$$

so I is expressed in terms of the conjugates $^{\overline{\tau}}\rho_l^T$ of ρ_l^T.

We denote the inverse image of $\Gamma_{F'} \subseteq \Gamma_{\mathbf{Q}}$ in $^L Res_{F/\mathbf{Q}}(GL(2))_l$ by

$$^L Res_{F/\mathbf{Q}}(GL(2))_l|_{\Gamma_{F'}}.$$

On this latter group is defined the irreducible representation r_B on $\overline{\mathbf{Q}_l}^{2^r}$. We review its construction. Recall that $J_{F,nc} = \{\tau_1, ..., \tau_r\}$ is an ordering of the set of real embeddings $J_{F,\mathbf{R}} \subseteq Hom(F,\mathbf{R})$ of F where B is split. Then on the connected component $GL(2, \overline{\mathbf{Q}_l})^{Hom(F,\mathbf{R})}$, and for $g = (g_\tau)_{\tau \in Hom(F,\mathbf{R})}$,

$$r_B(g) = \otimes_{i=1}^{i=r} g_{\tau_i}.$$

By definition of F', an $\eta \in \Gamma_{F'}$ on $Hom(F,\mathbf{R})$ defines a permutation of $J_{F,nc}$. If we define r_B^η by the rule,

$$r_B^\eta(g) = r_B(\eta(g)),$$

then r_B^η is isomorphic to r_B. Let

$$P \subset GL(2, \overline{\mathbf{Q}_l})^{Hom(F,\mathbf{R})}$$

be product of the groups of upper triangular matrices in each factor. Then (i) $\eta(P) = P$ and (ii) $r_B(P)$ fixes a unique line Λ in $\overline{\mathbf{Q}_l}^{2^r}$. If $i(\eta)$ is an isomorphism satisfying, for all g,

$$i(\eta)r_B(g) = r_B^\eta(g)i(\eta),$$

then $i(\eta)(\Lambda) = \Lambda$. The choice of $i(\eta)$ is, by Schur's Lemma, unique up to a scalar, and we define $r_B(\eta)$ to be the unique choice which leave L pointwise fixed. The rule $r_B((g,\eta)) = r_B(g)r_B(\eta)$ gives the sought representation.

Note that the restriction of R_l^T to the Galois closure L of F is just the representation $R_l(\pi)$ defined in Section 3. Hence R_l^T is semisimple and satisfies WMC at each finite place v of F' where $(v,l) = 1$. Furthermore, for such v, the representations (ρ_v, N_v) of WD_v defined by the Φ- semisimplification of the restriction of R_l^T to a decomposition group $D_{\overline{v}}$ at v coincide, since ρ_l^T satisfies the Langlands correspondence, with

$$r_B(\sigma(JL(\pi')_p)|WD_{F'_v}).$$

Now, at this point we know that, for all v prime to l, the representations

$$(\rho^*_{W,v}, N_{W,v})$$

of WD_v, defined by the restriction to $D_{\overline{v}}$ of the $\Gamma_{F'}$ module

$$H^r(Sh_{B,W}, \overline{\mathbf{Q}_l})(\pi'_{f,W})$$

satisfy-whatever they may be- WMC. Thus (see 1.12), for each v, the nilpotent data N_v and $N_{W,v}$ are uniquely determined by the semisimple representations ρ_v and $\rho^*_{W,v}$. Hence it will suffice to show that

$$m(\pi'_f, W)\rho_v = \rho^*_{W,v}.$$

Now, for almost all v, (i) $N_v = 0$ and $N_{W,v} = 0$, (ii) ρ_v and $\rho^*_{W,v}$ are unramified, and (iii) the computation ([**Re1**],[**L1**],[**BrLa**]) of the unramified zeta function shows exactly that this formula holds. Using the l-adic Cebotarev theorem again, we see that the semisimplified $\Gamma_{F'}$-module

$$H^r(Sh_{B,W}, \overline{\mathbf{Q}_l})(\pi'_{f,W})^{ss}$$

is isomorphic to

$$m(\pi'_f, W)R_l^T.$$

Now let v be any place of F' which is prime to l. Then evidently,

$$(H^r(Sh_{B,W}, \overline{\mathbf{Q}_l})(\pi'_{f,W})|D_{\overline{v}})^{ss}$$

is isomorphic to

$$m(\pi'_f, W)((R_l^T)|D_{\overline{v}})^{ss}.$$

Since the former gives rise to the parameter $\rho^*_{W,v}$ and the latter gives rise to $m(\pi'_f, W)\rho_v$, we are done.   □

Acknowledgements. I thank the SFB 478 at Münster and the Université Henri Pasteur at Strasbourg for their generous hospitality. I thank Ron Livné for asking me several years ago about the RC at *all* places, and Takeshi Saito for a stimulating conversation which led me to highlight WMC in the presentation of the results. Finally, I thank the referee for conscientious reading of the manuscript and the reporting of many misprints.

References

[AC]   Arthur, J.; Clozel, L., *Simple algebras, base change, and the advanced theory of the trace formula*. Annals of Mathematics Studies, **120**. Princeton University Press, Princeton, NJ, 1989.

[BR1]   Blasius, D.; Rogawski, J., *Motives for Hilbert modular forms*. Invent. Math. **114** (1993), no. 1, 55-87.

[BR2]   Blasius, Don; Rogawski, Jonathan D., *Zeta functions of Shimura varieties*. Motives (Seattle, WA, 1991), 525-571, Proc. Sympos. Pure Math., **55**, Part 2, Amer. Math. Soc., Providence, RI, 1994.

[BrLa] Brylinski, J.-L.; Labesse, J.-P., *Cohomologie d'intersection et fonctions L de certaines variétés de Shimura*. Ann. Sci. École Norm. Sup. (4) **17** (1984), no. 3, 361–412.

[Ca]   Carayol, Henri, *Sur les représentations l-adiques associées aux formes modulaires de Hilbert*. Ann. Sci. École Norm. Sup. (4) **19** (1986), no. 3, 409–468.

[Cl]   Clozel, Laurent, *Motifs et formes automorphes: applications du principe de fonctorialité*. Automorphic forms, Shimura varieties, and L-functions, Vol. I (Ann Arbor, MI, 1988), 77–159, Perspect. Math., **10**, Academic Press, Boston, MA, 1990.

[DJ]   De Jong, A. J., *Smoothness, semi-stability and alterations*. Inst. Hautes Études Sci. Publ. Math. No. **83** (1996), 51–93.

[DS]   De Shalit, E., *The p-adic monodromy-weight conjecture for p-adically uniformized varieties*. Comp. Math. **141** (2005), 101-120.

[D1]   Deligne, P., *Théorie de Hodge I*. Actes ICM Nice, t.I, pp. 425-430, Gauthier-Villars, 1970.

[D2]   Deligne, P., *Travaux de Shimura*. Séminaire Bourbaki 1970-71, Exposé 389, Lecture Notes in Mathematics, Vol. **244**. Springer-Verlag, Berlin-New York, 1971.

[D3]   Deligne, P., *Les constantes des équations fonctionnelles des fonctions L*. Modular functions of one variable, II (Proc. Internat. Summer School, Univ. Antwerp, Antwerp, 1972), pp. 501–597. Lecture Notes in Mathematics, Vol. **349**, Springer-Verlag, Berlin, 1973.

[D4]   Deligne, Pierre, *La conjecture de Weil II*. Inst. Hautes Études Sci. Publ. Math. No. **52** (1980), 137–252.

[DM]   Deligne, P., Milne, J., *Tannakian Categories*. Hodge cycles, motives, and Shimura varieties. Lecture Notes in Mathematics, Vol. **900**. Springer-Verlag, Berlin-New York, 1982.

[DMOS] Deligne, P.; Milne, J.; Ogus, A.; Shih, K., *Hodge cycles, motives, and Shimura varieties*. Lecture Notes in Mathematics, Vol. **900**. Springer-Verlag, Berlin-New York, 1982.

[HT]   Harris, M.; Taylor, R., *The Geometry and Cohomology of Some Simple Shimura Varieties*, Annals of Mathematics Studies, **151**. Princeton University Press, Princeton, NJ, 2001.

[SGA7-I] Grothendieck, A., *Groupes de monodromie en géométrie algébrique, I*. Lecture Notes in Mathematics, Vol. **288**, Springer Verlag, Berlin-New York, 1972.

[Il]   Illusie, Luc, *Autour du théoréme de monodromie locale*. Périodes p-adiques (Bures-sur-Yvette, 1988). Astérisque No. **223** (1994), 9-57.

[It1]   Ito, T., *Weight-monodromy conjecture for certain 3-folds in mixed characteristic*, math.NT/0212109, 2002.

[It2]   Ito, T., *Weight-monodromy conjecture for p-adically uniformized varieties*, MPI 2003-6, math.NT/ 0301202, 2003.

[JL]   Jacquet, H.; Langlands, R. P., *Automorphic forms on GL(2)*. Lecture Notes in Mathematics, Vol. **114**. Springer-Verlag, Berlin-New York, 1970.

[J]   Jarvis, Frazer, *On Galois representations associated to Hilbert modular forms*. J. Reine Angew. Math. **491** (1997), 199–216.

[Ku]   Kudla, Stephen S., *The local Langlands correspondence: the non-Archimedean case*. Motives (Seattle, WA, 1991), 365–391, Proc. Sympos. Pure Math., **55**, Part 2, Amer. Math. Soc., Providence, RI, 1994.

[L1]   Langlands, R. P., *On the zeta functions of some simple Shimura varieties*. Canad. J. Math. **31** (1979), no. 6, 1121–1216.

[L2]   Langlands, R.P., *Base change for GL(2)*. Annals of Mathematics Studies, **96**. Princeton University Press, Princeton, N.J.,1980.

[Oh]   Ohta, Masami, *On l-adic representations attached to automorphic forms*. Japan. J. Math. (N.S.) **8** (1982), no. 1, 1–47.

[RZ] Rapoport, M.; Zink, Th., *Ueber die lokale Zetafunktion von Shimuravarietaeten. Mon-odromiefiltration und verschwindende Zyklen in ungleicher Charakteristik.* Invent. Math. **68** (1982), no. 1, 21–101.

[Ra] Rapoport, M., *On the bad reduction of Shimura varieties.* Automorphic forms, Shimura varieties, and L-functions, Vol. II (Ann Arbor, MI, 1988), 253-321, Perspect. Math., **11**, Academic Press, Boston, MA, 1990.

[Re1] Reimann, Harry, *The semi-simple zeta function of quaternionic Shimura varieties.* Lecture Notes in Mathematics, Vol.**1657**. Springer-Verlag, Berlin, 1997. viii+143 pp.

[Re2] Reimann, Harry, *On the zeta function of quaternionic Shimura varieties.* Math. Ann. **317** (2000), no. 1, 41–55.

[RZ] Reimann, Harry; Zink, Thomas, *Der Dieudonnmodul einer polarisierten abelschen Mannig-faltigkeit vom CM-Typ.* Ann. of Math. (2) **128** (1988), no. 3, 461–482.

[RT] Rogawski, J. D.; Tunnell, J. B., *On Artin L-functions associated to Hilbert modular forms of weight one.* Invent. Math. **74** (1983), no. 1, 1–42.

[Roh] Rohrlich, David E., *Elliptic curves and the Weil-Deligne group.* Elliptic curves and related topics, 125–157, CRM Proc. Lecture Notes, **4**, Amer. Math. Soc., Providence, RI, 1994.

[Sa] Saito, Takeshi, *Weight-monodromy conjecture for l-adic representations associated to mod-ular forms.* A supplement to: "Modular forms and p-adic Hodge theory" [Invent. Math. **129** (1997), no. 3, 607-620] , in The arithmetic and geometry of algebraic cycles (Banff, AB, 1998), 427-431, NATO Sci. Ser. C Math. Phys. Sci., **548**, Kluwer Acad. Publ., Dordrecht, 2000.

[Se] Serre, Jean-Pierre, *Abelian l-adic Representations and Elliptic Curves.* Addison-Wesley, Redwood City, 1989.

[ST] Serre, Jean-Pierre; Tate, John, *Good reduction of abelian varieties.* Ann. of Math. (2) **88** (1968) 492-517.

[Sha] Shahidi, Freydoon, *On the Ramanujan conjecture and finiteness of poles for certain L-functions.* Ann. of Math. (2) **127** (1988), no. 3, 547–584.

[Shi] Shimura, Goro, *Construction of class fields and zeta functions of algebraic curves.* Ann. of Math. (2) **85** (1967), 58–159.

[Tad] Tadić, Marko, *Classification of unitary representations in irreducible representations of general linear group (non-Archimedean case).* Ann. Sci. École Norm. Sup. (4) **19** (1986), no. 3, 335–382.

[Ta] Tate, J., *Number theoretic background.* Automorphic forms, representations and L-functions (Corvallis, Oregon, 1977), 3-26, Proc. Sympos. Pure Math.,**XXXIII**, Part 2, Amer. Math. Soc., Providence, R.I., 1979.

[T1] Taylor, R., *On Galois representations associated to Hilbert modular forms.* Invent. Math. **98** (1989), no. 2, 265-280.

[T2] Taylor, R., *On Galois representations associated to Hilbert modular forms. II.* Elliptic curves, modular forms, & Fermat's last theorem (Hong Kong, 1993), 185–191, Ser. Number Theory, I, Internat. Press, Cambridge, MA, 1995.

[TY] Taylor, R.; Yoshida, Teruyoshi, *Compatibility of local and global Langlands correspondences* preprint, math.NT/0412357, December 2004.

[Y]   Yoshida, Hiroyuki, *On a conjecture of Shimura concerning periods of Hilbert modular forms.* Am. J. Math. **117** (1995), no. 4, 1019-1038.

[W]   Wiles, A., *On ordinary λ-adic representations associated to modular forms.* Invent. Math. **94** (1988), no. 3, 529–573.

Farey fractions and two-dimensional tori

Florin P. Boca and Alexandru Zaharescu

ABSTRACT. The Farey sequence gives a natural filtration with finite subsets of the set of rational numbers in $[0, 1]$. It is elementary to define this sequence and to prove that it is uniformly distributed on the interval $[0, 1]$. However, other aspects regarding its distribution appear to be quite intricate and are related with a number of important open problems in mathematics.

This survey paper revolves around three main themes. The first one is concerned with their spacing statistics, especially the existence and computation of consecutive level spacing measures. The second theme is the connection with some problems in geometric probability concerning the statistics of the linear flow in a punctured flat two-torus when the diameter of the puncture tends to zero. Finally some results and open problems about noncommutative two-tori, some of their subalgebras, and the spectral theory of almost Mathieu operators are reviewed. In particular, a connection between Farey fractions and the structure of gaps in the Hofstadter butterfly is discussed.

1. Farey fractions. Elementary properties

The *Farey sequence of order* n is the set $\mathcal{F}_n$ of fractions $\gamma = \frac{a}{q}$ with $1 \leq q \leq n$ in the interval $(0, 1]$, written in lowest terms and arranged in increasing order of magnitude. The number of elements of $\mathcal{F}_n$ is

$$N_n = \sum_{q=1}^{n} \varphi(q) = \frac{3n^2}{\pi^2} + O(n \log n).$$

Two reduced rational numbers $\gamma = \frac{a}{q}$ and $\gamma' = \frac{a'}{q'}$ are *adjacent* in $\mathcal{F}_n$ if and only if the following conditions hold:

$$\begin{cases} 0 < a \leq q \leq n \\ 0 < a' \leq q' \leq n \\ q + q' > n \\ a'q - aq' = 1. \end{cases}$$

Three Farey fractions of order n, $\gamma_j = \frac{a_j}{q_j} < \gamma_{j+1} = \frac{a_{j+1}}{q_{j+1}} < \gamma_{j+2} = \frac{a_{j+2}}{q_{j+2}}$, are adjacent in $\mathcal{F}_n$ if and only if

(1.1) $$q_{j+2} = \left[\frac{n + q_j}{q_{j+1}} \right] q_{j+1} - q_j.$$

The condition (1.1) is actually equivalent to

$$\left(\frac{q_{j+1}}{n}, \frac{q_{j+2}}{n}\right) = T^j\left(\frac{q_1}{n}, \frac{q_2}{n}\right),$$

where $\frac{a_j}{q_j}$ denote the elements of $\mathcal{F}_n$ written in increasing order of magnitude, and T is the bijective area-preserving transformation of the *Farey triangle*

$$\mathcal{T} = \{(x,y) \in \mathbb{R}^2 : 0 < x \le y \le 1, \ x + y > 1\}$$

defined ([**12**]) by

$$(1.2) \qquad T(x,y) = \left(y, \left[\frac{1+x}{y}\right]y - x\right), \qquad (x,y) \in \mathcal{T}.$$

The sets $\mathcal{T}_k = \{(x,y) \in \mathcal{T} : [\frac{1+x}{y}] = k\}$, $k = 1, 2, \dots$, give a partition of $\mathcal{T}$. On each set $\mathcal{T}_k$ the map T is linear and $T(x,y) = (y, ky - x)$, but $\mathcal{T}_k$ is not T-invariant (see Figure 1). The pair $(T, \mathcal{T})$ is a fibred system in the sense of [**53**]. In spite of its apparent resemblance with the Gauss and Jacobi-Perron maps, the transformation T seems to have a different type of behaviour.

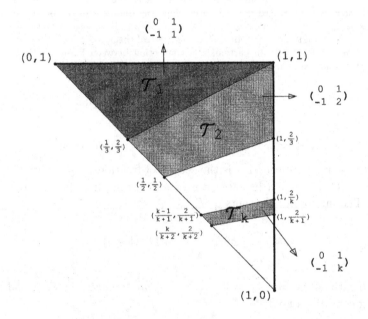

FIGURE 1. *The decomposition* $\mathcal{T} = \cup_{k=1}^{\infty}\mathcal{T}_k$.

REMARK 1.1. 1) *The T-orbit of any rational point in $\mathcal{T}$ is finite and*

$$\#\left\{T^j\left(\frac{r}{n}, \frac{s}{n}\right) : j \in \mathbb{N}\right\} \le \frac{N_n}{\gcd(r,s)}.$$

2) *The map T can be used to generate the Farey fractions of any order: if* $\gamma_1 = \frac{1}{n} < \gamma_2 = \frac{1}{n-1} < \cdots < \gamma_{N_n} = 1$ *denote the elements of* $\mathcal{F}_n$, $\gamma_0 = \gamma_{N_n} = 1$ *and*

$\gamma_j = \frac{a_j}{q_j}$ with $\gcd(a_j, q_j) = 1$, then $q_0 = 1$, $q_1 = n$, $q_2 = n - 1$, and

$$nT^j\left(\frac{1}{n}, 1\right) = (q_j, q_{j+1}), \qquad 1 \leq j \leq N_n.$$

The fractions γ_j can be retrieved from the pair (q_j, q_{j+1}) by solving for each j the equation $q_j a_{j+1} - q_{j+1} a_j = 1$, with $0 < a_j \leq q_j$ and $0 < a_{j+1} \leq q_{j+1}$.

2. Some statistical aspects

2.1. Uniform distribution. Denote by $\tau(n)$ the number of positive divisors of an integer n and by μ the Möbius function. The Landau symbols 'big oh' and 'little oh' will be denoted by O and o. The meaning of $f(x) \ll g(x)$ will be that $f(x) = O(g(x))$.

For each interval $I \subseteq (0, 1]$, consider the set $\mathcal{F}_n(I)$ of Farey fractions of order n from I. The number of elements of $\mathcal{F}_n(I)$ can in an elementary way be estimated as

$$N_n(I) = \sum_{q=1}^{n} \#\{a \in qI \,;\, \gcd(a, q) = 1\} = \sum_{q=1}^{n}\left(\frac{\varphi(q)}{q} \cdot q|I| + O(\tau(q))\right)$$

$$= |I| \sum_{q=1}^{n} \varphi(q) + O_\varepsilon(n^{1+\varepsilon}) = \frac{3|I|n^2}{\pi^2}\left(1 + \frac{\log n}{n}\right) + O_\varepsilon(n^{1+\varepsilon}).$$

In particular we have

$$\frac{N_n(I)}{N_n} = |I|\left(1 + O\left(\frac{\log n}{n}\right)\right) + O_\varepsilon(n^{-1+\varepsilon}) \qquad \text{as } n \to \infty,$$

which shows that Farey fractions are uniformly distributed on the unit interval.

The distribution of Farey fractions is closely related to the growth of the function

$$M(x) := \sum_{j \leq x} \mu(j),$$

as one can see from the equality

(2.1) $$\sum_{\gamma \in \mathcal{F}_n} f(\gamma) = \sum_{k=1}^{\infty}\sum_{j=1}^{k} f\left(\frac{j}{k}\right) M\left(\frac{n}{k}\right).$$

Taking $f_h(x) = e^{2\pi i h x}$ in (2.1) we find for every integer $h \geq 1$ that

$$\sum_{\gamma \in \mathcal{F}_n} e^{2\pi i h \gamma} = \sum_{d|h} dM\left(\frac{n}{d}\right),$$

and in particular that

$$\sum_{\gamma \in \mathcal{F}_n} e^{2\pi i \gamma} = M(n).$$

From the very definition of M we have $|M(n)| \leq n$. The precise order of growth of $M(n)$ as $n \to \infty$ is very important. Recall that $M(n) = o(n)$ implies the prime number theorem, and also that (see [28])

$$\text{Riemann hypothesis}$$

$$\Updownarrow \text{ (Littlewood 1912)}$$

$$M(n) = O_\varepsilon(n^{\frac{1}{2}+\varepsilon}) \quad \forall \varepsilon > 0$$

$$\Updownarrow \text{ (Franel 1924)}$$

$$\sum_{j=1}^{N_n} \left(\gamma_j - \frac{j}{N_n}\right)^2 = O_\varepsilon(n^{-1+\varepsilon}) \quad \forall \varepsilon > 0$$

$$\Updownarrow \text{ (Landau 1924)}$$

$$\sum_{j=1}^{N_n} \left|\gamma_j - \frac{j}{N_n}\right| = O_\varepsilon(n^{\frac{1}{2}+\varepsilon}) \quad \forall \varepsilon > 0.$$

Farey fraction partitions of the unit interval also played an important role in various applications of the circle method.

2.2. The definition of the consecutive level spacing measures. The spacing statistics measure the distribution of sequences of real numbers in a more subtle way than the classical Weyl uniform distribution. They are conveniently expressed in the convergence of certain measures, called level correlations and respectively level spacings, associated with the sequence.

DEFINITION 2.1. Let $A = (x_j)_{j=1}^N$ be a finite sequence of real numbers, arranged in increasing order of magnitude. This sequence is scaled to $\tilde{x}_j = \frac{N x_j}{x_N - x_1}$ whose mean spacing is $\frac{\tilde{x}_N - \tilde{x}_1}{N} = 1$. For each positive integer $h \geq 1$, consider the probability measure with finite support on $[0, \infty)^h$ defined by

$$\nu_A^{(h)} := \frac{1}{N-h} \sum_{j=1}^{N-h} \delta_{\tilde{x}_{j+1}-\tilde{x}_j} \otimes \delta_{\tilde{x}_{j+2}-\tilde{x}_{j+1}} \otimes \cdots \otimes \delta_{\tilde{x}_{j+h}-\tilde{x}_{j+h-1}},$$

where δ_x denotes the Dirac measure of mass one concentrated at x.

Equivalently, for any continuous function f with compact support on $[0, \infty)^h$,

$$\int_{[0,\infty)^h} f \, d\nu_A^{(h)} = \frac{1}{N-h} \sum_{j=1}^{N-h} f(\tilde{x}_{j+1} - \tilde{x}_j, \tilde{x}_{j+2} - \tilde{x}_{j+1}, \ldots, \tilde{x}_{j+h} - \tilde{x}_{j+h-1}).$$

DEFINITION 2.2. Suppose that $(A_n)_n$ is an increasing sequence of finite sets of real numbers such that the sequence of probability measures $(\nu_{A_n}^{(h)})_n$ is weakly convergent; that is, there exists a probability measure $\nu^{(h)}$ on $[0, \infty)^h$ such that, for every continuous function f with compact support on $[0, \infty)^h$,

$$\int_{[0,\infty)^h} f \, d\nu_{A_n}^{(h)} \xrightarrow{n} \int_{[0,\infty)^h} f \, d\nu^{(h)}.$$

The measure $\nu^{(h)}$ is called the *h-level consecutive spacing measure* of $(A_n)_n$.

For example, the h-level consecutive spacing measure of the sequence of sets $A_n = \{\frac{j}{n} : j = 1, \ldots, n\}$ is the Dirac measure concentrated at $(1, \ldots, 1) \in [0, \infty)^h$.

2.3. The consecutive level spacing measures of the Farey sequence.
We are interested in the situation where $A_n = \mathcal{F}_n$ is the set of Farey fractions of
order n. The first level spacings were computed by R. R. Hall.

THEOREM 2.1. ([**37**]) *The first consecutive spacing distribution exists and the
limit measure $\nu^{(1)}$ is given for every $0 < \alpha < \beta$ by*

$$\int_\alpha^\beta d\nu^{(1)} = 2\text{Area}\left\{ (x,y) \in \mathcal{T} \ : \ \frac{3}{\pi^2 \beta} < xy < \frac{3}{\pi^2 \alpha} \right\}.$$

The density g_1 of $\nu^{(1)}$ is given by

$$g_1(t) = \begin{cases} 0 & \text{if } t \in \left[0, \frac{3}{\pi^2}\right], \\ \frac{6}{\pi^2 t^2} \log \frac{\pi^2 t}{3} & \text{if } t \in \left[\frac{3}{\pi^2}, \frac{12}{\pi^2}\right], \\ \frac{12}{\pi^2 t^2} \log \frac{\pi^2 t}{6}\left(1 - \sqrt{1 - \frac{12}{\pi^2 t}}\right) & \text{if } t \in \left[\frac{12}{\pi^2}, \infty\right). \end{cases}$$

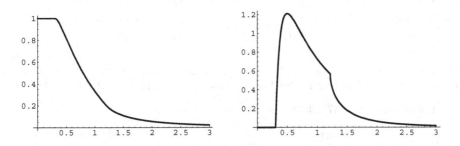

FIGURE 2. *The distribution and density functions of $\nu^{(1)}$.*

For the study of higher level consecutive spacings, it is advantageous to consider
the iterates

$$T^i(x,y) = \big(L_i(x,y), L_{i+1}(x,y)\big), \qquad i \geq 0,$$

of the area-preserving bijective transformation T of the triangle $\mathcal{T}$ defined by (1.2),
and also the map $\Phi_h : \mathcal{T} \to (0,\infty)^h$ defined by

$$\Phi_h(x,y) = \frac{3}{\pi^2}\left(\frac{1}{L_0(x,y)L_1(x,y)}, \frac{1}{L_1(x,y)L_2(x,y)}, \cdots, \frac{1}{L_{h-1}(x,y)L_h(x,y)} \right).$$

The higher level consecutive spacing measures are shown to exist and computed in

THEOREM 2.2. ([**3**]) *For every $h \geq 2$, the h-level consecutive spacing measure
$\nu^{(h)}$ of $(\mathcal{F}_n)_n$ exists and is given, for any box $\mathcal{B} \subset (0,\infty)^h$, by*

$$\nu^{(h)}(\mathcal{B}) = 2\text{Area}\,\Phi_h^{-1}(\mathcal{B}).$$

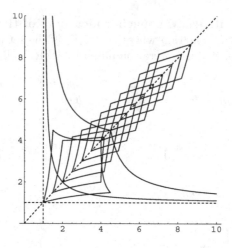

FIGURE 3. *The support of the measure* $\nu^{(2)}$.

2.4. Other results on the statistics of Farey fractions. Denote by $(\gamma_j)_{j=1}^{N_n}$ the elements of $\mathcal{F}_n$ arranged in increasing order of magnitude, and consider the sum

$$S_h(n) = \sum_{j=1}^{N_n-h} (\gamma_{j+h} - \gamma_j)^2.$$

The problem of estimating the first two terms in the asymptotic of these sums as $n \to \infty$ was considered and solved in the cases $h = 1$ and $h = 2$ by R. R. Hall.

THEOREM 2.3. ([**37**],[**38**])

(i) $S_1(n) = \dfrac{12\log n}{\pi^2 n^2} + \dfrac{12\left(\gamma - \frac{\zeta'(2)}{\zeta(2)} + \frac{1}{2}\right)}{\pi^2 n^2} + O_\varepsilon\left(\dfrac{\log^{\frac{5}{3}} n (\log\log n)^{1+\varepsilon}}{n^3}\right),$

(ii) $S_2(n) = \dfrac{36\log n}{\pi^2 n^2} + \dfrac{12\left(3\gamma - \frac{\zeta'(2)}{\zeta(2)} + 1 + B\right)}{\pi^2 n^2} + O\left(\dfrac{\log n}{n^{\frac{5}{2}}}\right),$

where

$$B = \frac{1}{2} + \log 2 + 2\sum_{k=1}^{\infty} \frac{\zeta(2k)-1}{2k-1} = 2.546277\ldots$$

Based on this result, Hall conjectured ([**38**]) that an asymptotic formula of type

$$S_h(n) = \frac{12(2h-1)\log n}{\pi^2 n^2} + \frac{B(h)}{n^2} + o\left(\frac{1}{n^2}\right) \quad \text{as } n \to \infty,$$

should be valid for any $h \geq 1$. An answer to this problem is given by

THEOREM 2.4. ([**12**]) *For any* $h \geq 3$,

$$S_h(n) = \frac{12(2h-1)\log n}{\pi^2 n^2} + \frac{B(h)}{n^2} + O_h\left(\frac{\log^{\frac{1}{h+2}} n}{n^{2+\frac{1}{h+2}}}\right),$$

where

$$B(h) = \frac{12}{\pi^2}\left((2h-1)\left(\gamma - \frac{\zeta'(2)}{\zeta(2)}\right) + \frac{h}{2} + (h-1)B + \sum_{r=2}^{h-1}(h-r)I_r\right)$$

and for $r \geq 2$ we take

$$I_r = \iint_{\mathcal{T}} \frac{dx\,dy}{xy L_r(x,y) L_{r+1}(x,y)} < \infty.$$

Other aspects of the distribution of the Farey sequence that were recently investigated include, for example, the distribution of the integers $\nu_n(\gamma) = \left[\frac{n+q}{q'}\right]$ ([39],[14]), the distribution of consecutive Farey fractions with odd denominators ([40],[13]), and the distribution of visible lattice points in n-homotheties of a bounded region in $\mathbb{R}^2$ as $n \to \infty$ ([11]).

In many of these problems the iterates of the area-preserving, piecewise linear transformation T of the triangle $\mathcal{T}$ do play a prominent role, both in proofs and statements of results. It is therefore natural to raise the following

Problem. Is T ergodic? weakly mixing? What is the entropy of T?

3. A technical device: estimates of Kloosterman sums

In these kinds of problems one can also study *short interval* versions, considering the set $\mathcal{F}_n(I)$ of Farey fractions of order n in an interval $I \subseteq (0,1]$. For applications, situations when $|I| \to 0$ and $n \to \infty$ in a prescribed way are important to consider.

In the short interval situations the main terms in the asymptotic expansions tend to reflect the uniform distribution of Farey fractions. However, it is more difficult to control the error terms because one also has to take into account the distribution of multiplicative inverses (mod q). The equality $a'q - aq' = 1$ says that a is the unique integer between 0 and q for which $aq' \equiv -1$ (mod q). This is achieved by using Weil type bounds on Kloosterman sums

$$K(r,s;q) := \sum_{x(\mathrm{mod}\ q)} \exp\frac{2\pi i(rx + s\bar{x})}{q},$$

where $\bar{x}$ denotes the multiplicative inverse of $x(\mathrm{mod}\ q)$. For q prime and $\gcd(r,s,q) = 1$, Weil proved ([58]) that

$$|K(r,s;q)| \leq 2q^{\frac{1}{2}}.$$

For a modulus $q \geq 2$ which is not necessarily prime, one has ([32],[44]) that

$$(3.1) \qquad |K(r,s;q)| \ll \gcd(r,s,q)^{\frac{1}{2}} \tau(q)^{\frac{1}{2}} q^{\frac{1}{2}}.$$

Using a standard argument one can derive upper bounds for incomplete Kloosterman sums

$$S_{\mathcal{I}}(r,s;q) := \sum_{\substack{x \in \mathcal{I} \\ \gcd(x,q)=1}} \exp\frac{2\pi i(rx + s\bar{x})}{q}.$$

One shows ([11]) for any integer $q \geq 2$ and any interval $\mathcal{I} \subseteq [1,q]$, that

$$(3.2) \qquad |S_{\mathcal{I}}(r,s;q)| \ll \tau(q) \gcd(s,q)^{\frac{1}{2}} q^{\frac{1}{2}} (2 + \log q).$$

The estimate (3.2) can be used to evaluate the asymptotic behaviour of the counting function for multiplicative inverses, defined by

$$N_q(\mathcal{I}_1, \mathcal{I}_2) := \#\{(x,y) \in \mathcal{I}_1 \times \mathcal{I}_2 : xy \equiv 1 \ (\mathrm{mod}\ q)\}.$$

LEMMA 3.1. ([11]) *Suppose that q is a large integer and $\mathcal{I}_1$ and $\mathcal{I}_2$ are intervals with $|\mathcal{I}_1|, |\mathcal{I}_2| < q$. Then*

$$N_q(\mathcal{I}_1, \mathcal{I}_2) = \frac{\varphi(q)}{q^2} \cdot |\mathcal{I}_1| \cdot |\mathcal{I}_2| + O\left(\tau^2(q) q^{\frac{1}{2}} \log^2 q\right).$$

Given an interval $I \subseteq (0, 1]$, a domain $\Omega \subset \mathcal{T}$, and a function f, we need to estimate sums of type

$$S_{f,\Omega,I}(n) := \sum_{\substack{\gamma \in \mathcal{F}_n(I) \\ (q,q') \in n\Omega}} f(q, q', a)$$

for large values of n. In many cases of interest we get an estimate of the form

$$\frac{6}{\pi^2} \iiint_{n\Omega \times I} f(v, w, v\gamma) \, dv \, dw \, d\gamma + \text{error},$$

where the order of magnitude of the error is smaller than that of the main term. The error term can be accurately estimated by using the following two lemmas.

LEMMA 3.2. ([15]) *Assume that $q \geq 1$ is an integer, $\mathcal{I}_1$ and $\mathcal{I}_2$ are intervals with $|\mathcal{I}_1|, |\mathcal{I}_2| < q$, and $g : \mathcal{I}_1 \times \mathcal{I}_2 \to \mathbb{R}$ is a C^1 function. Write $Dg = \left|\frac{\partial g}{\partial x}\right| + \left|\frac{\partial g}{\partial y}\right|$ and let $\|\cdot\|_\infty$ denote the L^∞ norm on $\mathcal{I}_1 \times \mathcal{I}_2$. Then for any integer $T > 1$ one has*

$$\sum_{\substack{a \in \mathcal{I}_1, b \in \mathcal{I}_2 \\ ab = 1 (\bmod q)}} g(a, b) = \frac{\varphi(q)}{q^2} \iint_{\mathcal{I}_1 \times \mathcal{I}_2} g(x, y) dx dy + E_{q,\mathcal{I}_1,\mathcal{I}_2,g,T},$$

where, for all $\delta > 0$,

$$E_{q,\mathcal{I}_1,\mathcal{I}_2,g,T} \ll_\delta T^2 q^{\frac{1}{2}+\delta} \|g\|_\infty + T q^{\frac{3}{2}+\delta} \|Dg\|_\infty + \frac{|\mathcal{I}_1| \, |\mathcal{I}_2| \, \|Dg\|_\infty}{T}.$$

LEMMA 3.3. ([11]) *Suppose that $0 < a < b$ are real numbers and that f is a C^1 function on $[a, b]$. Then*

$$\sum_{a < k \leq b} \frac{\varphi(k)}{k} f(k) = \frac{6}{\pi^2} \int_a^b f + O\left(\log b\left(\|f\|_\infty + \int_a^b |f'|\right)\right).$$

The proof of Lemma 3.2 uses an 'infinitesimal' application of Lemma 3.1. The proof of Lemma 3.3 is elementary and only uses elementary properties of the Möbius function.

4. Some applications to statistical mechanics

4.1. The two-dimensional periodic Lorentz gas and the free path length.
A periodic two-dimensional Lorentz gas (Sinai billiard) is a billiard system on the two-dimensional torus with one or more circular regions (scatterers) removed ([54], [21], [34], [8], [24]). Such systems were introduced by Lorentz in 1905 to study the dynamics of electrons in metals ([47]). The problem, which belongs to classical geometric probability, is to understand the statistics of the linear flow on a punctured flat two-dimensional torus.

Consider scatterers of small radius $\varepsilon > 0$ centered at all integer lattice points in the plane. By removing these discs, one gets the region

$$Z_\varepsilon = \{x \in \mathbb{R}^2 : \text{dist}(x, \mathbb{Z}^2) > \varepsilon\}$$

and the punctured disc

$$Y_\varepsilon = Z_\varepsilon/\mathbb{Z}^2.$$

Consider a point-like particle with initial position $x \in Z_\varepsilon$ and initial velocity $w \in \mathbb{T} = [0, 2\pi)$, which moves along the linear trajectory $t \mapsto x + wt$. In this situation the phase space – comprising the initial position and velocity – is the probability space $(Y_\varepsilon \times \mathbb{T}, \mu_\varepsilon)$, where μ_ε denotes the normalized Lebesgue measure of total mass one on $Y_\varepsilon \times \mathbb{T}$. The *free path length* (first exit time)

$$\tau_\varepsilon(x, w) = \inf\{\tau > 0 : x + \tau w \in \partial Z_\varepsilon\} \in (0, \infty]$$

measures the length of the path from the initial position to the first scatterer. When the trajectory does not reach any scatterer (which may well happen when the slope of w is rational) we take $\tau_\varepsilon(x, w) = \infty$. In this way one defines a Borel map $(x, w) \mapsto \tau_\varepsilon(x, w)$ on $Y_\varepsilon \times \mathbb{T}$. The problem of estimating the distribution of τ_ε is of compelling interest; it consists in finding the probability that the free path length is, say, larger than a given $t > 0$. For very small values of ε, the average value of $\tau_\varepsilon(x, w)$ is of order of magnitude $\frac{1}{\varepsilon}$. It is therefore appropriate to study the probability that $\varepsilon\tau_\varepsilon > t$, or perhaps $2\varepsilon\tau_\varepsilon > t$. This probability is defined by

$$\mathbb{P}_\varepsilon(t) = \int_{Y_\varepsilon \times \mathbb{T}} e_{(t,\infty)}(2\varepsilon\tau_\varepsilon)\, d\mu_\varepsilon = \mu_\varepsilon\{(x, w) \in Y_\varepsilon \times \mathbb{T} : 2\varepsilon\tau_\varepsilon(x, w) > t\}.$$

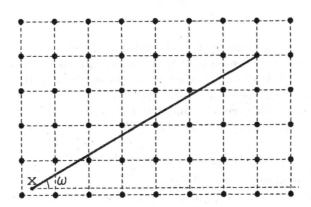

FIGURE 4. *The free path length* $\tau_\varepsilon(x, w)$, $(x, w) \in Y_\varepsilon \times \mathbb{T}$.

There is another possible choice for the phase space, motivated by the connection with dynamical systems and billiards: consider the probability space $(\Sigma_\varepsilon^+, \nu_\varepsilon)$, where $\Sigma_\varepsilon^+ := \{(x, w) \in \partial Y_\varepsilon \times \mathbb{T} : w \cdot n_x > 0\}$ with n_x being the inward unit normal to Y_ε at the point x, and where ν_ε denotes the probability measure obtained by normalizing the Liouville measure $w \cdot n_x\, dx\, dw$ on Σ_ε^+. The Liouville measure is invariant under the billiard map $T_\varepsilon : \Sigma_\varepsilon^+ \to \Sigma_\varepsilon^+$, which is defined as the return map on Σ_ε^+ of the billiard flow. In this case the free path is sometimes called the *geometric free path*. The probability that $2\varepsilon\tau_\varepsilon > t$ is this time given by

$$\mathbb{G}_\varepsilon(t) = \int_{\Sigma_\varepsilon^+} e_{(t,\infty)}(2\varepsilon\tau_\varepsilon)\, d\nu_\varepsilon = \nu_\varepsilon\{(x, w) \in \Sigma_\varepsilon^+ : 2\varepsilon\tau_\varepsilon(x, w) > t\}.$$

Another related quantity of interest is

$$C_\varepsilon := \log \int_{\Sigma_\varepsilon^+} \tau_\varepsilon \, d\nu_\varepsilon - \int_{\Sigma_\varepsilon^+} \log \tau_\varepsilon \, d\nu_\varepsilon.$$

4.2. The free path lengths in the small-scatterer limit. The distribution $\mathbb{P}_\varepsilon$ of the free path length was studied by Bourgain, Golse, and Wennberg ([20]), and by Caglioti and Golse ([22]). In [20] the existence of two universal constants $c_1, c_2 > 0$ such that

$$\frac{c_1}{t} < \mathbb{P}_\varepsilon(t) < \frac{c_2}{t} \qquad \text{for all } t > 2$$

was proved. In [22] the Cesaro lim sup and lim inf means were proved to exist and have the same asymptotic form for large t:

$$\limsup_{\delta \to 0^+} \frac{1}{|\log \delta|} \int_\delta^{1/4} \mathbb{P}_\varepsilon(t) \, d\varepsilon = \frac{2}{\pi^2 t} + O\left(\frac{1}{t^2}\right) = \liminf_{\delta \to 0^+} \frac{1}{|\log \delta|} \int_\delta^{1/4} \mathbb{P}_\varepsilon(t) \, d\varepsilon.$$

The free path length on the probability space $(\Sigma_\varepsilon^+, \nu_\varepsilon)$ was studied by Chernov, who proved relying on a formula of Santaló from integral geometry ([23]) that the first moment of τ_ε on this space is given (in closed form), for any $\varepsilon \in (0, \frac{1}{2})$, by

$$\int_{\Sigma_\varepsilon^+} \tau_\varepsilon \, d\nu_\varepsilon = \frac{1 - \pi\varepsilon^2}{2\varepsilon} .$$

He also proved that the quantity C_ε is bounded as $\varepsilon \to 0^+$.

A comprehensive discussion of the connection with the Kolmogorov-Sinai entropy $h(T_\varepsilon)$ of the billiard map T_ε on Σ_ε^+ is contained in [24]. Relations of recent results with the kinetic theory are discussed in [35].

A close connection between the quantities C_ε and $h(T_\varepsilon)$ was shown by Friedman, Kubo, and Oono ([33]). They proved that if $C = \lim_{\varepsilon \to 0^+} C_\varepsilon$ exists, then

$$h(T_\varepsilon) = -2 \log \varepsilon + 2 - C + o(1),$$

and also conjectured that $C = 0.44 \pm 0.01$.

The distribution $\mathbb{G}_\varepsilon(t)$ was studied by Dahlqvist ([27]) using the Farey tree. He exhibited the remarkable formula

$$-\mathbb{G}'(t) = \frac{6}{\pi^2} \cdot \begin{cases} 1 & \text{if } t \in [0, 1], \\ \frac{1}{t} + 2\left(1 - \frac{1}{t}\right)^2 \log\left(1 - \frac{1}{t}\right) - \frac{1}{2}\left(1 - \frac{2}{t}\right)^2 \log\left|1 - \frac{2}{t}\right| & \text{if } t \in [1, \infty], \end{cases}$$

for the density of the limit $\mathbb{G}(t)$ of $\mathbb{G}_\varepsilon(t)$ as $\varepsilon \to 0^+$. His approach however stopped short of providing a rigorous proof for the fact that the error term is smaller than the main term. Based on this formula for the limit distribution of the geometric free path, he found that one should expect $h(T_\varepsilon)$ to be of the form $-2 \log \varepsilon + 1 - 4 \log 2 + \frac{27\zeta(3)}{2\pi^2} + o(1)$ as $\varepsilon \to 0^+$.

Building upon the technique described in Section 3 and employing a certain partition of the torus $\mathbb{T}^2$ introduced by Blank and Krikorian ([7]), and used in conjunction with the free path length in [22], we proved the following result which provides complete answers to a number of open questions about the asymptotic behaviour of the free path lengths in the small-scatterer limit.

THEOREM 4.1. ([18]) *For all $t > 0$ and $\delta > 0$,*

$$\mathbb{P}_\varepsilon(t) = \mathbb{P}(t) + O_\delta(\varepsilon^{\frac{1}{8}-\delta}) \quad and \quad \mathbb{G}_\varepsilon(t) = \mathbb{G}(t) + O_\delta(\varepsilon^{\frac{1}{8}-\delta}) \qquad as \ \ \varepsilon \to 0^+,$$

where

$$\mathbb{P}(t) = \begin{cases} 1 - t + \dfrac{3t^2}{\pi^2} & \text{if } t \in [0,1], \\[2ex] \dfrac{6}{\pi^2}\displaystyle\int_0^{t-1}\psi(x,t)\,dx + \dfrac{6}{\pi^2}\displaystyle\int_{t-1}^1 \phi(x,t)\,dx & \text{if } t \in [1,2], \\[3ex] \dfrac{6}{\pi^2}\displaystyle\int_0^1 \psi(x,t)\,dx & \text{if } t \in [2,\infty), \end{cases}$$

and

$$\psi(x,t) = \frac{(1-x)^2}{x}\left(2\log\frac{t-x}{t-2x} - \frac{t}{x}\cdot\log\frac{(t-x)^2}{t(t-2x)} \right),$$

$$\phi(x,t) = \frac{(1-x)^2}{x}\left(2\log\frac{t-x}{1-x} - \frac{t}{x}\cdot\log\frac{t-x}{t(1-x)} \right)$$
$$+ \frac{1-t}{x}\cdot\log\frac{1}{t-x} + \frac{(t-x)(x-t+1)}{x}.$$

We also have

$$\mathbb{G}(t) = -\mathbb{P}'(t).$$

We notice that our formula for $g(t) = -\mathbb{G}'(t)$ coincides with that given by Dahlqvist, and that for $t \geq 2$ we have

$$\mathbb{P}(t) = \frac{24}{\pi^2}\sum_{n=1}^{\infty}\frac{2^n-1}{n^2(n+1)^2(n+2)t^n},$$

$$\mathbb{G}(t) = \frac{24}{\pi^2 t}\sum_{n=1}^{\infty}\frac{2^n-1}{n(n+1)^2(n+2)t^n},$$

$$g(t) = \frac{24}{\pi^2 t^2}\sum_{n=1}^{\infty}\frac{2^n-1}{n(n+1)(n+2)t^n}.$$

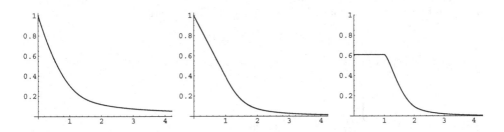

FIGURE 5. *The graphs of $\mathbb{P}(t)$, $\mathbb{G}(t)$, and $g(t)$.*

We also obtain the following corollary which answers the open problem raised in [24] concerning the existence of the second term in the asymptotic of the KS entropy $h(T_\varepsilon)$. Our answer is close to the prediction of Dahlqvist ([27]) (which seems to be off by a factor of two somewhere).

COROLLARY 4.2. ([**18**]) *The following two estimates hold as $\varepsilon \to 0^+$:*

(i) $C_\varepsilon = 3 \log 2 - \dfrac{9\zeta(3)}{4\zeta(2)} + o(1),$

(ii) $h(T_\varepsilon) = -2 \log \varepsilon + 2 - 3 \log 2 + \dfrac{9\zeta(3)}{4\zeta(2)} + o(1).$

REMARK 4.3. (i) Note that $3 \log 2 - \frac{9\zeta(3)}{4\zeta(2)} = 0.43522513609\ldots$ is indeed within 0.01 of 0.44, as predicted in [**33**].

(ii) No conjectures on the distribution of $\mathbb{P}_\varepsilon$ in the small-scatterer limit were available before [**18**].

4.3. The case when the initial position is at the origin. It is natural to try to estimate the length of the free path when the initial position x is fixed and the phase space is the range of the velocity ω. This is certainly a very difficult problem in general. The case that the trajectory starts at the origin is already nontrivial. The problem of estimating the moments of the average over ω of $\tilde{\tau}_\varepsilon(\omega) = \tau_\varepsilon(O, \omega)$ was raised by Ya. G. Sinai. An answer was given by Gologan and the authors.

THEOREM 4.4. ([**15**]) *For any interval $I \subseteq [0, 2\pi)$, any $r > 0$ and any $\delta > 0$,*

$$\int_I \left(\varepsilon \tilde{\tau}_\varepsilon\right)^r(\omega)\, d\omega = c_r |I| + \begin{cases} O_{r,\delta}(\varepsilon^{\frac{1}{8}-\delta}) & \text{if } r \neq 2 \\ O_{r,\delta}(\varepsilon^{\frac{1}{4}-\delta}) & \text{if } r = 2 \end{cases} \qquad \text{as } \varepsilon \to 0^+,$$

where[1]

$$c_r = \frac{12}{\pi^2} \int_0^{1/2} \left(x\left(x^{r-1} + (1-x)^{r-1}\right) + \frac{1-(1-x)^r}{rx(1-x)} - \frac{1-(1-x)^{r+1}}{(r+1)x(1-x)} \right) dx.$$

This result shows in particular that the sequence of probability measures $(\tilde{\mu}_\varepsilon)_\varepsilon$ defined by

$$\tilde{\mu}_\varepsilon(f) = \frac{1}{2\pi} \int_0^{2\pi} f\left(\varepsilon \tilde{\tau}_\varepsilon(\omega)\right) d\omega, \qquad f \in C_c\left([0, \infty)\right),$$

converges weakly, as $\varepsilon \to 0^+$, to a probability measure with compact support $\tilde{\mu}$ on $[0, \infty)$ and with moments of order n given by c_n. The distribution

$$H(t) := \tilde{\mu}\left([t, \infty)\right)$$

was shown ([**16**]) to be given by

(4.1) $$H(t) = \begin{cases} 1 - \dfrac{12t}{\pi^2} & \text{if } t \in \left[0, \frac{1}{2}\right], \\[2mm] \dfrac{12}{\pi^2} \displaystyle\int_t^1 \dfrac{1-x}{x}\left(1 + \log \dfrac{x}{1-x}\right) dx & \text{if } t \in \left[\frac{1}{2}, 1\right], \\[2mm] 0 & \text{if } t \in [1, \infty). \end{cases}$$

---

[1]Note that $\displaystyle\lim_{r \to 0^+} c_r = -\frac{12}{\pi^2} \int_0^{1/2} \frac{\log(1-x)}{x(1-x)}\, dx = 1.$

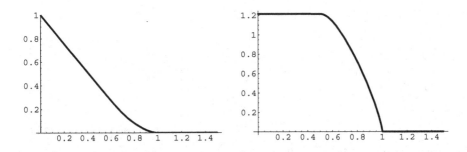

FIGURE 6. *The graphs of $H(t)$, and respectively of $h(t) = -H'(t)$.*

Equality (4.1) also implies that the limit measure $\tilde{\mu}$ is absolutely continuous with respect to the Lebesgue measure and its density is

$$h(t) = -H'(t) = \frac{12}{\pi^2} \cdot \begin{cases} 1 & \text{if } t \in [0, \frac{1}{2}], \\ \frac{1-t}{t}\left(1 + \log\frac{t}{1-t}\right) & \text{if } t \in [\frac{1}{2}, 1], \\ 0 & \text{if } t \in [1, \infty). \end{cases}$$

5. Gap labelling for almost Mathieu operators

In this final section we discuss some spectral properties of almost Mathieu operators and how Farey fractions are related to the gap labelling of these operators.

5.1. Almost Mathieu operators and rotation C^*-algebras.

For each number $\theta \in [0, 1]$, consider the *rotation C^*-algebra* (or *noncommutative two-torus*)

$$A_\theta = C^*(u_1, u_2 \text{ unitaries} : u_1 u_2 = e^{2\pi i\theta} u_2 u_1),$$

the *almost Mathieu operator* $(\lambda \in \mathbb{R})$

$$H_{\theta,\mu} = u_1 + u_1^* + \mu(u_2 + u_2^*) \in A_\theta,$$

and the *Harper operator*

$$H_\theta = H_{\theta,1} = u_1 + u_1^* + u_2 + u_2^*.$$

The mapping

$$(u_1, u_2) \mapsto (\sigma(u_1), \sigma(u_2)) = (u_2, u_1^*)$$

extends to an order four automorphism σ of A_θ which is called the *Fourier automorphism* of A_θ. An automorphism of a C^*-algebra preserves the spectrum and thus σ maps $H_{\theta,\mu}$ to $\mu H_{\theta,\frac{1}{\mu}}$, thus one gets the *Aubry duality*

$$\text{spec}(H_{\theta,\mu}) = \mu\,\text{spec}(H_{\theta,\frac{1}{\mu}}).$$

When $\theta = \frac{p}{q}$ is a (positive) rational number in lowest terms, one can exploit the fact that all irreducible representations have dimension q and are explicitly known, to prove that $\mathrm{spec}(H_{\frac{p}{q},\mu})$ coincides with the set

$$\bigcup_{t_1,t_2\in[0,\frac{2\pi}{q})} \mathrm{spec}\begin{pmatrix} 2\cos t_1 & \mu e^{it_2} & 0 & \cdots & \mu e^{-it_2} \\ \mu e^{-it_2} & 2\cos\left(t_1+\frac{2\pi p}{q}\right) & \mu e^{it_2} & \cdots & 0 \\ 0 & \mu e^{-it_2} & 2\cos\left(t_1+\frac{4\pi p}{q}\right) & \cdots & 0 \\ 0 & 0 & \mu e^{-it_2} & \cdots & 0 \\ \vdots & \vdots & \vdots & \ddots & \vdots \\ \mu e^{it_2} & 0 & 0 & \cdots & 2\cos\left(t_1+\frac{2\pi(q-1)p}{q}\right) \end{pmatrix}.$$

This set is contained in $[-2-2|\mu|, 2+2|\mu|]$, is symmetric about the origin, and consists in

- the union of q disjoint bands if q is odd,
- the union of $q-1$ bands if q even.

The tower of graphs of the Harper operators $H_{\frac{p}{q}}$ produces the Hofstadter butterfly ([43]).

FIGURE 7. *The Hofstadter butterfly.*

The spectral continuity of the spectrum of almost Mathieu operators with respect to θ has been thoroughly investigated. The following theorem features the best quantitative results on the continuity of the maps $\theta \mapsto \mathrm{spec}(H_{\theta,\mu})$, and respectively $\theta \mapsto \|H_{\theta,\mu}\|$.

THEOREM 5.1. (i) ([36]) *For any $\theta \in [0,1]$ there exists a faithful representation $\pi_\theta : A_\theta \to \mathcal{B}(\mathcal{H})$ of the C^*-algebra A_θ on the same separable Hilbert space $\mathcal{H}$ such that, for all $\theta, \theta' \in [0,1]$,*

$$\|\pi_\theta(u_j) - \pi_{\theta'}(u_j)\| \ll |\theta-\theta'|^{\frac{1}{2}}, \qquad j=1,2.$$

In particular, the Hausdorff distance between the spectra of $H_{\theta,\mu}$ and $H_{\theta',\mu}$ is at most $c|\theta - \theta'|^{\frac{1}{2}}$, for some universal constant c.

(ii) ([5]) $\left| \|H_{\theta,\mu}\| - \|H_{\theta',\mu}\| \right| \ll_{\mu} |\theta - \theta'|.$

The proofs of these two results make essential use of the unique irreducible unitary representation π of the real Heisenberg group on a separable Hilbert space $\mathcal{H}$ such that $\pi_{\lambda} = \lambda I_{\mathcal{H}}$, $\lambda \in \mathbb{T}$.

The C^*-algebra A_{θ} is endowed with the tracial state defined by $\tau(u_1^m u_2^n) = \delta_{m,0}\delta_{n,0}$. An explicit method of constructing nontrivial projections in A_{θ} was exhibited by Rieffel ([52]), who proved the inclusion $\{\{n\theta\} : n \in \mathbb{Z}\} \subseteq \tau(\mathcal{P}(A_{\theta}))$, where $\mathcal{P}(A_{\theta})$ denotes the set of (self-adjoint) projections in A_{θ}. The converse inclusion $\tau(\mathcal{P}(A_{\theta})) \subseteq \{\{n\theta\} : n \in \mathbb{Z}\} \cup \{1\}$ was proved by Pimsner and Voiculescu ([50]), by embedding an irrational rotation C^*-algebra A_{θ} into the Effros-Shen AF-algebra (see [29]) associated with the continued fraction decomposition of θ. If $\theta = [a_1, a_2, a_3, \dots]$ is the continued fraction decomposition of the irrational number θ, this AF-algebra is defined by the Bratteli diagram from Figure 8.

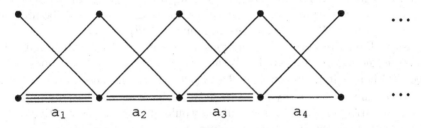

FIGURE 8. *The Bratteli diagram of the Effros-Shen AF-algebra.*

The equality $\tau(\mathcal{P}(A_{\theta})) = \{\{n\theta\} : n \in \mathbb{Z}\} \cup \{1\}$ also follows from the following fundamental theorem of Elliott and Evans concerning the structure of irrational rotation C^*-algebras.

THEOREM 5.2. ([30], see also [31] and [10]) *The irrational rotation C^*-algebra A_{θ} can be realized as an inductive limit*

$$A_{\theta} = \lim_{\rightarrow} \left(M_q(\mathbb{C}) \otimes C(\mathbb{T})\right) \oplus \left(M_{q'}(\mathbb{C}) \otimes C(\mathbb{T})\right),$$

of type I algebras, where $\frac{p}{q} < \theta < \frac{p'}{q'}$ are consecutive terms in the continued fraction decomposition of θ.

Since $H_{\theta,\mu}$ is a self-adjoint element in A_{θ}, it follows that if g is a gap in $\mathrm{spec}(H_{\theta,\mu})$, $t \in g$, and $e = e_{(-\infty,t)}(H_{\theta,\mu})$ is the corresponding spectral projection, then

$$\tau(e) = \{n\theta\} \qquad \text{for some } n \in \mathbb{Z}^*.$$

If $e \in \mathcal{P}(A_{\theta})$ and $\tau(e) = \{n\theta\}$, then n coincides with Connes's first Chern class $c_1(e)$ when θ is irrational, but we only have $n = c_1(e) \pmod{q}$ when $\theta = \frac{p}{q}$ is a rational number in lowest terms (see [26]).

Open problem. (M. Kac's 'Ten Martini problem') *Are all gaps open for irrational θ? That is, is it true, for irrational θ, that*

$$\forall n \in \mathbb{Z}^*, \ \exists e \in \mathcal{P}(H_{\theta}) \text{ such that } \tau(e) = \{n\theta\} ?$$

The answer was shown to be affirmative if θ is a Liouville number (cf. [25]). Various weaker conjectures and results on the Cantoriality and on the Lebesgue measure of the spectrum were proved in [2], [6], [41], [42], [45], [46], [56], and most recently in [1] and [51].

There are some very interesting fixed point subalgebras of A_θ which are related to almost Mathieu or Harper operators. The square σ_0 of the Fourier automorphism σ maps u_1 to u_1^* and u_2 to u_2^*, and is called the *flip automorphism*. Consider the C^*-subalgebras

$$A_\theta^\sigma = \{a \in A_\theta \,:\, \sigma(a) = a\} \subset A_\theta^{\sigma_0} = \{a \in A_\theta \,:\, \sigma_0(a) = a\} \subset A_\theta$$

of fixed points of σ, and respectively of σ_0. It is easily seen that $A_\theta^{\sigma_0}$ contains all almost Mathieu operators $H_{\theta,\mu}$, $\mu \in \mathbb{R}$, and that $H_\theta \in A_\theta^\sigma$. Moreover, when θ is irrational, A_θ^σ is generated by the Harper operators $u_1 + u_1^* + u_2 + u_2^*$ and $u_1^2 + u_1^{*2} + u_2^2 + u_2^{*2}$. Although each of the consecutive inclusions above has Jones index 2, the study of these fixed point C^*-algebras turned out to be quite challenging. By a result of Bratteli and Kishimoto ([19]) the C^*-algebra $A_\theta^{\sigma_0}$ is AF for any irrational θ. Manifest symmetry shows that Rieffel projections can be chosen to be invariant under the flip; thus $\tau\big(\mathcal{P}(A_\theta^{\sigma_0})\big)$ contains all numbers $\{n\theta\}$, $n \in \mathbb{Z}$.

However, Rieffel projections are no longer invariant under σ which makes the task of constructing projections in A_θ^σ and of describing $\tau\big(\mathcal{P}(A_\theta^\sigma)\big)$ more complicated. Although it is known that $\theta \in \tau\big(\mathcal{P}(A_\theta^\sigma)\big)$ ([9]), the question whether $\tau\big(\mathcal{P}(A_\theta^\sigma)\big)$ contains $\{n\theta\}$ for <u>any</u> integer n seems to be open and does not appear to be easy. A negative answer for a particular θ would also provide a negative answer to the ten Martini problem. Recent progress on the structure of A_θ^σ was achieved by Walters ([57]) and Phillips ([49]), who proved that this C^*-algebra is AF for a G_δ dense set of parameters θ.

5.2. Gap labelling and Farey fractions.

For each $\mu > 0$ consider the compact set

$$\mathcal{B}_\mu = \bigcup_{\theta \in [0,1]} \{(E,\theta) \,:\, E \in \mathrm{spec}(H_{\theta,\mu})\} \subset [-2-2\mu, 2+2\mu] \times [0,1].$$

To each connected component g of the complement of $\mathcal{B}_\mu$ one associates the first Chern class $c_1(g)$ of the spectral projection in A_θ of some operator $H_{\theta,\mu}$ which has a gap at the θ-section of g. This integer is independent of θ; thus $g \mapsto c_1(g)$ defines a integral-valued map on the connected components of the complement of $\mathcal{B}_\mu$.

It was noticed in [48] that

$$\#\{g \,:\, c_1(g) = n\} \geq \#\mathcal{F}_{2|n|} \qquad \text{for all } n \in \mathbb{Z}^*.$$

Equality would hold if the Ten Martini problem had a positive answer.

The butterflies $\mathcal{B}_\mu$ have a common shape, which is captured in their stylized form, obtained by letting $\mu \to 0^+$. This set $\mathcal{C}$ can be defined as the complement in the unit square $\{0 \leq E, \theta < 1\}$ of the 1-dimensional infinite simplex

$$\mathfrak{S} = \bigcup_{Q=1}^{\infty} \mathfrak{S}_Q,$$

where we set

$$\mathfrak{S}_Q = \bigcup_{n=1}^{Q} (S_n^+ \cup S_n^-),$$

$$S_n^+ = \{(\{n\theta\}, \theta) : \theta \in [0,1)\}$$

$$= \left\{\left(E, \frac{E+k}{n}\right) : E \in [0,1), \ k = 0,1,\ldots,n-1\right\},$$

$$S_n^- = \{(\{-n\theta\}, \theta) : \theta \in [0,1)\} = \{(\{n(1-\theta)\}, \theta) : \theta \in [0,1)\}$$

$$= \left\{\left(E, 1 - \frac{E+k}{n}\right) : E \in [0,1), \ k = 0,1,\ldots,n-1\right\}.$$

Each set $S_n^{\pm}$ is the union of n parallel segments of slope $\frac{1}{n}$. For each positive integer Q, consider the set K_Q of points where two distinct segments from $S_n^{\pm} \cap S_m^{\pm}$ or from $S_n^{\pm} \cap S_m^{\mp}$ cross, $1 \le n \neq m \le Q$.[2] An easy computation shows that

$$\{\theta : \exists E \text{ such that } (E, \theta) \in K_Q\} = \mathcal{F}_{2Q}.$$

The case $Q = 4$ is illustrated in Figure 9.

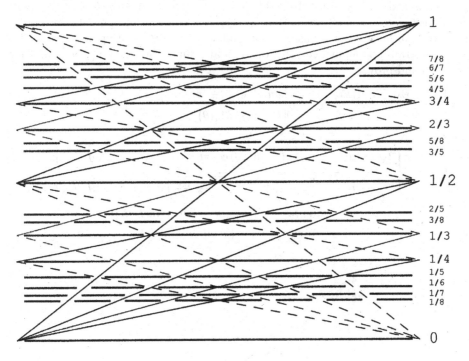

FIGURE 9. *The sets* $\mathfrak{S}_4$ *and* $\mathcal{F}_8$.

Note that for irrational θ, the θ-section of $\mathcal{C}$ is the Cantor set

$$\mathcal{C}_\theta = \{E : \exists\theta \text{ such that } (E, \theta) \in \mathcal{C}\} = [0,1) \setminus \{\{n\theta\} ; n \in \mathbb{Z}\}.$$

The first Chern class can be handily read on $\mathcal{C}$:

---

[2]Note that segments from $S_n^+ \cap S_n^-$ only cross at $E = \frac{1}{2}$, which corresponds to the point 0 in the spectrum of $H_{\theta,\mu}$, thus where there is a closed gap.

- Fix a positive integer Q and a rational number in lowest terms $\theta = \frac{a}{q} \in \mathcal{F}_{2Q}$.
- Draw the horizontal segment of ordinate θ which joins the vertical axis $E = 0$ with a point on one of the segments from $\mathfrak{S}_Q$. Let c be the smallest positive integer n with the property that the right endpoint belongs to $S_n^+ \cup S_n^-$.
- Then the first Chern class of the corresponding spectral projection in A_θ should be c or $-c$ according to whether the right endpoint belongs to S_c^+ or to S_c^-. In particular, c is a nonzero integer such that $-\frac{Q}{2} < c < \frac{Q}{2}$.

It would be interesting to find a closer connection between the simplex $\mathfrak{S}$, which contains a good deal of number-theoretical information and can be used for instance to find the multiplicative inverse of a unit in the finite ring $\mathbb{Z}/q\mathbb{Z}$, and the continuous field of C^*-algebras $(A_\theta)_{\theta \in [0,1)}$.

5.3. The label $n = 0$. Estimates for the norm. The label $n = 0$ corresponds to the external boundary of the butterfly. As a consequence of Theorem 5.1 (ii) this curve, which corresponds to the graph of the function $\theta \mapsto \|H_{\theta,\mu}\|$, is continuous and almost everywhere differentiable. The estimate $\|H_\theta\| < 4$ is non-trivial for $0 < \theta < 1$, and means that the gap $n = 0$ is always open. Lower bound estimates can be derived by using approximate eigenvectors. Exploiting cancellations in certain trigonometric sums one can also find an explicit upper bound for the norm and prove:

THEOREM 5.3. ([**17**]) (i) *For any $\theta \in [0, \frac{1}{2}]$ and any $\mu > 0$,*

$$\|H_{\theta,\mu}\| \leq M_\mu(\theta),$$

where

$$M_\mu(\theta) = \begin{cases} 2\sqrt{1 + \mu^2} + |\mu|(\cos \pi\theta - \sin \pi\theta) \cos \pi\theta & \text{if } \theta \in [0, \frac{1}{4}], \\ 2\sqrt{1 + \mu^2 - \left(1 - \frac{1}{\tan \pi\theta}\right)\left(1 - \sqrt{\frac{1 + \cos^2 4\pi\theta}{2}}\right) \min\{1, \mu^2\}} & \text{if } \theta \in [\frac{1}{4}, \frac{1}{2}]. \end{cases}$$

(ii) *There exists an elementary function $\theta \mapsto m(\theta)$ such that*

$$m(\theta) \leq \|H_\theta\| \qquad \text{for all } \theta \in [0, 1].$$

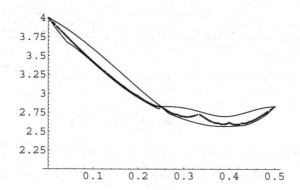

FIGURE 10. *The graphs of the functions $[0, \frac{1}{2}] \ni \theta \mapsto M_1(\theta), \|H_\theta\|, m(\theta)$.*

REMARK 5.4. For lack of space we do not give here the explicit form of the function m. We note however that in particular

$$\min_{\theta \in [0,0.5]} \|H_\theta\| \geq \min_{\theta \in [0,0.5]} m(\theta) \approx \sqrt{6.59303} \approx 2.56769.$$

This is a satisfactory estimate for $\min_{0.25 \leq \theta \leq 0.5} \|H_\theta\|$, which appears to be, by numerical computations, just fractionally larger than 2.59.

REMARK 5.5. The upper bound estimates in Theorem 5.3 significantly improve upon previous estimates obtained in [4] and in [55].

It would be interesting to prove or disprove the validity of the inequality

$$\min_{\theta \in [0,0.25]} \|H_\theta\| \geq 2\sqrt{2}.$$

References

[1] A. Avila, R. Krikorian, *Reducibility of non-uniform hyperbolicity for quasiperiodic Schrödinger cocycles*, preprint math.DS/0306382.

[2] J. Avron, P. H. M. van Mouche, B. Simon, *On the measure of the spectrum for the almost Mathieu operator*, Comm. Math. Phys. **132** (1990), 103–118.

[3] V. Augustin, F. P. Boca, C. Cobeli, A. Zaharescu, *The h-spacing distribution between Farey points*, Math. Proc. Camb. Phil. Soc. **131** (2001), 23–38.

[4] C. Béguin, A. Valette, A. Zuk, *On the spectrum of a random walk on the discrete Heisenberg group and the norm of Harper's operator*, J. Geom. Phys. **21** (1997), 337–356.

[5] J. Bellissard, *Lipschitz continuity of gap boundaries for Hofstadter-like spectra*, Comm. Math. Phys. **160** (1994), 599–613.

[6] J. Bellissard, B. Simon, *Cantor spectrum for the almost Mathieu equation*, J. Functional Anal. **48** (1982), 408–419.

[7] S. Blank, N. Krikorian, *Thom's problem on irrational flows*, Internat. J. Math. **4** (1993), 721–726.

[8] P. Blecher, *Statistical properties of two-dimensional periodic Lorentz with infinite horizon*, J. Stat. Phys. **66**, (1992), 315–373.

[9] F. P. Boca, *Projections in rotation algebras and theta functions*, Comm. Math. Phys. **202** (1999), 325–357.

[10] F. P. Boca, *Rotation C*-algebras and almost Mathieu operators*, Theta Series in Advanced Mathematics, 1. The Theta Foundation, Bucharest 2001.

[11] F. P. Boca, C. Cobeli, A. Zaharescu, *Distribution of lattice points visible from the origin*, Comm. Math. Phys. **213** (2000), 433–470.

[12] F. P. Boca, C. Cobeli and A. Zaharescu, *A conjecture of R. R. Hall on Farey points*, J. Reine Angew. Mathematik **535** (2001), 207–236.

[13] F. P. Boca, C. Cobeli, A. Zaharescu, *On the distribution of the Farey sequence with odd denominators*, Michigan Math. J. **51** (2003), 557–573.

[14] F. P. Boca, R. N. Gologan, A. Zaharescu, *On the index of Farey sequences*, Quart. J. Math. Oxford Ser. (2) **53** (2002), 377–391.

[15] F. P. Boca, R. N. Gologan, A. Zaharescu, *The average length of a trajectory in a certain billiard in a flat two-torus*, New York J. Math. **9** (2003), 303–330.

[16] F. P. Boca, R. N. Gologan, A. Zaharescu, *The statistics of a certain billiard in a flat two-torus*, Comm. Math. Phys. **240** (2003), 53–73.

[17] F. P. Boca, A. Zaharescu, *Norm estimates of almost Mathieu operators*, preprint arXiv math-ph/0201028.

[18] F. P. Boca, A. Zaharescu, *The distribution of the free path lengths in the periodic two-dimensional Lorentz gas in the small-scatterer limit*, preprint arXiv math.NT/0301270.

[19] O. Bratteli, A. Kishimoto, *Noncommutative spheres. III. Irrational rotations*, Comm. Math. Phys. **147** (1992), 605–624.

[20] J. Bourgain, F. Golse, B. Wennberg, *On the distribution of free path lengths for the periodic Lorentz gas*, Comm. Math. Phys. **190** (1998), 491–508.

[21] L. Bunimovich, *Billiards and other hyperbolic systems*, in Dynamical systems, ergodic theory and applications (Ya. G. Sinai and al. eds.), pp. 192–233, Encyclopedia Math. Sci. 100, 2nd edition, Springer-Verlag, Berlin, 2000.

[22] E. Caglioti, F. Golse, *On the distribution of free path lengths for the periodic Lorentz gas. III*, Comm. Math. Phys. **236** (2003), 199–221.

[23] N. Chernov, *New proof of Sinai's formula for the entropy of hyperbolic billiards. Application to Lorentz gases and Bunimovich stadium*, Funct. Anal. and Appl. **25** (1991), 204–219.

[24] N. Chernov, *Entropy values and entropy bounds*, in Hard ball systems and the Lorentz gas (D. Szász ed.), pp. 121–143, Encyclopaedia Math. Sci., 101, Springer-Verlag, Berlin, 2000.

[25] M. D. Choi, G. A. Elliott, N. Yui, *Gauss polynomials and the rotation algebras*, Invent. Math. **99** (1990), 225–246.

[26] A. Connes, *C*-algèbres et géométrie différentielle*, C. R. Acad. Sci. Paris Sér. A-B **290** (1980), 599–604.

[27] P. Dahlqvist, *The Lyapunov exponent in the Sinai billiard in the small scatterer limit*, Nonlinearity **10** (1997), 159-173.

[28] H. M. Edwards, *Riemann's zeta function*, Dover Publications, Inc., Mineola, New York, 2001.

[29] E. G. Effros, C. L. Shen, *Approximately finite C*-algebras and continued fractions*, Indiana Univ. Math. J. **29** (1980), 191–204.

[30] G. A. Elliott, D. E. Evans, *The structure of the irrational rotation C*-algebra*, Ann. Math. **138** (1993), 477–501.

[31] G. A. Elliott, Q. Lin, *Cut-down method in the inductive limit decomposition of noncommutative tori*, J. London Math. Soc. **54** (1996), 121–134.

[32] T. Estermann, *On Kloosterman's sum*, Mathematika **8** (1961), 83–86.

[33] B. Friedman, Y. Oono, I. Kubo, *Universal behaviour of Sinai billiard systems in the small-scatterer limit*, Phys. Rev. Letter **52** (1984), 709–712.

[34] G. Gallavotti, *Lectures on the billiard*, in Dynamical Systems, Theory and Applications (Battelle Seattle 1974 Rencontres, J. Moser ed.), Lecture Notes in Physics Vol. **38**, Springer-Verlag 1975.

[35] F. Golse, *On the statistics of free-path lengths for the periodic Lorentz gas*, Proceedings of the XIVth International Congress of Mathematical Physics, Lisbon, July, 2003, to appear.

[36] U. Haagerup, M. Rørdam, *Perturbations of the rotation C*-algebra and of the Heisenberg commutation relations*, Duke Math. J. **77** (1995), 627–656.

[37] R. R. Hall, *A note on Farey series*, J. London Math. Soc. **2** (1970), 139–148.

[38] R. R. Hall, *On consecutive Farey arcs II*, Acta Arith. **66** (1994), 1–9.

[39] R. R. Hall, P. Shiu, *The index of a Farey sequence*, Michigan Math. J. **51** (2003), 209–223.

[40] A. Haynes, *A note on Farey fractions with odd denominators*, J. Number Theory **98** (2003), 89–104.

[41] B. Helffer, J. Sjöstrand, *Analyse semi-classique pour l'équation de Harper. II. Comportement semi-classique près d'un rationnel*, Mém. Soc. Math. France No. 40, 1990; *Semi-classical analysis for Harper's equation. III. Cantor structure of the spectrum*, Mém. Soc. Math. France (N.S.), No. 39, 1989.

[42] B. Helffer, P. Kerdelhué, J. Sjöstrand, *Le papillon de Hofstadter revisité*, Mém. Soc. Math. de France, No. 43, 1990.

[43] D. R. Hofstadter, *Energy levels and wave functions of Bloch electrons in a rational or irrational magnetic field*, Phys. Rev. B **14** (1976), 2239–2249.

[44] C. Hooley, *An asymptotic formula in the theory of numbers*, Proc. London Math. Soc. **7** (1957), 396–413.

[45] Y. Last, *Zero measure spectrum for the almost Mathieu operator*, Comm. Math. Phys. **164** (1994), 421–432.

[46] Y. Last, M. Wilkinson, *A sum rule for the dispersion relations of the rational Harper equation*, J. Phys. A **25** (1992), 6123–6133.

[47] H. A. Lorentz, *Le mouvement des électrons dans les métaux*, Arch. Néerl. **10** (1905), 336, reprinted in *Collected papers*, Vol. 3, Martinus Nijhoff, The Hague, 1936.

[48] D. Osadchy, J. E. Avron, *Hofstadter butterfly as quantum phase diagram*, J. Math. Phys. **42** (2001), 5665–5671.

[49] N. C. Phillips, *Crossed products by finite cyclic group actions with the tracial Rokhlin property*, preprint math.OA/0306410.

[50] M. V. Pimsner, D. Voiculescu, *Imbedding the irrational rotation C^*-algebra into an AF-algebra*, J. Operator Theory **4** (1980), 201–210.

[51] J. Puig, *Cantor spectrum for the almost Mathieu operator*, Comm. Math. Phys. **244** (2004), 297–309.

[52] M. A. Rieffel, *C^*-algebras associated with irrational rotations*, Pacific J. Math. **93** (1981), 415–429.

[53] F. Schweiger, *Ergodic theory of fibred systems and metric number theory*, Oxford University Press, New York, 1995.

[54] Ya. G. Sinai, *Dynamical systems with ellastic reflections. Ergodic properties of dispersing billiards*, Russ. Math. Surveys **25** (1970), 137–189.

[55] R. Szwarc, *Norm estimates of discrete Schrödinger operators*, Coll. Math. **76** (1998), 153–160.

[56] D. Thouless, *Bandwidth for a quasiperiodic tight binding model*, Phys. Rev. B **28** (1983), 4272–4276.

[57] S. Walters, *The AF structure of non commutative toroidal $\mathbb{Z}/4\mathbb{Z}$ orbifolds*, J. Reine Angew. Mathematik **568**, 139–196, 2004.

[58] A. Weil, *On some exponential sums*, Proc. Nat. Acad. Sci. U.S.A. **34** (1948), 204–207.

FPB AND AZ: DEPARTMENT OF MATHEMATICS, UNIVERSITY OF ILLINOIS, 1409 W. GREEN STR., URBANA IL 61801, USA

INSTITUTE OF MATHEMATICS OF THE ROMANIAN ACADEMY, P.O. BOX 1-764, RO-014700 BUCHAREST, ROMANIA

FBOCA@MATH.UIUC.EDU, ZAHARESC@MATH.UIUC.EDU

Transgressions of the Godbillon-Vey Class and Rademacher functions

Alain Connes and Henri Moscovici

ABSTRACT. We construct, out of modular symbols, 1-traces that are invariant with respect to the actions of the Hopf algebra $\mathcal{H}_1$ on the crossed product $\mathcal{A}_\mathbb{Q}$ of the algebra of modular forms of all levels by $\mathrm{GL}^+(2, \mathbb{Q})$ investigated in earlier work. This provides a conceptual explanation for the construction of the Euler cocycle representing the image of the universal Godbillon-Vey class under the characteristic map of noncommutative Chern-Weil theory which we developed in our earlier work. We then refine the construction to produce secondary data by transgression. For the action determined by the Ramanujan connection the transgression takes place within the Euler class and the resulting cocycle coincides with the classical Rademacher function. The actions associated to cusp forms of higher weight produce transgressed cocycles that implement the Eichler-Shimura isomorphism. Finally, the actions corresponding to Eisenstein series give rise by transgression to the Eisenstein cocycle, expressed in terms of higher Dedekind sums and generalized Rademacher functions.

Introduction

In earlier work [8, 9] we investigated a surprising interconnection between the transverse geometry of codimension 1 foliations and modular forms. At the core of this interplay lies the Hopf algebra $\mathcal{H}_1$, the first in a series of Hopf algebras $\mathcal{H}_n$ that were found [6] to determine the affine transverse geometry of codimension n foliations. The periodic Hopf-cyclic cohomology of $\mathcal{H}_1$ is generated by two classes, $[\delta_1]$ for the odd component and $[RC_1]$ for the even component. The tautological action of $\mathcal{H}_1$ on the étale groupoid algebra $\mathcal{A}_\mathcal{G}$ associated to the frame bundle of a codimension 1 foliation preserves (up to a character) the canonical trace on $\mathcal{A}_\mathcal{G}$, and thus gives rise to a characteristic homomorphism in cyclic cohomology. This homomorphism maps $[\delta_1]$ to the Godbillon-Vey class and $[RC_1]$ to the transverse fundamental class.

The starting point of the investigation in [8] was the realization that the Hopf algebra $\mathcal{H}_1$ can be made to act on the crossed product $\mathcal{A}_\mathbb{Q}$ of the algebra of modular forms of all levels by $\mathrm{GL}^+(2, \mathbb{Q})$, via a natural connection provided by the Ramanujan operator on modular forms, thus conferring a symmetry structure to the space of lattices modulo the action of the Hecke correspondences. Although the algebra $\mathcal{A}_\mathbb{Q}$ no longer has an invariant trace, we used an ad hoc pairing with modular symbols to convert the Godbillon-Vey class $[\delta_1] \in HC^1(\mathcal{H}_1)$ into the Euler class of $\mathrm{GL}^+(2, \mathbb{Q})$.

In this paper we provide a completely conceptual explanation for the above pairing and at the same time extend it to the higher weight case. This is achieved by constructing out of modular symbols $\mathcal{H}_1$-invariant 1-traces that support characteristic maps for certain actions of $\mathcal{H}_1$ on $\mathcal{A}_\mathbb{Q}$, canonically associated to modular forms. Moreover, we show that the image of the Godbillon-Vey class through these characteristic homomorphisms, obtained by the cup product between $[\delta_1]$ and the invariant 1-traces, transgresses to secondary data. For the projective action determined by the Ramanujan connection the transgression takes place within the Euler class, in a manner that ressembles the K-homological transgression in the context of $SU_q(2)$ [5], and leads to the classical Rademacher function [23]. For the actions associated to cusp forms of higher weight the transgressed classes implement the Eichler-Shimura isomorphism. The actions corresponding to Eisenstein series give rise by transgression to higher Dedekind sums and generalized Rademacher functions ([28, 20]), or equivalently to the Eisenstein cocycle of [28].

Generalized Dedekind sums have been related to special values of L-functions in the work of C. Meyer on the class-number formula [18, 19], and higher Dedekind sums appear in the work of Siegel [25, 26] and Zagier [29] on the values at non-positive integers of partial zeta functions over real quadratic fields. Eisenstein cocycles were employed by Stevens [28] and by Sczech [24] in order to compute these values more efficiently. The fact that these notions can be interpreted as secondary invariants is reminiscent of the secondary nature of the regulator invariants (cf. e.g. [10]) that are involved in the expression of the special values at non-critical points of L-functions associated to number fields (see e.g. [16], [30]).

1. The standard modular Hopf action

In this preliminary section, we briefly review the basic facts ([6], [8]) concerning the Hopf algebra $\mathcal{H}_1$ and its standard Hopf action on the crossed product $\mathcal{A}_\mathbb{Q}$ of the algebra of modular forms of all levels by $GL^+(2,\mathbb{Q})$, associated to the Ramanujan connection.

1.1. The Hopf algebra $\mathcal{H}_1$ and its cyclic classes. We start by recalling the definition of the Hopf algebra $\mathcal{H}_1$. As an algebra, it coincides with the universal enveloping algebra of the Lie algebra with basis $\{X, Y, \delta_n\,; n \geq 1\}$ and brackets

$$[Y,X] = X\,, \quad [Y,\delta_n] = n\,\delta_n\,, \quad [X,\delta_n] = \delta_{n+1}\,, \quad [\delta_k,\delta_\ell] = 0\,, \quad n,k,\ell \geq 1\,.$$

As a Hopf algebra, the coproduct $\Delta : \mathcal{H}_1 \to \mathcal{H}_1 \otimes \mathcal{H}_1$ is determined by

$$\begin{aligned}
\Delta Y &= Y \otimes 1 + 1 \otimes Y\,, & \Delta\,\delta_1 &= \delta_1 \otimes 1 + 1 \otimes \delta_1 \\
\Delta X &= X \otimes 1 + 1 \otimes X + \delta_1 \otimes Y
\end{aligned}$$

and the multiplicativity property

$$\Delta(h^1 h^2) = \Delta h^1 \cdot \Delta h^2\,, \quad h^1, h^2 \in \mathcal{H}_1\,;$$

the antipode is determined by

$$S(Y) = -Y\,, \quad S(X) = -X + \delta_1 Y\,, \quad S(\delta_1) = -\delta_1$$

and the anti-isomorphism property

$$S(h^1 h^2) = S(h^2)\,S(h^1)\,, \quad h^1, h^2 \in \mathcal{H}_1\,;$$

finally, the counit is

$$\varepsilon(h) = \text{constant term of} \quad h \in \mathcal{H}_1 \,.$$

The modular character $\delta \in \mathcal{H}_1^*$, determined by

$$\delta(Y) = 1, \quad \delta(X) = 0, \quad \delta(\delta_n) = 0 \,,$$

together with the unit of $1 \in \mathcal{H}_1$ forms a modular pair in involution $(\delta, 1)$, and thus the Hopf-cyclic cohomology $HC^1_{(\delta,1)}(\mathcal{H}_1)$ is well-defined (for definitions, see [6, 7]). The element $\delta_1 \in \mathcal{H}_1$ is a Hopf-cyclic cocycle, which gives a nontrivial class

$$[\delta_1] \in HC^1_{(\delta,1)}(\mathcal{H}_1)$$

in the Hopf-cyclic cohomology of $\mathcal{H}_1$ with respect to the modular pair $(\delta, 1)$. Its periodic image generates the periodic group $HP^1(\mathcal{H}_1; \delta, 1)$, and represents the universal Godbillon-Vey class (cf. [8, Prop. 3]).

The even component of the periodic cyclic cohomology group $HP^0(\mathcal{H}_1; \delta, 1)$ is generated by the "transverse fundamental class", represented by the Hopf-cyclic 2-cocycle

$$RC_1 := X \otimes Y - Y \otimes X - \delta_1 Y \otimes Y \,.$$

(See [9] for the explanation of the notation.)

There is one other Hopf-cyclic 1-cocycle, intimately related to the classical Schwarzian, which plays a prominent role in the transverse geometry of modular Hecke algebras ([8, 9]). It is given by the primitive element

$$\delta_2' := \delta_2 - \frac{1}{2}\delta_1^2 \in \mathcal{H}_1 \,.$$

Its periodic class vanishes because $\delta_2' = B(c)$, where c is the following Hochschild 2-cocycle:

$$c := \delta_1 \otimes X + \frac{1}{2}\delta_1^2 \otimes Y \,.$$

1.2. Standard modular action of $\mathcal{H}_1$. The notation being as in [8, §1], we form the crossed product algebra

$$\mathcal{A}_{G^+(\mathbb{Q})} = \mathcal{M} \ltimes \text{GL}^+(2, \mathbb{Q}) \,,$$

where $\mathcal{M}$ is the algebra of (holomorphic) modular forms of all levels. The product of two elements in $\mathcal{A}_{G^+(\mathbb{Q})}$,

$$a^0 = \sum_\alpha f_\alpha^0 U_\alpha \quad \text{and} \quad a^1 = \sum_\beta f_\beta^1 U_\beta \,,$$

is given by the convolution rule

$$a^0 a^1 = \sum_{\alpha,\beta} f_\alpha^0 \, f_\beta^1|\alpha^{-1} \, U_{\alpha\beta} \,.$$

We recall (see [8, Prop. 7]) that there is a unique Hopf action of the Hopf alge-
bra $\mathcal{H}_1$ on $\mathcal{A}_{G+(\mathbb{Q})}$ determined by letting the generators $\{Y, X, \delta_1\}$ of $\mathcal{H}_1$ act on
monomials $fU_\gamma^* \in \mathcal{A}_{G+(\mathbb{Q})}$ as follows:

$$(1.1) \qquad Y(fU_\gamma^*) = Y(f)U_\gamma^*, \quad \text{where} \quad Y(f) = \frac{w(f)}{2} f, \quad w(f) = \text{weight}(f);$$

$$(1.2) \qquad X(fU_\gamma^*) = X(f)U_\gamma^*, \quad \text{where} \quad X = \frac{1}{2\pi i}\frac{d}{dz} - \frac{1}{2\pi i}\frac{d}{dz}(\log \eta^4) Y$$

and η stands for the Dedekind η-function,

$$\eta^{24}(z) = q \prod_{n=1}^{\infty}(1-q^n)^{24}, \qquad q = e^{2\pi i z};$$

lastly,

$$(1.3) \qquad\qquad\qquad \delta_1(fU_\gamma^*) = \mu_\gamma \, fU_\gamma^*,$$

with the factor μ_γ given by the expression

$$(1.4) \qquad \mu_\gamma(z) = \frac{1}{12\pi i}\frac{d}{dz}\log\frac{\Delta|\gamma}{\Delta} = \frac{1}{2\pi i}\frac{d}{dz}\log\frac{\eta^4|\gamma}{\eta^4};$$

equivalently,

$$(1.5) \qquad \mu_\gamma(z) = \frac{1}{2\pi^2}\left(G_2|\gamma(z) - G_2(z) + \frac{2\pi i c}{cz+d}\right),$$

where

$$G_2(z) = 2\zeta(2) + 2\sum_{m\geq 1}\sum_{n\in\mathbb{Z}}\frac{1}{(mz+n)^2} = \frac{\pi^2}{3} - 8\pi^2\sum_{m,n\geq 1}me^{2\pi imnz}$$

is the quasimodular holomorphic Eisenstein series of weight 2. The factor μ_γ can
further be expressed as the difference

$$(1.6) \qquad \mu_\gamma = 2(\phi_0|\gamma - \phi_0), \quad \phi_0 = \frac{1}{4\pi^2}G_0,$$

where G_0 is the modular (but nonholomorphic) weight 2 Eisenstein series

$$G_0(z) = G_0(z,0)$$

obtained by taking the value at $s = 0$ of the analytic continuation of the series

$$G_0(z,s) = \sum_{(m,n)\in\mathbb{Z}^2\backslash 0}(mz+n)^{-2}|mz+n|^{-s}$$

$$= 2\zeta(2+s) + 2\sum_{m\geq 1}\sum_{n\in\mathbb{Z}}(mz+n)^{-2}|mz+n|^{-s}, \quad \text{Re}\, s > 0.$$

It is related to G_2 by the identity

$$G_2(z) = G_0(z) + \frac{2\pi i}{z-\bar{z}}.$$

The equation (1.6) shows that the range of μ is contained in the space $\mathcal{E}_2(\mathbb{Q})$ of
weight 2 Eisenstein series whose constant term in the q-expansion at each cusp is
rational. We recall (following [27, §2.4]) Hecke's construction [14] of a lattice of
generators for the $\mathbb{Q}$-vector space $\mathcal{E}_2(\mathbb{Q})$.

For $\mathbf{a} = (a_1, a_2) \in (\mathbb{Q}/\mathbb{Z})^2$ and $z \in \mathbb{H}$ fixed, the series

$$G_{\mathbf{a}}(z, s) := \sum_{\mathbf{m} \in \mathbb{Q}^2 \backslash 0, \, \mathbf{m} \equiv \mathbf{a} \,(\mathrm{mod}\ 1)} (m_1 z + m_2)^{-2} |m_1 z + m_2|^{-s}, \quad \mathrm{Re}\, s > 0 \,;$$

defines a function that can be analytically continued beyond $\mathrm{Re}\, s = 0$, which allows to define

$$G_{\mathbf{a}}(z) := G_{\mathbf{a}}(z, 0)\,.$$

Furthermore, one has

$$G_{\mathbf{a}} | \gamma = G_{\mathbf{a} \cdot \gamma}, \quad \forall \gamma \in \Gamma(1)\,,$$

which shows that $G_{\mathbf{a}}(z)$ behaves like a weight 2 modular form of some level N. In fact the difference

$$\wp_{\mathbf{a}}(z) = G_{\mathbf{a}}(z) - G_0(z)$$

is the $\mathbf{a}$–division value of the Weierstrass $\wp$-function. The collection of functions

$$\left\{ \wp_{\mathbf{a}} \; ; \; \mathbf{a} \in \left(\frac{1}{N} \mathbb{Z}/\mathbb{Z} \right)^2 \backslash \mathbf{0} \right\}$$

generates the space of weight 2 Eisenstein series of level N.

In order to obtain a set of generators for $\mathcal{E}_2(\mathbb{Q})$, one considers the additive characters $\chi_{\mathbf{x}} : \left(\frac{1}{N} \mathbb{Z}/\mathbb{Z} \right)^2 \to \mathbb{C}^\times$ defined by

(1.7) $$\chi_{\mathbf{x}} \left(\frac{\mathbf{a}}{N} \right) := e^{2\pi i \, (a_2 x_1 - a_1 x_2)}\,,$$

for each $\mathbf{x} = (x_1, x_2) \in \left(\frac{1}{N} \mathbb{Z}/\mathbb{Z} \right)^2$, and one forms the series

(1.8) $$\phi_{\mathbf{x}}(z) := (2\pi N)^{-2} \sum_{\mathbf{a} \in \left(\frac{1}{N} \mathbb{Z}/\mathbb{Z} \right)^2} \chi_{\mathbf{x}}(\mathbf{a}) \cdot G_{\mathbf{a}}(z)\,.$$

The definition is independent of N and, for each $\mathbf{x} = (x_1, x_2) \in \left(\frac{1}{N} \mathbb{Z}/\mathbb{Z} \right)^2 \backslash \mathbf{0}$ then $\phi_{\mathbf{x}}$, gives a weight 2 Eisenstein series of level N.

To account for the special case when $\mathbf{x} = \mathbf{0}$, one adjoins the non-holomorphic but modular function ϕ_0 defined in (1.6).

All the linear relations among the functions $\phi_{\mathbf{x}}$, $\mathbf{x} \in (\mathbb{Q}/\mathbb{Z})^2$ are encoded in the *distribution property*

(1.9) $$\phi_{\mathbf{x}} = \sum_{\mathbf{y} \cdot \check{\gamma} = \mathbf{x}} \phi_{\mathbf{y}} | \gamma\,.$$

where

$$\check{\gamma} = \det \gamma \cdot \gamma^{-1}\,.$$

This allows to equip the extended Eisenstein space

$$\mathcal{E}_2^*(\mathbb{Q}) = \mathcal{E}_2(\mathbb{Q}) \oplus \mathbb{Q} \cdot \phi_0\,,$$

with a linear $\mathrm{PGL}^+(2, \mathbb{Q})$-action, as follows. Denoting

$$\mathcal{S} := (\mathbb{Q}/\mathbb{Z})^2, \quad \mathrm{resp.} \quad \mathcal{S}' := \mathcal{S} \backslash \mathbf{0}$$

and identifying in the obvious way

$$\mathrm{PGL}^+(2, \mathbb{Q}) \cong M_2^+(\mathbb{Z}) / \{\mathrm{scalars}\}\,,$$

where $M_2^+(\mathbb{Z})$ stands for the set of integral 2×2-matrices of determinant > 0, one defines the action of $\gamma \in M_2^+(\mathbb{Z})$ by:

$$\mathbf{x}|\gamma := \sum_{\mathbf{y}\cdot\tilde{\gamma}=\mathbf{x}} \mathbf{y} \in \mathbb{Q}[\mathcal{S}].$$

With this definition one has

$$\phi_\mathbf{x}|\gamma = \phi_{\mathbf{x}|\gamma}, \qquad \gamma \in M_2^+(\mathbb{Z}).$$

Modulo the subspace of 'distribution relations'

$$\mathcal{R} := \mathbb{Q} - \text{span of } \left\{ \mathbf{x} - \mathbf{x}\Big| \begin{pmatrix} n & 0 \\ 0 & n \end{pmatrix} ; \mathbf{x} \in \mathcal{S}, n \in \mathbb{Z} \setminus 0 \right\},$$

the assignment $\mathbf{x} \in \mathcal{S} \longmapsto \phi_\mathbf{x}$ induces an isomorphism of $\mathrm{PGL}_2^+(2,\mathbb{Q})$-modules

$$\mathbb{Q}[\mathcal{S}]/\mathcal{R} \cong \mathcal{E}_2^*(\mathbb{Q}).$$

In view of the above the identity (1.6) can be completed as follows:

$$(1.10) \qquad \mu_\gamma = 2(\phi_\mathbf{0}|\gamma - \phi_\mathbf{0}) = 2\left(\sum_{\mathbf{y}\cdot\tilde{\gamma}=\mathbf{0}} \phi_\mathbf{y} - \phi_\mathbf{0} \right), \qquad \forall \gamma \in M_2^+(\mathbb{Z}).$$

2. Characteristic map for the standard action

In this section we provide the conceptual explanation for the period pairing which was employed in [8] to obtain the Euler class out of the universal Godbillon-Vey class $[\delta_1] \in HC^1_{(\delta,1)}(\mathcal{H}_1)$, by showing that it is in fact the by-product of a characteristic map associated to a 1-trace which is invariant with respect to the standard action of $\mathcal{H}_1$ on $\mathcal{A}$.

Extending a classical 'splitting formula' for the restriction to $\mathrm{SL}(2,\mathbb{Z})$ of the 2-cocycle that gives the universal cover of $\mathrm{SL}(2,\mathbb{R})$, we shall then obtain a new formula for the rational 2-cocycle representing the Euler class found in [8], showing that it differs from the Petersson cocycle ([1], [21]) by precisely the coboundary of the classical Rademacher function.

2.1. Characteristic map and cup products.
In [6] (see also [7]) we defined a characteristic map associated to a Hopf module algebra with invariant trace. The construction has been subsequently extended to higher traces [11] and turned into a cup product in Hopf-cyclic cohomology, [15]. A predecessor of these constructions is the contraction of a cyclic n-cocycle by the generator of a 1-parameter group of automorphisms that fixes the cocycle, [4, *Chap.III.*6.β]. We shall apply the latter to a specific 1-trace $\tau_0 \in ZC^1(\mathcal{A})$, which will be described in details in the next subsection. Further on, it will also be applied in the context of cyclic cohomology with coefficients, to 1-traces $\tau_W \in ZC^1(\mathcal{A},W)$, where W denotes an algebraically irreducible $\mathrm{GL}(2,\mathbb{Q})$-module.

Let us assume that $\tau \in C^1(\mathcal{A})$ is a cyclic cocycle which satisfies, with respect to a given Hopf action of $\mathcal{H}_1$ on $\mathcal{A}$, the invariance property

$$(2.1) \qquad \tau(h_{(1)}(a^0), h_{(2)}(a^1)) = \delta(h)\,\tau(a^0, a^1), \qquad \forall h \in \mathcal{H}_1,\ a^0, a^1 \in \mathcal{A}.$$

The simplest expression for the cup product

$$gv = \delta_1 \# \tau \in ZC^2(\mathcal{A})$$

is given by the contraction formula in [4, $Chap.III.6.\beta$] mentioned above, which (in the non-normalized form, [11, §3]) takes the expression:

$$(2.2) \qquad gv\,(a^0, a^1, a^2) = \tau(a^0\,\delta_1(a^1), a^2), \qquad a^0, a^1, a^2 \in \mathcal{A}.$$

For the convenience of the reader, let us check directly that this formula gives a cocycle in the (b, B)-bicomplex of the algebra $\mathcal{A}$.

LEMMA 1. Let $\tau \in ZC^1(\mathcal{A})$ be a cyclic cocycle satisfying the $\mathcal{H}_1$-invariance property (2.1). Then $b(gv) = 0$ and $B(gv) = 0$.

PROOF. Using the fact that δ_1 acts as a derivation, one has

$$b(gv)\,(a^0, a^1, a^2, a^3) = \tau\,(a^0 a^1 \delta_1(a^2), a^3) - \tau\,(a^0 \delta_1(a^1 a^2), a^3)$$

$$+ \quad \tau\,(a^0 \delta_1(a^1), a^2 a^3) - \tau\,(a^3 a^0 \delta_1(a^1), a^2)$$

$$= \quad -\tau\,(a^0 \delta_1(a^1) a^2, a^3) + \tau\,(a^0 \delta_1(a^1), a^2 a^3)$$

$$- \quad \tau\,(a^3 a^0 \delta_1(a^1), a^2) = -b\tau\,(a^0 \delta_1(a^1), a^2, a^3) = 0\,.$$

Passing to B, since $\tau(a, 1) = 0$ for any $a \in \mathcal{A}$, one has

$$B(gv)\,(a^0, a^1) = gv\,(1, a^0, a^1) - gv\,(1, a^1, a^0)$$

$$= \quad \tau\,(\delta_1(a^0), a^1) - \tau\,(\delta_1(a^1), a^0) = \tau\,(\delta_1(a^0), a^1) + \tau\,(a^0, \delta_1(a^1)) = 0\,,$$

the vanishing taking place because $\tau \in ZC^1(\mathcal{A})$ is $\mathcal{H}_1$-invariant.   □

2.2. The basic invariant 1-cocycles. The modular symbol cocycle of weight 2 associated to a base point $z_0 \in \mathbb{H}$, $\tau_0 \in ZC^1(\mathcal{A})$, is defined as follows. For monomials $f^0 U_{\gamma_0}, f^1 U_{\gamma_1} \in \mathcal{A}$, $\tau_0(f^0 U_{\gamma_0}, f^1 U_{\gamma_1}) = 0$, unless they satisfy the condition

$$(2.3) \qquad w(f^0) + w(f^1) = 2 \qquad \text{and} \qquad \gamma_0 \gamma_1 = 1\,,$$

in which case it is given by the integral

$$(2.4) \qquad \tau_0(f^0 U_{\gamma_0}, f^1 U_{\gamma_1}) = \int_{z_0}^{\gamma_0 z_0} f^0\, f^1|_{\gamma_1}\, dz\,.$$

The fact that $\tau_0 \in C^1(\mathcal{A})$ is indeed an $\mathcal{H}_1$-invariant cyclic cocycle is the content of the following result.

PROPOSITION 2. The cochain $\tau_0 \in C^1(\mathcal{A})$ is a cyclic cocycle which satisfies the $\mathcal{H}_1$-invariance property (2.1) with respect to the standard action.

PROOF. Let $f^0 U_{\gamma_0}, f^1 U_{\gamma_1}, f^2 U_{\gamma_2} \in \mathcal{A}$ be such that

(2.5) $\qquad w(f^0) + w(f^1) + w(f^2) = 2 \qquad$ and $\qquad \gamma_0 \gamma_1 \gamma_2 = 1$

One has

$$b\tau_0(f^0 U_{\gamma_0}, f^1 U_{\gamma_1}, f^2 U_{\gamma_2}) =$$

$$= \tau_0(f^0 f^1|_{\gamma_0^{-1}} U_{\gamma_0 \gamma_1}, f^2 U_{\gamma_2}) - \tau_0(f^0 U_{\gamma_0}, f^1 f^2|_{\gamma_1^{-1}} U_{\gamma_1 \gamma_2})$$

$$+ \tau_0(f^2 f^0|_{\gamma_2^{-1}} U_{\gamma_2 \gamma_0}, f^1 U_{\gamma_1})$$

$$= \int_{z_0}^{\gamma_0 \gamma_1 z_0} f^0 f^1|_{\gamma_0^{-1}} f^2|_{\gamma_1^{-1} \gamma_0^{-1}} dz - \int_{z_0}^{\gamma_0 z_0} f^0 f^1|_{\gamma_0^{-1}} f^2|_{\gamma_1^{-1} \gamma_0^{-1}} dz$$

$$+ \int_{z_0}^{\gamma_2 \gamma_0 z_0} f^2 f^0|_{\gamma_2^{-1}} f^1|_{\gamma_0^{-1} \gamma_2^{-1}} dz$$

$$= \int_{\gamma_0 z_0}^{\gamma_0 \gamma_1 z_0} f^0 f^1|_{\gamma_0^{-1}} f^2|_{\gamma_2} dz + \int_{z_0}^{\gamma_2 \gamma_0 z_0} f^2 f^0|_{\gamma_2^{-1}} f^1|_{\gamma_0^{-1} \gamma_2^{-1}} dz$$

$$= \int_{\gamma_0 z_0}^{\gamma_2^{-1} z_0} f^0 f^1|_{\gamma_0^{-1}} f^2|_{\gamma_2} dz + \int_{\gamma_2^{-1} z_0}^{\gamma_0 z_0} f^2|_{\gamma_2} f^0 f^1|_{\gamma_0^{-1}} dz \quad = \quad 0,$$

and so τ_0 is a Hochschild cocycle.
It is also cyclic, because for $f^0 U_{\gamma_0}, f^1 U_{\gamma_1} \in \mathcal{A}$ satisfying (2.3) one has

$$\lambda_1 \tau_0(f^0 U_{\gamma_0}, f^1 U_{\gamma_1}) \quad = \quad -\tau_0(f^1 U_{\gamma_1}, f^0 U_{\gamma_0}) = -\int_{z_0}^{\gamma_1 z_0} f^1 f^0|_{\gamma_1^{-1}} dz$$

$$= \quad \int_{z_0}^{\gamma_0 z_0} f^1|_{\gamma_1} f^0 dz \quad = \quad \tau_0(f^0 U_{\gamma_0}, f^1 U_{\gamma_1}).$$

In view of its multiplicative nature, it suffices to check the $\mathcal{H}_1$-invariance property
(2.1) on the algebra generators $\{Y, X, \delta_1\}$. Starting with Y, and with $f^0 U_{\gamma_0}, f^1 U_{\gamma_1}$
satisfying (2.3), one has

$$\tau_0(Y(f^0 U_{\gamma_0}), f^1 U_{\gamma_1}) + \tau_0(f^0 U_{\gamma_0}, Y(f^1 U_{\gamma_1})) =$$

$$= \int_{z_0}^{\gamma_0 z_0} Y(f^0 f^1|_{\gamma_0^{-1}}) dz = \frac{w(f^0) + w(f^1)}{2} \int_{z_0}^{\gamma_0 z_0} f^0 f^1|_{\gamma_0^{-1}} dz$$

$$= \int_{z_0}^{\gamma_0 z_0} f^0 f^1|_{\gamma_0^{-1}} dz \quad = \quad \delta(Y) \tau_0(f^0 U_{\gamma_0}, f^1 U_{\gamma_1}).$$

Passing to X, the identity (2.1) is nontrivial only if $f^0 U_{\gamma_0}, f^1 U_{\gamma_1} \in \mathcal{A}$ satisfy

(2.6) $\qquad w(f^0) + w(f^1) = 0 \qquad$ and $\qquad \gamma_0 \gamma_1 = 1,$

which actually implies that f^0 and f^1 are constants. One gets

$$\tau(X(f^0)U_{\gamma_0}, f^1 U_{\gamma_1}) + \tau(f^0 U_{\gamma_0}, X(f^1)U_{\gamma_1}) + \tau(\delta_1(f^0 U_{\gamma_0}), Y(f^1)U_{\gamma_1}) = 0$$

since $X(f^j) = 0$ and $Y(f^1) = 0$.
Finally, with $f^0 U_{\gamma_0}, f^1 U_{\gamma_1}$ as above, one has

$$\tau(\delta_1(f^0 U_{\gamma_0}), f^1 U_{\gamma_1}) + \tau(f^0 U_{\gamma_0}, \delta_1(f^1 U_{\gamma_1})) =$$

$$= \int_{z_0}^{\gamma_0 z_0} (\mu_{\gamma_1} + \mu_{\gamma_0}|_{\gamma_1}) f^0 f^1|_{\gamma_1} dz = 0,$$

because of the cocycle property of μ.   □

Allowing the base point z_0 to belong to the 'arithmetic' boundary of the upper
half plane $P^1(\mathbb{Q})$ requires some regularization of the integral. This can be achieved
by the standard procedure of removing the poles of Eisenstein series, [**27**]. In the
case at hand, it amounts to a coboundary modification which we proceed now to
describe.

To obtain it, we start from the observation that the derivative of τ_0 with respect
to the base point $z_0 \in \mathbb{H}$ is a coboundary:

$$\frac{d}{dz_0}\tau_0(f^0 U_{\gamma_0}, f^1 U_{\gamma_1}) = (f^0|_{\gamma_0} f^1)(z_0) - (f^0 f^1|_{\gamma_1})(z_0) = -b\epsilon(f^0 U_{\gamma_0}, f^1 U_{\gamma_1}),$$

where ϵ is the evaluation map at z_0:

$$\epsilon(f\, U_\gamma) = \begin{cases} f(z_0) & \text{if } \gamma = 1,\ f \in \mathcal{M}_2 \\ 0 & \text{otherwise}. \end{cases}$$

Taking the base point at the cusp ∞, we split the evaluation functional ϵ into two
other functionals, the constant term at ∞

$$\mathbf{a}_0(f\, U_\gamma) = \begin{cases} a_0 & \text{if } \gamma = 1,\ f \in \mathcal{M}_2 \\ 0 & \text{otherwise} \end{cases}$$

and the evaluation of the remainder

$$\tilde{\epsilon}(f\, U_\gamma) = \begin{cases} \tilde{f}(z_0) & \text{if } \gamma = 1,\ f \in \mathcal{M}_2 \\ 0 & \text{otherwise}, \end{cases}$$

where for $f \in \mathcal{M}_2$ of level N,

$$f(z) = \sum_{n=0}^{\infty} a_n\, e^{\frac{2\pi i n z}{N}}$$

represents its Fourier expansion at ∞, and

$$\tilde{f}(z) := f(z) - \mathbf{a}_0(f).$$

Both functionals are well-defined, because a_0 is independent of the level. To obtain
a cohomologous cocycle independent of z_0, it suffices to add to τ_0 the sum of
coboundaries of suitable anti-derivatives for the two components. We therefore
define

$$\tilde{\tau}_0 = \tau_0 + z_0\, b\, \mathbf{a}_0 - \int_{z_0}^{i\infty} b\tilde{\epsilon}\, dz,$$

that is, for $f^0 U_{\gamma_0}, f^1 U_{\gamma_1} \in \mathcal{A}$ as in (2.3)

$$\tilde{\tau}_0(f^0 U_{\gamma_0}, f^1 U_{\gamma_1}) = \int_{z_0}^{\gamma_0 z_0} f^0 f^1|_{\gamma_1}\, dz + z_0\, \mathbf{a}_0(f^0 f^1|_{\gamma_1} - f^0|_{\gamma_0} f^1)$$

(2.7) $$- \int_{z_0}^{i\infty} \tilde{\epsilon}(f^0 f^1|_{\gamma_1} - f^0|_{\gamma_0} f^1)\, dz.$$

The fact that $\tilde{\tau}_0 \in C^1(\mathcal{A})$ still satisfies Proposition 2 can be checked by an obvious
adaptation of its proof.

2.3. Euler class and the transgression formula. As a first example, we now specialize the construction of the cup product to the invariant cocycle $\tau_0 \in ZC^1(\mathcal{A})$. Let $f^0 U_{\gamma_0}, f^1 U_{\gamma_1}, f^2 U_{\gamma_2} \in \mathcal{A}$ be such that $\gamma_0 \gamma_1 \gamma_2 = 1$. Then

$$gv\left(f^0 U_{\gamma_0}, f^1 U_{\gamma_1}, f^2 U_{\gamma_2}\right) = \tau_0\left(f^0 U_{\gamma_0} \mu_{\gamma_1 - 1} f^1 U_{\gamma_1}, f^2 U_{\gamma_2}\right) =$$

$$= \tau_0(f^0 f^1 |\gamma_0^{-1} \mu_{\gamma_1 - 1}|\gamma_0^{-1} U_{\gamma_0 \gamma_1}, f^2 U_{\gamma_2})$$

$$= \int_{z_0}^{\gamma_0 \gamma_1 z_0} f^0 f^1 |\gamma_0^{-1} \mu_{\gamma_1 - 1}|\gamma_0^{-1} f^2 |\gamma_1^{-1} \gamma_0^{-1} \, dz$$

$$= \int_{\gamma_0^{-1} z_0}^{\gamma_1 z_0} f^0 |\gamma_0 \, f^1 \, \mu_{\gamma_1 - 1} \, f^2 |\gamma_1^{-1} \, dz = \int_{\gamma_2 z_0}^{z_0} f^0 |\gamma_0 \gamma_1 \, f^1 |\gamma_1 \, \mu_{\gamma_1 - 1} |\gamma_1 \, f^2 \, dz$$

$$(2.8) \quad = \int_{z_0}^{\gamma_2 z_0} f^0 |\gamma_0 \gamma_1 \, f^1 |\gamma_1 \, f^2 \, \mu_{\gamma_1} \, dz \, .$$

In particular, the restriction to $\mathcal{A}_0 = \mathbb{C}[\mathrm{GL}^+(2, \mathbb{Q})]$ is the group cocycle

$$(2.9) \qquad GV\left(\gamma_1, \gamma_2\right) := gv\left(U_{\gamma_0}, U_{\gamma_1}, U_{\gamma_2}\right) = \int_{z_0}^{\gamma_2 z_0} \mu_{\gamma_1} \, dz \, ,$$

whose real part

$$(2.10) \qquad \mathrm{Re}\, GV\left(\gamma_1, \gamma_2\right) := \mathrm{Re} \int_{z_0}^{\gamma_2 z_0} \mu_{\gamma_1} \, dz \, , \qquad \gamma_1, \gamma_2 \in \mathrm{GL}^+(2, \mathbb{Q})$$

represents a generator of $H^2(\mathrm{SL}(2, \mathbb{Q}), \mathbb{R})$, hence a multiple of the Euler class (cf. [8, Thm. 16]).

For a more precise identification, we shall be very specific about the choice of the Euler class. Namely, we take it as the class $\mathbf{e} \in H^2_{\mathrm{bor}}(\mathbb{T}, \mathbb{Z})$ defined by the extension

$$0 \to \mathbb{Z} \to \mathbb{R} \to \mathbb{T} \to 1 \, ;$$

via the canonical isomorphisms

$$H^2_{\mathrm{bor}}(\mathbb{T}, \mathbb{Z}) \simeq H^2(B\mathbb{T}, \mathbb{Z}) \simeq H^2(B\,\mathrm{SL}(2, \mathbb{R}), \mathbb{Z}) \simeq H^2_{\mathrm{bor}}(\mathrm{SL}(2, \mathbb{R}), \mathbb{Z})$$

followed by the succession of natural map

$$H^2_{\mathrm{bor}}(\mathrm{SL}(2, \mathbb{R}), \mathbb{Z}) \to H^2(\mathrm{SL}(2, \mathbb{Q}), \mathbb{Z}) \to H^2(\mathrm{SL}(2, \mathbb{Q}), \mathbb{R})$$

we regard it as a class $\mathbf{e} \in H^2(\mathrm{SL}(2, \mathbb{Q}), \mathbb{R})$.

PROPOSITION 3. *The 2-cocycle* $\mathrm{Re}\, GV \in Z^2(\mathrm{SL}(2, \mathbb{Q}), \mathbb{R})$ *represents the class* $-2\mathbf{e} \in H^2(\mathrm{SL}(2, \mathbb{Q}), \mathbb{R})$, *while* $\mathrm{Im}\, GV$ *is a coboundary.*

PROOF. In view of the definition (2.9) and using (1.4), one has for $\gamma_1, \gamma_2 \in \mathrm{GL}^+(2, \mathbb{Q})$,

$$12\pi i \, GV(\gamma_1, \gamma_2) = 12\pi i \int_{z_0}^{\gamma_2 z_0} \mu_{\gamma_1}(z) \, dz = \int_{z_0}^{\gamma_2 z_0} \frac{d}{dz} \log \frac{\Delta|\gamma_1}{\Delta} \, dz$$

$$= \int_{z_0}^{\gamma_2 z_0} (d \log \Delta|\gamma_1 - d \log \Delta)$$

$$= \log \Delta|\gamma_1(\gamma_2 z_0) - \log \Delta|\gamma_1(z_0)$$

$$(2.11) \qquad \qquad - (\log \Delta(\gamma_2 z_0) - \log \Delta(z_0))$$

where, since both Δ and $\Delta|\gamma_1$ don't have zeros in $\mathbb{H}$ one lets $\log \Delta$ and $\log \Delta|\gamma_1$ be holomorphic determinations of the logarithm whose choice is unimportant at this stage, since the additive constant which depends only on γ_1 cancels out. Let

$$j(\gamma, z) = cz + d, \qquad \gamma = \begin{pmatrix} a & b \\ c & d \end{pmatrix} \in \mathrm{GL}^+(2, \mathbb{Q}),$$

be the automorphy factor. Since it has no zero in $\mathbb{H}$ one can choose for each γ a holomorphic determination $\log j^2(\gamma, z)$ of its logarithm (for instance using the principal branch for $\mathbb{C} \setminus [0, \infty)$ of the logarithm when $c \neq 0$ and taking $\log d^2$ when $c = 0$). One then has,

$$\log \Delta|\gamma(z) = \log \Delta(\gamma z) - 6 \ \mathrm{Log}\, j^2(\gamma, z) + 2\pi i\, k(\gamma), \quad \forall z \in \mathbb{H}$$

for some $k(\gamma) \in \mathbb{Z}$. Thus (2.11) can be continued as follows:

$$12\pi i GV(\gamma_1, \gamma_2) = \log \Delta(\gamma_1 \gamma_2 z_0) - \log \Delta(\gamma_1 z_0) - \log \Delta(\gamma_2 z_0)$$

$$(2.12) \qquad + \log \Delta(z_0) - 6 \left(\log j^2(\gamma_1, \gamma_2 z_0) - \log j^2(\gamma_1, z_0) \right)$$

again after the cancelation of the additive constants. The equality

$$(2.13) \qquad \log j^2(\gamma_1 \gamma_2, z_0) = \log j^2(\gamma_1, \gamma_2 z_0) + \log j^2(\gamma_2, z_0) - 2\pi i\, c(\gamma_1, \gamma_2),$$

determines a cocycle $c \in Z^2(\mathrm{PSL}(2, \mathbb{R}), \mathbb{Z})$ (which is precisely the cocycle discussed in [2, §B-2], and whose cohomology class is independent of the choices of the branches $\log j^2(\gamma, z)$ of the logarithm). Inserting (2.13) into (2.12) one obtains

$$12\pi i GV(\gamma_1, \gamma_2) = \log \Delta(z_0) + \log \Delta(\gamma_1 \gamma_2 z_0) - \log \Delta(\gamma_1 z_0) - \log \Delta(\gamma_2 z_0)$$

$$- 6 \left(\log j^2(\gamma_1 \gamma_2, z_0) - \log j^2(\gamma_2, z_0) - \log j^2(\gamma_1, z_0) \right)$$

$$(2.14) \qquad - 12\pi i\, c(\gamma_1, \gamma_2).$$

This identity shows that the cocycles $\mathrm{Re}\, GV$ and $-c$ are cohomologous in $Z^2(\mathrm{PSL}(2, \mathbb{Q}), \mathbb{R})$, and also that $\mathrm{Im}\, GV$ is a coboundary.

On the other hand, the restriction of $c \in Z^2_{\mathrm{bor}}(\mathrm{SL}(2, \mathbb{R}), \mathbb{Z})$ to $\mathbb{T} = SO(2)$,

$$c(\gamma(\theta_1), \gamma(\theta_2)) = \frac{1}{2\pi i} \left(\mathrm{Log}\, e^{2i\theta_1} + \mathrm{Log}\, e^{2i\theta_2} - \mathrm{Log}\, e^{2i(\theta_1 + \theta_2)} \right),$$

$$\text{where} \quad \gamma(\theta) = \begin{pmatrix} \cos\theta & -\sin\theta \\ \sin\theta & \cos\theta \end{pmatrix}, \qquad \theta \in [0, 2\pi).$$

evidently represents the class $2\mathbf{e} \in H^2_{\mathrm{bor}}(\mathbb{T}, \mathbb{Z})$, which concludes the proof. $\square$

The Euler class $\mathbf{e} \in H^2(\mathrm{SL}(2, \mathbb{Q}), \mathbb{Z})$ occurs naturally in the context of the Chern character in K-homology [3], as the Chern character of a natural Fredholm module given by the "dual Dirac" operator relative to a base point $z_0 \in \mathbb{H}$. When moving the base point to the cusp $i\infty \in P^1(\mathbb{Q})$, it coincides with the restriction of the 2-cocycle $e \in Z^2(\mathrm{SL}(2, \mathbb{R}), \mathbb{Z})$ introduced by Petersson, [21], and investigated in detail by Asai, [1], where it is denoted w). It is defined, for $g_1, g_2 \in \mathrm{SL}(2, \mathbb{R})$, by the formula

$$(2.15) \qquad e(g_1, g_2) = \frac{1}{2\pi i} \left(\log j(g_2, z) + \log j(g_1, g_2 z) - \log j(g_1 g_2, z) \right),$$

with the logarithm chosen so that $\mathrm{Im} \log \in [-\pi, \pi)$; the above definition is independent of $z \in \mathbb{H}$.

Asai [1, §1-4] has shown that it can be given a simple expression, analogous to Kubota's cocycles [17] for coverings over local fields, which is as follows:

(2.16) $e(g_1, g_2) = -(x(g_1)|x(g_2)) + (-x(g_1)x(g_2)|x(g_1 g_2))$,

where

$$x(g) = \begin{cases} c & \text{if } c > 0, \\ d & \text{if } c = 0, \end{cases} \qquad \forall g = \begin{pmatrix} a & b \\ c & d \end{pmatrix} \in \mathrm{SL}(2, \mathbb{R})$$

and for any two numbers $x_1, x_2 \in \mathbb{R}$, the (Hilbert-like) symbol $(x_1|x_2)$ is defined as

$$(x_1|x_2) = \begin{cases} 1 & \text{iff } x_1 < 0 \text{ and } x_2 < 0, \\ 0 & \text{otherwise}. \end{cases}$$

Replacing τ_0 by $\tilde{\tau}_0$ one obtains, as in (2.8) and (2.9), the cohomologous group 2-cocycle on $\mathrm{GL}^+(2, \mathbb{Q})$

$$\widetilde{GV}(\gamma_1, \gamma_2) = \tilde{\tau}_0(U_{\gamma_0} \delta_1(U_{\gamma_1}), U_{\gamma_2}) = \tilde{\tau}_0(\mu_{\gamma_1^{-1}}|\gamma_0^{-1} U_{\gamma_0 \gamma_1}, U_{\gamma_2})$$

$$(2.17) \qquad \int_{z_0}^{\gamma_2 z_0} \mu_{\gamma_1} dz + z_0 \, \mathbf{a}_0(\mu_{\gamma_1} - \mu_{\gamma_1}|\gamma_2) + \int_{z_0}^{i\infty} (\widetilde{\mu_{\gamma_1}|\gamma_2} - \widetilde{\mu_{\gamma_1}}) dz.$$

In [8, §4] we found an explicit rational formula for $\operatorname{Re}\widetilde{GV}$, in terms of Rademacher-Dedekind sums, which we proceed now to recall.

Since by its very definition $\operatorname{Re}\widetilde{GV}$ descends to a 2-cocycle on $\mathrm{PGL}^+(2, \mathbb{Q})$, it suffices to express it for pairs of matrices with integer entries

$$(2.18) \qquad \gamma_1 = \begin{pmatrix} a_1 & b_1 \\ c_1 & d_1 \end{pmatrix}, \quad \gamma_2 = \begin{pmatrix} a_2 & b_2 \\ c_2 & d_2 \end{pmatrix} \in M_2^+(\mathbb{Z}).$$

When $\gamma_2 \in B^+(\mathbb{Z})$, that is $c_2 = 0$, then

$$(2.19) \qquad \operatorname{Re}\widetilde{GV}(\gamma_1, \gamma_2) = \frac{b_2}{d_2} \sum_{\mathbf{x}\cdot\vec{\gamma}_1 = 0, \mathbf{x} \neq 0} \mathbf{B}_2(x_1),$$

where $\mathbf{B}_2(x) := (x - [x])^2 - (x - [x]) + \frac{1}{6}$, with $[x] =$ greatest integer $\leq x$. When $c_2 > 0$, then

$$\operatorname{Re}\widetilde{GV}(\gamma_1, \gamma_2) = \frac{a_2}{c_2} \sum_{\mathbf{x}\cdot\vec{\gamma}_1 = 0, \mathbf{x} \neq 0} \mathbf{B}_2(x_1) + \frac{d_2}{c_2} \sum_{\mathbf{x}\cdot\vec{\gamma}_2 \vec{\gamma}_1 = 0, \mathbf{x} \neq 0} \mathbf{B}_2(x_1)$$

$$(2.20) \quad - 2 \sum_{\mathbf{x}\cdot\vec{\gamma}_1 = 0, \mathbf{x} \neq 0} \sum_{j=0}^{c_2'-1} \mathbf{B}_1\left(\frac{x_1 + j}{c_2'}\right) \mathbf{B}_1\left(\frac{a_2'(x_1 + j)}{c_2'} + x_2\right)$$

where $\frac{a_2'}{c_2'} = \frac{a_2}{c_2}$, $(a_2', c_2') = 1$, and $\mathbf{B}_1(x) := x - [x] - \frac{1}{2}$, for any $x \in \mathbb{R}$.

We shall obtain below a simpler expression for $\operatorname{Re}\widetilde{GV} \in Z^2(\mathrm{SL}(2, \mathbb{Q}), \mathbb{Q})$, through a transgression formula which involves only the classical Dedekind sums, through the

Rademacher function. Let us recall that, for a pair of integers m, n with $(m, n) = 1$ and $n \geq 1$, the Dedekind sum is given by the formula

$$(2.21) \qquad s\left(\frac{m}{n}\right) = \sum_{j=1}^{n-1} \mathbf{B}_1\left(\frac{j}{n}\right) \mathbf{B}_1\left(\frac{mj}{n}\right).$$

The Rademacher function $\Phi : \mathrm{SL}(2, \mathbb{Z}) \to \mathbb{Z}$ is uniquely characterized [23] by the coboundary relation

$$(2.22) \qquad \Phi(\sigma_1 \sigma_2) = \Phi(\sigma_1) - \Phi(\sigma_2) - 3\,\mathrm{sign}(c_1 c_2 c_3), \qquad \sigma_1, \sigma_2 \in \mathrm{SL}(2, \mathbb{Q})$$

where $\sigma_3 = \sigma_1 \sigma_2$, $\sigma_i = \begin{pmatrix} a_i & b_i \\ c_i & d_i \end{pmatrix} \in \mathrm{SL}(2, \mathbb{Z})$, $i = 1, 2, 3$, and is explicitly given by the following formula, [22]:

$$(2.23) \qquad \Phi(\sigma) = \begin{cases} \frac{b}{d} & \text{if } c = 0, \\[2mm] \frac{a+d}{c} - 12\,\mathrm{sign}(c) s\left(\frac{a}{|c|}\right) & \text{if } c \neq 0 \end{cases}$$

for any $\sigma = \begin{pmatrix} a & b \\ c & d \end{pmatrix} \in \mathrm{SL}(2, \mathbb{Z})$.

We extend it, in a slightly modified version, to a function $\widetilde{\Phi} : \mathrm{GL}^+(2, \mathbb{Q}) \to \mathbb{Q}$, as follows. First, for any $\sigma = \begin{pmatrix} a & b \\ c & d \end{pmatrix} \in \mathrm{SL}(2, \mathbb{Z})$, we define

$$(2.24) \qquad \widetilde{\Phi}(\sigma) = \begin{cases} \frac{b}{12d} + \frac{1-\mathrm{sign}(d)}{4} & \text{if } c = 0, \\[2mm] \frac{a+d}{12c} - \mathrm{sign}(c)\left(\frac{1}{4} + s\left(\frac{a}{|c|}\right)\right) & \text{if } c \neq 0, \end{cases}$$

while for any $\beta \in B(\mathbb{Q})$, where $B(\mathbb{Q}) = \left\{\beta = \begin{pmatrix} a & b \\ 0 & d \end{pmatrix} \in \mathrm{GL}^+(2, \mathbb{Q})\right\}$, we set

$$(2.25) \qquad \widetilde{\Phi}(\beta) = \frac{b}{12d} + \frac{1 - \mathrm{sign}(d)}{4}.$$

Now given $\gamma \in \mathrm{GL}^+(2, \mathbb{Q})$, after factoring it in the form

$$\gamma = \sigma \cdot \beta \qquad \text{with} \qquad \sigma \in \mathrm{SL}(2, \mathbb{Z}) \quad \text{and} \quad \beta \in B(\mathbb{Q}),$$

we define

$$(2.26) \qquad \widetilde{\Phi}(\gamma) = \widetilde{\Phi}(\sigma) + \widetilde{\Phi}(\beta);$$

one easily checks, by elementary calculations, that the definition is consistent.

The transgression formula within the Euler class can now be stated as follows.

THEOREM 4. *The function* $\widetilde{\Phi} : \mathrm{SL}(2, \mathbb{Q}) \to \mathbb{Q}$ *is uniquely characterized by the identity*

$$(2.27) \qquad \frac{1}{2}\,\mathrm{Re}\,\widetilde{GV}(\gamma_1, \gamma_2) + e(\gamma_1, \gamma_2) = \widetilde{\Phi}(\gamma_1 \gamma_2) - \widetilde{\Phi}(\gamma_1) - \widetilde{\Phi}(\gamma_2),$$

for any $\gamma_1, \gamma_2 \in \mathrm{SL}(2, \mathbb{Q})$.

PROOF. By Proposition 3, $[\frac{1}{2}\operatorname{Re}\widetilde{GV} + e] = 0$ in $H^2(\mathrm{SL}(2,\mathbb{Q}),\mathbb{R})$. Since

$$H^1(\mathrm{SL}(2,\mathbb{Q}),\mathbb{R}) = 0\,,$$

there exists a unique function $\Psi : \mathrm{SL}(2,\mathbb{Q}) \to \mathbb{R}$ such that

$$(2.28) \qquad \frac{1}{2}\operatorname{Re}\widetilde{GV}(\gamma_1,\gamma_2) + e(\gamma_1,\gamma_2) = \Psi(\gamma_1\gamma_2) - \Psi(\gamma_1) - \Psi(\gamma_2)\,.$$

Restricting to $\mathrm{SL}(2,\mathbb{Z})$, and taking into account that $\mu_\sigma = 0$ for all $\sigma \in \mathrm{SL}(2,\mathbb{Z})$, one obtains

$$e(\sigma_1,\sigma_2) = \Psi(\sigma_1\sigma_2) - \Psi(\sigma_1) - \Psi(\sigma_2)\,, \qquad \sigma_1,\sigma_2 \in \mathrm{SL}(2,\mathbb{Z})\,.$$

This is the splitting formula (2.22) which uniquely characterizes the restriction (2.24) of $\widetilde{\Phi}$ to $\mathrm{SL}(2,\mathbb{Z})$ (see [**1**, Thm. 3]), so that

$$(2.29) \qquad \Psi(\sigma) = \widetilde{\Phi}(\sigma)\,, \qquad \forall\,\sigma \in \mathrm{SL}(2,\mathbb{Z})\,.$$

Furthermore, taking in (2.28) $\gamma_1 = \sigma \in \mathrm{SL}(2,\mathbb{Z})$ and

$$\gamma_2 = \beta \in B_1^+(\mathbb{Q}) = \left\{\beta = \begin{pmatrix} a & b \\ 0 & d \end{pmatrix} \in B(\mathbb{Q})\,;\, a > 0\,,\, a\,d = 1\right\}\,,$$

one obtains

$$(2.30) \qquad \Psi(\sigma\beta) = \widetilde{\Phi}(\sigma) + \Psi(\beta)\,, \qquad \forall\,\sigma \in \mathrm{SL}(2,\mathbb{Z})\,,\, \beta \in B_1^+(\mathbb{Q})\,,$$

because $e(\sigma,\beta) = 0$, by [**1**, Lemma 3]. In particular, for $\sigma = -I$, one has

$$(2.31) \qquad \Psi(-\beta) = \widetilde{\Phi}(-I) + \Psi(\beta) = \Psi(\beta) + \frac{1}{2}\,, \qquad \forall\,\beta \in B_1^+(\mathbb{Q})\,,$$

since $\widetilde{\Phi}(-I) = \frac{1}{2}$, cf. [**1**, Lemma 4].

It remains to prove that $\Psi(\beta) = \widetilde{\Phi}(\beta)$ for any $\beta \in B_1^+(\mathbb{Q})$. Specializing (2.28) to $B_1^+(\mathbb{Q})$, and recalling that e vanishes on $B_1^+(\mathbb{Q})$, one obtains

$$\Psi(\beta_1\beta_2) - \Psi(\beta_1) - \Psi(\beta_2) = \frac{1}{2}\operatorname{Re}\widetilde{GV}(\beta_1,\beta_2)$$

and by (2.17) and (1.6) this can be computed as the real part of

$$\int_{z_0}^{\beta_2 z_0} (\phi_0|\beta_1 - \phi_0)dz + z_0\,\mathbf{a}_0(\phi_0|\beta_1 - \phi_0 - \phi_0|\beta_1\beta_2 + \phi_0|\beta_2)$$

$$+ \int_{z_0}^{i\infty} (\widetilde{\phi_0|\beta_1\beta_2} - \widetilde{\phi_0|\beta_2} - \widetilde{\phi_0|\beta_1} + \widetilde{\phi_0})dz$$

$$= \int_{z_0}^{\beta_2 z_0} (\widetilde{\phi_0|\beta_1} - \widetilde{\phi_0})dz + (\beta_2 z_0 - z_0)\,\mathbf{a}_0(\phi_0|\beta_1 - \phi_0)$$

$$+ z_0\,\mathbf{a}_0(\phi_0|\beta_1 - \phi_0 - \phi_0|\beta_1\beta_2 + \phi_0|\beta_2)$$

$$+ \int_{z_0}^{i\infty} (\widetilde{\phi_0|\beta_1\beta_2} - \widetilde{\phi_0|\beta_2} - \widetilde{\phi_0|\beta_1} + \widetilde{\phi_0})dz$$

$$= \int_{z_0}^{\beta_2 z_0} (\widetilde{\phi_0|\beta_1} - \widetilde{\phi_0})dz + \int_{z_0}^{i\infty} (\widetilde{\phi_0|\beta_1\beta_2} - \widetilde{\phi_0|\beta_2} - \widetilde{\phi_0|\beta_1} + \widetilde{\phi_0})dz$$

$$+ \beta_2 z_0\,\mathbf{a}_0(\phi_0|\beta_1 - \phi_0) - z_0\,\mathbf{a}_0((\phi_0|\beta_1 - \phi_0)|\beta_2)\,.$$

Since obviously

$$(2.32) \qquad \mathbf{a}_0(f|\beta) = \frac{a}{d}\,\mathbf{a}_0(f), \quad \text{for} \quad \beta = \begin{pmatrix} a & b \\ 0 & d \end{pmatrix} \in B(\mathbb{Q}),$$

the last line contributes

$$\left(\frac{a_2\,z_0 + b_2}{d_2} - \frac{a_2}{d_2}\,z_0\right)\mathbf{a}_0(\phi_0|\beta_1 - \phi_0) = \frac{b_2}{d_2}\,\mathbf{a}_0(\phi_0|\beta_1 - \phi_0)$$

which can be further computed as

$$= \frac{1}{12}\frac{b_2}{d_2}\left(\frac{a_1}{d_1} - 1\right),$$

by using (2.19) or more directly the Fourier expansion of ϕ_0 (cf. e.g. [**27**, Prop. 2.4.2]). On the other hand, when $z_0 \to i\infty$ then $\beta_2\,z_0 \to i\infty$ too, so that both integrals converge to 0. We conclude that

$$(2.33) \qquad \Psi(\beta_1\beta_2) - \Psi(\beta_1) - \Psi(\beta_2) = \frac{1}{2}\operatorname{Re}\widetilde{GV}(\beta_1,\beta_2) = \frac{1}{12}\frac{b_2}{d_2}\left(\frac{a_1}{d_1} - 1\right).$$

Using (2.25) it is elementary to check that, for $\beta_1,\beta_2 \in B_1^+(\mathbb{Q})$,

$$\widetilde{\Phi}(\beta_1\beta_2) - \widetilde{\Phi}(\beta_1) - \widetilde{\Phi}(\beta_2) = \frac{1}{12}\frac{b_2}{d_2}\left(\frac{a_1}{d_1} - 1\right).$$

Thus, $\Psi|B_1^+(\mathbb{Q})$ and $\widetilde{\Phi}|B_1^+(\mathbb{Q})$ can only differ by a character of $B_1^+(\mathbb{Q})$. To show that they coincide, it suffices to prove that Ψ vanishes on the torus

$$T = \left\{\delta = \begin{pmatrix} a & 0 \\ 0 & d \end{pmatrix}; \delta \in B_1^+(\mathbb{Q})\right\}.$$

For any $\delta \in T$, one has

$$\sigma_0\,\delta = \delta^{-1}\,\sigma_0, \quad \text{where} \quad \sigma_0 = \begin{pmatrix} 0 & -1 \\ 1 & 0 \end{pmatrix},$$

hence

$$\Psi(\sigma_0\,\delta) = \Psi(\delta^{-1}\,\sigma_0).$$

From (2.28) it follows that

$$\Psi(\sigma_0) + \Psi(\delta) + \frac{1}{2}\operatorname{Re}\widetilde{GV}(\sigma_0,\delta) = \Psi(\delta^{-1}) + \Psi(\sigma_0) + \frac{1}{2}\operatorname{Re}\widetilde{GV}(\delta^{-1},\sigma_0),$$

therefore, since $\operatorname{Re}\widetilde{GV}(\sigma_0,\delta) = 0$,

$$\Psi(\delta) - \Psi(\delta^{-1}) = \frac{1}{2}\operatorname{Re}\widetilde{GV}(\delta^{-1},\sigma_0).$$

On the other hand, by (2.33)

$$\Psi(\delta) + \Psi(\delta^{-1}) = 0.$$

Hence

$$\Psi(\delta) = \frac{1}{4}\operatorname{Re}\widetilde{GV}(\delta^{-1},\sigma_0),$$

and it remains to show that the right hand side vanishes. We shall apply formula (2.20), which allows us to replace δ^{-1}, after multiplication by a scalar, with a diagonal matrix with positive integer entries $\rho = \begin{pmatrix} m & 0 \\ 0 & n \end{pmatrix} \in M_2^+(\mathbb{Z})$. In this case, it simply gives

$$\frac{1}{2} \operatorname{Re} \widetilde{GV}(\rho, \sigma_0) = -\sum_{\mathbf{x} \cdot \tilde{\rho} = 0} \mathbf{B}_1(x_1) \mathbf{B}_1(x_2) + \mathbf{B}_1(0)^2$$

$$= -\sum_{j=0}^{m-1} \mathbf{B}_1(\frac{j}{m}) \sum_{k=0}^{n-1} \mathbf{B}_1(\frac{k}{n}) + \mathbf{B}_1(0)^2 = 0,$$

because of the distribution property of the first Bernoulli function.   □

REMARK 5. Note that while the cocycle e admits a K-homological interpretation as a Chern character, the cohomologous cocycle $\frac{1}{2} \operatorname{Re} \widetilde{GV}$ obviously should also have such an interpretation. This in turn would allow to put the above transgression on the same K-homological footing as in [**5**].

3. Modular symbol cyclic cocycles of higher weight

We begin in this section to extend the above results to the case of higher weight modular symbols. This will involve introducing the cyclic cohomology with coefficients $HC^*(\mathcal{A}, W)$ and describing the higher weight analogues of the basic invariant cocycles. They will be used in the next section to define the characteristic maps corresponding to the degenerate actions of weight $m \geq 2$ of $\mathcal{H}_1$ on $\mathcal{A} = \mathcal{A}_{G+(\mathbb{Q})}$.

3.1. Cyclic cohomology with coefficients. We need to introduce the *cyclic cohomology with coefficients* $HC^*(\mathcal{A}, W)$, where $\mathcal{A}$ is the crossed product algebra $\mathcal{A}_{G+(\mathbb{Q})}$ and W is a $\mathrm{GL}^+(2, \mathbb{Q})$-module. It is a special case of the Hopf-cyclic cohomology with coefficients, cf. [**12**, §3], in which the 'gauge' Hopf algebra is the group ring

$$\mathcal{G} = \mathbb{C}[G^+(\mathbb{Q})], \qquad G^+(\mathbb{Q}) := \mathrm{GL}^+(2, \mathbb{Q}),$$

equipped with its usual Hopf algebra structure; W is viewed as a left $\mathcal{G}$-module and as a trivial $\mathcal{G}$-comodule, and $\mathcal{A}$ is regarded as a left $\mathcal{G}$-comodule algebra with respect to the intrinsic coaction

$$a = f U_\gamma \longmapsto a_{(-1)} \otimes a_{(0)} := U_\gamma \otimes f U_\gamma \in \mathcal{G} \otimes \mathcal{A}.$$

Thus, by definition, $HC^*(\mathcal{A}, W)$ is the cohomology associated to the cyclic module

$$(3.1) \qquad\qquad C^*(\mathcal{A}, W) := \operatorname{Hom}^{\mathcal{G}}(\mathcal{A}^{*+1}, W),$$

whose cyclic structure is defined by the operators

$$\partial_i \phi(a^0 \otimes \cdots \otimes a^n) = \phi(a^0 \otimes \cdots a^i a^{i+1} \otimes \cdots \otimes a^n), \, 0 \leq i < n,$$

$$\partial_n \phi(a^0 \otimes \cdots \otimes a^n) = S(a^n_{(-1)}) \, \phi(a^n_{(0)} a^0 \otimes \cdots \otimes a^{n-1}),$$

$$\sigma_i \phi(a^0 \otimes \cdots \otimes a^n) = \phi(a^0 \otimes \cdots a^i \otimes a^{i+1} \otimes \cdots \otimes a^n), \, 0 \leq i < n,$$

$$\tau_n \phi(a^0 \otimes \cdots \otimes a^n) = S(a^n_{(-1)}) \, \phi(a^n_{(0)} \otimes a^0 \otimes \cdots \otimes a^{n-1}).$$

For each $m \in \mathbb{N}$ we denote by W_m the simple $\mathrm{SL}(2, \mathbb{C})$-module of dimension $m+1$, realized as the the space

$$W_m = \{P(T_1, T_2) \in \mathbb{C}[T_1, T_2] \, ; \, P \text{ is homogeneous of degree } m\}$$

and we let $GL^+(2,\mathbb{R})$ act on W_m by

$$(g \cdot P)(T_1, T_2) = \det(g)^{-\frac{m}{2}} P(aT_1 + cT_2, bT_1 + dT_2), \quad g = \begin{pmatrix} a & b \\ c & d \end{pmatrix}.$$

Note that as a $GL^+(2,\mathbb{R})$-module, W_m is the complexification of

$$W_m(\mathbb{R}) = \{P(T_1, T_2) \in \mathbb{R}[T_1, T_2] \,;\, P \text{ is homogeneous of degree } m\},$$

and we denote by $\mathrm{Re} : W_m \to W_m(\mathbb{R})$ the projection obtained by taking the real parts of the coefficients. We also note that as a $GL^+(2,\mathbb{Q})$-module $W_m(\mathbb{R})$ has an obvious rational structure

$$W_m(\mathbb{Q}) = \{P(T_1, T_2) \in \mathbb{Q}[T_1, T_2] \,;\, P \text{ is homogeneous of degree } m\}.$$

We denote by $F_m : \mathbb{H} \to W_m$ the polynomial function

(3.2) $$F_m(z) = (zT_1 + T_2)^m,$$

and note that it satisfies, for any $g \in GL^+(2,\mathbb{R})$, the covariance property

(3.3) $$F_m|g(z) \equiv \det(g)^{-\frac{m}{2}}(cz+d)^m F_m(gz) = g \cdot F_m(z).$$

3.2. Cyclic 1-cocycles with coefficients: base point in $\mathbb{H}$. With these ingredients at hand, and after making the additional choice of a 'base point' $z_0 \in \mathbb{H}$, we proceed to define the *invariant cocycles* which will support characteristic maps associated to Hopf actions in the degenerate case.

Regarding $\mathrm{Hom}_{\mathcal{G}}(\mathcal{A} \otimes \mathcal{A}, W_m)$ as a graded linear space with respect to the weight filtration inherited from $\mathcal{A}$, we define the weight 2 element $\tau_m \in \mathrm{Hom}_{\mathcal{G}}(\mathcal{A} \otimes \mathcal{A}, W_m)$ as follows. Let

$$a^0 = \sum_\alpha f_\alpha^0 U_\alpha \in \mathcal{A}_{w(a^0)} \quad \text{and} \quad a^1 = \sum_\beta f_\beta^1 U_\beta \in \mathcal{A}_{w(a^1)},$$

be two homogeneous elements in $\mathcal{A}$; by definition,

$$\tau_m(a^0, a^1) = \begin{cases} \displaystyle\sum_\alpha \int_{z_0}^{\alpha z_0} F_m \, f_\alpha^0 \, f_{\alpha^{-1}}^1 |\alpha^{-1} \, dz & \text{if} \quad w(a^0) + w(a^1) = m+2, \\[3mm] 0 & \text{otherwise}. \end{cases}$$

LEMMA 6. *For each $m \geq 2$, $\tau_{m-2} \in C^1(\mathcal{A}, W_{m-2})$ is a cyclic cocycle.*

PROOF. For notational convenience, we shall omit the subscript $m-2$ in the ensuing calculations, which are of course similar to those in the proof of Proposition 2.

Let $f^0 U_{\gamma_0}, f^1 U_{\gamma_1}, f^2 U_{\gamma_2} \in \mathcal{A}$ be such that

$$w(f^0) + w(f^1) + w(f^2) = m \quad \text{and} \quad \gamma_0\gamma_1\gamma_2 = 1.$$

Then

$$b\tau(f^0 U_{\gamma_0}, f^1 U_{\gamma_1}, f^2 U_{\gamma_2}) =$$

$$= \quad \tau(f^0 f^1|\gamma_0{}^{-1}U_{\gamma_0\gamma_1}, f^2 U_{\gamma_2}) - \tau(f^0 U_{\gamma_0}, f^1 f^2|\gamma_1{}^{-1}U_{\gamma_1\gamma_2})$$

$$+ \quad \gamma_2{}^{-1}\tau(f^2 f^0|\gamma_2{}^{-1}U_{\gamma_2\gamma_0}, f^1 U_{\gamma_1})$$

$$= \quad \int_{z_0}^{\gamma_0\gamma_1 z_0} F f^0 f^1|\gamma_0{}^{-1}f^2|\gamma_1{}^{-1}\gamma_0{}^{-1}dz - \int_{z_0}^{\gamma_0 z_0} F f^0 f^1|\gamma_0{}^{-1}f^2|\gamma_1{}^{-1}\gamma_0{}^{-1}dz$$

$$+ \quad \gamma_2{}^{-1}\int_{z_0}^{\gamma_2\gamma_0 z_0} F f^2 f^0|\gamma_2{}^{-1}f^1|\gamma_0{}^{-1}\gamma_2{}^{-1}dz$$

$$= \quad \int_{\gamma_0 z_0}^{\gamma_0\gamma_1 z_0} F f^0 f^1|\gamma_0{}^{-1}f^2|\gamma_2 dz + \int_{z_0}^{\gamma_2\gamma_0 z_0} F|\gamma_2{}^{-1}f^2 f^0|\gamma_2{}^{-1}f^1|\gamma_0{}^{-1}\gamma_2{}^{-1}dz$$

$$= \quad \int_{\gamma_0 z_0}^{\gamma_2{}^{-1}z_0} F f^0 f^1|\gamma_0{}^{-1}f^2|\gamma_2 dz + \int_{\gamma_2{}^{-1}z_0}^{\gamma_0 z_0} F f^2|\gamma_2 f^0 f^1|\gamma_0{}^{-1}dz \quad = \quad 0 \,,$$

and so $b\tau = 0$. It is also cyclic, because for $f^0 U_{\gamma_0}, f^1 U_{\gamma_1} \in \mathcal{A}$ such that

(3.4) $$\qquad\qquad w(f^0) + w(f^1) = m \qquad \text{and} \qquad \gamma_0\gamma_1 = 1$$

one has

$$\lambda_1\tau(f^0 U_{\gamma_0}, f^1 U_{\gamma_1}) = -\gamma_1{}^{-1}\tau(f^1 U_{\gamma_1}, f^0 U_{\gamma_0}) =$$

$$= \quad -\gamma_1{}^{-1}\int_{z_0}^{\gamma_1 z_0} F f^1 f^0|\gamma_1{}^{-1}dz = -\int_{z_0}^{\gamma_1 z_0} F|\gamma_1{}^{-1}f^1 f^0|\gamma_1{}^{-1}dz$$

$$= \quad \int_{z_0}^{\gamma_0 z_0} F f^1|\gamma_1 f^0 dz = \tau(f^0 U_{\gamma_0}, f^1 U_{\gamma_1}) \,.$$

$$\square$$

3.3. Cyclic 1-cocycles with coefficients: base point at cusps. We now extend the construction of subsection 2.2, allowing the base point z_0 to belong to the 'arithmetic' boundary of the upper half plane $P^1(\mathbb{Q})$, to the general case of weight $m \geq 2$. To this end we shall just apply the same procedure to the cocycle $\tau = \tau_{m-2}$. For $f^0 U_{\gamma_0}, f^1 U_{\gamma_1} \in \mathcal{A}$ such that

(3.5) $$\qquad\qquad w(f^0) + w(f^1) = m \qquad \text{and} \qquad \gamma_0\gamma_1 = 1$$

it is given by the integral

$$\tau(f^0 U_{\gamma_0}, f^1 U_{\gamma_1}) = \int_{z_0}^{\gamma_0 z_0} F f^0 f^1|\gamma_1 dz \,,$$

where $F = F_{m-2}$. Taking its derivative with respect to $z_0 \in \mathbb{H}$ gives

$$\frac{d}{dz_0}\tau(f^0 U_{\gamma_0}, f^1 U_{\gamma_1}) \quad = \quad F|\gamma_0(z_0)\,(f^0|\gamma_0 f^1)(z_0) - F(z_0)\,(f^0 f^1|\gamma_1)(z_0)$$

$$= \quad \gamma_0 \cdot F(z_0)\,(f^0|\gamma_0 f^1)(z_0) - F(z_0)\,(f^0 f^1|\gamma_1)(z_0)$$

$$= \quad -b\epsilon_F\,(f^0 U_{\gamma_0}, f^1 U_{\gamma_1}),$$

where

$$\epsilon_F = F(z_0)\,\epsilon \,.$$

As above, we split it into two functionals

$$\epsilon_F = F(z_0)\,\mathbf{a}_0 + F(z_0)\,\widetilde{\epsilon},$$

and regularize the cocycle τ by adding the coboundaries of suitable anti-derivatives of the two components:

$$\widetilde{\tau} = \tau + b\left(\check{F}(z_0)\,\mathbf{a}_0\right) - \int_{z_0}^{i\infty} b\left(F(z)\,\widetilde{\epsilon}\right)\,dz\,,$$

where

(3.6) $\qquad \check{F}_{m-2}(z_0) = \int_0^{z_0} F(z)dz = \sum_{k=0}^{m-2} \frac{(m-2)!}{(k+1)!(m-k-2)!}\,z_0^{k+1}\,T_1^k\,T_2^{m-k-2}\,.$

Explicitly, for $f^0 U_{\gamma_0}, f^1 U_{\gamma_1} \in \mathcal{A}$ satisfying (3.5),

$$\widetilde{\tau}_{m-2}(f^0 U_{\gamma_0}, f^1 U_{\gamma_1}) =$$

$$= \int_{z_0}^{\gamma_0 z_0} F f^0\, f^1|_{\gamma_1}\,dz + \check{F}(z_0)\,\mathbf{a}_0(f^0\, f^1|_{\gamma_1}) - \gamma_0 \cdot \check{F}(z_0)\,\mathbf{a}_0(f^0|_{\gamma_0} f^1)$$

(3.7) $\qquad - \int_{z_0}^{i\infty} F\,\widetilde{\epsilon}(f^0\, f^1|_{\gamma_1})\,dz + \int_{z_0}^{i\infty} \gamma_0 \cdot F\,\widetilde{\epsilon}(f^0|_{\gamma_0} f^1)\,dz\,.$

Since by its very definition $\widetilde{\tau}_{m-2} \in C^1(\mathcal{A}, W_{m-2})$ differs from the cocycle $\tau_{m-2} \in ZC^1(\mathcal{A}, W_{m-2})$ by a coboundary, it is itself a cyclic cocycle.

4. Transgression for degenerate actions

The higher weight counterparts of the above results involve cyclic cocycles with coefficients that are invariant under 'degenerate' actions of $\mathcal{H}_1$, associated to modular forms of arbitrary weight. In the case of cusp forms, the corresponding cup products by the universal Godbillon-Vey cocycle $\delta_1 \in ZC^1(\mathcal{H}_1; \delta, 1)$ transgress to 1-dimensional cohomology classes of congruence subgroups, implementing the Eichler-Shimura isomorphism, while the degenerate actions corresponding to Eisenstein series give rise to generalized functions of Rademacher type, [20], as well as to the Eisenstein cocycle of Stevens [28].

4.1. Degenerate actions of $\mathcal{H}_1$ of higher weight. We recall a few basic facts about cocycle perturbations of Hopf actions. Given a Hopf algebra $\mathcal{H}$ and an algebra $\mathcal{A}$ endowed with a Hopf action of $\mathcal{H}$, a 1-cocycle $u \in Z^1(\mathcal{H}, \mathcal{A})$ is an invertible element of the convolution algebra $\mathcal{L}(\mathcal{H}, \mathcal{A})$ of linear maps from $\mathcal{H}$ to $\mathcal{A}$, satisfying

(4.1) $\qquad u(h\,h') = \sum u(h_{(1)})\,h_{(2)}(u(h'))\,, \quad \forall h \in \mathcal{H}\,.$

The conjugate under $u \in Z^1(\mathcal{H}, \mathcal{A})$ of the original action of $\mathcal{H}$ on $\mathcal{A}$ is given by

(4.2) $\qquad \tilde{h}(a) := \sum u(h_{(1)})\,h_{(2)}(a)\,u^{-1}(h_{(3)})$

The standard action of $\mathcal{H}_1$ commutes with the natural coaction of $G^+(\mathbb{Q})$ on $\mathcal{A}_{G^+(\mathbb{Q})}$. It is natural to only consider cocycles u with the same property. The values of such a 1-cocycle u on generators must belong to the subalgebra $\mathcal{M} \subset \mathcal{A}_{G^+(\mathbb{Q})}$, and have to be of the following form:

$$u(X) = \theta \in \mathcal{M}_2, \ u(Y) = \lambda \in \mathbb{C}, \ u(\delta_1) = \omega \in \mathcal{M}_2.$$

By [8, Prop. 11], for each such data $\theta \in \mathcal{M}_2, \lambda \in \mathbb{C}, \omega \in \mathcal{M}_2$,

1^0 there exists a unique 1-cocycle $u \in \mathcal{L}(\mathcal{H}_1, \mathcal{A}_{G^+(\mathbb{Q})})$ such that

$$u(X) = \theta, \ u(Y) = \lambda, \ u(\delta_1) = \omega;$$

2^0 the conjugate under u of the action of $\mathcal{H}_1$ is given on generators as follows:

(4.3)
$$\begin{aligned}
\tilde{Y} &= Y, & \tilde{X}(a) &= X(a) + [(\theta - \lambda\omega), a] - \lambda\,\delta_1(a) + \omega Y(a), \\
& & \tilde{\delta}_1(a) &= \delta_1(a) + [\omega, a], & a \in \mathcal{A}_{G^+(\mathbb{Q})};
\end{aligned}$$

3^0 The conjugate under u of δ_2' is given by the operator

$$\tilde{\delta}_2'(a) = [X(\omega) + \frac{\omega^2}{2} - \Omega_4, \, a], \qquad a \in \mathcal{A}_{G^+(\mathbb{Q})}$$

and there is no choice of u for which $\tilde{\delta}_2' = 0$.

The actions described above were called *projective* because $\tilde{\delta}_2'$ acts by an inner transformation.

For our purposes, it is the difference $\tilde{\delta}_1 - \delta_1$ which matters. In view of (4.3), we may as well start from the trivial action and may also assume $\theta = 0, \lambda = 0$.

We call the *trivial action of weight m* of $\mathcal{H}_1$ on $\mathcal{A}_{G^+(\mathbb{Q})}$ the action defined by

(4.4)
$$Y(a) = \frac{w(f)}{m}\, a, \qquad X(a) = 0, \qquad \delta_n(a) = 0, \quad n \geq 1$$

To any modular form $\omega \in \mathcal{M}_m$ we shall now associate a 'degenerate' action of the Hopf algebra $\mathcal{H}_1$ on the crossed product algebra $\mathcal{A}_{G^+(\mathbb{Q})}$ as follows.

PROPOSITION 7. *Let $\omega \in \mathcal{M}$ be a modular of weight $w(\omega) = m$.*

1^0. *There exists a unique 1-cocycle $u = u_\omega \in Z^1(\mathcal{H}_1, \mathcal{A}_{G^+(\mathbb{Q})})$ such that*

$$u(Y) = 0, \qquad u(X) = 0, \qquad u(\delta_1) = \omega.$$

2^0. *The conjugate under u_ω of the trivial action of weight m of $\mathcal{H}_1$ on $\mathcal{A} = \mathcal{A}_{G^+(\mathbb{Q})}$ is determined by*

(4.5)
$$\tilde{Y}(a) = \frac{w(a)}{m}\, a, \quad \tilde{X}(a) = \omega\,\tilde{Y}(a), \quad \tilde{\delta}_1(a) = [\omega, a], \quad \forall a \in \mathcal{A}.$$

Under this action, one has, for any $n \geq 1$,

(4.6)
$$\tilde{\delta}_n(a) = (n-1)!\,\omega^{n-1}\,[\omega, a], \qquad \forall a \in \mathcal{A},$$

(4.7)
$$resp. \quad \delta_n(f\,U_\gamma^*) = X^{n-1}(\omega - \omega|\gamma)\,f\,U_\gamma^*, \quad \forall f\,U_\gamma^* \in \mathcal{A}.$$

PROOF. The first statement is a variant of [8, Proposition 11, 1^0]. Its proof gives the following explicit description of the 1-cocycle $u \in Z^1(\mathcal{H}_1, \mathcal{A}_{G^+(\mathbb{Q})})$. First, recall that any element of $\mathcal{H}_1$ can be uniquely written as a linear combination of monomials of the form $P(\delta_1, \delta_2, ..., \delta_\ell) X^n Y^m$. We then let $\omega^{(k)}$ be defined by induction by

$$\omega^{(1)} := \omega, \qquad \omega^{(k+1)} := \omega Y(\omega^{(k)}) = k! \omega^{k+1}, \quad \forall k \geq 1.$$

Then

$$u\left(P(\delta_1, \delta_2, ..., \delta_\ell) X^n Y^m\right) = \begin{cases} P(\omega^{(1)}, \omega^{(2)}, ..., \omega^{(\ell)}) & \text{if} \quad n = m = 0, \\ 0 & \text{otherwise}, \end{cases}$$

while its inverse u^{-1} is given by

$$u^{-1}\left(P(\delta_1, \delta_2, ..., \delta_\ell) X^n Y^m\right) = \begin{cases} P(-\omega, 0, ..., 0) & \text{if} \quad n = m = 0, \\ 0 & \text{otherwise}. \end{cases}$$

The formulae (4.5) are obtained from 1^0 above and the definitions (4.2), (4.4) (cf. also [8, Proposition 11, 2^0]). Finally (4.6), and hence (4.7), follows by straightforward computation. $\square$

The actions as in 2^0 above are *degenerate*, in the sense that $\tilde{\delta}_1$ acts by an inner transformation.

LEMMA 8. *With respect to the degenerate action of weight m, one has*

$$(4.8) \qquad \tau(h_{(1)}(a^0), h_{(2)}(a^1)) = \delta(h)\,\tau(a^0, a^1), \qquad \forall h \in \mathcal{H}_1, \ a^0, a^1 \in \mathcal{A},$$

for $\tau = \tau_{m-2}$ or $\tau = \tilde{\tau}_{m-2}$.

PROOF. We shall check the $\mathcal{H}_1$-invariance property (4.8) for $\tau = \tau_{m-2}$. As in the proof of Proposition 2, it suffices to verify it on the generators. Starting with $\tilde{Y}$, for $f^0 U_{\gamma_0}, f^1 U_{\gamma_1}$ satisfying (3.4) one has

$$\tau(\tilde{Y}(f^0 U_{\gamma_0}), f^1 U_{\gamma_1}) + \tau(f^0 U_{\gamma_0}, \tilde{Y}(f^1 U_{\gamma_1})) =$$

$$= \int_{z_0}^{\gamma_0 z_0} F \tilde{Y}(f^0 f^1 |\gamma_0^{-1}) dz = \frac{w(f^0) + w(f^1)}{m} \int_{z_0}^{\gamma_0 z_0} F f^0 f^1 |\gamma_0^{-1} dz$$

$$= \int_{z_0}^{\gamma_0 z_0} F f^0 f^1 |\gamma_0^{-1} dz = \delta(Y)\,\tau(f^0 U_{\gamma_0}, f^1 U_{\gamma_1}).$$

Passing to $\tilde{X}$, the identity (4.8) is nontrivial only if $f^0 U_{\gamma_0}, f^1 U_{\gamma_1} \in \mathcal{A}$ satisfy

$$(4.9) \qquad w(f^0) + w(f^1) = 0 \qquad \text{and} \qquad \gamma_0 \gamma_1 = 1.$$

One then has

$$\tau(\tilde{X}(f^0)U_{\gamma_0}, f^1 U_{\gamma_1}) + \tau(f^0 U_{\gamma_0}, \tilde{X}(f^1)U_{\gamma_1}) + \tau(\delta_1(f^0 U_{\gamma_0}), \tilde{Y}(f^1)U_{\gamma_1}) =$$

$$= \int_{z_0}^{\gamma_0 z_0} F \omega \tilde{Y}(f^0) f^1 |\gamma_0^{-1} dz + \int_{z_0}^{\gamma_0 z_0} F f^0 \omega |\gamma_0^{-1} \tilde{Y}(f^1) |\gamma_0^{-1} dz$$

$$+ \int_{z_0}^{\gamma_0 z_0} F (\omega - \omega |\gamma_0^{-1}) f^0 \tilde{Y}(f^1) |\gamma_0^{-1} dz = \int_{z_0}^{\gamma_0 z_0} F \omega \tilde{Y}(f^0 f^1 |\gamma_0^{-1}) dz$$

$$= \frac{w(f^0) + w(f^1)}{m} \int_{z_0}^{\gamma_0 \, z_0} F f^0 f^1 |\gamma_0{}^{-1} dz = 0,$$

the vanishing being a consequence of (4.9).

Finally, in the case of $\tilde{\delta}_1$ and with $f^0 U_{\gamma_0}, f^1 U_{\gamma_1}$ as in (4.9), if the action is degenerate, one has

$$\tau(\tilde{\delta}_1(f^0 U_{\gamma_0}), f^1 U_{\gamma_1}) + \tau(f^0 U_{\gamma_0}, \tilde{\delta}_1(f^1 U_{\gamma_1})) =$$

$$= \int_{z_0}^{\gamma_0 \, z_0} F(\omega - \omega|\gamma_1) f^0 f^1 |\gamma_1 dz + \int_{z_0}^{\gamma_0 \, z_0} F f^0 (\omega - \omega|\gamma_0)|\gamma_1 f^1 |\gamma_1 dz = 0.$$

$\square$

4.2. Transgression in the higher weight case.

We now consider the degenerate action associated to a modular form $\omega \in \mathcal{M}_m$ with $m \geq 2$ and define the cup product of δ_1 with the invariant 1-trace τ as in Lemma 8 by the formula

$$gv_\omega (a^0, a^1, a^2) = \tau(a^0 \, \tilde{\delta}_1(a^1), \, a^2) = \tau(a^0 \, \omega a^1 - a^1 \omega, \, a^2), \quad a^0, a^1, a^2 \in \mathcal{A}$$

LEMMA 9. One has $gv_\omega \in ZC^2(\mathcal{A}, W_{m-2})$, that is

$$b(gv_\omega) = 0 \quad \text{and} \quad B(gv_\omega) = 0.$$

PROOF. As in the proof of Lemma 1, using the primitivity of δ_1, one has

$$b(gv_\omega) (a^0, a^1, a^2, a^3) = \tau (a^0 a^1 \delta_1(a^2), a^3) - \tau (a^0 \delta_1(a^1 a^2), a^3)$$

$$+ \quad \tau (a^0 \delta_1(a^1), a^2 a^3) - S(a^3_{(-1)}) \tau (a^3_{(0)} a^0 \delta_1(a^1), a^2)$$

$$= \quad -\tau (a^0 \delta_1(a^1) a^2, a^3) + \tau (a^0 \delta_1(a^1), a^2 a^3)$$

$$- \quad S(a^3_{(-1)}) \tau (a^3_{(0)} a^0 \delta_1(a^1), a^2) = -b\tau (a^0 \delta_1(a^1), a^2, a^3) = 0.$$

Next, since $\tau(a, 1) = 0$ for any $a \in \mathcal{A}$, one has

$$B(gv_\omega) (a^0, a^1) = g_\omega v (1, a^0, a^1) - S(a^1_{(-1)}) gv_\omega (1, a^1_{(0)}, a^0)$$

$$= \tau (\delta_1(a^0), a^1) - S(a^1_{(-1)}) \tau (\delta_1(a^1_{(0)}), a^0).$$

We now use the fact that the action of δ_1 commutes with the coaction of $\mathrm{GL}^+(2, \mathbb{Q})$, to continue

$$B(gv_\omega) (a^0, a^1) = \tau (\delta_1(a^0), a^1) - S(\delta_1(a^1)_{(-1)}) \tau (\delta_1(a^1)_{(0)}, a^0)$$

$$= \tau (\delta_1(a^0), a^1) + \tau (a^0, \delta_1(a^1)) = 0,$$

the vanishing being a consequence of the $\mathcal{H}_1$-invariance property of $\tau \in ZC^1(\mathcal{A}, W_{m-2})$.

$\square$

Because degenerate actions are perturbations of the trivial action of weight m, one expects the cup product

$$[gv_\omega] = [\delta_1] \# [\tau_{m-2}]$$

to vanish. This is indeed the case, the vanishing being a consequence of a general transgression formula for degenerate actions. To state it, we introduce the cochain $tgv_\omega \in C^1(\mathcal{A}, W_{m-2})$,

(4.10) $$tgv_\omega(a^0, a^1) = -\tau(a^0, \omega\, a^1), \qquad a^0, a^1 \in \mathcal{A}$$

PROPOSITION 10. *For any $\omega \in \mathcal{M}_m$,*

$$gv_\omega = b\,(tgv_\omega) \quad and \quad B\,(tgv_\omega) = 0.$$

PROOF. One has

$$b(tgv_\omega)(a^0, a^1, a^2) = -\tau(a^0 a^1, \omega a^2) + \tau(a^0, \omega a^1 a^2) - a^2_{(-1)}\tau(a^2 a^0, \omega a^1);$$

after replacing the first term in the right hand side using

$$0 \;=\; b\tau(a^0, a^1, \omega a^2) =$$

$$= \; \tau(a^0 a^1, \omega a^2) - \tau(a^0, a^1 \omega a^2) + a^2_{(-1)}\tau(\omega a^2 a^0, a^1),$$

one obtains

$$b(tgv_\omega)\,(a^0, a^1, a^2) = \tau(a^0, \omega a^1 a^2) - \tau(a^0, a^1 \omega a^2) +$$

$$+\; a^2_{(-1)}\left(\tau(\omega a^2 a^0, a^1) - \tau(a^2 a^0, \omega a^1)\right)$$

$$= \; \tau(a^0, \widetilde{\delta}_1(a^1)a^2) + a^2_{(-1)}\left(\tau(\omega a^2 a^0, a^1) - \tau(a^2 a^0, \omega a^1)\right).$$

We now use the identity

$$0 = b\tau(a^0, \widetilde{\delta}_1(a^1), a^2) = \tau(a^0 \widetilde{\delta}_1(a^1), a^2) - \tau(a^0, \widetilde{\delta}_1(a^1)a^2) + a^2_{(-1)}\tau(a^2 a^0, \widetilde{\delta}_1(a^1))$$

to replace the term $\tau(a^0, \widetilde{\delta}_1(a^1)a^2)$, thus obtaining

$$b(tgv_\omega)\,(a^0, a^1, a^2) = \tau(a^0 \widetilde{\delta}_1(a^1), a^2) +$$

$$+\; a^2_{(-1)}\left(\tau(a^2 a^0, \widetilde{\delta}_1(a^1)) + \tau(\omega a^2 a^0, a^1) - \tau(a^2 a^0, \omega a^1)\right)$$

(4.11) $$= \tau(a^0 \widetilde{\delta}_1(a^1), a^2) + a^2_{(-1)}\left(-\tau(a^2 a^0, a^1 \omega) + \tau(\omega a^2 a^0, a^1)\right).$$

Next, using

$$0 = b\tau(c^0, c^1, \omega) \;=\; \tau(c^0 c^1, \omega) - \tau(c^0, c^1 \omega) + \tau(\omega c^0, c^1)$$

$$=\; -\tau(c^0, c^1 \omega) + \tau(\omega c^0, c^1),$$

for $c^0 = a^2 a^0$ and $c^1 = a^1$, one sees that the term in paranthesis vanishes, and therefore (4.11) reduces to

$$b(tgv_\omega)\,(a^0, a^1, a^2) = \tau(a^0 \widetilde{\delta}_1(a^1), a^2) = gv_\omega\,(a^0, a^1, a^2).$$

On the other hand, by the very definition of $\tau = \tau_{m-2}$, one has

$$B(tgv_\omega)\,(a^0) = -\tau(1, \omega a^0) - \tau(a^0, \omega) = 0.$$

$\square$

Let $\omega \in \mathcal{M}_m(\Gamma)$, for some congruence subgroup $\Gamma \subset \mathrm{SL}(2,\mathbb{Z})$. The transgression proper occurs when gv_ω is restricted to the subalgebra

$$\mathcal{A}^\Gamma = \mathcal{M} \ltimes \Gamma,$$

on which $\tilde{\delta}_1$ vanishes. Then tgv_ω becomes a cocycle in $ZC^2(\mathcal{A}^\Gamma, W_{m-2})$ and its restriction to $\mathcal{A}_0^\Gamma = \mathbb{C}[\Gamma]$

$$tgv_\omega(U_{\gamma_0}, U_{\gamma_1}) = -\tau(U_{\gamma_0}, \omega\, U_{\gamma_1}) = -\int_{z_0}^{\gamma_0 z_0} F\omega\, dz\,, \qquad \gamma_0\gamma_1 = 1,$$

gives a group cocycle in $Z^1(\Gamma, W_{m-2})$. After a change of sign, we denote this cocycle

$$(4.12) \qquad TGV_\omega(\gamma) := \int_{z_0}^{\gamma z_0} F\omega\, dz\,, \qquad \forall\, \gamma \in \Gamma.$$

Taking its real part, one obtains the linear map

$$(4.13) \qquad \omega \in \mathcal{M}_m(\Gamma) \longmapsto ES(\omega) := [\mathrm{Re}\, TGV_\omega] \in H^1(\Gamma, W_{m-2}(\mathbb{R}))\,,$$

whose restriction to the cuspidal subspace $\mathcal{M}_m^0(\Gamma)$ gives the Eichler-Shimura embedding of $\mathcal{M}_m^0(\Gamma)$ into $H^1(\Gamma, W_{m-2}(\mathbb{R}))$.

In the remainder of this section we shall look closer at the restriction of the assignment (4.13) to the Eisenstein subspace $\mathcal{E}_m(\Gamma) \subset \mathcal{M}_m(\Gamma)$. To this end, we proceed to recall the construction of the Hecke lattice for higher weight Eisenstein series. For $\mathbf{a} = (a_1, a_2) \in (\mathbb{Q}/\mathbb{Z})^2$, the holomorphic Eisenstein series $G_{\mathbf{a}}^{(m)}$ of weight $m > 2$ is defined by the absolutely convergent series

$$G_{\mathbf{a}}^{(m)}(z) := \sum_{\mathbf{k} \in \mathbb{Q}^2 \setminus 0,\, \mathbf{k} \equiv \mathbf{a}\,(\mathrm{mod}\,1)} (k_1 z + k_2)^{-m}\,;$$

averaging over $\left(\frac{1}{N}\mathbb{Z}/\mathbb{Z}\right)^2$, and using as weights the additive characters (see (1.7)) $\left\{\chi_{\mathbf{x}}\,;\, \mathbf{x} \in \left(\frac{1}{N}\mathbb{Z}/\mathbb{Z}\right)^2\right\}$, gives rise to the series

$$(4.14) \qquad \phi_{\mathbf{x}}^{(m)}(z) := \frac{(m-1)!}{(2\pi i N)^m} \sum_{\mathbf{a} \in \left(\frac{1}{N}\mathbb{Z}/\mathbb{Z}\right)^2} \chi_{\mathbf{x}}(\mathbf{a}) \cdot G_{\mathbf{a}}^{(m)}(z)\,.$$

We now apply Proposition 10 for the case of the degenerate action defined by an Eisenstein series $\phi_{\mathbf{x}}^{(m)}$, $\mathbf{x} \in \left(\frac{1}{N}\mathbb{Z}/\mathbb{Z}\right)^2$, and with respect to the $\mathcal{H}_1$-invariant cocycle $\tilde{\tau} = \tilde{\tau}_{m-2}$ (see (3.7)). Restricting to $\mathbb{C}[\Gamma(N)]$ one obtains the transgressed group cocycle $\widetilde{TGV}_{\phi_{\mathbf{x}}} \in Z^1(\Gamma(N), W_{m-2})$. We then take its real part $\Phi_{\mathbf{x}}^{(m)} :=$ $\mathrm{Re}\,\widetilde{TGV}_{\phi_{\mathbf{x}}} \in Z^1(\Gamma(N), W_{m-2}(\mathbb{R}))$, which still satisfies the cocycle relation

$$(4.15) \qquad \Phi_{\mathbf{x}}^{(m)}(\alpha\beta) = \Phi_{\mathbf{x}}^{(m)}(\alpha) + \alpha \cdot \Phi_{\mathbf{x}}^{(m)}(\beta), \qquad \alpha, \beta \in \Gamma(N).$$

We shall show that $\Phi_{\mathbf{x}}^{(m)}$ coincides with the generalized Rademacher function of [20, §2], and in particular $\Phi_{\mathbf{x}}^{(m)} \in Z^1(\Gamma(N), W_{m-2}(\mathbb{Q}))$. As a matter of fact, we shall explicitly compute its extension to $\mathrm{GL}^+(2, \mathbb{Q})$,

$$\Phi_{\mathbf{x}}^{(m)} = \mathrm{Re}\, \Psi_{\mathbf{x}}^{(m)}\,.$$

In turn,

$$\Psi_{\mathbf{x}}^{(m)}(\gamma) = \widetilde{TGV}_{\phi_{\mathbf{x}}}\,, \qquad \gamma \in \mathrm{GL}^+(2, \mathbb{Q}),$$

is obtained by specializing the formula (3.7) to $f^0 = \phi_{\mathbf{x}}^{(m)}$, $f^1 = 1$ and thus is given by the expression

$$
\begin{aligned}
\Psi_{\mathbf{x}}^{(m)}(\gamma) \;=\;& \int_{z_0}^{\gamma z_0} F(z)\,\phi_{\mathbf{x}}^{(m)}(z)dz + \check{F}(z_0)\,\mathbf{a}_0(\phi_{\mathbf{x}}^{(m)}) - \gamma \cdot \check{F}(z_0)\,\mathbf{a}_0(\phi_{\mathbf{x}}^{(m)}|\gamma) \\
(4.16) \qquad & - \int_{z_0}^{i\infty} F(z)\,\widetilde{\epsilon}(\phi_{\mathbf{x}}^{(m)})(z)dz + \int_{z_0}^{i\infty} \gamma \cdot F(z)\,\widetilde{\epsilon}(\phi_{\mathbf{x}}^{(m)}|\gamma)(z)dz \,,
\end{aligned}
$$

which is independent of $z_0 \in \mathbb{H}$. The coboundary relation in Proposition 10 is equivalent with the following cocycle property

$$
(4.17) \qquad \Psi_{\mathbf{x}}^{(m)}(\alpha\beta) \;=\; \Psi_{\mathbf{x}}^{(m)}(\alpha) + \alpha \cdot \Psi_{\mathbf{x}|\alpha}^{(m)}(\beta), \qquad \alpha,\beta \in \mathrm{GL}^+(2,\mathbb{Q}),
$$

and the collection $\{\Phi_{\mathbf{x}}^{(m)} \,;\, \mathbf{x} \in \mathbb{Q}^2/\mathbb{Z}^2\}$ is equivalent to the distribution valued *Eisenstein cocycle* of Stevens [28].

To state the precise result, we need to recall one more definition, that of a *generalized higher Rademacher-Dedekind sum* ([23, 13, 28, 20]); given $a, c \in \mathbb{Z}$ with $(a, c) = 1$ and $c \geq 1$, $0 < k < m \in \mathbb{N}$ and $\mathbf{x} \in \left(\frac{1}{N}\mathbb{Z}/\mathbb{Z}\right)^2$, it is defined by the expression

$$
(4.18) \qquad S_{\mathbf{x}}^{(m-k,k)}\left(\frac{a}{c}\right) = \sum_{r=0}^{c-1} \frac{\mathbf{B}_{m-k}\left(\frac{x_1+r}{c}\right)}{m-k} \, \frac{\mathbf{B}_k\left(x_2 + a\frac{x_1+r}{c}\right)}{k} \,,
$$

where $\mathbf{B}_j : \mathbb{R} \to \mathbb{R}$ is the j-th periodic Bernoulli function

$$
\mathbf{B}_j(x) \;=\; B_j(x - [x]), \qquad x \in \mathbb{R} \,.
$$

THEOREM 11. *Let $m \geq 2$ and let $\mathbf{x} \in \mathbb{Q}^2/\mathbb{Z}^2$, with $\mathbf{x} \neq 0$ if $m = 2$.*

1^0. *For $\beta = \begin{pmatrix} a & b \\ 0 & d \end{pmatrix} \in B(\mathbb{Q})$,*

$$
\Phi_{\mathbf{x}}^{(m)}(\beta) \;=\; -\frac{\mathbf{B}_m(x_1)}{m} \int_0^{\frac{b}{d}} (tT_1 + T_2)^{m-2}\, dt \,.
$$

2^0. *For $\sigma = \begin{pmatrix} a & b \\ c & d \end{pmatrix} \in \mathrm{SL}(2,\mathbb{Z})$, with $c > 0$,*

$$
\Phi_{\mathbf{x}}^{(m)}(\sigma) = \begin{cases} - \dfrac{\mathbf{B}_m(x_1)}{m} \displaystyle\int_0^{\frac{a}{c}} (tT_1 + T_2)^{m-2}\, dt \\[3mm] - \dfrac{\mathbf{B}_m(ax_1 + cx_2)}{m} \displaystyle\int_{-\frac{d}{c}}^{0} (t(aT_1 + cT_2) + bT_1 + dT_2)^{m-2}\, dt \\[3mm] + \displaystyle\sum_{k=0}^{m-2} (-1)^k \binom{m-2}{k} S_{\mathbf{x}}^{(m-k-1,k+1)}\left(\frac{a}{c}\right) T_1^k (aT_1 + cT_2)^{m-k-2}. \end{cases}
$$

PROOF. 1^0. Restricting $\Psi_{\mathbf{x}}^{(m)}$ to $B(\mathbb{Q})$, one has

$$
\Psi_{\mathbf{x}}^{(m)}(\beta) \;=\; \int_{z_0}^{\beta z_0} F(z)\,\widetilde{\epsilon}(\phi_{\mathbf{x}}^{(m)})(z)dz + \mathbf{a}_0(\phi_{\mathbf{x}}^{(m)}) \int_{z_0}^{\beta z_0} F(z)dz
$$

$$+ \quad \check{F}(z_0)\, \mathbf{a}_0(\phi_{\mathbf{x}}^{(m)}) - \beta \cdot \check{F}(z_0)\, \mathbf{a}_0(\phi_{\mathbf{x}}^{(m)}|\beta)$$

$$- \quad \int_{z_0}^{i\infty} F(z)\, \widetilde{\epsilon}(\phi_{\mathbf{x}}^{(m)})(z)\, dz + \int_{z_0}^{i\infty} \beta \cdot F(z)\, \widetilde{\epsilon}(\phi_{\mathbf{x}}^{(m)}|\beta)(z)\, dz$$

$$= \quad \int_{z_0}^{\beta z_0} F(z)\, \widetilde{\epsilon}(\phi_{\mathbf{x}}^{(m)})(z)\, dz + \mathbf{a}_0(\phi_{\mathbf{x}}^{(m)})\, (\check{F}(\beta z_0) - \check{F}(z_0))$$

$$+ \quad \check{F}(z_0)\, \mathbf{a}_0(\phi_{\mathbf{x}}^{(m)}) - \beta \cdot \check{F}(z_0)\, \mathbf{a}_0(\phi_{\mathbf{x}}^{(m)}|\beta)$$

$$- \quad \int_{z_0}^{i\infty} F(z)\, \widetilde{\epsilon}(\phi_{\mathbf{x}}^{(m)})(z)\, dz + \int_{z_0}^{i\infty} \beta \cdot F(z)\, \widetilde{\epsilon}(\phi_{\mathbf{x}}^{(m)}|\beta)(z)\, dz$$

$$= \quad \int_{z_0}^{\beta z_0} F(z)\, \widetilde{\epsilon}(\phi_{\mathbf{x}}^{(m)})(z)\, dz + \mathbf{a}_0(\phi_{\mathbf{x}}^{(m)})\check{F}(\beta z_0) - \beta \cdot \check{F}(z_0)\, \mathbf{a}_0(\phi_{\mathbf{x}}^{(m)}|\beta)$$

$$- \quad \int_{z_0}^{i\infty} F(z)\, \widetilde{\epsilon}(\phi_{\mathbf{x}}^{(m)})(z)\, dz + \int_{z_0}^{i\infty} \beta \cdot F(z)\, \widetilde{\epsilon}(\phi_{\mathbf{x}}^{(m)}|\beta)(z)\, dz\,.$$

When $z_0 \to i\infty$ all three integrals vanish. For the remaining terms we note that, using the weight m analogue of (2.32), one has

$$\mathbf{a}_0(\phi_{\mathbf{x}}^{(m)})\check{F}(\beta z_0) - \beta \cdot \check{F}(z_0)\mathbf{a}_0(\phi_{\mathbf{x}}^{(m)}|\beta) = \mathbf{a}_0(\phi_{\mathbf{x}}^{(m)}) \left(\check{F}(\beta z_0) - \left(\frac{a}{d}\right)^{\frac{m}{2}} \beta \cdot \check{F}(z_0) \right).$$

Since

$$\frac{d}{dz_0} \left(\check{F}(\beta z_0) - \left(\frac{a}{d}\right)^{\frac{m}{2}} \beta \cdot \check{F}(z_0) \right) = 0\,,$$

as one can easily check employing (3.3), the expression is constant in z_0, hence equal to its value at $z_0 = 0$

$$\check{F}(\beta z_0) - \left(\frac{a}{d}\right)^{\frac{m}{2}} \beta \cdot \check{F}(z_0) = \check{F}(\frac{b}{d})\,.$$

From the known Fourier transform of $\phi_{\mathbf{x}}^{(m)}$ ([20], formula (2.1)), one reads that

$$(4.19) \qquad\qquad \mathbf{a}_0(\phi_{\mathbf{x}}^{(m)}) = -\frac{\mathbf{B}_m(x_1)}{m}\,.$$

It follows that

$$(4.20) \qquad\qquad \Phi_{\mathbf{x}}^{(m)}(\beta) = -\frac{\mathbf{B}_m(x_1)}{m} \int_0^{\frac{b}{d}} (tT_1 + T_2)^{m-2}\, dt$$

which proves the first statement.

2^0. Let us now compute the value of $\Psi_{\mathbf{x}}^{(m)}$ at $\sigma_0 = \begin{pmatrix} 0 & -1 \\ 1 & 0 \end{pmatrix}$ by choosing in (4.16) $z_0 = i$, which is a fixed point of σ_0. One has

$$\Psi_{\mathbf{x}}^{(m)}(\sigma_0) = \mathbf{a}_0(\phi_{\mathbf{x}}^{(m)})\, \check{F}(i) - \sigma_0 \cdot \check{F}(i)\, \mathbf{a}_0(\phi_{\mathbf{x}}^{(m)}|\sigma_0)$$

$$- \int_i^{i\infty} F(z)\, \widetilde{\epsilon}(\phi_{\mathbf{x}}^{(m)})(z)\, dz + \int_i^{i\infty} \sigma_0 \cdot F(z)\, \widetilde{\epsilon}(\phi_{\mathbf{x}}^{(m)}|\sigma_0)(z)\, dz\,.$$

This can be related to the Mellin transform of $F\phi_{\mathbf{x}}^{(m)}$,

$$D(F\phi_{\mathbf{x}}^{(m)}, s) = \int_0^{i\infty} F(z)\, \widetilde{\epsilon}(\phi_{\mathbf{x}}^{(m)})(z)\, y^{s-1}\, dz\,, \qquad z = x + iy\,.$$

More precisely, as in the proof of [27, Prop. 2.3.3], one shows that

$$\Psi_{\mathbf{x}}^{(m)}(\sigma_0) = -D(F\phi_{\mathbf{x}}^{(m)}, 1),$$

which in turn can be computed, as in [27, Prop. 2.2.1], from the Fourier transform of f $\phi_{\mathbf{x}}^{(m)}$. Taking the real part one obtains, [28, Thm. 6.9 (a)],

$$(4.21) \qquad \Phi_{\mathbf{x}}^{(m)}(\sigma_0) = \sum_{k=0}^{m-2} (-1)^k \binom{m-2}{k} \frac{\mathbf{B}_{m-k-1}(x_1)}{m-k-1} \frac{\mathbf{B}_{k+1}(x_2)}{k+1} T_1^k T_2^{m-k-2}$$

In view of the Bruhat decomposition $\mathrm{GL}^+(2,\mathbb{Q}) = B(\mathbb{Q}) \cup B(\mathbb{Q})\sigma_0 B(\mathbb{Q})$, the expressions (4.20) and (4.21), together with the cocycle relation (4.17), uniquely determine $\Phi_{\mathbf{x}}^{(m)}$ and allow its explicit calculation (comp. [27, Ch. 2], [28, §5], where the weight 2 case is treated in detail). This calculation has in fact been done by Nakamura [20], for a cocycle defined in a similar manner but using an alternate construction of the Eichler-Shimura period integrals. Since both (4.20) and (4.21) agree with Nakamura's formula [20, (2.8)] for the restriction of $\Phi_{\mathbf{x}}^{(m)}$ to $\mathrm{SL}(2,\mathbb{Z})$, one can conclude that $\Phi_{\mathbf{x}}^{(m)}(\sigma)$ is given by the cited formula for any $\sigma \in \mathrm{SL}(2,\mathbb{Z})$.   □

A similar construction can be performed starting with any holomorphic function $F : \mathbb{H} \to W$ that satisfies the covariance law (3.3),

$$F|g(z) \equiv \det(g)^{-\frac{m-2}{2}} (cz+d)^{m-2} F(gz) = g \cdot F(z), \qquad g \in \mathrm{GL}^+(2,\mathbb{R}).$$

Of particular interest, in view of its connection to special values at non-positive integers of partial zeta functions for real quadratic fields (see [28, §7]), is the following choice ([28, (6.4) (b)]). For $m = 2n \in 2\mathbb{N}$, one takes

$$W_{n-1,n-1} = (W_{n-1} \otimes W_{n-1})^{\mathrm{sym}} \subset W_{n-1} \otimes W_{n-1},$$

i.e. the subspace fixed by the involution $P_1 \otimes P_2 \mapsto P_2 \otimes P_1$; its elements may be identified with the homogeneous polynomials $P \in \mathbb{C}[T_1, T_2, T_3, T_4]$ of degree $2(n-1)$ such that

$$P(T_1, T_2, T_3, T_4) = P(T_3, T_4, T_1, T_2).$$

$\mathrm{GL}^+(2,\mathbb{R})$ acts on $W_{n-1,n-1}$ by the tensor product of the natural representations on the two factors W_{n-1}, and the function $F = F_{n-1,n-1} : \mathbb{H} \to W_{n-1,n-1}$ is defined by the formula

$$F_{n-1,n-1}(z) = (zT_1 + T_2)^{n-1} (zT_3 + T_4)^{n-1}.$$

For this specific choice, taking into account [28, Theorem 6.9], the statement of Theorem 11 becomes modified as follows.

THEOREM 12. *Let* $m = 2n \geq 2$ *and let* $\mathbf{x} \in \mathbb{Q}^2/\mathbb{Z}^2$, *with* $\mathbf{x} \neq 0$ *if* $m = 2$.
1^0. *For* $\beta = \begin{pmatrix} a & b \\ 0 & d \end{pmatrix} \in B(\mathbb{Q})$,

$$\Phi_{\mathbf{x}}^{(m)}(\beta) = -\frac{\mathbf{B}_m(x_1)}{m} \int_0^{\frac{b}{d}} (tT_1 + T_2)^{n-1} (tT_3 + T_4)^{n-1} \, dt.$$

2^0. For $\sigma = \begin{pmatrix} a & b \\ c & d \end{pmatrix} \in \mathrm{SL}(2, \mathbb{Z})$, with $c > 0$,

$$
\Phi_{\mathbf{x}}^{(m)}(\sigma) = \begin{cases}
-\dfrac{\mathbf{B}_m(x_1)}{m} \displaystyle\int_0^{\frac{a}{c}} (tT_1 + T_2)^{n-1} (tT_3 + T_4)^{n-1} \, dt \\[2ex]
-\dfrac{\mathbf{B}_m(ax_1 + cx_2)}{m} \displaystyle\int_{-\frac{d}{c}}^0 (t(aT_1 + cT_2) + bT_1 + dT_2)^{n-1} \\[1ex]
\qquad\qquad\qquad\qquad \cdot (t(aT_3 + cT_4) + bT_3 + dT_4)^{n-1} \, dt \\[2ex]
+ \displaystyle\sum_{k=0}^{n-1} \sum_{\ell=0}^{n-1} (-1)^{k+\ell} \binom{n-1}{k} \binom{n-1}{\ell} S_{\mathbf{x}}^{(m-k-\ell-1, k+\ell+1)}(\tfrac{a}{c}) \\[1ex]
\qquad\qquad\qquad\qquad \cdot T_1^k (aT_1 + cT_2)^{n-k-1} T_3^\ell (aT_3 + cT_4)^{n-\ell-1}.
\end{cases}
$$

REMARK 13. Reformulating a result of Siegel [**25**, **26**], Stevens has shown [**28**, §7] that the Eisenstein cocycle Φ of Theorem 12 can be used to calculate the values at nonpositive integers of partial zeta functions over a real quadratic field. This is achieved by specializing Φ to a certain Eisenstein series E, then evaluating it at the element $\sigma \in \mathrm{SL}(2, \mathbb{Z})$ that represents the action of a certain unit, and finally computing the polynomial $\Phi_E(\sigma)$ on the basis elements and their conjugates. It is intriguing to observe that the above transgressive construction confers Φ the secondary status reminiscent of the Borel regulator invariants that enter in the expression of the special values at non-critical points of L-functions associated to number fields ([**16**, **29**, **30**]).

Acknowledgment. The second author is supported by the National Science Foundation award no. DMS-0245481.

References

[1] Asai, T., *The reciprocity of Dedekind sums and the factor set for the universal covering group of* SL$(2, \mathbb{R})$, Nagoya Math. J. **37** (1970), 67-80.

[2] Barge, J. and Ghys, E., *Cocycles d'Euler et de Maslov*, Math. Ann. **294** (1992), 235–265.

[3] Connes, A., *Noncommutative differential geometry*, Inst. Hautes Etudes Sci. Publ. Math. **62** (1985), 257-360.

[4] Connes, A., *Noncommutative Geometry*, Academic Press, 1994.

[5] Connes, A., *Cyclic cohomology, quantum group symmetries and the local index formula for* $SU_q(2)$, J. Inst. Math. Jussieu, **3** (2004), 17-68.

[6] Connes, A. and Moscovici, H., *Hopf algebras, cyclic cohomology and the transverse index theorem*, Commun. Math. Phys. **198** (1998), 199-246.

[7] Connes, A. and Moscovici, H., *Cyclic cohomology and Hopf algebra symmetry*, Letters Math. Phys. **52** (2000), 1–28.

[8] Connes, A. and Moscovici, H., *Modular Hecke algebras and their Hopf symmetry*, Moscow Math. J. **4** (2004).

[9] Connes, A. and Moscovici, H., *Rankin-Cohen brackets and the Hopf algebra of transverse geometry*, Moscow Math. J. **4** (2004), 111–130.

[10] Dupont, J., Hain, R., Zucker, S., *Regulators and characteristic classes of flat bundles*, "The arithmetic and geometry of algebraic cycles" (Banff, AB, 1998), 47–92, CRM Proc. Lecture Notes, 24, Amer. Math. Soc., Providence, RI, 2000.

[11] Gorokhovsky, A., *Secondary classes and cyclic cohomology of Hopf algebras*, Topology, **41** (2002), 993 – 1016.

[12] Hajac, P. M., Khalkhali, M., Rangipour, B., Sommerhäuser, Y., *Hopf-cyclic homology and cohomology with coefficients*, C. R. Math. Acad. Sci. Paris, **338** (2004), 667–672.

[13] Hall, R. R., Wilson, J. C., Zagier, D., *Reciprocity formulae for general Dedekind-Rademacher sums*, Acta Arith. **73** (1995), 389–396.

[14] Hecke, E., *Theorie der Eisensteinschen Reihen höherer Stufe und ihre Anwendung auf Funktionentheorie und Arithmetik*, Abh. Math. Sem. Univ. Hamburg, **5** (1927), 199-224 ; also in *Mathematische Werke*, Third edition, Vandenhoeck & Ruprecht, Göttingen, 1983.

[15] Khalkhali, M., Rangipour, B., *Cup products in Hopf-cyclic cohomology*, to appear in C. R. Math. Acad. Sci. Paris.

[16] Kontsevich M. and Zagier, D., *Periods*, "Mathematics unlimited—2001 and beyond", 771–808, Springer, Berlin, 2001.

[17] Kubota, T., *Topological covering of* SL(2) *over a local field*, J. Math. Soc. Japan, **19** (1967), 231-267.

[18] Meyer, C., *Die Berechnung der Klassenzahl Abelscher Korper über quadratischen Zahlkorpern*, Akademie-Verlag, Berlin, 1957.

[19] Meyer, C., *Über einige Anwendungen Dedekindsche Summen*, J. Reine Angew. Math. **198** (1957), 143-203.

[20] Nakamura, H., *Generalized Rademacher functions and some congruence properties*, "Galois Theory and Modular Forms", Kluwer Academic Publishers, 2003, 375-394.

[21] Petersson, H., *Zur analytischen Theorie der Grenzkreisgruppen I*, Math. Ann. **115** (1938), 23-67.

[22] Rademacher, H., *Zur Theorie der Modulfunktionen*, J. Reine Angew. Math. **167** (1931), 312-366.

[23] Rademacher, H. and Grosswald, E., *Dedekind Sums*, The Carus Mathematical Monographs, No. **16**, Mathematical Association of America, 1972.

[24] Sczech, R., *Eisenstein cocycles for* GL$_2$(ℚ) *and values of L-functions in real quadratic fields*, Comment. Math. Helvetici, **67** (1992), 363-382.

[25] Siegel, C. L., *Bernoullische Polynome und quadratische Zahlkörper*, Nachr. Akad. Wiss. Göttingen Math.-physik, **2** (1968), 7-38.

[26] Siegel, C. L., *Über die Fourierschen Koeffizienten von Modulformen*, Nachr. Akad. Wiss. Göttingen Math.-physik, **3** (1970), 15-56.

[27] Stevens, G., *Arithmetic on modular curves*, Progress in Mathematics, **20**, Birkhäuser, Boston, MA, 1982.

[28] Stevens, G., *The Eisenstein measure and real quadratic fields*, in "Proceedings of the International Conference on Number Theory" (Quebec, 1987), 887-927, de Gruyter, Berlin, 1987.

[29] Zagier, D., *Valeurs des fonctions zeta des corps quadratiques réels aux entiers négatifs*, Astérisque, **41-42** (1977), 135-151.

[30] Zagier, D., *Polylogarithms, Dedekind zeta functions and the algebraic K-theory of fields*, In: "Arithmetic Algebraic Geometry", Progr. Math. **89**, Birkhäuser (1991), 391-430.

ALAIN CONNES: COLLÈGE DE FRANCE, 3 RUE D'ULM, 75005 PARIS, FRANCE

HENRI MOSCOVICI: DEPARTMENT OF MATHEMATICS, THE OHIO STATE UNIVERSITY, COLUMBUS, OH 43210, USA

Archimedean cohomology revisited

Caterina Consani and Matilde Marcolli

ABSTRACT. Archimedean cohomology provides a cohomological interpretation
for the calculation of the local L-factors at archimedean places as zeta regularized determinant of a log of Frobenius. In this paper we investigate further
the properties of the Lefschetz and log of monodromy operators on this cohomology. We use the Connes-Kreimer formalism of renormalization to obtain
a fuchsian connection whose residue is the log of the monodromy. We also
present a dictionary of analogies between the geometry of a tubular neighborhood of the "fiber at arithmetic infinity" of an arithmetic variety and the
complex of nearby cycles in the geometry of a degeneration over a disk, and we
recall Deninger's approach to the archimedean cohomology through an interpretation as global sections of a analytic Rees sheaf. We show that action of the
Lefschetz, the log of monodromy and the log of Frobenius on the archimedean
cohomology combine to determine a spectral triple in the sense of Connes. The
archimedean part of the Hasse-Weil L-function appears as a zeta function of
this spectral triple. We also outline some formal analogies between this cohomological theory at arithmetic infinity and Givental's homological geometry
on loop spaces.

1. Introduction

C. Deninger produced a unified description of the local factors at arithmetic infinity
and at the finite places where the local Frobenius acts semi-simply, in the form of a
Ray–Singer determinant of a "logarithm of Frobenius" Φ on an infinite dimensional
vector space (the *archimedean cohomology* $H^\cdot_{ar}(X)$ at the archimedean places, [11]).
The first author gave a cohomological interpretation of the space $H^\cdot_{ar}(X)$, in terms
of a double complex $K^{\cdot,\cdot}$ of real differential forms on a smooth projective algebraic
variety X (over $\mathbb{C}$ or $\mathbb{R}$), with Tate-twists and suitable cutoffs, together with an
endomorphism N, which represents a "logarithm of the local monodromy at arithmetic infinity". Moreover, in this theory the cohomology of the complex $\text{Cone}(N)^\cdot$
computes real Deligne cohomology of X, [9]. The construction of[9] is motivated
by a dictionary of analogies between the geometry of the tubular neighborhoods
of the "fibers at arithmetic infinity" of an arithmetic variety X and the geometric
theory of the limiting mixed Hodge structure of a degeneration over a disk. Thus,
the formulation and notation used in [9] for the double complex and archimedean
cohomology mimics the definition, in the geometric case, of a resolution of the
complex of nearby cycles and its cohomology([28]).

In Section 2 and 3 we give an equivalent description of Consani's double complex,
which allows us to investigate further the structure induced on the complex and

the archimedean cohomology by the operators N, Φ, and the Lefschetz operator $\mathbb{L}$. In Section 4 we illustrate the analogies between the complex and archimedean cohomology and a resolution of the complex of nearby cycles in the classical geometry of an analytic degeneration with normal crossings over a disk. In Section 5 we show that, using the Connes–Kreimer formalism of renormalization, we can identify the endomorphism N with the residue of a Fuchsian connection, in analogy to the log of the monodromy in the geometric case. In Section 6 we recall Deninger's approach to the archimedean cohomology through an interpretation as global sections of a real analytic Rees sheaf over $\mathbb{R}$. In Section 7 we show how the action of the endomorphisms N and $\mathbb{L}$ and the Frobenius operator Φ define a noncommutative manifold (a spectral triple in the sense of Connes), where the algebra is related to the $\mathrm{SL}(2,\mathbb{R})$ representation associated to the Lefschetz $\mathbb{L}$, the Hilbert space is obtained by considering Kernel and Cokernel of powers of N, and the log of Frobenius Φ gives the Dirac operator. The archimedean part of the Hasse-Weil L-function is obtained from a zeta function of the spectral triple. In Section 8 we outline some formal analogies between the complex and cohomology at arithmetic infinity and the equivariant Floer cohomology of loop spaces considered in Givental's homological geometry of mirror symmetry.

2. Cohomology at arithmetic infinity

Let X be a compact Kähler manifold of (complex) dimension n. Consider the complex of $\mathbb{C}$-vector spaces

$$(2.1) \qquad C^{\cdot} = \Omega_X^{\cdot} \otimes \mathbb{C}[U, U^{-1}] \otimes \mathbb{C}[\hbar, \hbar^{-1}],$$

where $\Omega_X^{\cdot} = \oplus_{p,q} \Omega_X^{p,q}$ is the complex of global sections of the sheaves of (p,q)-forms on X, $\hbar$ and U are formal independent variables, with U of degree two. Our choice of notation wants to be suggestive of [15], in view of the analogies illustrated in the last section of this paper. On $C^{\cdot}$ we consider the total differential $\delta_C = d'_C + d''_C$, where $d'_C = \hbar\, d$, with $d = \partial + \bar\partial$ the usual de Rham differential and $d''_C = \sqrt{-1}(\bar\partial - \partial)$. The hypercohomology $\mathbb{H}^{\cdot}(C^{\cdot}, \delta_C)$ is then simply given by the infinite dimensional vector space $H^{\cdot}(X;\mathbb{C}) \otimes \mathbb{C}[U, U^{-1}] \otimes \mathbb{C}[\hbar, \hbar^{-1}]$.

We also consider the positive definite inner product

$$(2.2) \qquad \langle \alpha \otimes U^r \otimes \hbar^k, \eta \otimes U^s \otimes \hbar^t \rangle := \langle \alpha, \eta \rangle\, \delta_{r,s}\delta_{k,t},$$

where $\langle \alpha, \eta \rangle$ denotes the Hodge inner product on forms $\Omega_X^{\cdot}$, given by

$$(2.3) \qquad \langle \alpha, \eta \rangle := \int_X \alpha \wedge * C(\bar\eta),$$

with $C(\eta) = (\sqrt{-1})^{p-q}$, for $\eta \in \Omega_X^{p,q}$, and $\delta_{a,b}$ the Kronecker delta.

We then introduce certain cutoffs on $C^{\cdot}$, which will allow us to recover the complex at arithmetic infinity of [9] from $C^{\cdot}$.

To fix notation, for fixed $p, q \in \mathbb{Z}_{\geq 0}$ with $m = p + q$, let

$$(2.4) \qquad \lambda(q, r) := \max\{0, 2r + m, r + q\},$$

where $2r + m$ is the total degree of the complex. Let $\tilde\Lambda_{p,q} \subset \mathbb{Z}^2$ be the set of lattice points satisfying

$$(2.5) \qquad \tilde\Lambda_q = \{(r, k) \in \mathbb{Z}^2 :\ k \geq \lambda(q, r)\},$$

for $\lambda(q, r)$ as in (2.4).

For fixed (p,q) with $m = p+q$, let $\mathfrak{C}^{m,2r}_{p,q} \subset C^{\cdot}$ be the complex linear subspace given by the span of the elements of the form

$$(2.6) \qquad \alpha \otimes U^r \otimes \hbar^k,$$

where $\alpha \in \Omega^{p,q}_X$ and $(r,k) \in \tilde{\Lambda}_q$. We regard $\mathfrak{C}^{m,*}_{p,q}$ as a $2\mathbb{Z}$-graded complex vector space.

Let $\mathfrak{C}^{\cdot}$ be the direct sum of all the $\mathfrak{C}^{m,*}_{p,q}$, for varying (p,q). We regard it as a $\mathbb{Z}$-graded complex vector space with total degree $2r + m$.

In the cutoff (2.4), while the integer $2r+m$ is just the total degree in $\Omega^{\cdot}_X \otimes \mathbb{C}[U, U^{-1}]$, the constraint $k \geq r + q$ can be explained in terms of the Hodge filtration.

Let $\gamma^{\cdot} = F^{\cdot} \cap \bar{F}^{\cdot}$, where $F^{\cdot}$ and $\bar{F}^{\cdot}$ are the Hodge filtrations

$$(2.7) \qquad F^p \Omega^m_X := \bigoplus_{\substack{p'+q=m \\ p' \geq p}} \Omega^{p',q}_X,$$

$$(2.8) \qquad \bar{F}^q \Omega^m_X := \bigoplus_{\substack{p+q''=m \\ q'' \geq q}} \Omega^{p,q''}_X.$$

The condition defining $\mathfrak{C}^{\cdot}$ can be rephrased in the following way.

LEMMA 2.1. *The complex $\mathfrak{C}^{\cdot}$ has an equivalent description as $\mathfrak{C}^i = \bigoplus_{i=m+2r} \mathfrak{C}^{m,2r}$,* with

$$(2.9) \qquad \mathfrak{C}^{m,2r} = \bigoplus_{\substack{p+q=m \\ k \geq \max\{0, 2r+m\}}} \left(F^{m+r-k} \Omega^m_X \right) \otimes U^r \otimes \hbar^k.$$

with the filtration $F^{\cdot}$ as in (2.7).

PROOF. This follows immediately by

$$(2.10) \qquad F^{m+r-k} \Omega^m_X = \bigoplus_{\substack{p+q=m \\ k \geq r+q}} \Omega^{p,q}_X.$$

$\square$

Let c denote the complex conjugation operator acting on complex differential forms. We set $\mathcal{T}^{\cdot} := (\mathfrak{C}^{\cdot})^{c=id}$. This is the real complex

$$(C^{\cdot})^{c=id} = \Omega^{\cdot}_{X,\mathbb{R}} \otimes \mathbb{R}[U, U^{-1}] \otimes \mathbb{R}[\hbar, \hbar^{-1}].$$

Here $\Omega^m_{X,\mathbb{R}}$ is the $\mathbb{R}$-vector space of *real differential forms* of degree m, spanned by forms $\alpha = \xi + \bar{\xi}$, with $\xi \in \Omega^{p,q}_X$ and such that $p + q = m$, namely

$$(2.11) \qquad \Omega^m_{X,\mathbb{R}} = \bigoplus_{p+q=m} (\Omega^{p,q}_X + \Omega^{q,p}_X).$$

We have then the following equivalent description of $\mathcal{T}^{\cdot}$.

LEMMA 2.2. *The complex $\mathcal{T}^{\cdot} = (\mathfrak{C}^{\cdot})^{c=id}$ has the equivalent description $\mathcal{T}^i = \bigoplus_{i=m+2r} \mathcal{T}^{m,2r}$ with*

$$(2.12) \qquad \mathcal{T}^{m,2r} = \bigoplus_{\substack{p+q=m \\ k \geq \max\{0, 2r+m\}}} \left(\gamma^{m+r-k} \Omega^m_X \right) \otimes U^r \otimes \hbar^k,$$

with the filtration $\gamma^{\cdot} = F^{\cdot} \cap \bar{F}^{\cdot}$.

PROOF. We have

$$F^{m+r-k}\Omega_X^m = \bigoplus_{\substack{p+q=m \\ k \geq r+q}} \Omega_X^{p,q}$$

$$\bar{F}^{m+r-k}\Omega_X^m = \bigoplus_{\substack{p+q=m \\ k \geq r+p}} \Omega_X^{p,q},$$

hence one obtains

$$\gamma^{m+r-k}\Omega_X^m = \bigoplus_{\substack{p+q=m \\ k \geq \frac{|p-q|+2r+m}{2}}} \Omega_X^{p,q},$$

where $(|p-q|+2r+m)/2 = r + \max\{p,q\}$.

$\square$

Notice that the inner product (2.3) is real valued on real forms, hence it induces an inner product on $\mathcal{T}^{\cdot}$.

Lemma 2.2 suggests the following convenient description of $\mathcal{T}^{\cdot}$, which we shall use in the following.

For fixed (p,q) with $m = p+q$, let

$$(2.13) \qquad \kappa(p,q,r) := \max\left\{0, 2r+m, \frac{|p-q|+2r+m}{2}\right\}.$$

Then let $\Lambda_{p,q} \subset \mathbb{Z}^2$ be the set

$$(2.14) \qquad \Lambda_{p,q} = \{(r,k) \in \mathbb{Z}^2 : k \geq \kappa(p,q,r)\},$$

with $\kappa(p,q,r)$ as in (2.13).

PROPOSITION 2.3. *The elements of the $\mathbb{Z}$-graded real vector space $\mathcal{T}^{\cdot}$ are linear combinations of elements of the form*

$$(2.15) \qquad \alpha \otimes U^r \otimes \hbar^k,$$

with $\alpha \in (\Omega_X^{p,q} + \Omega_X^{q,p})$, $\alpha = \xi + \bar{\xi}$, for some (p,q) with $p+q = m$, and (r,k) in the corresponding $\Lambda_{p,q}$.

PROOF. By Lemma 2.1, we have seen that the cutoff $\lambda(q,r)$ of (2.4) corresponds to the Hodge filtration $F^{\cdot}$, while Lemma 2.2 shows that, when we impose $c = id$ we can describe $(\mathbb{C}^{\cdot})^{c=id}$ in terms of the $\gamma^{\cdot}$-filtration as in (2.12). For fixed (p,q) with $p+q = m$, this corresponds to the fact that

$$(2.16) \qquad \mathcal{T}_{p,q}^{m,2r} := (\mathbb{C}_{p,q}^{m,2r} \oplus \mathbb{C}_{q,p}^{m,2r})^{c=id}$$

is the real vector space generated by elements of the form (2.15), where $\alpha = \xi + \bar{\xi} \in (\Omega_X^{p,q} + \Omega_X^{q,p})$ is a real form and the indices (r,k) satisfy the conditions $k \geq 0$, $k \geq 2r+m$ and $k \geq r + \max\{p,q\}$. Equivalently, $(r,k) \in \Lambda_{p,q}$. Notice that, since $\kappa(p,q,r) = \kappa(q,p,r)$, we have $\Lambda_{p,q} = \Lambda_{q,p}$. In fact $\Lambda_{p,q}$ depends on (p,q) only through $|p-q|$ and $m = p+q$.

$\square$

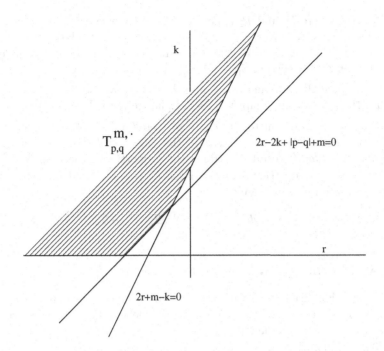

FIGURE 1. *The region* $\Lambda_{p,q} \subset \mathbb{Z}^2$ *defining* $\mathcal{T}_{p,q}^{m,\cdot}$

Figure 1 describes, for fixed values of m and $|p - q|$, the effect of the cutoff (2.13) on the varying indices (r, k). Namely, for fixed (p, q) with $p + q = m$, the region $\Lambda_{p,q} \subset \mathbb{Z}^2$ defined in (2.14) is the shaded region in Figure 1. The graph of the function $k = \kappa(p, q, r)$ of (2.13) is the boundary of the shaded region in the Figure. By construction, the real vector space $\mathcal{T}_{p,q}^{m,2r}$ is the linear span of the (2.15) with $(r, k) \in \Lambda_{p,q}$ and $\mathcal{T}^{\cdot}$ is the direct sum of all the $\mathcal{T}_{p,q}^{m,*}$, for varying (p, q), viewed as a $\mathbb{Z}$-graded real vector space with total degree $2r + m$. Namely, we can think of a single $\mathcal{T}_{p,q}^{m,*}$ as a "slice" of $\mathcal{T}^{\cdot}$ for fixed (p, q), namely, for each (p, q) there is a corresponding Figure 1 and $\mathcal{T}^{\cdot}$ is obtained when considering the union of all of them.

The differentials d'_C and d''_C induce corresponding differentials d' and d'' on $\mathcal{T}^{\cdot}$, where $d' = d'_C = \hbar d$ and $d'' = P^{\perp} d''_C$, with $P^{\perp}$ the orthogonal projection of $(C^{\cdot})^{c=id}$ onto $\mathcal{T}^{\cdot}$. Notice that, since d' and d'' change the values of (p, q), the differentials move from one "slice" $\mathcal{T}_{p,q}^{m,*} \subset \mathcal{T}^{\cdot}$ to another.

2.1. Operators. In the formulation introduced above, we then obtain the very simple description of the operators N and Φ of [9] as

(2.17) $$N = U\hbar \qquad \Phi = -U\frac{\partial}{\partial U}.$$

In [9] these represent, respectively, a logarithm of the local monodromy and a logarithm of Frobenius at arithmetic infinity. We consider the Hilbert space completion of $\mathcal{T}^{\cdot}$ in the inner product induced by (2.2). With a slight abuse of notation, we still denote this Hilbert space by $\mathcal{T}^{\cdot}$. The linear operator N satisfies $N^*N = P_1$

and $NN^* = P_2$, where P_1 and P_2 denote, respectively, the orthogonal projections onto the closed subspaces $\mathrm{Ker}(N)^\perp$ and $\mathrm{Coker}(N)^\perp$ of $\mathcal{T}^\cdot$. The operator N also has the property that $[N, d'] = [N, d''] = 0$. The operator Φ is an unbounded, self adjoint operator with spectrum $\mathrm{Spec}(\Phi) = \mathbb{Z}$. It also satisfies $[\Phi, d'] = [\Phi, d''] = 0$.

Notice that, unlike the differentials d' and d'' that move between different slices $\mathcal{T}_{p,q}^{m,*}$ of $\mathcal{T}^\cdot$, the monodromy map N does not change the values (p, q).

This means that, in addition to the result of Corollary 4.4 of [9] on the "*global*" properties of injectivity and surjectivity of the map $N : \mathcal{T}^\cdot \to \mathcal{T}^{\cdot+2}$, we can also give an analogous "*local*" result describing the properties of the map $N : \mathcal{T}_{p,q}^{m,*} \to \mathcal{T}_{p,q}^{m,*+2}$, restricted to an action on a fixed "slice" (*i.e.* for fixed p and q). In this case, we obtain the following result.

PROPOSITION 2.4. *The endomorphism* $N : \mathcal{T}_{p,q}^{m,2r} \to \mathcal{T}_{p,q}^{m,2(r+1)}$ *has the following properties:*

(1) N *is surjective iff* r *is in the range* $r > -\max\{p, q\}$
(2) N *is injective iff* r *is in the range* $r < -\min\{p, q\}$.

PROOF. (1) For fixed (p, q) with $p + q = m$, let $\Lambda_{p,q} \subset \mathbb{Z}^2$ denote the shaded region in Figure 1, as in (2.14). Let $Z_{p,q} \subset \mathbb{Z}^2$ denote the set of lattice points $Z_{p,q} = \{(r, k) \in \mathbb{Z}^2 : r > -\max\{p, q\}\}$.

The point $(-\max\{p, q\}, 0) \in \Lambda_{p,q}$ is the intersection point of the lines $k = 0$ and $2r - 2k + m + |p - q| = 0$ in the boundary of $\Lambda_{p,q}$. Thus, one sees that the only points in $(r, k) \in \Lambda_{p,q}$ such that $(r-1, k-1) \notin \Lambda_{p,q}$ are those of the form $(r, 0)$ with $r \leq -\max\{p, q\}$. This shows that every point $(r, k) \in \Lambda_{p,q} \cap Z_{p,q}$ has the property that $(r-1, k-1) \in \Lambda_{p,q}$, hence N is surjective in the range $r > -\max\{p, q\}$. It also shows that, for every $r \leq -\max\{p, q\}$, the point $(r, 0)$ in $\Lambda_{p,q}$ is such that $(r-1, -1) \notin \Lambda_{p,q}$, so that N cannot be surjective in the range $r \leq -\max\{p, q\}$.

(2) The case of injectivity is proved similarly. Let

$$W_{p,q} = \{(r, k) \in \mathbb{Z}^2 : r < -\min\{p, q\}\}.$$

Notice that the only points $(r, k) \in \Lambda_{p,q}$ such that $(r+1, k+1) \notin \Lambda_{p,q}$ are those on the boundary line $k = 2r + m$. The point $(-\min\{p, q\}, |p - q|)$ is the intersection point of the lines $k = 2r + m$ and $2r - 2k + m + |p - q| = 0$ in the boundary of $\Lambda_{p,q}$. Thus, we see that every point $(r, k) \in \Lambda_{p,q} \cap W_{p,q}$ is such that $(r+1, k+1) \in \Lambda_{p,q}$, and conversely, for all $r \geq -\min\{p, q\}$ there exists a point $(r, k = 2r + m) \in \Lambda_{p,q}$ such that $(r+1, k+1) \notin \Lambda_{p,q}$, hence N is injective in the range $r < -\min\{p, q\}$, while it cannot be injective for $r \geq -\min\{p, q\}$. $\qquad \square$

The complex $(\mathcal{T}^\cdot, \delta)$ has another important structure, given by the Lefschetz operator, which, together with the polarization and the monodromy, endows $(\mathcal{T}^\cdot, \delta)$ with the structure of a polarized Hodge–Lefschetz module, in the sense of Deligne and Saito ([9], [18], [24]). The Lefschetz endomorphism $\mathbb{L}$ is given by

(2.18) $\mathbb{L} = (\cdot \wedge \omega)\, U^{-1}$

where ω is the canonical real closed $(1,1)$-form determined by the Kähler structure. The Lefschetz operator satisfies $[\mathbb{L}, d'] = [\mathbb{L}, d''] = 0$.

Notice that, unlike the monodromy operator N that preserves the "slices" $\mathcal{T}_{p,q}^{m,*}$, the Lefschetz moves between different slices, namely

$$\mathbb{L} : \mathcal{T}_{p,q}^{m,*} \to \mathcal{T}_{p+1,q+1}^{m+2,*-1} .$$

2.2. Dualities. There are two important duality maps on the complex $\mathcal{T}^{\cdot}$. The first is defined on forms by

(2.19) $$S : \alpha \otimes U^r \otimes \hbar^{2r+m+\ell} \mapsto \alpha \otimes U^{-(r+m)} \otimes \hbar^\ell,$$

for $\alpha \in \Omega_X^m$, and it induces, at the level of cohomology, the duality map of Proposition 4.8 of [**9**]. The map S induces, in particular, the duality between kernel and cokernel of the monodromy map in cohomology, described in [**9**] and [**10**] in terms of powers of the monodromy (*cf.* Proposition 2.5 below).

The other duality is given by the map

(2.20) $$\tilde{S} : \alpha \otimes U^r \otimes \hbar^k \mapsto C(*\alpha) \otimes U^{r-(n-m)} \otimes \hbar^k.$$

PROPOSITION 2.5. *Let S and $\tilde{S}$ be the maps defined in* (2.19) *and* (2.20).

(1) *The map $S : \mathcal{T}^{\cdot} \to \mathcal{T}^{\cdot}$ is an involution, namely $S^2 = 1$. It gives a collection of linear isomorphisms*

$$S = N^{-(2r+m)} : \operatorname{span}\{\alpha \otimes U^r \otimes \hbar^{2r+m+\ell}\} \to \operatorname{span}\{\alpha \otimes U^{-(r+m)} \otimes \hbar^\ell\},$$

realized by powers $N^{-(2r+m)}$ of the monodromy.

(2) *The map $\tilde{S} : \mathcal{T}^{\cdot} \to \mathcal{T}^{\cdot}$ is an involution, $\tilde{S}^2 = 1$. The map induced by $\tilde{S}$ on the primitive part of the cohomology, with respect to the Lefschetz decomposition, agrees (up to a non-zero real constant) with the power $\mathbb{L}^{n-m}$ of the Lefschetz operator.*

PROOF. (1) The result for S follows directly from the definition, in fact, we have

$$S^2(\alpha \otimes U^r \otimes \hbar^{2r+m+\ell}) = S(\alpha \otimes U^{-(r+m)} \otimes \hbar^\ell)$$
$$= \alpha \otimes U^{r+m-m} \otimes \hbar^{\ell+2(r+m)-m} = \alpha \otimes U^r \otimes \hbar^{2r+m+\ell}.$$

This means that the duality S preserve the "slices" $\mathcal{T}_{p,q}^{m,*}$ and on them it can be identified with the symmetry of $\mathcal{T}^{\cdot}$ obtained by reflection of the shaded area of Figure 1 along the line illustrated in Figure 2. The elements of the form $\alpha \otimes U^{-m/2} \otimes \hbar^{|p-q|}$, for $\alpha \in \Omega_X^m$ are fixed by the involution S since

$$S(\alpha \otimes U^{-m/2} \otimes \hbar^{|p-q|}) = \alpha \otimes U^{\frac{m}{2}-m} \otimes \hbar^{|p-q|+\frac{2m}{2}-m} = \alpha \otimes U^{-m/2} \otimes \hbar^{|p-q|}.$$

We prove (2). The map $\tilde{S}$ preserves $\mathcal{T}^{\cdot}$, since the cutoffs described by the conditions $k \geq 0$, $2r + m - 2k + |p - q| \leq 0$ and $k \geq 2r + m$ are preserved by mapping $r \mapsto r - (n - m)$, $k \mapsto k$, $m \mapsto 2n - m$, $(p, q) \mapsto (n - q, n - p)$. We have $\tilde{S}^2 = 1$, since

$$\tilde{S}^2(\alpha \otimes U^r \otimes \hbar^k) = (\sqrt{-1})^{p-q} \tilde{S}(*\alpha \otimes U^{r-(n-m)} \otimes \hbar^k)$$
$$= (\sqrt{-1})^{p-q}(\sqrt{-1})^{n-q-(n-p)} *^2 \alpha \otimes U^{r-(n-m)-(n-(2n-m))} \otimes \hbar^k$$
$$= (\sqrt{-1})^{2(p-q)}(-1)^m \alpha \otimes U^r \otimes \hbar^k,$$

where we used $*^2 = (-1)^{m(2n-m)} = (-1)^m$.

Let $P^m(X)$ be the primitive part of the cohomology $H^m(X, \mathbb{C})$, with respect to the Lefschetz decomposition ([**29**] §V.6). Let J be the operator induced on $P^{p,q}(X)$ by

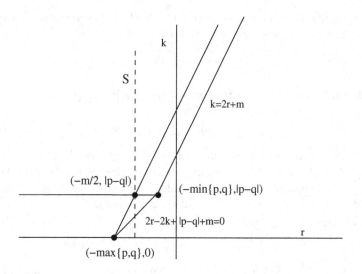

FIGURE 2. *The duality S*

$C(\eta) = (\sqrt{-1})^{p-q} \eta$. On the primitive cohomology one has the identification (up to multiplication by a non-zero real constant)

(2.21) $\mathbb{L}^{n-m} J\eta = *\eta$.

In particular, (2.21) implies that the map $\tilde{S}$ agrees (up to a normalization factor) with $\mathbb{L}^{n-m}$, on the primitive cohomology.

$\square$

Thus, we can think of the two dualities S and $\tilde{S}$ as related, respectively, to the action of N^{2r+m} and $\mathbb{L}^{n-m}$, *i.e.* of powers of the monodromy and Lefschetz.

2.3. Representations. The nilpotent endomorphisms $\mathbb{L}$ and N of $\mathcal{T}^{\cdot}$ introduced above define two representations of $SL(2,\mathbb{R})$ on $\mathcal{T}^{\cdot}$ as follows. With the notation
(2.22)
$$\chi(\lambda) := \begin{pmatrix} \lambda & 0 \\ 0 & \lambda^{-1} \end{pmatrix} \lambda \in \mathbb{R}^*, \qquad u(s) := \begin{pmatrix} 1 & s \\ 0 & 1 \end{pmatrix} s \in \mathbb{R}, \qquad w := \begin{pmatrix} 0 & 1 \\ -1 & 0 \end{pmatrix},$$
we define σ^L and σ^R by

(2.23) $\sigma^L(\chi(\lambda)) = \lambda^{-n+m}, \qquad \sigma^L(u(s)) = \exp(s\,\mathbb{L}), \qquad \sigma^L(w) = (\sqrt{-1})^n\, C\, \tilde{S}.$

(2.24) $\sigma^R(\chi(\lambda)) = \lambda^{2r+m}, \qquad \sigma^R(u(s)) = \exp(s\,N), \qquad \sigma^R(w) = C\, S.$

Here C is the operator on forms $C(\eta) = (\sqrt{-1})^{p-q}$ for $\eta \in \Omega_X^{p,q}$, and S and $\tilde{S}$ the dualities on $\mathcal{T}^{\cdot}$, as in Proposition 2.5. The results of [**18**], [**9**] and [**10**] yield the following.

PROPOSITION 2.6. *The operators* (2.23) *and* (2.24) *on* $\mathcal{T}^{\cdot}$ *define a representation* $\sigma = (\sigma^L, \sigma^R) : SL(2,\mathbb{R}) \times SL(2,\mathbb{R}) \to \mathrm{Aut}(\mathcal{T}^{\cdot})$. *In the representation* σ^L, *the group* $SL(2,\mathbb{R})$ *acts by bounded operators on the completion of* $\mathcal{T}^{\cdot}$ *in the inner product (2.2). Both* $\mathrm{SL}(2,\mathbb{R})$ *actions commute with the Laplacian* $\square = \delta\delta^* + \delta^*\delta$, *hence they define induced representations on the cohomology* $\mathbb{H}^{\cdot}(\mathcal{T}^{\cdot},\delta)$.

PROOF. For completeness, we give here a simple proof of the proposition. In order to show that we have representations of $\mathrm{SL}(2,\mathbb{R})$ it is sufficient ([19] §XI.2) to check that (2.23) and (2.24) satisfy the relations

(2.25)
$$\sigma(w)^2 = \sigma(\chi(-1))$$
$$\sigma(\chi(\lambda))\sigma(u(s))\sigma(\chi(\lambda^{-1})) = \sigma(u(s\lambda^2))$$

We show it first for σ^R. We show that we have $\sigma^R(w) = (-1)^m$, as in [10]. This follows directly from the fact that $S^2 = 1$, since $\sigma^R(w)^2 = CSCS = (\sqrt{-1})^{2(p-q)} = (-1)^m$. Thus, we have $\sigma^R(w)^2 = (-1)^m = \lambda^{2r+m}|_{\lambda=-1}$ and the first relation of (2.25) is satisfied. To check the second relation notice that, on an element $\alpha \otimes U^r \otimes \hbar^k$ with $\alpha \in \Omega_X^m$ we have

$$\sigma^R(\chi(\lambda))\,\sigma^R(u(s))\,\sigma^R(\chi(\lambda^{-1}))\,\alpha \otimes U^r \otimes \hbar^k =$$

$$\sigma^R(\chi(\lambda))\left(1 + sN + \frac{s^2}{2}N^2 + \cdots\right)\lambda^{-(2r+m)}\,\alpha \otimes U^r \otimes \hbar^k =$$

$$\left(1 + \lambda^{2(r+1)+m}sN\lambda^{-(2r+m)} + \lambda^{2(r+2)+m}\frac{s^2}{2}N^2\lambda^{-(2r+m)} + \cdots\right)\alpha \otimes U^r \otimes \hbar^k =$$

$$\exp(s\lambda^2 N)\,\alpha \otimes U^r \otimes \hbar^k,$$

hence the second relation is satisfied.

We show that σ^L also satisfies the relations (2.25). Again, we first show that $\sigma^L(w) = (-1)^{n+m}$, as in [9]. We have

$$C\tilde{S}(\alpha \otimes U^r \otimes \hbar^k) = (\sqrt{-1})^{n-q-n+p}(\sqrt{-1})^{p-q} * \alpha \otimes U^{r-(n-m)} \otimes \hbar^k,$$

hence

$$\sigma^L(w)^2\,(\alpha \otimes U^r \otimes \hbar^k) = (-1)^n(-1)^m\,C\tilde{S}(*\alpha \otimes U^{r-(n-m)} \otimes \hbar^k)$$

$$= (-1)^n(-1)^m(-1)^m(\sqrt{-1})^{p-q}(\sqrt{-1})^{n-q-n+p}\alpha \otimes U^r \otimes \hbar^k = (-1)^n(-1)^m\alpha \otimes U^r \otimes \hbar^k$$

where, in the left-hand side, we used $*^2 = (-1)^m$.

Thus, we have $\sigma^L(w)^2 = (-1)^{n+m} = \lambda^{m-n}|_{\lambda=-1}$. Moreover, we have

$$\sigma^L(\chi(\lambda))\,\sigma^L(u(s))\,\sigma^L(\chi(\lambda^{-1}))\,\alpha \otimes U^r \otimes \hbar^k =$$

$$\sigma^L(\chi(\lambda))\left(1 + s\mathbb{L} + \frac{s^2}{2}\mathbb{L}^2 + \cdots\right)\lambda^{n-m}\,\alpha \otimes U^r \otimes \hbar^k =$$

$$\left(1 + \lambda^{-n+m+2}s\mathbb{L}\lambda^{n-m} + \lambda^{-n+m+4}\frac{s^2}{2}\mathbb{L}^2\lambda^{n-m} + \cdots\right)\alpha \otimes U^r \otimes \hbar^k =$$

$$\exp(s\lambda^2\,\mathbb{L})\,\alpha \otimes U^r \otimes \hbar^k,$$

hence the second relation is also satisfied.

The fact that $\sigma^L(\chi(\lambda))$ is a bounded operator in the inner product induced by (2.2), while for $\lambda \neq \pm 1$ the operators $\sigma^R(\chi(\lambda))$ are unbounded is clear from the fact that the index $2r + m$ ranges over all of $\mathbb{Z}$, while $-n \leq m - n \leq n$. For the fact that $[\square, \sigma^L] = [\square, \sigma^R] = 0$ we refer to [9].

$\square$

2.4. Ring of differential operators. Let $\mathcal{D}$ denote the algebra of differential operators on a 1-dimensional complex torus $T_{\mathbb{C}}$, generated by the operators $Q = e^z$ and $P = \frac{\partial}{\partial z}$ satisfying the commutation relation

$$(2.26) \qquad PQ - QP = Q.$$

Let $\mathcal{R}$ be the ring of functions defining the coefficients of the differential operators in $\mathcal{D}$. This is a subring of the ring of functions on $\mathbb{C}^*$. For $\mathcal{R} = \mathbb{C}[Q]$ we obtain $\mathcal{D} = \mathbb{C}[P, Q]/(PQ - QP = Q)$.

Since the operators N and Φ satisfy the commutation relation $[\Phi, N] = -N$ and the operators $\mathbb{L}$ and Φ satisfy the commutation relation $[\Phi, \mathbb{L}] = \mathbb{L}$, the pairs of operators $(N, -\Phi)$ and $(\mathbb{L}, \Phi)$ define actions π^R and π^L of $\mathcal{D}$ on the complex $\mathfrak{C}^{\cdot}$ and on its cohomology, by setting

$$(2.27) \qquad \begin{aligned} \pi^L(P) &= \Phi & \pi^L(Q) &= \mathbb{L} \\ \pi^R(P) &= -\Phi & \pi^R(Q) &= N. \end{aligned}$$

There is an induced action of the ring $\mathcal{D} = \mathbb{R}[P, Q]/(PQ - QP = Q)$ on the complex $\mathcal{T}^{\cdot}$ and on its cohomology.

2.5. Weil–Deligne group at arithmetic infinity. On $\mathcal{T}^{\cdot}$ consider the "Frobenius flow"

$$(2.28) \qquad F_t = e^{t\Phi}, \quad \forall t \in \mathbb{R},$$

generated by the operator Φ. We write $F = F_1$. This satisfies

$$(2.29) \qquad F N F^{-1} = e^{-1} N.$$

Thus, the operators F and N can be thought of as defining an analog at arithmetic infinity of the Weil–Deligne group $\mathbb{G}_a \rtimes W_K$, which acts on the finite dimensional vector space associated to the étale cohomology of the geometric generic fiber of a local geometric degeneration for K a non-archimedean local field. In fact, in that case, the action of the Frobenius $\varphi \in W_K$ on $\mathbb{G}_a$ is given by

$$(2.30) \qquad \varphi \, x \, \varphi^{-1} = q^{-1} x,$$

where q is the cardinality of the residue field. The formal replacement of q by e and of φ by F determines (2.29) from (2.30). In the archimedean case, this "Weil–Deligne group" acts directly at the level of the complex, not just on the cohomology.

3. Archimedean cohomology

We now describe the relation between the complex $(\mathcal{T}^{\cdot}, \delta = d' + d'')$ defined in the first section and the cohomology theory at arithmetic infinity developed in [9].

On a smooth projective algebraic variety X of dimension n over $\mathbb{C}$ or $\mathbb{R}$, the complex of Tate-twisted real differential forms introduced in (4.1) of [9] is defined as

$$(3.1)$$

$$K^{i,j,k} = \begin{cases} \displaystyle\bigoplus_{\substack{p+q=j+n \\ |p-q|\leq 2k-i}} (\Omega_X^{p,q} \oplus \Omega_X^{q,p})_{\mathbb{R}} \otimes_{\mathbb{R}} \mathbb{R}\left(\dfrac{n+j-i}{2}\right) & \text{if } \begin{aligned} j+n-i &\equiv 0(2), \\ k &\geq \max(0, i) \end{aligned} \\ 0 & \text{otherwise,} \end{cases}$$

for $i, j, k \in \mathbb{Z}$. Here $\mathbb{R}(r)$ denotes the real Hodge structure $\mathbb{R}(r) := (2\pi\sqrt{-1})^r \mathbb{R}$, and the differentials d' and d'' are given by

(3.2) $$d' : K^{i,j,k} \to K^{i+1,j+1,k+1}, \qquad d'(\alpha) = d(\alpha)$$

(3.3)
$$d'' : K^{i,j,k} \to K^{i+1,j+1,k}, \qquad d''(\alpha) = \sqrt{-1}(\bar{\partial}-\partial)(\alpha) \quad \text{(projected onto } K^{i+1,j+1,k}\text{).}$$

The inner product on $\mathbb{H}^{\cdot}(K^{\cdot}, d' + d'')$ is defined in terms of the bilinear form

(3.4) $$Q(\alpha, \eta) = \int_X \mathbb{L}^{n-m} \alpha \wedge J\bar{\eta},$$

for η, α in the primitive part $P^m(X)$ of the de Rham cohomology $H^m(X, \mathbb{R})$ with respect to the Lefschetz decomposition.

The relation between the complex $\mathcal{T}^{\cdot}$ and the total complex $K^{\cdot}$ of (3.1) is described by the following result, which shows that the complex $\mathcal{T}^{\cdot}$ is identified with $K^{\cdot}$ after applying the simple change of variables

(3.5) $$i = m + 2r, \qquad r = -\frac{n+j-i}{2}, \qquad j = -n + m$$

to the indices of $\mathcal{T}^{\cdot}$, taking fixed points under complex conjugation ($c = id$) and replacing the variable U by a Tate twist.

PROPOSITION 3.1. *Upon identifying the formal variable U^{-1} with the Tate twist given by multiplication by $2\pi\sqrt{-1}$, we obtain an isomorphism of complexes*

(3.6) $$(\mathcal{T}^{\cdot}|_{U=(2\pi\sqrt{-1})^{-1}}, \delta) \cong (K^{\cdot}, d' + d''),$$

given by a reparameterization of the indices. Up to a normalization factor, the inner product induced by (2.2) on the cohomology $\mathbb{H}^{\cdot}(\mathcal{T}^{\cdot}|_{U=(2\pi\sqrt{-1})^{-1}}, \delta)$ agrees with the one defined by (3.4) on $\mathbb{H}^{\cdot}(K^{\cdot}, d' + d'')$.

PROOF. We define a homomorphism

$$I : \mathcal{T}^{\cdot} \to K^{\cdot}$$

as follows. For fixed p, q with $p + q = m$, consider the region $\Lambda_{p,q} \subset \mathbb{Z}^2$ as in (2.14), with $\kappa(p, q, r)$ as in (2.13). For every $\alpha \in (\Omega_X^{p,q} \oplus \Omega_X^{q,p})_{\mathbb{R}}$ a real form, $\alpha = \xi + \bar{\xi}$, with $\xi \in \Omega_X^{p,q}$, and for every point $(r, k) \in \Lambda_{p,q}$, we have

$$\alpha \otimes U^r \otimes \hbar^k \in \mathcal{T}_{p,q}^{m,2r},$$

and, by Proposition 2.3, every element of $\mathcal{T}^{\cdot}$ is a linear combination of elements of this form, for varying (p, q) and corresponding $(r, k) \in \Lambda_{p,q}$.
We now define the map I in the following way. To an element

(3.7) $$(\xi + \bar{\xi}) \otimes U^r \otimes \hbar^k,$$

with $\xi \in \Omega_X^{p,q}$, with $p + q = m$, and with $k \geq \kappa(p, q, r)$, we assign an element $I(\eta) \in K^{i,j,k}$, with the same index k and with

(3.8) $$i = m + 2r, \qquad \text{and} \qquad j = -n + m,$$

by setting

(3.9) $$I(\eta) = (2\pi\sqrt{-1})^{-r} (\xi + \bar{\xi}).$$

In fact, for (i, j) as in (3.8), the index $r \in \mathbb{Z}$ can be written in the form

$$r = -\frac{n+j-i}{2}, \qquad \text{where } n + j - i = 0 \mod 2.$$

Thus, the element (3.9) can be written as

$$(3.10) \qquad\qquad (2\pi\sqrt{-1})^{\frac{n+j-i}{2}} (\xi + \bar{\xi}).$$

By the definition (3.1), to check that this is an element in $K^{i,j,k}$, it is sufficient to verify that $p + q = j + n$, and that the conditions $|p - q| \le 2k - i$ and $k \ge \max\{0, i\}$ are satisfied. Since $j = -n + m$ and $p + q = m$ we have $p + q = j + n$. The index k is the same as in (3.7), hence it satisfies $k \ge \kappa(p, q, r)$. This means that $k \ge 0$ and that $k \ge 2r + m = i$, so that $k \ge \max\{0, i\}$ satisfied. Since $k \ge \kappa(p, q, r)$ also implies $k \ge (|p - q| + 2r + m)/2$, which, by $i = 2r + m$ is the condition $|p - q| \le 2k - i$. It is clear that the map I defined this way is injective.

Thus, we have shown that, for every real form $\alpha \in (\Omega_X^{p,q} \oplus \Omega_X^{q,p})_{\mathbb{R}}$ and for every lattice point $(r, k) \in \Lambda_{p,q}$ we have a unique corresponding element in the complex $K^{i,j,k}$.

The map I is also surjective, hence a linear isomorphism. In fact, every element in $K^{\cdot}$ is a linear combination of elements of the form (3.10) in $K^{i,j,k}$, for $\xi \in \Omega_X^{p,q}$, and indices $(i, j, k) \in \mathbb{Z}^3$ satisfying $n + j - i = 0 \bmod 2$, $p + q = j + n$, $k \ge \max\{i, 0\}$ and $|p - q| \le 2k - i$. It is sufficient to show that, for any such element, there exists a point $(r, k) \in \Lambda_{p,q}$ such that

$$I((\xi + \bar{\xi}) \otimes U^r \otimes \hbar^k) = (2\pi\sqrt{-1})^{\frac{n+j-i}{2}} (\xi + \bar{\xi}) \in K^{i,j,k}.$$

This is achieved by taking the point

$$(3.11) \qquad\qquad \left(r = -\frac{n+j-i}{2}, k \right).$$

Since $n + j - i = 0 \bmod 2$ this point is in $\mathbb{Z}^2$, and since under the change of variables (3.8) the conditions $k \ge \max\{i, 0\}$ and $|p - q| \le 2k - i$ are equivalent to the condition $k \ge \kappa(p, q, r)$ of (2.13), the point (3.11) is in $\Lambda_{p,q}$.

The map I is compatible with the differentials, namely

$$I(\delta\eta) = (d' + d'')I(\eta),$$

where on the left hand side $\delta = d_C' + P^\perp d_C''$ is the differential on $\mathcal{T}^{\cdot}$ and on the right hand side $d' + d''$ is as in (3.2) (3.3). To see this, first notice that the differential d' of (3.2) satisfies

$$d' = Id_C' I^{-1} = I\hbar d I^{-1} : K^{2r+m,-n+m,k} \to K^{2r+(m+1),-n+(m+1),k+1}.$$

The analogous statement $d'' = IP^\perp d_C'' I^{-1}$ for the differential d'' of (3.3) also involves the fact that the orthogonal projection onto $K^{i+1,j+1,k}$ in (3.3), induced by the inner product (3.4) agrees with the corresponding orthogonal projection $P^\perp$ in $\mathcal{T}^{\cdot}$ induced by the inner product (2.2).

The identification (2.21) implies that the inner product (3.4) on $\mathbb{H}^{\cdot}(K^{\cdot}, d' + d'')$ considered in [9] and the inner product induced by (2.2) agree up to a normalization factor.

$\square$

In particular, the 'weight type' condition $|a - b| \le 2k - i$ on the real forms in (3.1) describes, as in (2.7) (2.8), the filtration $\gamma^{\cdot} := F^{\cdot} \cap \bar{F}^{\cdot}$ on the complex of real differential forms on X. It follows that the complex $K^{i,j,k}$ has a real analytic type, even when X defined over $\mathbb{C}$ does not have a real structure.

To the abelian group $K^{i,j,k}$ we assign the weight: $-n - j + i \in \mathbb{Z}$. Keeping in mind that $\mathbb{R}(\frac{n+j-i}{2})$ is the real Hodge structure of rank one and pure bi-degree $(-\frac{n+j-i}{2}, -\frac{n+j-i}{2})$, we obtain the following description in terms of the filtration γ:

$$(3.12) \quad K^{i,j,k} = \gamma^{\frac{n+j+i}{2}-k}\Omega_X^{n+j} \otimes_{\mathbb{R}} \mathbb{R}(\frac{n+j-i}{2}) = \gamma^{\frac{n+j+i}{2}-k}\Omega_X^{n+j} \otimes_{\mathbb{R}} \gamma^{-\frac{n+j-i}{2}}\mathbb{R}.$$

When considering the tensor product of the two structures one sees that the index of the γ-filtration on the product (*i.e.* on $K^{i,j,k}$) is $i - k$.

3.1. Deligne cohomology.
By Proposition 3.1, the complex $\mathcal{T}^{\cdot}$ is related to the real Deligne cohomology $H_{\mathcal{D}}^*(X, \mathbb{R}(r))$. These groups can be computed via the Deligne complex $(C_{\mathcal{D}}^*(r), d_{\mathcal{D}})$ ([3], [9]). The relation of $(C_{\mathcal{D}}^*(r), d_{\mathcal{D}})$ to the complex $(\mathcal{T}^{\cdot}, \delta)$ is given by the following result of [9] (Prop. 4.1), which, for convenience, we reformulate here in our notation.

PROPOSITION 3.2. *For N acting on $(\mathcal{T}^{\cdot}, \delta)$, consider the complex $(Cone(N)^{\cdot}, D)$ with differential $D(\alpha, \beta) = (\delta(\alpha), N(\beta) - \delta(\beta))$.*

(1) *For $2r + m > 0$, the map $N^{-(2r+m)}$ gives an isomorphism between the cohomology group in $\mathbb{H}^{\cdot}(\text{Ker}(N)^{\cdot})$, which lies over the point of coordinates $(2r, 2r + m)$ in Figure 1, and the cohomology group in $\mathbb{H}^{\cdot}(\text{Coker}(N)^{\cdot})$ that lies over the point of coordinates $(-2(r + m), 0)$ in Figure 1.*

(2) *In the range $2r + m < -1$, the cohomology $\mathbb{H}^{\cdot}(Cone(N)^{\cdot}, D)$ is identified with $\mathbb{H}^{\cdot+1}(\text{Coker}(N), \delta)$.*

(3) *Upon identifying the variable U^{-1} with the Tate twist by $2\pi\sqrt{-1}$, and for a for fixed $r \in \mathbb{Z}_{\leq 0}$, we obtain quasi isomorphic complexes*

$$(3.13) \qquad (Cone(N)^{\cdot}, D)|_{U^r=(2\pi\sqrt{-1})^{-r}} \simeq (C_{\mathcal{D}}^*(-r), d_{\mathcal{D}})$$

3.2. Local factors.
The "archimedean factor" (*i.e.* the local factor at arithmetic infinity) $L_{\kappa}(H^m, s)$ is a product of powers of shifted Gamma functions, with exponents and arguments that depend on the Hodge structure on $H^m = H^m(X, \mathbb{C}) = \oplus_{p+q=m}H^{p,q}$. More precisely, it is given by ([25])
$$(3.14)$$

$$L_{\kappa}(H^m, s) = \begin{cases} \prod_{p,q}\Gamma_{\mathbb{C}}(s - \min(p,q))^{h^{p,q}} & \kappa = \mathbb{C} \\ \\ \prod_{p<q}\Gamma_{\mathbb{C}}(s - p)^{h^{p,q}} \prod_p \Gamma_{\mathbb{R}}(s - p)^{h^{p+}}\Gamma_{\mathbb{R}}(s - p + 1)^{h^{p-}} & \kappa = \mathbb{R}, \end{cases}$$

where the $h^{p,q}$, with $p + q = m$, are the Hodge numbers, $h^{p,\pm}$ is the dimension of the $\pm(-1)^p$-eigenspace of de Rham conjugation on $H^{p,p}$, and

$$\Gamma_{\mathbb{C}}(s) := (2\pi)^{-s}\Gamma(s) \qquad \Gamma_{\mathbb{R}}(s) := 2^{-1/2}\pi^{-s/2}\Gamma(s/2).$$

It is shown in [11] that the local factor (3.14) can be computed as a Ray–Singer determinant

$$(3.15) \qquad L_{\kappa}(H^m, s) = \det_{\infty}\left(\frac{1}{2\pi}(s - \Phi)|_{H_{ar}^m(X)}\right)^{-1},$$

where the "archimedean cohomology" $H_{ar}^m(X)$ is an infinite-dimensional real vector space, and the zeta regularized determinant of an unbounded self adjoint operator T is defined as

$$\det_{\infty}(s - T) = \exp(-\frac{d}{dz}\zeta_T(s, z)|_{z=0}).$$

In [9] (*cf.* §5) the archimedean cohomology is identified with the "inertia invariants"

(3.16) $$H_{ar}^{\cdot}(X) = \mathbb{H}^{\cdot}(K^{\cdot}, d' + d'')^{N=0},$$

where $\mathbb{H}^{\cdot}(K^{\cdot}, d' + d'')$ is the hypercohomology of the complex (3.1) and $\mathbb{H}^{\cdot}(K^{\cdot}, d' + d'')^{N=0}$ is the kernel of the map induced in hypercohomology by the monodromy N. This follows the expectation that the fiber over arithmetic infinity has semi-stable reduction.

At arithmetic infinity, the alternating product of the local factors of X can be described in terms of the operators $\sigma^L(w)^2$ and Φ ([10] par. 3.4). Let Φ_0 denote the restriction of the operator Φ to $\mathbb{H}^{\cdot}(K^{\cdot})^{N=0}$. For a a bounded operator on $\mathbb{H}^{\cdot}(K^{\cdot})^{N=0}$, let $\zeta_{a,\Phi_0}(s,z)$ denote the two variable zeta function

(3.17) $$\zeta_{a,\Phi_0}(s,z) = \sum_{\lambda \in \mathrm{Spec}(\Phi_0)} \mathrm{Tr}(a\Pi_\lambda)(s-\lambda)^{-z},$$

where Π_λ are the spectral projections of Φ_0. Let $\det_{\infty,a,\Phi_0}$ denote the zeta regularized determinant

(3.18) $$\det_{\infty,a,\Phi_0}(s) = \exp\left(-\frac{d}{dz}\zeta_{a,\Phi_0}(s,z)|_{z=0}\right).$$

PROPOSITION 3.3. *The two variable zeta function* $\zeta_{\sigma^L(w)^2,\Phi_0}(s,z)$ *satisfies*

(3.19) $$\det_{\infty,\sigma^L(w)^2,\Phi_0/(2\pi)}\left(\frac{s}{2\pi}\right)^{-1} = \prod_{m=0}^{2n} L_{\mathbb{C}}(H^m, s)^{(-1)^{m+n}}.$$

PROOF. The operator Φ_0 has spectrum

$$\mathrm{Spec}(\Phi_0) = \{\lambda_{\ell,p,q} = \min\{p,q\} - \ell : \ell \in \mathbb{Z}_{\geq 0}\},$$

with eigenspaces $E_{\ell,p,q} = H^{p,q}(X) \otimes U^r \otimes \hbar^{2r+m}$, with $r = \ell - \min\{p,q\}$ and $m = p + q$. In fact, for $\mathbb{H}^{\cdot}(K)^{N=0}$ we have $2r + m = k$, and $k \geq |p-q|$ so that $r \geq -\min\{p,q\}$. The result then follows as in [10] §3.3.

$\square$

4. Archimedean cohomology and nearby cycles

The definition of the complex (3.1) was inspired by an analogy with the resolution of the complex of nearby cycles associated to an analytic degeneration with normal crossings over a disk, [28]. In this section we recall this classical construction and we relate it to its archimedean counterpart by making the analogy more explicit.

4.1. The complex of nearby cycles. Let $\mathfrak{X}$ and Δ be complex analytic manifolds, of complex dimensions $\dim \mathfrak{X} = n+1$ and $\dim \Delta = 1$, and let $f : \mathfrak{X} \to \Delta$ be a flat, proper morphism with projective fibers. For $0 \in \Delta$, we write $Y = f^{-1}(0)$, $\mathfrak{X}^* = \mathfrak{X} \smallsetminus Y$, and $\Delta^* = \Delta \smallsetminus \{0\}$. We assume that the map f is smooth on $\mathfrak{X}^*$ and that Y is a divisor with normal crossings on $\mathfrak{X}$. Under these hypotheses, the relative de Rham complex of sheaves of differential forms with logarithmic poles along Y is well defined and of the form

$$\Omega_{\mathfrak{X}/\Delta}^m(\log Y) := \wedge^m \Omega_{\mathfrak{X}/\Delta}^1(\log Y),$$

where we use the inclusion $f^*\Omega_\Delta^1(\log 0) \subset \Omega_{\mathfrak{X}}^1(\log Y)$ to define

$$\Omega_{\mathfrak{X}/\Delta}^1(\log Y) := \Omega_{\mathfrak{X}}^1(\log Y)/f^*\Omega_\Delta^1(\log 0).$$

Consider local coordinates $\{z_0, \ldots z_n\}$ at $P \in Y$, with t the local coordinate at $0 \in \Delta$, so that $t \circ f = z_0^{e_0} \cdots z_k^{e_k}$ for some $0 \leq k \leq n$ and $e_i \in \mathbb{N}$. The stalk $\Omega^1_{\mathfrak{X}/\Delta}(\log Y)_P$ is the $\mathcal{O}_{\mathfrak{X},P}$-module with generators $\{dz_0/z_0, \ldots, dz_k/z_k, dz_{k+1}, \ldots dz_n\}$ satisfying the relation $\sum_{i=0}^{k} e_i dz_i/z_i = 0$. Thus, $\Omega^1_{\mathfrak{X}/\Delta}(\log Y)$ is a locally free sheaf of rank $n = \dim \mathfrak{X} - 1$ endowed with differential d given by the composite

$$d : \mathcal{O}_{\mathfrak{X}} \to \Omega^1_{\mathfrak{X}} \hookrightarrow \Omega^1_{\mathfrak{X}}(\log Y) \to \Omega^1_{\mathfrak{X}/\Delta}(\log Y),$$

where $\Omega^1_{\mathfrak{X}}(\log Y)$ is the free sheaf of $\mathcal{O}_{\mathfrak{X}}$-modules on the same generators of $\Omega^1_{\mathfrak{X}/\Delta}(\log Y)$.

The relative hypercohomology sheaves $\mathbb{R}^m f_* \Omega^{\cdot}_{\mathfrak{X}/\Delta}(\log Y)$ are locally free $\mathcal{O}_{\Delta}$-modules of finite rank ([**28**]). This follows from the fact that the restriction $f : \mathfrak{X}^* \to \Delta^*$, which is a smooth fiber bundle, determines, for all $s \in \Delta^*$, a canonical isomorphism of complexes

$$\Omega^{\cdot}_{\mathfrak{X}/\Delta}(\log Y) \otimes_{\mathcal{O}_{\mathfrak{X}}} \mathcal{O}_{\mathfrak{X}_s} \xrightarrow{\simeq} \Omega^{\cdot}_{\mathfrak{X}_s}$$

on $\mathfrak{X}_s = f^{-1}(s)$. This implies that $\Omega^{\cdot}_{\mathfrak{X}/\Delta}(\log Y) \otimes_{\mathcal{O}_{\mathfrak{X}}} \mathcal{O}_{\mathfrak{X}_s}$ is a resolution of the constant sheaf $\mathbb{C}$ on $\mathfrak{X}_s$, hence

$$\mathbb{H}^m(\mathfrak{X}_s, \Omega^{\cdot}_{\mathfrak{X}/\Delta}(\log Y) \otimes_{\mathcal{O}_{\mathfrak{X}}} \mathcal{O}_{\mathfrak{X}_s}) \simeq H^m(\mathfrak{X}_s, \mathbb{C}), \quad \forall s \in \Delta^*,$$

hence the complex dimension of $\mathbb{H}^m(\mathfrak{X}_s, \Omega^{\cdot}_{\mathfrak{X}/\Delta}(\log Y) \otimes_{\mathcal{O}_{\mathfrak{X}}} \mathcal{O}_{\mathfrak{X}_s})$ is a locally constant function. One obtains

$$(4.1) \qquad \mathbb{R}^m f_* \Omega^{\cdot}_{\mathfrak{X}/\Delta}(\log Y) \otimes_{\mathcal{O}_{\Delta^*}} k(s) \simeq H^m(\mathfrak{X}_s, \mathbb{C}), \quad \forall s \in \Delta^*,$$

so that $\mathbb{R}^m f_* \Omega^{\cdot}_{\mathfrak{X}/\Delta}(\log Y)$ is a locally free $\mathcal{O}_{\Delta^*}$-module of finite rank. Moreover, if Δ is a small disk, there exists an isomorphism

$$(4.2) \qquad H^m(\tilde{\mathfrak{X}}^*, \mathbb{C}) \xrightarrow{\simeq} \mathbb{H}^m(Y, \Omega^{\cdot}_{\mathfrak{X}/\Delta}(\log Y) \otimes_{\mathcal{O}_{\mathfrak{X}}} \mathcal{O}_Y)$$

where $\tilde{\mathfrak{X}}^* = \mathfrak{X} \times_{\Delta} \tilde{\Delta}^*$ and $\tilde{\Delta}^* \to \Delta^*$ is the universal covering space of $\Delta^* = \Delta \setminus \{0\}$, so that $\dim \mathbb{H}^m(\mathfrak{X}_s, \Omega^{\cdot}_{\mathfrak{X}/\Delta}(\log Y) \otimes_{\mathcal{O}_{\mathfrak{X}}} \mathcal{O}_{\mathfrak{X}_s})$ is locally constant on Δ.

A stronger result ([**28**], §9) shows that the Gauss–Manin connection

$$(4.3) \qquad \nabla : \mathbb{R}^m f_* \Omega^{\cdot}_{\mathfrak{X}/\Delta}(\log Y) \to \Omega^1_{\Delta}(\log 0) \otimes_{\mathcal{O}_{\Delta}} \mathbb{R}^m f_* \Omega^{\cdot}_{\mathfrak{X}/\Delta}(\log Y)$$

has logarithmic singularities at $0 \in \Delta$ and in case of an algebraic morphism f admits an algebraic description. In particular the residue of ∇ at zero is a well defined operator

$$(4.4) \qquad N := \mathrm{Res}_0(\nabla)$$

in fact it is an endomorphism of $H^m(\tilde{\mathfrak{X}}^*, \mathbb{C})$. The eigenvalues of $\mathrm{Res}_0(\nabla)$ are rational numbers α with $0 \leq \alpha < 1$. The monodromy transformation T induces an automorphism T_0 of (4.1) and it can be shown that

$$(4.5) \qquad T_0 = \exp(-2\pi\sqrt{-1}\,\mathrm{Res}_0(\nabla)).$$

It follows that the eigenvalues of T_0 are roots of unity, so that a power T_0^d is unipotent. In fact, up to a base change $\Delta \to \Delta$, $z \mapsto z^d$, one can assume that T_0 is already unipotent, this means that the residue (4.4) is nilpotent.

There are two important filtrations on the cohomology $H^m(\tilde{\mathfrak{X}}^*, \mathbb{C})$. One is the Hodge filtration $F^p H^m(\tilde{\mathfrak{X}}^*, \mathbb{C})$ determined by the isomorphism (4.2) and the 'naive filtration' on $\Omega^{\cdot}_{\mathfrak{X}/\Delta}(\log Y) \otimes_{\mathcal{O}_{\mathfrak{X}}} \mathcal{O}_Y$. The other is the Picard–Lefschetz filtration

$L_\ell H^m(\tilde{\mathfrak{X}}^*, \mathbb{C})$, $\ell \in \mathbb{Z}$, which is a canonical, finite, increasing filtration associated to the map N and defined by induction. The properties and behavior of this filtration are a priori rather mysterious as its definition is given via an "indirect method". The main result in [**28**] shows that the data $(H^\cdot(\tilde{\mathfrak{X}}^*, \mathbb{C}), L_\cdot, F^\cdot)$ determine a mixed $\mathbb{Q}$-Hodge structure on $H^\cdot(\tilde{\mathfrak{X}}^*, \mathbb{Q})$.

This result is obtained by studying "explicitly" the Picard–Lefschetz filtration $L_\cdot$ on a resolution of the complex of sheaves $\Omega^\cdot_{\mathfrak{X}/\Delta}(\log Y) \otimes_{\mathcal{O}_{\mathfrak{X}}} \mathcal{O}_{Y^{\mathrm{red}}}$ of the form

$$(4.6) \qquad A^{s,k} := \Omega^{s+k+1}_{\mathfrak{X}}(\log Y)/W_k \Omega^{s+k+1}_{\mathfrak{X}}(\log Y), \qquad s, k \in \mathbb{Z}_{\geq 0},$$

where the weight filtration $W_\cdot$ on $\Omega^\cdot_{\mathfrak{X}}(\log Y)$ is defined as

$$W_k \Omega^m_{\mathfrak{X}}(\log Y) := \Omega^k_{\mathfrak{X}}(\log Y) \wedge \Omega^{m-k}_{\mathfrak{X}},$$

and the differentials

$$(4.7) \qquad d' : A^{s,k} \to A^{s+1,k}, \qquad d'' : A^{s,k} \to A^{s,k+1}.$$

on (4.6) are the usual differential d' on $\Omega^\cdot_{\mathfrak{X}}(\log Y)$ and

$$(4.8) \qquad d''(\alpha) = (-1)^s \alpha \wedge \theta, \quad \text{for} \quad \theta = f^*\left(\frac{dt}{t}\right).$$

Notice that θ can be seen as an element of

$$(4.9) \qquad H^1(\tilde{\mathfrak{X}}^*, \mathbb{Q})(1) = 2\pi\sqrt{-1}\, H^1(\tilde{\mathfrak{X}}^*, \mathbb{Q}),$$

because the form dt/t on Δ^* has period $2\pi\sqrt{-1}$, so that

$$(4.10) \qquad (2\pi\sqrt{-1})^{-1} dt/t \in H^1(\Delta^*, \mathbb{Z}) \subset H^1(\Delta^*, \mathbb{Q}).$$

Wedging with θ provides an injective map

$$\wedge\theta : \Omega^m_{\mathfrak{X}/\Delta}(\log Y) \hookrightarrow \Omega^{m+1}_{\mathfrak{X}}(\log Y)$$

and an induced morphism of complexes of sheaves $\phi : \Omega^\cdot_{\mathfrak{X}/\Delta}(\log Y) \otimes_{\mathcal{O}_{\mathfrak{X}}} \mathcal{O}_{Y^{\mathrm{red}}} \to A^\cdot$, which defines a resolution of the unipotent factor of the complex of nearby cycles, with $(A^\cdot, \delta = d' + d'')$ the total complex of (4.6). The endomorphism $\nu^{s,k} : A^{s,k} \to A^{s-1,k+1}$ given by the natural projection on forms is non-trivial because of the presence of the cutoff by the weight filtration W_k on $\Omega^m_{\mathfrak{X}}(\log Y)$ and it plays a central role in this theory as it describes the local monodromy map on the resolution $A^\cdot$.

It can be shown (see [**18**]) that the map induced in hypercohomology by

$$(4.11) \qquad \tilde{\nu} = (-1)^s \nu^{s,k} : A^{s,k} \to A^{s-1,k+1},$$

satisfying $\tilde{\nu}\delta + \delta\tilde{\nu} = 0$, is the residue (4.4) of the Gauss–Manin connection,

$$N : H^m(\tilde{\mathfrak{X}}^*, \mathbb{C}) \to H^m(\tilde{\mathfrak{X}}^*, \mathbb{C}).$$

More precisely, N is obtained as the connecting homomorphism in the long exact sequence of hypercohomology associated to the exact sequence of complexes of sheaves on Y

$$0 \to A^\cdot[-1] \xrightarrow{\epsilon} \mathrm{Cone}^\cdot(\tilde{\nu}) \xrightarrow{\eta} A^\cdot \to 0.$$

The complex $(\mathrm{Cone}^\cdot(\tilde{\nu}), D)$, $D = (\delta + \tilde{\nu}, \delta)$ is quasi-isomorphic to $\Omega^\cdot_{\mathfrak{X}}(\log Y) \otimes_{\mathcal{O}_{\mathfrak{X}}} \mathcal{O}_{Y^{\mathrm{red}}}$. The map $\tilde{\nu}$ measures the difference between differentiation in $\mathrm{Cone}^\cdot(\tilde{\nu})$ and in $A^\cdot$ which appears when one considers the section of $\eta: A^{s,k} \to \mathrm{Cone}^{s,k}(\tilde{\nu}) = A^{s-1,k} \oplus A^{s,k}$.

4.2. Nearby cycles at arithmetic infinity. The main idea that motivates the definition of the complex (3.1) is to "transfer" these results to the archimedean setting, where one deals with a smooth, projective algebraic variety X over $\mathbb{C}$ or $\mathbb{R}$, interpreted as the generic fiber of a degeneration "around" infinity. The main aspects of the above construction that one wishes to retain are the fact that the cutoff by W_k on $\Omega^m_{\mathfrak{X}}(\log Y)$ is introduced because the complex of the nearby cycles is the restriction to Y of the complex $\Omega^m_{\mathfrak{X}/\Delta}(\log Y)$. It is the presence of this cutoff that makes the morphism $\nu^{s,k}$ non-trivial on $A^{s,k}$. Moreover, another essential observation is that translates of the weight filtration

$$(4.12) \qquad L_\ell A^{s,k} = W_{2k+\ell+1}\Omega^{s+k+1}_{\mathfrak{X}}(\log Y)/W_k\Omega^{s+k+1}_{\mathfrak{X}}(\log Y); \qquad \ell \in \mathbb{Z}$$

and the corresponding graded spaces

$$(4.13) \qquad gr^L_\ell A^{s,k} = \begin{cases} gr^W_{2k+\ell+1}\Omega^{s+k+1}_{\mathfrak{X}}(\log Y) & \ell + k \geq 0 \\ 0 & \ell + k < 0 \end{cases}$$

describe the strata of the special fiber Y, through the Poincaré residue map

$$(4.14) \qquad \mathrm{Res} : W_k\Omega^m_{\mathfrak{X}}(\log Y) \to (a_k)_*\Omega^{m-k}_{\tilde{Y}^{(k)}},$$

where

$$\tilde{Y}^{(k)} := \coprod_{1\leq i_1<\ldots<i_k\leq M} Y_{i_1} \cap \ldots \cap Y_{i_k}$$

is the k-th stratum of $Y = Y_1 \cup \ldots \cup Y_M$ and $(a_k)_* : \tilde{Y}^{(k)} \to \mathfrak{X}$ is the canonical projection. The Poincaré residue (4.14) is given by

$$(4.15) \quad \mathrm{Res}\left(\sum_{1\leq i_1<\ldots<i_k\leq K} \omega_{i_1\ldots i_k}\frac{dz_{i_1}}{z_{i_1}} \wedge \ldots \wedge \frac{dz_{i_k}}{z_{i_k}}\right) = \sum_{1\leq i_1<\ldots<i_k\leq K} \mathrm{res}(\omega_{i_1\ldots i_k}),$$

where $z_1 \cdots z_K = 0$ is the local description of Y^{red} (the closed subset Y of $\mathfrak{X}$ with its reduced scheme structure), $\omega_{i_1\ldots i_k}$ is a section of $\Omega^{m-k}_{\mathfrak{X}}$, and res $: \Omega^{m-k}_{\mathfrak{X}} \to (a_k)_*\Omega^{m-k}_{\tilde{Y}^{(k)}}$ is the restriction to the stratum $\tilde{Y}^{(k)}$.

This means that it is sufficient to provide a version at arithmetic infinity of the weight filtration $W.$ and the Picard–Lefschetz filtration $L.$, as these are sufficient to characterize the geometry of the singular fiber $Y = f^{-1}(0)$, which is strictly related to the behavior of the monodromy map.

Notice that the period $2\pi\sqrt{-1}$ of the form dt/t on Δ^* (cf. (4.8), (4.9), (4.10)) corresponds to a Tate twist on the (rational) cohomology of the generic fiber $\tilde{\mathfrak{X}}^*$. It is important to stress the fact that this 'detects' the presence of the singular fiber through an operation that does not involve Y explicitly, and therefore can be transported at arithmetic infinity (where the description of the fiber 'over infinity' is still mysterious) on a complex of real differential forms. Moreover, the fact that the cutoff by W_k on $\Omega^m_{\mathfrak{X}}(\log Y)$ implies the non-triviality of the monodromy map $\nu^{s,k}$ suggests a definition of the monodromy operator N "at infinity" in terms of an analogous weight filtration on a complex of Tate–twisted real differential forms.

By *weight* of a real m-form

$$\alpha \in \bigoplus_{p+q=m} (\Omega^{p,q}_X + \Omega^{q,p}_X)_{\mathbb{R}}$$

on X we mean the non-negative integer $|p-q|$. At arithmetic infinity, the operations of taking the residues and hence considering holomorphic differential forms on the strata of Y can be rephrased in a form suitable to be included in the definition of the complex $K^{\cdot}$ of (3.1).

In fact, one should interpret the archimedean complex (3.1) as a "filtered copy" of $A^{s,k}$, with an additional condition characterizing the graded pieces $gr^L A^{\cdot}$. In fact, in the case of the complex of the nearby cycles, there is a graded isomorphism

$$(4.16) \qquad gr_\ell^L A^{\cdot} \simeq \bigoplus_{k \geq \max(0,-\ell)} (a_{2k+\ell+1})_* \Omega_{\tilde{Y}(2k+\ell+1)}^{\cdot+1}[-\ell - 2k - 1]$$

under the condition $\ell + k \geq 0$. Furthermore, in $gr_\ell^L A^{\cdot}$ the second differential is trivial, $d'' = 0$. The induced *weights spectral sequence*

$$(4.17) \qquad E_1^{i,m-i} = \bigoplus_{k \geq \max(0,i)} H^{m+i-2k}(\tilde{Y}^{(2k-i+1)}, \mathbb{Q})(i - k) \Rightarrow H^m(\tilde{\mathfrak{X}}^*, \mathbb{Q})$$

degenerates at the E_2 term. The $E_1^{i,m-i}$ term is a pure Hodge structure of weight $m - i$. A major result in the theory shows that the filtration induced on the abutment coincides with the Picard-Lefschetz filtration.

At arithmetic infinity, in terms of the complex (3.1), for

$$K^{i,j} = \bigoplus_{k \geq \max\{0,i\}} K^{i,j,k},$$

the terms
(4.18)

$$K^{i,m-n,k} = \begin{cases} \displaystyle\bigoplus_{\substack{p+q=m \\ |p-q| \leq 2k-i}} (\Omega_X^{p,q} \oplus \Omega_X^{q,p})_{\mathbb{R}}(\frac{m-i}{2}) & \text{if } m - i \equiv 0(2), \ k \geq \max(0,i) \\ 0 & \text{otherwise} \end{cases}$$

give the archimedean analog, at the level of the real differential forms, of the $E_1^{i,m-i}$-term of the spectral sequence (4.17). At arithmetic infinity the weight is $i - m$.

These analogies suggest a geometric interpretation of the indices involved in the definition of (3.1). The first index is associated to the ℓ-th piece of an "archimedean monodromy filtration" L_ℓ on $\oplus_{p+q=m}(\Omega_X^{p,q} \oplus \Omega_X^{q,p})_{\mathbb{R}}$, with $\ell = -i$. This explains our previous comment that the archimedean complex should be thought of as a filtered copy of $A^{s,k}$. The cutoffs $|p - q| \leq 2k - i$ and $k \geq \max\{0,i\}$ correspond to considering the filtered piece $L_\ell A^{s,k} = W_{2k+\ell+1} A^{s,k}$ in Steenbrink's theory with the cutoff by W_k on $\Omega_X^{s+k+1}(\log Y)$, justified by restricting (relative) differential forms to the special fiber. In this identification the third index k of (3.1) plays the role of the index k of the anti-holomorphic forms in the double complex (4.6). The archimedean theory is a weighted "even theory" since $i - m = 2r$. The second index $j = m - n$ detects the total degree m of the differential forms.

Finally, we remark that there is a fundamental difference in the definition of the differentials (3.2) (3.3) in the complex (3.1) at infinity and their geometric analogs (4.7). In fact, while the differential d' is similar in both theories, in the geometric case, the action of d'' by the wedging with the form $\theta = f^*(dt/t)$ involves a Tate twist on the rational version of the complex $A^{s,k}$, the differential d'' in the archimedean case does not involve any twist.

5. Monodromy and the renormalization group

In the construction at arithmetic infinity we obtain an analog of the formula (4.4) for the logarithm of the monodromy N as the residue of a connection, in terms of a Birkhoff decomposition of loops and a Riemann–Hilbert problem analogous to those underlying the theory of renormalization in QFT, as developed by Connes and Kreimer ([7] [5], also [8]).

In our case, the Birkhoff decomposition will take place in the group G of automorphisms of the complex $(\mathcal{T}_{\mathbb{C}}^{\cdot}, \delta)$, where $\mathcal{T}_{\mathbb{C}}^{\cdot}$ is the complexification of $\mathcal{T}^{\cdot}$. The Birkhoff decomposition will determine a one parameter family of principal G-bundles $\mathcal{P}_\mu$ on $\mathbb{P}^1(\mathbb{C})$, with trivializations

$$(5.1) \qquad \phi_\mu(z) = \phi_\mu^-(z)^{-1}\, \phi_\mu^+(z), \quad z \in \partial\Delta \subset \mathbb{P}^1(\mathbb{C}), \quad \mu \in \mathbb{C}^*,$$

where ϕ_μ^+ is a holomorphic function on a disk Δ around $z = 0$ and ϕ_μ^- is holomorphic on $\mathbb{P}^1(\mathbb{C}) \smallsetminus \Delta$, normalized by $\phi_\mu^-(\infty) = 1$. The parameter μ is related to a scaling action by $\mathbb{R}_+^*$, by $\mu \mapsto \lambda\mu$. We shall construct the data (5.1) in such a way that the negative part of the Birkhoff decomposition $\phi_\mu^- = \phi^-$ is in fact independent of μ, as in the theory of renormalization. As part of the data, one also considers a one parameter group of automorphisms θ, and the corresponding infinitesimal generator $\Upsilon = \frac{d}{dt}\theta_t|_{t=0}$, so that

$$(5.2) \qquad \phi_{\lambda\mu}(\epsilon) = \theta_{t\epsilon}\,\phi_\mu(\epsilon), \quad \forall \lambda = e^t \in \mathbb{R}_+^*, \quad \epsilon \in \partial\Delta.$$

The corresponding *renormalization group* is the one parameter group

$$(5.3) \qquad \rho(\lambda) = \lim_{\epsilon \to 0} \phi^-(\epsilon)\, \theta_{t\epsilon}(\phi^-(\epsilon)^{-1}), \quad \forall \lambda = e^t \in \mathbb{R}_+^*.$$

Following [7], one can write $\phi^-(z)$ in the form

$$(5.4) \qquad \phi^-(z)^{-1} = 1 + \sum_{k=1}^\infty \frac{d_k}{z^k},$$

with coefficients

$$(5.5) \qquad d_k = \int_{s_1 \geq \cdots \geq s_k \geq 0} \theta_{-s_1}(\beta) \cdots \theta_{-s_k}(\beta)\, ds_1 \cdots ds_k.$$

Here β is the beta-function of renormalization, related to the residue at zero of ϕ by

$$(5.6) \qquad \beta = \Upsilon \operatorname{Res} \phi,$$

where Υ is the generator of θ_t and the residue is defined as

$$(5.7) \qquad \operatorname{Res} \phi = \frac{d}{dz}\left(\phi^-(1/z)^{-1}\right)|_{z=0}.$$

In the construction at arithmetic infinity, the grading operator Φ induces a time evolution on the complex $(\mathcal{T}_{\mathbb{C}}^{\cdot}, \delta)$ given by the "Frobenius flow" (2.28),

$$(5.8) \qquad F_t = e^{t\Phi}, \quad t \in \mathbb{R},$$

and we denote by θ_t the induced time evolution on $\operatorname{End}(\mathcal{T}_{\mathbb{C}}^{\cdot}, \delta)$,

$$(5.9) \qquad \theta_t(a) = e^{-t\Phi}\, a\, e^{t\Phi},$$

with the corresponding Υ given by

$$(5.10) \qquad \Upsilon(a) = \frac{d}{dt}\theta_t(a)|_{t=0} = [a, \Phi].$$

The analogy with the complex of nearby cycles of a geometric degeneration over a disk suggests to make the following prescription for the residue of ϕ:

$$(5.11) \qquad\qquad\qquad \text{Res}\,\phi = N.$$

As in the case of Connes–Kreimer, the residue uniquely determines ϕ^- via (5.5) and (5.4). We obtain the following result.

THEOREM 5.1. *There is a unique holomorphic map $\phi^- : \mathbb{P}^1(\mathbb{C}) \setminus \{0\} \to \text{Aut}(\mathcal{T}_{\mathbb{C}})$ satisfying (5.4), with coefficients (5.5) and residue (5.11), and it is of the form*

$$(5.12) \qquad\qquad\qquad \phi^-(z) = \exp(-N/z).$$

PROOF. First notice that the time evolution (5.9) satisfies

$$(5.13) \qquad\qquad\qquad \theta_t(N) = e^t\,N.$$

Moreover, by (5.10) and (5.11), we have

$$(5.14) \qquad\qquad\qquad \beta = [N, \Phi] = N$$

Thus, we can write the coefficients d_k of (5.5) in the form

$$d_k = N^k \int_{s_1 \geq \cdots \geq s_k \geq 0} e^{-(s_1 + \cdots + s_k)}\, ds_1 \cdots ds_k.$$

It is easy to see by induction that

$$u_k(t) := \int_0^t \int_0^{s_1} \cdots \int_0^{s_{k-1}} e^{-(s_1 + \cdots + s_k)}\, ds_1 \cdots ds_k = \frac{(1 - e^{-t})^k}{k!},$$

satisfying the recursion $u'_{k+1}(t) = e^{-t} u_k(t)$, so that

$$\int_{s_1 \geq \cdots \geq s_k \geq 0} e^{-(s_1 + \cdots + s_k)}\, ds_1 \cdots ds_k = \frac{1}{k!}.$$

This implies that $d_k = N^k/k!$, hence we obtain that the series (5.4) is just

$$\phi^-(z)^{-1} = \sum_{k=0}^{\infty} \frac{z^{-k}}{k!} N^k,$$

and $\phi^-(z) = \exp(-N/z)$. $\qquad\qquad\qquad\qquad\qquad\qquad\qquad\qquad\square$

Correspondingly, we see that the renormalization group (5.3) is given by

$$(5.15) \qquad\qquad \rho(\lambda) = \lambda^N = \exp(tN), \qquad \forall \lambda = e^t \in \mathbb{R}_+^*,$$

since we have

$$(5.16) \qquad\qquad \theta_{t\epsilon}(\phi^-(\epsilon)) = \exp\left(-\frac{\lambda^\epsilon}{\epsilon} N\right) \qquad \forall \lambda = e^t \in \mathbb{R}_+^*.$$

We now show that the other term of the Birkhoff decomposition (5.1) is determined by the requirement (5.2) of compatibility between the scaling and the time evolution.

THEOREM 5.2. *Consider the holomorphic map $\phi_\mu^+ : \mathbb{P}^1(\mathbb{C}) \setminus \{\infty\} \to \text{Aut}(\mathcal{T}_{\mathbb{C}})$ given by*

$$(5.17) \qquad\qquad\qquad \phi_\mu^+(z) = \exp\left(\frac{\mu^z - 1}{z} N\right).$$

The loop $\phi_\mu(z) = \phi^-(z)^{-1}\phi_\mu^+(z)$ with $\phi^-(z) = \exp(-N/z)$ and ϕ_μ^+ as in (5.17) satisfies the relation (5.2).

PROOF. First notice that (5.17) is indeed holomorphic at $z = 0$ with $\phi_\mu^+(0) = \exp(\log(\mu)N) = \mu^N$. By (5.1) and (5.16), the relation (5.2) is equivalent to requiring that, for all $\lambda = e^t \in \mathbb{R}_+^*$, the function ϕ_μ^+ satisfies

$$(5.18) \qquad \phi_{\lambda\mu}^+(\epsilon) = \exp\left(\frac{\lambda^\epsilon - 1}{\epsilon} N\right) \theta_{t\epsilon}(\phi_\mu^+(\epsilon)).$$

For ϕ_μ^+ as in (5.17) we have

$$\theta_{t\epsilon}(\phi_\mu^+(\epsilon)) = \exp\left(\frac{\mu^\epsilon - 1}{\epsilon}\lambda^\epsilon N\right) = \exp\left(\frac{(\lambda\mu)^\epsilon - 1}{\epsilon} N\right)\exp\left(\frac{1 - \lambda^\epsilon}{\epsilon} N\right),$$

so that (5.18) is satisfied. $\qquad\square$

Notice that, via the representation (2.24), it is possible to lift the Birkhoff decomposition (5.1) to a Birkhoff decomposition of the form

$$(5.19) \qquad g_\mu(z) = g^-(z)^{-1} g_\mu^+(z),$$

where, with the notation of (2.22), we have

$$g^-(z) = u(1/z), \quad g_\mu(z) = u(\mu^z/z).$$

The renormalization group (5.15) then becomes simply the horocycle flow

$$(5.20) \qquad \rho(\lambda) = u(t) = \begin{pmatrix} 1 & t \\ 0 & 1 \end{pmatrix}.$$

There is a Riemann–Hilbert problem associated to the Birkhoff decomposition considered in the theory of renormalization ([8]). Namely, for γ a generator of the fundamental group $\pi_1(\Delta^*) = \mathbb{Z}$ of the punctured disk, consider a complex linear representation $\pi : \mathbb{Z} \to G$. Under the assumption that the eigenvalues of $\pi(\gamma)$ satisfy

$$0 \le \mathrm{Re}\,\frac{\lambda}{2\pi\sqrt{-1}} < 1,$$

we can take the logarithm

$$(5.21) \qquad \frac{1}{2\pi\sqrt{-1}} \log \pi(\gamma).$$

By the Riemann–Hilbert correspondence ([1]), a representation $\pi : \mathbb{Z} \to G$ determines a bundle with connection $(\mathcal{E}, \nabla)$, where the Fuchsian connection ∇ has local gauge potential on the disk Δ of the form

$$(5.22) \qquad -\phi^+(z)^{-1}\frac{\log \pi(\gamma)\,dz}{z}\phi^+(z) + \phi^+(z)^{-1}\,d\phi^+(z),$$

with $\phi^+(z)$ the local trivialization of $\mathcal{E}$ over Δ. The data $(\mathcal{E}, \nabla)$ correspond to a linear differential system $f'(z) = A(z)f(z)$, with $\nabla f = df - A(z)f\,dz$, for sections $f \in \Gamma(\Delta^*, \mathcal{E})$.

In the case of the cohomological theory at arithmetic infinity, this amounts to a vector bundle $\mathcal{E}_\mu$ over $\mathbb{P}^1(\mathbb{C})$, with fiber $\mathcal{T}_\mathbb{C}$, associated to the principal G-bundles $\mathcal{P}_\mu$, with transition function the loop $\phi_\mu(z)$ and local trivializations given by the $\phi^-(z)$ and $\phi_\mu^+(z)$. We use the representation specified by

$$(5.23) \qquad \pi(\gamma) := \exp(-2\pi\sqrt{-1}\,N),$$

which is the analog of (4.5). This determines a one parameter family $(\mathcal{E}_\mu, \nabla_\mu)$ of linear differential systems over the disk Δ, of the form

$$(5.24) \qquad\qquad \nabla_\mu : \mathcal{E}_\mu^{\cdot} \to \mathcal{E}_\mu^{\cdot} \otimes_{\mathcal{O}_\Delta} \Omega_\Delta^{\cdot}(\log 0),$$

where the connection on the restriction of $\mathcal{E}_\mu$ over Δ is given by

$$(5.25) \qquad\qquad \nabla_\mu = N \left(\frac{1}{z} + \frac{d}{dz} \frac{\mu^z - 1}{z} \right) dz.$$

Using the induced representation (2.24) in cohomology, this determines a corresponding linear differential system on the bundle of hypercohomologies $\mathbb{H}^{\cdot}(\mathcal{E}_\mu^{\cdot})$ with fiber $\mathbb{H}^{\cdot}(\mathcal{T}_{\mathbb{C}}^{\cdot})$,

$$(5.26) \qquad\qquad \nabla_\mu : \mathbb{H}^{\cdot}(\mathcal{E}_\mu^{\cdot}) \to \mathbb{H}^{\cdot}(\mathcal{E}_\mu^{\cdot}) \otimes_{\mathcal{O}_\Delta} \Omega_\Delta^{\cdot}(\log 0),$$

which is the analog of the Gauss–Manin connection (4.3) in the geometric case. The connections (5.25) form an isomonodromic family, with

$$(5.27) \qquad\qquad \mathrm{Res}_{z=0} \nabla_\mu = N.$$

6. Rees sheaves at arithmetic infinity

In the description (3.15) ([**11**]) of the archimedean factor of the Hasse-Weil L-function of the "motive" $H^m(X)$, for X a smooth projective variety over a number field, the definition of the archimedean cohomology $H_{ar}^m(X)$ is motivated by previous work of J. M. Fontaine and it is expressed in terms of an additive functor $\mathbb{D}$ (derivation) from the (abelian) category of pure Hodge structures over $\kappa = \mathbb{C}, \mathbb{R}$ to the additive category whose objects are free modules of finite rank over the $\mathbb{R}$-algebra of polynomials in one variable endowed with a $\mathbb{R}$-linear endomorphism and satisfying certain properties. More precisely one sets

$$H_{ar}^m(X) = \begin{cases} Fil^0(H_B^m(X, \mathbb{C}) \otimes_{\mathbb{C}} \mathbb{C}[z^{\pm 1}])^{c=id} & \text{if } \kappa = \mathbb{C} \\ Fil^0(H_B^m(X, \mathbb{C}) \otimes_{\mathbb{C}} \mathbb{C}[z^{\pm 1}])^{c=id, F_\infty = id} & \text{if } \kappa = \mathbb{R}. \end{cases}$$

Here, Fil^q denotes the filtration on the tensor product $H_B^m(X(\mathbb{C}), \mathbb{C}) \otimes_{\mathbb{C}} \mathbb{C}[z^{\pm 1}]$, which is obtained from the Hodge filtration $F^{\cdot}$ on the Betti cohomology $H_B^m(X) := H_B^m(X(\mathbb{C}), \mathbb{C})$ and the one on the ring of Laurent polynomials $\mathbb{C}[z^{\pm 1}]$ given by $F^q \mathbb{C}[z^{\pm 1}] := z^{-q} \mathbb{C}[z^{-1}]$, $\forall q \in \mathbb{Z}$. By c one denotes the conjugate linear involution (complex conjugation) and F_∞ is a $\mathbb{C}$-linear involution (the infinite Frobenius).

The expectation is that one should obtain a description of the archimedean cohomology together with the linear "Frobenius flow" generated by Φ directly by some natural homological construction on a suitable non-linear dynamical system. In this direction, a more geometric construction of the archimedean cohomology was given in [**13**] (*cf.* par. 3), where it was interpreted (for instance when $\kappa = \mathbb{C}$) as the space of global sections of a real analytic sheaf (Rees sheaf) $\zeta_{\mathbb{C}}^\omega(H^m(X), \gamma^{\cdot})$ over $\mathbb{R}$. A similar description holds when $\kappa = \mathbb{R}$. Here, as before, $\gamma^{\cdot}$ denotes the descending filtration $F^{\cdot} \cap \bar{F}^{\cdot}$ on $H^m(X^{\mathrm{an}}, \mathbb{R})$ endowed with its real Hodge structure. The locally-free sheaf $\zeta_{\mathbb{C}}^\omega(H^m(X), \gamma^{\cdot})$ has the remarkable description ([**13**] Thm. 4.4)

$$(6.1) \quad \zeta_{\mathbb{C}}^\omega(H^m(X), \gamma^{\cdot}) \simeq \mathrm{Ker} \left(\mathbb{R}^m \pi_*(\Omega_{X_{\mathbb{C}}^{\mathrm{an}} \times \mathbb{R}/\mathbb{R}}^{\cdot}, sd) \to (\mathbb{R}^m \pi_* \mathcal{DR}_{X/\mathbb{C}})/\mathcal{T}_{X/\mathbb{C}} \right)$$

$$H_{ar}^m(X) = \Gamma(\mathbb{R}, \zeta_{\mathbb{C}}^\omega(H^m(X), \gamma^{\cdot})) \quad \text{if} \quad \kappa = \mathbb{C}$$

where $\pi : X_{\mathbb{C}}^{\mathrm{an}} \times \mathbb{R} \to \mathbb{R}$ is the projection, s is a standard coordinate on $\mathbb{R}$, $\mathcal{A}_{\mathbb{R}}$ denotes the sheaf of real-analytic functions on the real analytic manifold $\mathbb{R}$, $\mathcal{DR}_{X/\mathbb{C}}$ is the cokernel of the natural inclusion of complexes of $\pi^{-1}\mathcal{A}_{\mathbb{R}}$-modules on $X_{\mathbb{C}}^{\mathrm{an}} \times \mathbb{R}$

$$\pi^{-1}\mathcal{A}_{\mathbb{R}} \hookrightarrow (\Omega^{\cdot}_{X_{\mathbb{C}}^{\mathrm{an}} \times \mathbb{R}/\mathbb{R}}, sd)$$

and $\mathcal{T}_{X/\mathbb{C}}$ is the $\mathcal{A}_{\mathbb{R}}$-torsion in $\mathbb{R}^m\pi_*\mathcal{DR}_{X/\mathbb{C}}$. $\Omega^{\cdot}_{X_{\mathbb{C}}^{\mathrm{an}} \times \mathbb{R}/\mathbb{R}}$ is the complex of $\mathbb{C}$-valued smooth relative differential forms on $X^{\mathrm{an}} \times \mathbb{R}/\mathbb{R}$ which are holomorphic in the X^{an}-coordinates and real analytic in the $\mathbb{R}$-variable. One considers the scaling flow on $\mathbb{R}$ given by the map $\phi_{\mathbb{C}}^t(s) = se^{-t}$. It induces an action of this group on the relative differential complex by means of: $\psi^t(\omega) = e^{t\deg\omega} \cdot (\mathrm{id} \times \phi_{\mathbb{C}}^t)^*\omega$. This action defines in turn a $\mathcal{A}_{\mathbb{R}}$-linear action on the complex of higher direct image sheaves on $\mathbb{R}$. A similar result holds when $\kappa = \mathbb{R}$: in this case one gets a $\mathcal{A}_{\mathbb{R}^{\geq 0}}$-action. In the description of $H_{\mathrm{ar}}^m(X)$ given in (6.1), the Hodge theoretic notions required in the definition of the archimedean cohomology have been replaced by using suitably deformed complexes of sheaves of modules on $\mathbb{R}$ (on $\mathbb{R}^{\geq 0}$ when $\kappa = \mathbb{R}$). The deformed complex of locally free sheaves of relative differentials $(\Omega^{\cdot}_{X_{\mathbb{C}} \times \mathbb{A}^1/\mathbb{A}^1}, zd)$, filtered by the Hodge filtration $F^{\cdot}$, for z coordinate on $\mathbb{A}^1$, was firstly studied by Simpson in [27]. He introduced the algebraic version $\zeta_{\mathbb{C}}(H^m(X^{\mathrm{an}}, \mathbb{C}), F^{\cdot})$ of the real analytic Rees sheaf and proved that $\zeta_{\mathbb{C}}(H^m(X^{\mathrm{an}}, \mathbb{C}), F^{\cdot}) \simeq \mathbb{R}^m\pi_*(\Omega^{\cdot}_{X_{\mathbb{C}} \times \mathbb{A}^1/\mathbb{A}^1}, zd)$. Following this viewpoint, (6.1) is the analogue of Simpson's formula for the non-algebraic filtration $\gamma^{\cdot}$. A very interesting fact ([13], §4) is that in the real analytic setting, the higher direct image sheaves fit into short exact sequences of coherent $\mathcal{A}_{\mathbb{R}}$-modules

$$(6.2) \qquad 0 \to \mathbb{R}^m\pi_*(\pi^{-1}\mathcal{A}_{\mathbb{R}}) \to \mathbb{R}^m\pi_*(\Omega^{\cdot}_{X_{\mathbb{C}}^{\mathrm{an}} \times \mathbb{R}/\mathbb{R}}, sd) \xrightarrow{\alpha} \mathbb{R}^m\pi_*\mathcal{DR}_{X/\mathbb{C}} \to 0$$

where $\mathbb{R}^m\pi_*(\pi^{-1}\mathcal{A}_{\mathbb{R}}) = H^m(X^{\mathrm{an}}, \mathbb{R}) \otimes \mathcal{A}_{\mathbb{R}}$. Here we like to think of the sheaf $\mathbb{R}^m\pi_*(\Omega^{\cdot}_{X_{\mathbb{C}}^{\mathrm{an}} \times \mathbb{R}/\mathbb{R}}, sd)$ as the archimedean real analytic analog of the relative analytic hypercohomology $\mathbb{R}^m f_*\Omega^{\cdot}_{\mathfrak{X}/\Delta}(\log Y)$ over a small disk Δ centered at the origin, whose algebraic description has been recalled in § 3.3. It turns out that $\zeta_{\mathbb{C}}^{\omega}(H^m(X), \gamma^{\cdot})$ is also canonically isomorphic to a twisted dual of $\mathbb{R}^m\pi_*\mathcal{DR}_{X/\mathbb{C}}$. More precisely, if $d, m \in \mathbb{Z}_{\geq 0}$ with $m + d = 2n$ ($n = \dim X$), then there are isomorphisms of $\mathcal{A}_{\mathbb{R}}$-modules ([13] Thm. 4.2)

$$(6.3) \qquad \zeta_{\mathbb{C}}^{\omega}(H^m(X), \gamma^{\cdot}) \simeq (2\pi\sqrt{-1})^{1-n}\mathcal{H}om_{\mathcal{A}_{\mathbb{R}}}(\mathbb{R}^d\pi_*\mathcal{DR}_{X/\mathbb{C}}, \mathcal{A}_{\mathbb{R}}(-n)).$$

Dualization detects the $\gamma^{\cdot}$-filtration from the Hodge filtration on $\mathbb{R}^d\pi_*\mathcal{DR}_{X/\mathbb{C}}$. From (6.1) and (6.3) one gets isomorphisms respecting the $\mathcal{A}_{\mathbb{R}}$-module structures and the flow

$$(6.4) \qquad \begin{aligned} \mathrm{Ker}\left(\mathbb{R}^m\pi_*(\Omega^{\cdot}_{X_{\mathbb{C}}^{\mathrm{an}} \times \mathbb{R}/\mathbb{R}}, sd) \to (\mathbb{R}^m\pi_*\mathcal{DR}_{X/\mathbb{C}})/\mathcal{T}_{X/\mathbb{C}}\right) \\ \simeq (2\pi\sqrt{-1})^{1-n}\mathcal{H}om_{\mathcal{A}_{\mathbb{R}}}(\mathbb{R}^d\pi_*\mathcal{DR}_{X/\mathbb{C}}, \mathcal{A}_{\mathbb{R}}(-n)). \end{aligned}$$

This statement is the dynamical sheaf-theoretic analog of the duality isomorphisms between the hypercohomology of the complexes $\mathrm{Ker}(N)^{\cdot}$ and $\mathrm{Coker}(N)^{\cdot}$ of vector spaces of [9] (Prop. 4.13) induced by powers of the 'local monodromy at arithmetic infinity', that is, of the duality S of (2.19).

This way, one can reinterpret the archimedean cohomology as the space of global sections on $\mathbb{R}$ ($\mathbb{R}^{\geq 0}$ for $\kappa = \mathbb{R}$) of the sheaf inverse image in $\mathbb{R}^m\pi_*(\Omega^{\cdot}_{X^{\mathrm{an}} \times \mathbb{R}/\mathbb{R}}, sd)$ of the maximal $\mathcal{A}_{\mathbb{R}}$-submodule of $\mathbb{R}^m\pi_*\mathcal{DR}_{X/\mathbb{C}}$ with support in $0 \in \mathbb{R}$. This statement is in accord with the classical description of the inertia invariants as the kernel of the local monodromy map N, viewed as the residue at zero of the Gauss-Manin connection

$$\mathrm{Ker}(\mathrm{Res}_0 \nabla : \mathbb{R}^m f_* \Omega^{\cdot}_{\mathfrak{X}/\Delta}(\log Y) \otimes_{\mathcal{O}_\Delta} k(0) \to \mathbb{R}^m f_* \Omega^{\cdot}_{\mathfrak{X}/\Delta}(\log Y) \otimes_{\mathcal{O}_\Delta} k(0))$$

cf. § 3.3 and (5.24), (5.25), (5.26) in the archimedean case.

In the archimedean setting (e.g. for $\kappa = \mathbb{C}$), the exact sequences

$$(6.5) \quad 0 \to \zeta_{\mathbb{C}}^{\omega}(H^m(X), \gamma) \to \mathbb{R}^m \pi_*(\Omega^{\cdot}_{X_{\mathbb{C}}^{\mathrm{an}} \times \mathbb{R}/\mathbb{R}}, sd) \to (\mathbb{R}^m \pi_* \mathcal{DR}_{X/\mathbb{C}})/\mathcal{T}_{X/\mathbb{C}} \to 0$$

are the sheaf theoretic analogs of the hypercohomology exact sequences associated to the nearby cycles complex as defined by Deligne in [26] (Exp. XIII, § 1.4: (1.4.2.2)). At arithmetic infinity, the hypercohomology of the complex of vanishing cycles is replaced by the hypercohomology sheaf $(\mathbb{R}^m \pi_* \mathcal{DR}_{X/\mathbb{C}})/\mathcal{T}_{X/\mathbb{C}}$, whereas the archimedean analog of the variation map in the formalism of the nearby cycles produces the duality isomorphisms (6.4).

The sequence (6.5) has also interesting analogies with the exact sequence of C_k-modules (C_k = idèle class group of a global field k)

$$(6.6) \qquad\qquad 0 \to L_\delta^2(X)_0 \xrightarrow{E} L_\delta^2(C_k) \to \mathcal{H} \to 0$$

studied by A. Connes in [4], §III (33). It is tempting to connect the role played by the space $L_\delta^2(C_k)$ to that of the cohomology $(\mathbb{R}^{\cdot}\pi_*(\Omega^{\cdot}_{X_{\mathbb{C}}^{\mathrm{an}} \times \mathbb{R}/\mathbb{R}}, sd))$, hence to recast the spaces $\mathcal{H}$ and $L_\delta^2(X)_0$ of Connes' theory, respectively, in the role of invariant and vanishing cycles. A connection with singularity theory in [4] is motivated by the "bad" behavior (at zero) of the action of k^* on the adeles space $\mathbb{A}_k$. This expected connection suggests, in turn, a singular behavior of $\overline{Spec(\mathbb{Z})}$ around infinity.

One of the interesting questions related to the archimedean cohomology and the archimedean factor is that of writing the logarithm $\log L_\kappa(H^m(X), s)$ of the regularized determinant (3.15) via a Lefschetz trace formula for the Frobenius operator. We shall return to these topics in future work.

7. "Arithmetic" spectral triples

In this section we present a refined version of the proposed construction of an "arithmetic spectral triple" in [10]. This version has the advantage that it holds at the level of differential forms and for X of any dimension. We first introduce natural subcomplexes and quotient complexes of $\mathcal{T}^{\cdot}$.

7.1. Inertia invariants and coinvariants.

We consider certain complexes of vector spaces related to the action of the endomorphism N on $(\mathcal{T}^{\cdot}, \delta)$. We introduce the notation $\mathcal{T}_\ell^{\cdot} \subset \mathcal{T}^{\cdot}$ for the $\mathbb{Z}$-graded linear subspace obtained as follows. For fixed (p, q) with $p+q = m$, let $(T_{p,q}^{m,*})_\ell \subset T_{p,q}^{m,*}$ be the $2\mathbb{Z}$-graded real vector space spanned by elements of the form $\alpha \otimes U^r \otimes \hbar^k$, with $\alpha \in (\Omega_X^{p,q} \oplus \Omega_X^{q,p})_{\mathbb{R}}$ and $(r, k) \in \Lambda_{p,q}$ lying on the line $k = 2r + m + \ell$ (cf. Figure 3). We let $\mathcal{T}_\ell^{\cdot} = \oplus_{p,q}(T_{p,q}^{m,*})_\ell$.

Each $\mathcal{T}_\ell^{\cdot}$ is a subcomplex with respect to the differential $d' = \hbar d$, while the second differential satisfies $d'' : \mathcal{T}_\ell^{\cdot} \to \mathcal{T}_{\ell-1}^{\cdot+1}$.

Similarly, we denote by $\check{\mathcal{T}}_\ell^{\cdot}$ the linear subspace of $\mathcal{T}^{\cdot}$, which is the direct sum of the subspaces $(\check{T}_{p,q}^{m,*})_\ell \subset T_{p,q}^{m,*}$ spanned by elements $\alpha \otimes U^r \otimes \hbar^k$, with $\alpha \in (\Omega_X^{p,q} \oplus \Omega_X^{q,p})_{\mathbb{R}}$

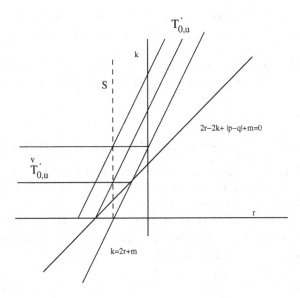

FIGURE 3. *The complexes* $\check{\mathcal{T}}_{0,u}^{\cdot}$ *and* $\mathcal{T}_{0,u}^{\cdot}$ *and the duality* S

and with $(r,k) \in \Lambda_{p,q}$ lying on the horizontal line $k = \ell$ (*cf.* Figure 3). Each $\check{\mathcal{T}}_{\ell}^{\cdot}$ is a subcomplex with respect to the differential d'', while $d' : \check{\mathcal{T}}_{\ell}^{\cdot} \to \check{\mathcal{T}}_{\ell+1}^{\cdot+1}$.

In particular, for $\ell = 0$ and $k = 2r + m \geq 0$, we obtain the subcomplex of "inertia invariants"

$$\mathcal{T}_0^{\cdot} := \mathrm{Ker}(N)^{\cdot}.$$

This complex agrees with the complex that computes the $\mathbb{H}^{\cdot}(\mathcal{T}^{\cdot}, \delta)^{N=0}$, for all $k = 2r + m > 0$, so that the map

$$\mathbb{H}^*(\mathrm{Ker}(N)^{\cdot}, d) \to \mathbb{H}^*(\mathcal{T}^{\cdot}, d)^{N=0}$$

is almost always a bijection.

Similarly, for $u \in \mathbb{N}$, one can consider subcomplexes $\mathcal{T}_{0,u}^{\cdot} \subset \mathcal{T}^{\cdot}$

$$(7.1) \qquad \mathcal{T}_{0,u}^{\cdot} := \oplus_{\ell=0}^{u} \mathcal{T}_{\ell}^{\cdot} = \mathrm{Ker}(N^{u+1})^{\cdot}.$$

These satisfy $\mathcal{T}^{\cdot} = \varinjlim_u \mathcal{T}_{0,u}^{\cdot}$. We also consider the quotient complexes

$$(7.2) \qquad \check{\mathcal{T}}_{0,u}^{\cdot} := \oplus_{\ell=0}^{u} \check{\mathcal{T}}_{\ell}^{\cdot} = \mathrm{Coker}(N^{u+1})^{\cdot},$$

where we denote by $\check{\delta}$ the induced differential on $\check{\mathcal{T}}_{0,u}^{\cdot}$. We obtain this way $\mathcal{T}^{\cdot} = \varprojlim_u \check{\mathcal{T}}_{0,u}^{\cdot}$. We call $\mathrm{Coker}(N)^{\cdot}$ the complex of "inertia coinvariant" and we refer to the $\mathrm{Ker}(N^{u+1})^{\cdot}$ and $\mathrm{Coker}(N^{u+1})^{\cdot}$ as higher inertia co/invariants.

7.2. Spectral triple. We consider the complex

$$(7.3) \qquad \mathcal{T}_u^{\cdot} = \mathcal{T}_{0,u}^{\cdot} \oplus \check{\mathcal{T}}_{0,u}^{\cdot}[+1],$$

with induced differential $\delta_u = \delta \oplus \check{\delta}$, where $(\mathcal{T}_{0,u}^{\cdot}, \delta)$ and $(\check{\mathcal{T}}_{0,u}^{\cdot}, \check{\delta})$ are the complexes of higher invariants and coinvariants of (7.1) and (7.2).

We still denote by Φ the linear operator on $\mathcal{T}_u^{\cdot}$ which agrees with the operator $\Phi = -U\partial_U$ on the subspaces $0 \oplus \check{\mathcal{T}}_{0,u}^{\cdot}$ and $\mathcal{T}_{0,u}^{\cdot} \oplus 0$, identified with the corresponding linear subspaces of $\mathcal{T}^{\cdot}$.

On the compact Kähler manifold X consider the operator $d+d^*$ on real forms $\Omega_{X,\mathbb{R}}^{\cdot}$. For the next result we need to assume the following conditions on the spectrum of $d+d^*$:

$$\#\{(r,\lambda) \in \mathbb{Z} \times \mathrm{Spec}(d+d^*) : t = r + \lambda\} < \infty \quad \forall t \in \mathbb{R}$$

(7.4)

$$\{r + \lambda : r \in \mathbb{Z}, \lambda \in \mathrm{Spec}(d+d^*)\} \subset \mathbb{R} \qquad \text{is discrete.}$$

THEOREM 7.1. *Let $U(\mathfrak{g})$ be the universal enveloping algebra of the Lie algebra $\mathfrak{g} = \mathrm{sl}(2,\mathbb{R})$. Let $\mathcal{H}^{\cdot}$ denote the completion of $\mathcal{T}_u^{\cdot}$ with respect to the inner product induced by (2.2). Let $\mathcal{D}$ be the linear operator $\mathcal{D} = \Phi + \delta_u + \delta_u^*$. Let $\mathcal{A} = C^\infty(X,\mathbb{R}) \otimes U(\mathfrak{g})$. If X has the property (7.4), then the data $(\mathcal{A}, \mathcal{H}^{\cdot}, \mathcal{D})$ satisfy the properties:*

- *The representation (2.23) determines an action of the algebra $\mathcal{A}$ by bounded operators on the Hilbert space $\mathcal{H}^{\cdot}$.*
- *The commutators $[\mathcal{D}, \sigma^L(a)]$, for $a \in \mathcal{A}$, are bounded operators on $\mathcal{H}^{\cdot}$.*
- *The operator $\mathcal{D}$ is a densely defined unbounded self-adjoint operator on $\mathcal{H}^{\cdot}$, such that $(1 + \mathcal{D}^2)^{-1}$ is a compact operator.*

PROOF. The representation σ^L of $\mathrm{SL}(2,\mathbb{R})$ on $\mathcal{T}^{\cdot}$ defined in (2.23) preserves the subspaces $\mathcal{T}_{0,u}^{\cdot}$ and $\check{\mathcal{T}}_{0,u}^{\cdot}$. In fact, the operator $\mathbb{L}$ changes $m \mapsto m+2$, $r \mapsto r-1$ and $k \mapsto k$, so that the constraint $k \leq u$ defining $\check{\mathcal{T}}_{0,u}^{\cdot}$ and the constraint $k = 2r+m+\ell$ with $0 \leq \ell \leq u$ defining $\mathcal{T}_{0,u}^{\cdot}$ are preserved by the action of $\mathbb{L}$. Similarly, the involution $\tilde{S}$ changes $m \mapsto 2n-m$, $r \mapsto r-(n-m)$ and $k \mapsto k$, so that again both constraints $k \leq u$ and $k = 2r+m+\ell$, for $0 \leq \ell \leq u$, are preserved. Thus, we can consider on $\mathcal{T}_u^{\cdot}$ the corresponding derived representation $d\sigma^L$ of the Lie algebra $\mathfrak{g} = sl(2,\mathbb{R})$,

$$d\sigma^L(v) = \frac{d}{ds}\sigma^L(\exp(sv))|_{s=0}.$$

Let $\{v_\pm, v_0\}$ be the basis of $\mathfrak{g}$ with $[v_0, v_+] = 2v_+$, $[v_0, v_-] = -2v_-$, and $[v_+, v_-] = v_0$. We have $d\sigma^L(v_+) = \mathbb{L}$, while $d\sigma^L(v_0)$ is the linear operator that multiplies elements $\alpha \otimes U^r \otimes \hbar^k$ with $\alpha \in \Omega_{X,\mathbb{R}}^m$ by $-n+m$. Thus, we obtain an action of the algebra $U(\mathfrak{g})$ on $\mathcal{H}^{\cdot}$ by bounded linear operators.

We have $[\sigma^L(\gamma), \delta_u] = [\sigma^L(\gamma), \delta_u^*] = 0$ hence $[\mathcal{D}, d\sigma^L(v)] = [\Phi, d\sigma^L(v)]$. Using $[\Phi, \mathbb{L}] = \mathbb{L}$ and $[\Phi, \tilde{S}] = (-n+m)\tilde{S}$, we obtain that $[\mathcal{D}, d\sigma^L(v)]$ is a bounded operator for all $v \in U(\mathfrak{g})$. The algebra of real valued smooth functions $C^\infty(X,\mathbb{R})$ acts on $\mathcal{T}^{\cdot}$ and on $\mathcal{T}_u^{\cdot}$ by the usual action on real forms $\Omega_{X,\mathbb{R}}^{\cdot}$. This action commutes with Φ so that $[\mathcal{D}, f] = [\delta_u + \delta_u^*, f]$, for all $f \in C^\infty(X,\mathbb{R})$, and we can estimate $\|[\mathcal{D}, f]\| \leq C \sup |df|$ for some $C > 0$.

For $\mathcal{D}_u = \delta_u + \delta_u^*$, we have $\mathcal{D} = \Phi + \mathcal{D}_u$, with $[\Phi, \mathcal{D}_u] = 0$. The operator Φ on $\mathcal{T}_u^{\cdot}$ has spectrum $\mathbb{Z}$. The eigenspace E_r with eigenvalue $r \in \mathbb{Z}$ is the span of the elements

$$(\alpha_1 \otimes U^r \otimes \hbar^{k_1}, \alpha_2 \otimes U^r \otimes \hbar^{k_2}) \in \mathcal{T}_{0,u}^{\cdot} \oplus \check{\mathcal{T}}_{0,u}^{\cdot},$$

namely, elements with $\alpha_i \in \Omega_{X,\mathbb{R}}^{m_i}$ and k_i satisfying $0 \leq k_1 - 2r - m_1 \leq u$ and $0 \leq k_2 \leq u$. The operator $\mathcal{D}_u$ restricted to the eigenspace E_r has discrete spectrum. The multiplicity m_λ of an eigenvalue λ of $\mathcal{D}_u|_{E_r}$ is bounded by $2u\, n_\lambda$, where n_λ is the multiplicity of λ as an eigenvalue of the operator $d+d^*$ on real differential forms $\Omega'_{X,\mathbb{R}}$. Condition (7.4) then implies that $\mathcal{D}$ has discrete spectrum with finite multiplicities and that $(1 + \mathcal{D}^2)^{-1}$ is compact.

$\square$

When passing to cohomology, we obtain an induced structure of the following form.

COROLLARY 7.2. *Let now $\mathcal{H}^{\cdot}$ denote the Hilbert completion of the cohomology $\mathbb{H}^{\cdot}(\mathcal{T}_u^{\cdot}, \delta_u)$ and let $\mathcal{D} = \Phi$, the operator induced in cohomology. For $\mathcal{A} = U(\mathfrak{g})$, the triple $(\mathcal{A}, \mathcal{H}^{\cdot}, \mathcal{D})$ satisfies:*

- *The representation (2.23) induces an action of algebra $\mathcal{A}$ by bounded operators on $\mathcal{H}^{\cdot}$.*
- *The commutators $[\mathcal{D}, \sigma^L(a)]$, for $a \in \mathcal{A}$, are bounded operators on $\mathcal{H}^{\cdot}$.*
- *The operator $\mathcal{D}$ is a densely defined unbounded self-adjoint operator on $\mathcal{H}^{\cdot}$, such that $(1 + \mathcal{D}^2)^{-1}$ is a compact operator.*

PROOF. The statement follows as in the case of Theorem 7.1. Notice that now we have $\mathrm{Spec}(\Phi) = \mathbb{Z}$ with multiplicites

$$m_r = \dim \frac{\mathrm{Ker}(d' : \mathcal{T}_{0,u}^{\cdot,2r} \to \mathcal{T}_{0,u}^{\cdot+1,2r})}{\mathrm{Im}(d' : \mathcal{T}_{0,u}^{\cdot-1,2r} \to \mathcal{T}_{0,u}^{\cdot,2r})} + \dim \frac{\mathrm{Ker}(d'' : \check{\mathcal{T}}_{0,u}^{\cdot+1,2r} \to \check{\mathcal{T}}_{0,u}^{\cdot+2,2r})}{\mathrm{Im}(d'' : \check{\mathcal{T}}_{0,u}^{\cdot,2r} \to \check{\mathcal{T}}_{0,u}^{\cdot+1,2r})}.$$

This cohomological result no longer depends on the property (7.4).

$\square$

Recall that $(\mathcal{A}, \mathcal{H}^{\cdot}, \mathcal{D})$ is a spectral triple in the sense of Connes ([6]) if the three properties listed in the statement of Theorem 7.1 and Corollary 7.2 hold and the algebra $\mathcal{A}$ is a dense involutive subalgebra of a C^*-algebra.

In our case, the adjoints of elements in $\mathcal{A}$ with respect to the inner product on $\mathcal{H}^{\cdot}$ are again contained in $\mathcal{A}$. In fact, one can see that the adjoint of the Lefschetz $\mathbb{L}$ is given by

(7.5) $$\mathbb{L}^* = (\cdot \rfloor \omega)\, U,$$

where $\rfloor$ is the interior product and ω is the Kähler form, and we obtain (7.5) from

$$\sigma^L(w)^{-1}\mathbb{L}\sigma^L(w) = *((*\cdot) \wedge \omega)\, U.$$

Moreover, a choice of the Kähler form (of the Kähler class in the cohomological case) determines a corresponding representation of the involutive algebra $\mathcal{A}$ on the Hilbert space $\mathcal{H}^{\cdot}$. The choice of the Kähler class ranges over the Kähler cone

(7.6) $$\mathcal{K} = \{c \in H^{1,1}(X) : \int_M c^k > 0\}$$

for all $M \subset X$ complex submanifolds of dimension $1 \le k \le X$. Thus, in the case of Corollary 7.2 for instance we can consider a norm

(7.7) $$\|a\| := \sup_{c \in \overline{\mathcal{K}}:\, \|c\| = 1} \|\sigma_c^L(a)\|,$$

where $\overline{\mathcal{K}}$ is the nef cone and σ_c^L is the representation of $\mathcal{A}$ determined by the choice of the class c. Thus, the data $(\mathcal{A}, \mathcal{H}^{\cdot}, \mathcal{D})$ of Theorem 7.1 and Corollary 7.2 determine a spectral triple.

An interesting arithmetic aspect of spectral triples is that they have an associated family of zeta functions, the basic one being the zeta function of the Dirac operator,

$$\zeta_{\mathcal{D}}(z) = \mathrm{Tr}(|\mathcal{D}|^{-z}) = \sum_{\lambda} \mathrm{Tr}(\Pi(\lambda, |\mathcal{D}|))\lambda^{-z},$$

where $\Pi(\lambda, |\mathcal{D}|)$ denotes the orthogonal projection onto the eigenspace $E(\lambda, |\mathcal{D}|)$. More generally, one considers for $a \in \mathcal{A}$ the corresponding zeta function

$$\zeta_{a,\mathcal{D}}(z) = \mathrm{Tr}(a|\mathcal{D}|^{-z}) = \sum_{\lambda} \mathrm{Tr}(a\,\Pi(\lambda, |\mathcal{D}|))\lambda^{-z}.$$

These provide a refined notion of dimension for noncommutative spaces, the dimension spectrum, which is the complement in $\mathbb{C}$ of the set where all the $\zeta_{a,\mathcal{D}}$ extend holomorphically.

One can also consider the associated two-variable zeta functions,

$$\zeta_{a,\mathcal{D}}(s, z) := \sum_{\lambda} \mathrm{Tr}(a\,\Pi(\lambda, |\mathcal{D}|))(s - \lambda)^{-z}$$

and the corresponding regularized determinants,

$$\det_{\infty\, a,\mathcal{D}}(s) := \exp\left(-\frac{d}{dz}\zeta_{a,\mathcal{D}}(s, z)|_{z=0}\right).$$

Proposition 3.3 then shows that the archimedean factor of the Hasse-Weil L-function is given by the regularized determinant of a zeta function $\zeta_{a,\mathcal{D}}$ of the spectral triple of Corollary 7.2, namely the one for $a = \sigma^L(w)^2$. One advantage of this point of view is that one can now see the archimedean factor of the Hasse-Weil L-function as an element in the family $\det_{\infty\, a,\mathcal{D}}$ associated to the noncommutative geometry $(\mathcal{A}, \mathcal{H}, \mathcal{D})$.

Different representations of the Lie algebra $\mathfrak{g} = \mathrm{sl}(2)$, for different choice of the Kähler class and the corresponding Lefschetz operators, were considered also in [20]. It would be interesting to see if one can use this formalism of spectral triples in that context to further investigate the structure of the resulting Kähler Lie algebra (or of the Neron–Severi Lie algebra of projective varieties considered in [20]).

In the special case of arithmetic surfaces considered in [10], where X is a compact Riemann surface, the result of Corollary 7.2 can be related to the "arithmetic spectral triple" of [10].

Consider the complex $\mathrm{Cone}(N)^{\cdot} = \mathcal{T}^{\cdot} \oplus \mathcal{T}^{\cdot}[+1]$ with differential

$$d_{Cone} = \begin{pmatrix} \delta & N \\ 0 & -\delta \end{pmatrix}.$$

Proposition 2.23 of [10] and §4 of [9] show that, for X a compact Riemann surface, we have

$$\mathbb{H}^{\cdot}(\mathrm{Cone}(N)^{\cdot}, d_{Cone}) \simeq \mathbb{H}^{\cdot}(\mathrm{Ker}(N)^{\cdot}, d') \oplus \mathbb{H}^{\cdot+1}(\mathrm{Coker}(N)^{\cdot}, d'').$$

This is isomorphic to $\mathbb{H}^{\cdot}(\mathcal{T}_u^{\cdot}, \delta_u)|_{u=0}$. Moreover, under this identification, the operator Φ on the cohomology $\mathbb{H}^{\cdot}(\mathrm{Cone}(N)^{\cdot}, d_{Cone})$ considered in [10] agrees with the operator Φ on $\mathbb{H}^{\cdot}(\mathcal{T}_u^{\cdot}, \delta_u)|_{u=0}$.

8. Analogies with loop space geometry

Besides the original motivating analogy with the case of a geometric degeneration and the resolution (4.6) of the complex of nearby cycles, the cohomology theory at arithmetic infinity defined by the complex $(T^{\cdot}, \delta)$ also bears some interesting formal analogies with Givental's homological geometry on the loop space of a Kähler manifold ([15]).

Let X be a compact Kähler manifold, with the symplectic form ω representing an integral class in cohomology, such that the morphism $\omega : \pi_2(X) \to \mathbb{Z}$ is onto. Let LX denote the space of contractible loops on X, and $\widetilde{LX}$ the cyclic cover with group of deck transformations

$$(8.1) \qquad \pi_2(X)/\mathrm{Ker}\{\omega : \pi_2(X) \to \mathbb{Z}\} \cong \mathbb{Z},$$

and with the S^1-action that rotates loops. This covering makes the action functional

$$(8.2) \qquad \mathcal{A}(\phi) = \int_\Delta \phi^* \omega, \qquad \forall \phi \in \widetilde{LX}$$

single valued, for $\partial \Delta = S^1$. In fact, if γ denotes the generator of (8.1), we have $\gamma^* \mathcal{A} = \mathcal{A} + 1$. The critical manifold $Crit(\mathcal{A}) = Fix(S^1)$ consists of a trivial cyclic cover of the submanifold of constant loops X. Formally, one can consider an equivariant Floer complex for the functional $\mathcal{A}$, which is the complex (2.1) with a differential $d_{S^1} \pm \pi_* \pi^*$, which combines the equivariant differential on each component of $Crit(\mathcal{A})$ with a pullback–pushforward along gradient flow lines of $\mathcal{A}$ between different components of the critical manifold.

More precisely, a *formal* setting for the construction of equivariant Floer cohomologies can be described as follows. On a configuration space $\mathcal{C}$, which is an infinite dimensional manifold with an S^1 action, consider an S^1-invariant functional $\mathcal{A} : \mathcal{C} \to \mathbb{R}$, satisfying the following assumptions: *(i)* The critical point equation $\nabla \mathcal{A} = 0$ cuts out a finite dimensional smooth compact S^1-manifold $\mathrm{Crit}(\mathcal{A}) \subset \mathcal{C}$. *(ii)* The Hessian $H(\mathcal{A})$ on $\mathrm{Crit}(\mathcal{A})$ is non-degenerate in the normal directions. *(iii)* For any $x, y \in \mathrm{Crit}(\mathcal{A})$, there is a well defined locally constant relative index $\mathrm{ind}(x) - \mathrm{ind}(y) \in \mathbb{Z}$. *(iv)* For any two S^1-orbits $\mathcal{O}^\pm$ in $\mathrm{Crit}(\mathcal{A})$, the set $\mathcal{M}(\mathcal{O}^+, \mathcal{O}^-)$ of solutions to the flow equation

$$(8.3) \qquad \frac{d}{dt}u(t) + \nabla \mathcal{A}(u(t)) = 0, \qquad \lim_{t \to \pm\infty} u(t) \in \mathcal{O}^\pm,$$

modulo reparameterizations by translations, is either empty or a smooth manifold of dimension $\mathrm{ind}(\mathcal{O}^+) - \mathrm{ind}(\mathcal{O}^-) + \dim \mathcal{O}^+ - 1$. *(v)* The manifolds $\mathcal{M}(\mathcal{O}^+, \mathcal{O}^-)$ admit a compactification to smooth manifolds with corners, with codimension one boundary strata $\cup_{\mathrm{ind}(\mathcal{O}^+) \geq \mathrm{ind}(\mathcal{O}) \geq \mathrm{ind}(\mathcal{O}^-)} \mathcal{M}(\mathcal{O}^+, \mathcal{O}) \times_\mathcal{O} \mathcal{M}(\mathcal{O}, \mathcal{O}^-)$, and with compatible endpoint fibrations $\pi_\pm : \mathcal{M}(\mathcal{O}^+, \mathcal{O}^-) \to \mathcal{O}^\pm$.

For each component $\mathcal{O} \subset \mathrm{Crit}(\mathcal{A})$ one can then consider the S^1-equivariant de Rham complex

$$(8.4) \qquad \Omega^{\cdot,\infty}_{S^1}(\mathcal{O}) := \Omega^{\cdot}_{inv}(\mathcal{O}) \otimes \mathbb{C}[U, U^{-1}],$$

where U is of degree two, so that the total degree of $\alpha \otimes U^r$, with $\alpha \in \Omega^m_\mathcal{O}$ is $i = m + 2r$. The differential is of the form

$$(8.5) \qquad d_{S^1}(\alpha \otimes U^r) = d\alpha \otimes U^r + \iota_V(\alpha) \otimes U^{r+1}, \qquad d_{S^1}U = 0,$$

where ι_V denotes contraction with the vector field generated to the S^1-action on $\mathcal{O}$. The complex (8.4) computes the periodic S^1-equivariant cohomology of $\mathcal{O}$,

$$H^{\cdot}(\Omega^{\cdot,\infty}_{S^1}(\mathcal{O}), d_{S^1}) = H^{\cdot}_{S^1, per}(\mathcal{O}; \mathbb{C}),$$

which is the localization of $H^{\cdot}_{S^1}(\mathcal{O}; \mathbb{C})$ obtained by inverting U. The Floer complex is then defined as

$$(8.6) \qquad CF^{\ell,\infty}_{S^1} = \oplus_{\ell=\mathrm{ind}(\mathcal{O})+m+2r} \Omega^{m+2r,\infty}_{S^1}(\mathcal{O}),$$

with the relative index $\mathrm{ind}(\mathcal{O})$ computed with respect to a fixed base point in $\mathrm{Crit}(\mathcal{A})$, and with the Floer differential given by

$$(8.7) \qquad D_{\mathcal{O}^+,\mathcal{O}^-}(\alpha \otimes U^r) = \begin{cases} d_{S^1}(\alpha \otimes U^r) & \mathcal{O}^+ = \mathcal{O}^- \\ (-1)^m(\pi_{+*}\pi^*_-\,\alpha) \otimes U^r & \mathrm{ind}(\mathcal{O}^+) \geq \mathrm{ind}(\mathcal{O}^-) \\ 0 & \text{otherwise,} \end{cases}$$

where $\pi_{\pm} : \mathcal{M}(\mathcal{O}^+, \mathcal{O}^-) \to \mathcal{O}^{\pm}$ are the endpoint projections. The (periodic) equivariant Floer cohomology is the $\mathbb{C}[U, U^{-1}]$ module

$$(8.8) \qquad HF^{\cdot,\infty}_{S^1}(\mathcal{C}; \mathcal{A}) := H^{\cdot}(CF^{\infty}_{S^1}, D).$$

The property $D^2 = 0$ for the Floer differential holds because of the structure of the compactification of the spaces $\mathcal{M}(\mathcal{O}^+, \mathcal{O}^-)$ and its compatibility with the endpoint fibrations ([2]). The periodic equivariant Floer cohomology is related to equivariant Floer cohomology and homology via a natural exact sequence of complexes ([22]), of the form

$$(8.9) \qquad 0 \to CF^{*,+}_{S^1} \to CF^{*,\infty}_{S^1} \to CF^{*,-}_{S^1} \to 0,$$

where $CF^{*,+}_{S^1}$ is defined as in (8.6), but with $\mathbb{C}[U]$ instead of $\mathbb{C}[U, U^{-1}]$ in (8.4). The equivariant Floer cohomology is defined as

$$(8.10) \qquad HF^*_{S^1}(\mathcal{C}; \mathcal{A}) = HF^{*,+}_{S^1}(\mathcal{C}; \mathcal{A}) = H^*(CF^{*,+}_{S^1}, D),$$

while the quotient complex $CF^{*,-}_{S^1}$, with the induced Floer differential D^-, computes the equivariant Floer homology

$$(8.11) \qquad HF_{*,S^1}(\mathcal{C}; -\mathcal{A}) = HF^{*,-}_{S^1}(\mathcal{C}; \mathcal{A}).$$

There is also, in Floer theory, an analog of the weight filtration $W_{\cdot}$ given by the increasing filtration of the complex (8.6) by index of critical orbits, $\mathrm{ind}(\mathcal{O}) \geq k$,

$$(8.12) \qquad W_k CF^{\ell,\infty}(\mathcal{C}, \mathcal{A}) = \oplus_{\mathrm{ind}(\mathcal{O})\geq k,\ i+\mathrm{ind}(\mathcal{O})=\ell} \Omega^i_{S^1}(\mathcal{O}).$$

In the cases where the components of the boundary corresponding to flow lines in $\mathcal{M}(\mathcal{O}^+, \mathcal{O}^-)$ vanish, the exact sequence collapses and the Floer cohomology is the equivariant cohomology of the critical set, [15] and [16]. We have then an analog, in this context, of the Picard–Lefschetz filtration in the form (4.12), by setting

$$(8.13) \qquad AF^{s,k} := CF^{s+k+1,\infty}_{S^1}/W_k CF^{s+k+1,\infty}_{S^1},$$

as the analog of (4.6), and

$$(8.14) \qquad L_{\ell} AF^{s,k} = W_{2k+\ell+1} CF^{s+k+1,\infty}_{S^1}/W_k CF^{s+k+1,\infty}_{S^1},$$

with

$$(8.15) \qquad gr^L_{\ell} AF^{s,k} = \begin{cases} gr^W_{2k+\ell+1} CF^{s+k+1,\infty}_{S^1} & \ell+k \geq 0 \\ 0 & \ell+k < 0 \end{cases}$$

where

$$gr^W_{2k+\ell+1} CF^{s+k+1,\infty}_{S^1} = \oplus_{\text{ind}(\mathcal{O})=2k+\ell+1,s-i=\ell+k} \Omega^i_{S^1}(\mathcal{O}).$$

In the case of the loop space of a smooth compact symplectic manifold X, the action functional (8.2) is degenerate, the transversality conditions fail and the setup of Floer theory becomes more delicate [23]. The argument of [16] shows that, in the case of the loop space of a Kähler manifold, the components $\mathcal{M}(\mathcal{O}^+, \mathcal{O}^-)$ contribute trivially to the Floer differential, hence the equivariant Floer cohomology is computed by the E_1 term of the spectral sequence associated to the filtration $W.$, namely by the infinite dimensional vector space $H^{\cdot}(X;\mathbb{C}) \otimes \mathbb{C}[U, U^{-1}] \otimes \mathbb{C}[\hbar, \hbar^{-1}]$, where $\hbar$ implements the action of $\mathbb{Z}$ in (8.1).

Thus, one can see a formal analogy between the complex $\mathfrak{C}^{\cdot}$ of (2.9) and an equivariant Floer complex on the loop space. The cutoff $k \geq 0$ is then interpreted as a filtration by sublevel sets, corresponding to considering HF_{S^1} of $\mathcal{C}_0 = \mathcal{A}^{-1}(\mathbb{R}_{\geq 0})$. The cutoff $2r + m \geq 0$ instead corresponds in this analogy to a cutoff on the total degree of the equivariant differential forms, $HF^{\cdot \geq 0}_{S^1}$. Finally, the cutoff $k \geq r + q$ in (2.4) can be compared to a splitting of the form (8.9), adapted to the Hodge filtration. As in the case of the analogy with the resolution of the complex of nearby cycles, also in this analogy with Floer theory on loop spaces there is however a fundamental difference in the differentials. In fact, the term d_{S^1} of the Floer differential is replaced in the theory at arithmetic infinity by $\hbar d + d''$, which would not give a degree one differential on the Floer complex (except in special cases, like hyperkähler manifolds, where all components of $\text{Crit}(\mathcal{A})$ have relative index zero).

By the results of Givental, the equivariant Floer cohomology of the loop space also has an action of the ring $\mathcal{D}$ of differential operators on $\mathbb{C}^*$, where Q acts as $\hbar = \gamma^*$ and P as a combination of the symplectic form and action functional on the loop space, [15]. This structure on the Floer cohomology plays an important role in the phenomenon of mirror symmetry. It is interesting to remark that there is a conjectural mirror relation between the monodromy and the Lefschetz operators ([17] [21]). Thus, the question of developing a more precise relation between the complex $(\mathcal{T}^{\cdot}, \delta)$ and Floer theory, with the actions (2.27) of the ring of differential operators on $\mathbb{C}^*$, may be interesting in this respect, in view of the possibility of addressing such mirror symmetry questions in the context of arithmetic geometry, by adapting to arithmetic cohomological constructions the setting of homological geometry on loop spaces.

Acknowledgment. The first author is partially supported by NSERC grants 72016789, 72024520. The second author is partially supported by Humboldt Foundation Sofja Kovalevskaja Award.

References

[1] D.V. Anosov, A.A. Bolibruch, *The Riemann–Hilbert problem*, Aspects of Mathematics Vol.**22**, Vieweg, 1994.

[2] D.M. Austin, P.J. Braam, *Morse-Bott theory and equivariant cohomology*, Floer Memorial Volume, Birkhäuser 1995.

[3] J. Burgos, *Arithmetic Chow rings and Deligne-Beilinson cohomology*, J.Alg.Geom. **6** (1997) N.2 335–377.

[4] A. Connes, *Trace formula in noncommutative geometry and zeros of the Riemann zeta function*, Selecta Math. (N.S.) **5** (1999), no. 5, 29–106.

[5] A. Connes, *Symetries Galoisiennes et Renormalisation*, preprint math.QA/0211199.

[6]   A. Connes, *Geometry from the spectral point of view.* Lett. Math. Phys. **34** (1995), no. 3, 203–238.

[7]   A. Connes, D. Kreimer, *Renormalization in quantum field theory and the Riemann-Hilbert problem. II. The β-function, diffeomorphisms and the renormalization group.* Comm. Math. Phys. **216** (2001), no. 1, 215–241.

[8]   A. Connes, M. Marcolli, *From physics to number theory via noncommutative geometry,* in preparation.

[9]   C. Consani, *Double complexes and Euler L-factors,* Compositio Math. **111** (1998), 323–358.

[10] C. Consani, M. Marcolli, *Noncommutative geometry, dynamics, and ∞-adic Arakelov geometry,* to appear in Selecta Mathematica.

[11] C. Deninger *On the Γ-factors attached to motives,* Invent. Math. **104** (1991) 245–261.

[12] C. Deninger *Local L-factors of motives and regularized determinants,* Invent. Math. **107** (1992) 135–150.

[13] C. Deninger *On the Γ-factors of motives II,* Doc. Math. **6** (2001), 69–97.

[14] S.I. Gelfand, Yu.I. Manin, *Homological algebra,* EMS Vol.**38**, Springer Verlag, 1999.

[15] A.B. Givental, *Homological geometry. I. Projective hypersurfaces.* Selecta Math. (N.S.) **1** (1995), no. 2, 325–345.

[16] A.B. Givental, B. Kim, *Quantum cohomology of flag manifolds and Toda lattices.* Comm. Math. Phys. **168** (1995), no. 3, 609–641.

[17] V. Golyshev, V. Lunts, D. Orlov, *Mirror symmetry for abelian varieties,* J. Algebraic Geom. **10** (2001), no. 3, 433–496.

[18] F. Guillén, V. Navarro Aznar, *Sur le théorème local des cycles invariants.* Duke Math. J. **61** (1990), no. 1, 133–155.

[19] S. Lang, $SL_2(\mathbb{R})$, Addison–Wesley, 1975.

[20] E. Looijenga, V.A. Lunts, *A Lie algebra attached to a projective variety,* Invent. Math. **129** (1997) 361–412.

[21] Yu.I. Manin, *Moduli, motives, mirrors,* Progress in Matematics Vol. **201**, Birkhäuser 2001, pp. 53–73.

[22] M. Marcolli, B.L. Wang, *Variants of equivariant Seiberg–Witten Floer homology,* preprint math.GT/0211238.

[23] S.Piunikhin, D.Salamon, M.Schwarz, *Symplectic Floer-Donaldson theory and quantum cohomology.* Contact and symplectic geometry (Cambridge, 1994), 171–200, Publ. Newton Inst., **8**, Cambridge Univ. Press, Cambridge, 1996.

[24] M. Saito, *Modules de Hodge Polarisable.* Publ. Res. Inst. Math. Sci. **24** (1988) 849–995.

[25] J. P. Serre, *Facteurs locaux des fonctions zêta des variétés algébriques (définitions et conjectures).* Sém. Delange-Pisot-Poitou, exp. 19, 1969/70.

[26] P. Deligne, *Groupes de Monodromie en Géométrie Algébrique.* Lecture Notes in Mathematics **340**, Springer-Verlag, New York 1973.

[27] C. Simpson, *The Hodge filtration on nonabelian cohomology,* Proc. Symp. Pure Math. **62** (2) (1997) 217–281.

[28] J. Steenbrink, *Limits of Hodge structures.* Invent. Math. **31** (1976), 229–257.

[29] R.O. Wells, *Differential analysis on complex manifolds,* Springer Verlag, 1980.

C. CONSANI: UNIVERSITY OF TORONTO   CANADA
E-mail address: kc@math.toronto.edu

M. MARCOLLI: MAX–PLANCK INSTITUT FÜR MATHEMATIK BONN   GERMANY
E-mail address: marcolli@mpim-bonn.mpg.de

A twisted Burnside theorem for countable groups and Reidemeister numbers

Alexander Fel'shtyn and Evgenij Troitsky

ABSTRACT. The purpose of the present paper is to prove for finitely generated groups of type I the following conjecture of A. Fel'shtyn and R. Hill [8], which is a generalization of the classical Burnside theorem.

Let G be a countable discrete group, ϕ one of its automorphisms, $R(\phi)$ the number of ϕ-conjugacy classes, and $S(\phi) = \#\operatorname{Fix}(\widehat{\phi})$ the number of ϕ-invariant equivalence classes of irreducible unitary representations. If one of $R(\phi)$ and $S(\phi)$ is finite, then it is equal to the other.

This conjecture plays a important role in the theory of twisted conjugacy classes (see [12], [6]) and has very important consequences in Dynamics, while its proof needs rather sophisticated results from Functional and Non-commutative Harmonic Analysis.

We begin a discussion of the general case (which needs another definition of the dual object). It will be the subject of a forthcoming paper.

Some applications and examples are presented.

1. Introduction and formulation of results

DEFINITION 1.1. Let G be a countable discrete group and $\phi : G \to G$ an endomorphism. Two elements $x, x' \in G$ are said to be ϕ-conjugate or twisted conjugate iff there exists $g \in G$ with

$$x' = gx\phi(g^{-1}).$$

We shall write $\{x\}_\phi$ for the ϕ-conjugacy or twisted conjugacy class of the element $x \in G$. The number of ϕ-conjugacy classes is called the Reidemeister number of an endomorphism ϕ and is denoted by $R(\phi)$. If ϕ is the identity map then the ϕ-conjugacy classes are the usual conjugacy classes in the group G.

If G is a finite group, then the classical Burnside theorem (see e.g. [13, p. 140]) says that the number of classes of irreducible representations is equal to the number of conjugacy classes of elements of G. Let $\widehat{G}$ be the unitary dual of G, i.e. the set of equivalence classes of unitary irreducible representations of G.

REMARK 1.2. If $\phi : G \to G$ is an epimorphism, it induces a map $\widehat{\phi} : \widehat{G} \to \widehat{G}$, $\widehat{\phi}(\rho) = \rho \circ \phi$ (because a representation is irreducible if and only if the scalar operators in the space of representation are the only ones which commute with all operators of the representation). This is not the case for a general endomorphism ϕ, because $\rho\phi$ can be reducible for an irreducible representation ρ, and $\widehat{\phi}$ can be defined only as a multi-valued map. But nevertheless we can define the set of fixed points $\operatorname{Fix}\widehat{\phi}$ of $\widehat{\phi}$ on $\widehat{G}$.

Therefore, by the Burnside's theorem, if ϕ is the identity automorphism of any finite group G, then we have $R(\phi) = \#\operatorname{Fix}(\widehat{\phi})$.

To formulate our theorem for the case of a general endomorphism we first need an appropriate definition of the $\operatorname{Fix}(\widehat{\phi})$.

DEFINITION 1.3. Let $\operatorname{Rep}(G)$ be the space of equivalence classes of finite dimensional unitary representations of G. Then the corresponding map $\widehat{\phi}_R : \operatorname{Rep}(G) \to \operatorname{Rep}(G)$ is defined in the same way as above: $\widehat{\phi}_R(\rho) = \rho \circ \phi$.

Let us denote by $\operatorname{Fix}(\widehat{\phi})$ the set of points $\rho \in \widehat{G} \subset \operatorname{Rep}(G)$ such that $\widehat{\phi}_R(\rho) = \rho$.

THEOREM 1.4 (Main Theorem). *Let G be a finitely generated discrete group of type I, ϕ one of its endomorphism, $R(\phi)$ the number of ϕ-conjugacy classes, and $S(\phi) = \#\operatorname{Fix}(\widehat{\phi})$ the number of ϕ-invariant equivalence classes of irreducible unitary representations. If one of $R(\phi)$ and $S(\phi)$ is finite, then it is equal to the other.*

Let $\mu(d)$, $d \in \mathbb{N}$, be the *Möbius function*, i.e.

$$\mu(d) = \begin{cases} 1 & \text{if } d = 1, \\ (-1)^k & \text{if } d \text{ is a product of } k \text{ distinct primes}, \\ 0 & \text{if } d \text{ is not square} - \text{free}. \end{cases}$$

THEOREM 1.5 (Congruences for the Reidemeister numbers). *Let $\phi : G \to G$ be an endomorphism of a countable discrete group G such that all numbers $R(\phi^n)$ are finite and let H be a subgroup of G with the properties*

$$\phi(H) \subset H$$

$$\forall x \in G \; \exists n \in \mathbb{N} \text{ such that } \phi^n(x) \in H.$$

If the pair (H, ϕ^n) satisfies the conditions of Theorem 1.4 for any $n \in \mathbb{N}$, then one has for all n,

$$\sum_{d|n} \mu(d) \cdot R(\phi^{n/d}) \equiv 0 \mod n.$$

These theorems were proved previously in a special case of Abelian finitely generated plus finite group [8, 9].

The interest in twisted conjugacy relations has its origins, in particular, in the Nielsen-Reidemeister fixed point theory (see, e.g. [12, 6]), in Selberg theory (see, eg. [14, 1]), and Algebraic Geometry (see, e.g. [11]).

Concerning some topological applications of our main results, they are already obtained in the present paper (Theorem 8.5). The congruences give some necessary conditions for the realization problem for Reidemeister numbers in topological dynamics. The relations with Selberg theory will be presented in a forthcoming paper.

Let us remark that it is known that the Reidemeister number of an endomorphism of a finitely generated Abelian group is finite iff 1 is not in the spectrum of the restriction of this endomorphism to the free part of the group (see, e.g. [12]). The Reidemeister number is infinite for any automorphism of a non-elementary Gromov hyperbolic group [5].

To make the presentation more transparent we start from a new approach (E.T.) for Abelian (Section 2) and compact (Section 3) groups. Only after that we develop this approach and prove the main theorem for finitely generated groups of type I in Section 5. A discussion of some examples leading to conjectures is the subject of Section 6. Then we prove the congruences theorem (Section 7) and describe some topological applications (Section 8).

Acknowledgement. We would like to thank the Max Planck Institute for Mathematics in Bonn for its kind support and hospitality while the most part of this work has been completed. We are also indebted to MPI and organizers of the Workshop on Noncommutative Geometry and Number Theory (Bonn, August 18–22, 2003) where the results of this paper were presented.

The first author thanks the Research Institute in Mathematical Sciences in Kyoto for the possibility of the present research during his visit there.

The authors are grateful to V. Balantsev, B. Bekka, R. Hill, V. Kaimanovich, V. Manuilov, A. Mishchenko, A. Rosenberg, A. Shtern, L. Vainerman, A. Vershik for helpful discussions.

They are also grateful to the referee for valuable remarks and suggestions.

2. Abelian case

Let ϕ be an automorphism of an Abelian group G.

LEMMA 2.1. *The twisted conjugacy class H of e is a subgroup. The other ones are cosets gH.*

PROOF. The first statement follows from the equalities
$$h\phi(h^{-1})g\phi(g^{-1}) = gh\phi((gh)^{-1}), \quad (h\phi(h^{-1}))^{-1} = \phi(h)h^{-1} = h^{-1}\phi(h).$$
For the second statement suppose $a \sim b$, i.e. $b = ha\phi(h^{-1})$. Then
$$gb = gha\phi(h^{-1}) = h(ga)\phi(h^{-1}), \qquad gb \sim ga.$$
$\square$

LEMMA 2.2. *Suppose, $u_1, u_2 \in G$, χ_H is the characteristic function of H as a set. Then*
$$\chi_H(u_1 u_2^{-1}) = \begin{cases} 1, & \text{if } u_1, u_2 \text{ are in one coset}, \\ 0, & \text{otherwise}. \end{cases}$$

PROOF. Suppose, $u_1 \in g_1 H$, $u_2 \in g_2 H$, hence, $u_1 = g_1 h_1$, $u_2 = g_2 h_2$. Then
$$u_1 u_2^{-1} = g_1 h_1 h_2^{-1} g_2^{-1} \in g_1 g_2^{-1} H.$$
Thus, $\chi_H(u_1 u_2^{-1}) = 1$ if and only if $g_1 g_2^{-1} \in H$ and u_1 and u_2 are in the same class. Otherwise it is 0.
$\square$

The following Lemma is well known.

LEMMA 2.3. *For any subgroup H the function χ_H is of positive type.*

PROOF. Let us take arbitrary elements $u_1, u_2, \ldots, u_n$ of G. Let us reenumerate them in such a way that some first are in $g_1 H$, the next ones are in $g_2 H$, and so on, till $g_m H$, where $g_j H$ are different cosets. By the previous Lemma the matrix $\|p_{it}\| := \|\chi_H(u_i u_t^{-1})\|$ is block-diagonal with square blocks formed by units. These blocks, and consequently the whole matrix are positively semi-defined.
$\square$

LEMMA 2.4. *In the Abelian case characteristic functions of twisted conjugacy classes belong to the Fourier-Stieltjes algebra* $B(G) = (C^*(G))^*$.

PROOF. In this case the characteristic functions of twisted conjugacy classes are the shifts of the characteristic function of the class H of e. Indeed, we have the following sequence of equivalent properties:

$$a \sim b, \qquad b = ha\phi(h^{-1}) \text{ for some } h, \qquad gb = gha\phi(h^{-1}) \text{ for some } h,$$

$$gb = hga\phi(h^{-1}) \text{ for some } h, \qquad ga \sim gb.$$

Hence, by Corollary (2.19) of [4], these characteristic functions are in $B(G)$. $\qquad \square$

Let us remark that there exists a natural isomorphism (Fourier transform)

$$u \mapsto \widehat{u}, \qquad C^*(G) = C_r^*(G) \cong C(\widehat{G}), \qquad \widehat{g}(\rho) := \rho(g),$$

(this is a number because irreducible representations of an Abelian group are 1-dimensional). In fact, it is better to look (for what follows) at an algebra $C(\widehat{G})$ as an algebra of continuous sections of a bundle of 1-dimensional matrix algebras. over $\widehat{G}$.

Our characteristic functions, being in $B(G) = (C^*(G))^*$ in this case, are mapped to the functionals on $C(\widehat{G})$ which, by the Riesz-Markov-Kakutani theorem, are measures on $\widehat{G}$. Which of these measures are invariant under the induced (twisted) action of G? Let us remark, that an invariant non-trivial functional gives rise to at least one invariant space – its kernel.

Let us remark, that convolution under the Fourier transform becomes pointwise multiplication. More precisely, the twisted action, for example, is defined as

$$g[f](\rho) = \rho(g)f(\rho)\rho(\phi(g^{-1})), \qquad \rho \in \widehat{G}, \quad g \in G, \quad f \in C(\widehat{G}).$$

There are 2 possibilities for the twisted action of G on the representation algebra $A_\rho \cong \mathbb{C}$: 1) the linear span of the orbit of $1 \in A_\rho$ is equal to all A_ρ, 2) and the opposite case (the action is trivial).

The second case means that the space of interviewing operators between A_ρ and $A_{\widehat{\phi}\rho}$ equals $\mathbb{C}$, and ρ is a fixed point of the action $\widehat{\phi} : \widehat{G} \to \widehat{G}$. In the first case this is the opposite situation.

If we have a finite number of such fixed points, then the space of twisted invariant measures is just the space of measures concentrated in these points. Indeed, let us describe the action of G on measures in more detail.

LEMMA 2.5. *For any Borel set E one has* $g[\mu](E) = \int_E g[1] \, d\mu$.

PROOF. The restriction of measure to any Borel set commutes with the action of G, since the last is point wise on $C(\widehat{G})$. For any Borel set E one has

$$g[\mu](E) = \int_E 1 \, dg[\mu] = \int_E g[1] \, d\mu.$$

$$\square$$

Hence, if μ is twisted invariant, then for any Borel set E and any $g \in G$ one has

$$\int_E (1 - g[1]) \, d\mu = 0.$$

LEMMA 2.6. *Suppose, $f \in C(X)$, where X is a compact Hausdorff space, and μ is a regular Borel measure on X, i.e. a functional on $C(X)$. Suppose, for any Borel set $E \subset X$ one has $\int_E f \, d\mu = 0$. Then $\mu(h) = 0$ for any $h \in C(X)$ such that $f(x) = 0$ implies $h(x) = 0$. I.e. μ is concentrated off the interior of $\operatorname{supp} f$.*

PROOF. Since the functions of the form fh are dense in the space of the refered to above h's, it is sufficient to verify the statement for fh. Let us choose an arbitrary $\varepsilon > 0$ and a simple function $h' = \sum\limits_{i=1}^{n} a_i \chi_{E_i}$ such that $|\mu(fh') - \mu(fh)| < \varepsilon$. Then

$$\mu(fh') = \sum_{i=1}^{n} \int_{E_i} a_i f \, d\mu = \sum_{i=1}^{n} a_i \int_{E_i} f \, d\mu = 0.$$

Since ε is an arbitrary one, we are done.   □

Applying this lemma to a twisted invariant measure μ and $f = 1 - g[1]$ we obtain that μ is concentrated at our finite number of fixed points of $\widehat{\phi}$, because outside of them $f \neq 0$. If we have an infinite number of fixed points, then the space is infinite–dimensional (we have an infinite number of measures concentrated in finite number of points, each time different) and Reidemeister number is infinite as well. So, we are done.

3. Compact case

Let G be a compact group, hence $\widehat{G}$ is a discrete space. Then $C^*(G) = \oplus M_i$, where M_i are the matrix algebras of irreducible representations. The infinite sum is in the following sense:

$$C^*(G) = \{f_i\}, i \in \{1, 2, 3, \ldots\} = \widehat{G}, f_i \in M_i, \|f_i\| \to 0 (i \to \infty).$$

When G is finite and $\widehat{G}$ is finite this is exactly Peter-Weyl theorem.

A characteristic function of a twisted class is a functional on $C^*(G)$. For a finite group it is evident, for a general compact group it is necessary to verify only the measurability of the twisted class with the respect to Haar measure, i.e. that twisted class is Borel. For a compact G, the twisted conjugacy classes being orbits of twisted action are compact and hence closed. Then its complement is open, hence Borel, and the class is Borel too.

Under the identification it passes to a sequence $\{\varphi_i\}$, where φ_i is a functional on M_i (the properties of convergence can be formulated, but they play no role at the moment). The conditions of invariance are the following: for each $\rho_i \in \widehat{G}$ one has $g[\varphi_i] = \varphi_i$, i.e. for any $a \in M_i$ and any $g \in G$ one has $\varphi_i(\rho_i(g) a \rho_i(\phi(g^{-1}))) = \varphi_i(a)$.

Let us recall the following well-known fact.

LEMMA 3.1. *Each functional on matrix algebra has form $a \mapsto \operatorname{Tr}(ab)$ for a fixed matrix b.*

PROOF. One has $\dim(M(n, \mathbb{C}))' = \dim(M(n, \mathbb{C})) = n \times n$ and looking at matrices as at operators in V, $\dim V = n$, with base e_i, one can remark that functionals $a \mapsto \langle a e_i, e_j \rangle$, $i, j = 1, \ldots, n$, are linearly independent. Hence, any functional takes form

$$a \mapsto \sum_{i,j} b_j^i \langle a e_i, e_j \rangle = \sum_{i,j} b_j^i a_i^j = \operatorname{Tr}(ba), \qquad b := \|b_j^i\|.$$

□

Now we can study invariant ones:

$$\mathrm{Tr}(b\rho_i(g)a\rho_i(\phi(g^{-1}))) = \mathrm{Tr}(ba), \qquad \forall\, a, g,$$
$$\mathrm{Tr}((b - \rho_i(\phi(g^{-1}))b\rho_i(g))a) = 0, \qquad \forall\, a, g,$$

hence,

$$b - \rho_i(\phi(g^{-1}))b\rho_i(g) = 0, \qquad \forall\, g.$$

Since ρ_i is irreducible, the dimension of the space of such b is 1 if ρ_i is a fixed point of $\widehat{\phi}$ and 0 in the opposite case. So, we are done.

REMARK 3.2. In fact we are only interested in finite discrete case. Indeed, for a compact G, the twisted conjugacy classes being orbits of twisted action are compact and hence closed. If there is a finite number of them, then are open too. Hence, the situation is more or less reduced to a discrete group: quotient by the component of unity.

4. Extensions and Reidemeister classes

Consider a group extension respecting homomorphism ϕ:

$$
\begin{array}{ccccccccc}
0 & \longrightarrow & H & \xrightarrow{\;i\;} & G & \xrightarrow{\;p\;} & G/H & \longrightarrow & 0 \\
& & \downarrow{\scriptstyle \phi'} & & \downarrow{\scriptstyle \phi} & & \downarrow{\scriptstyle \overline{\phi}} & & \\
0 & \longrightarrow & H & \xrightarrow{\;i\;} & G & \xrightarrow{\;p\;} & G/H & \longrightarrow & 0,
\end{array}
$$

where H is a normal subgroup of G. The following argument has partial intersection with [10].

First of all let us notice that the Reidemeister classes of ϕ in G are mapped epimorphically on classes of $\overline{\phi}$ in G/H. Indeed,

$$p(\widetilde{g})p(g)\overline{\phi}(p(\widetilde{g}^{-1})) = p(\widetilde{g}g\phi(\widetilde{g}^{-1})).$$

Suppose, $R(\phi) < \infty$. Then the previous remark implies $R(\overline{\phi}) < \infty$. Consider a class $K = \{h\}_{\tau_g\phi'}$, where $\tau_g(h) := ghg^{-1}$, $g \in G$, $h \in H$. The corresponding equivalence relation is

(1) $$h \sim \widetilde{h}hg\phi'(\widetilde{h}^{-1})g^{-1}.$$

Since H is normal, the automorphism $\tau_g : H \to H$ is well defined. We will denote by K the image iK as well. By (1) the shift Kg is a subset of Hg is characterized by

(2) $$hg \sim \widetilde{h}(hg)\phi'(\widetilde{h}^{-1}).$$

Hence it is a subset of $\{hg\}_\phi \cap Hg$ and the partition $Hg = \cup(\{h\}_{\tau_g\phi'})g$ is a sub-partition of $Hg = \cup(Hg \cap \{hg\}_\phi)$.

LEMMA 4.1. Suppose, $|G/H| = N < \infty$. Then $R(\tau_g\phi') \leq NR(\phi)$. More precisely, the mentioned subpartition is not more than in N parts.

PROOF. Consider the following action of G on itself: $x \mapsto gx\phi(g^{-1})$. Then its orbits are exactly classes $\{x\}_\phi$. Moreover it maps classes (2) onto each other. Indeed,

$$\widetilde{g}\widetilde{h}(hg)\phi'(\widetilde{h}^{-1})\phi(\widetilde{g}^{-1}) = \widehat{h}\widetilde{g}(hg)\phi(\widetilde{g}^{-1})\phi'(\widehat{h}^{-1})$$

using normality of the H. This map is invertible ($\widetilde{g} \leftrightarrow \widetilde{g}^{-1}$), hence bijection. Moreover, $\widetilde{g}$ and $\widetilde{g}\widehat{h}$, for any $\widehat{h} \in H$, act in the same way. Or in the other words, H

is in the stabilizer of this permutation of classes (2). Hence, the cardinality of any orbit $\leq N$.   □

Hence, for any finite G/H the number of classes of the form (2) is finite: it is $\leq NR(\phi)$.

LEMMA 4.2. *Suppose, H satisfies the following property: for any automorphism of H with finite Reidemeister number the characteristic functions of Reidemeister classes of ϕ are linear combinations of matrix elements of some finite number of irreducible finite dimensional representations of H. Then the characteristic functions of classes (2) are linear combinations of matrix elements of some finite number of irreducible finite dimensional representations of G.*

PROOF. Let $\rho_1, \rho_2, \ldots, \rho_k$ be the above irreducible representations of H, ρ its direct sum acting on V, and π the regular (finite dimensional) representation of G/H. Let $\rho_1^I, \ldots, \rho_k^I, \rho^I$ be the corresponding induced representations of G. Let the characteristic function of K be represented under the form $\chi_K(h) = \langle \rho(h)\xi, \eta \rangle$. Let $\xi^I \in L^2(G/H, V)$ be defined by the formulas $\xi^I(\bar{e}) = \xi \in V$, $\xi^I(\bar{g}) = 0$ if $\bar{g} \neq \bar{e}$. Define similarly η^I. Then for $h \in iH$ we have

$$\rho^I(h)\xi^I(\bar{g}) = \rho(s(\bar{g})hs(\bar{g}h)^{-1})\xi(\bar{g}h) = \rho(hs(\bar{g})s(\bar{g})^{-1})\xi(\bar{g}) = \begin{cases} \rho(h)\xi, & \text{if } \bar{g} = \bar{e}, \\ 0, & \text{otherwise.} \end{cases}$$

Hence, $\langle \rho^I(h)\xi^I, \eta^I \rangle|_{iH}$ is the characteristic function of iK. Let $u, v \in L^2(G/H)$ be such vectors that $\langle \pi(\bar{g})u, v \rangle$ is the characteristic function of $\bar{e}$. Then

$$\langle (\rho^I \otimes \pi)(\xi^I \otimes u, \eta^I \otimes v) \rangle$$

is the characteristic function of iK. Other characteristic functions of classes (2) are shifts of this one. Hence matrix elements of the representation $\rho^I \otimes \pi$. It is finite dimensional. Hence it can be decomposed in a finite direct sum of irreducible representations.   □

COROLLARY 4.3 (of previous two lemmata). *Under the assumptions of the previous lemma, the characteristic functions of Reidemeister classes of ϕ are linear combinations of matrix elements of some finite number of irreducible finite dimensional representations of G.*

5. The case of groups of type I

THEOREM 5.1. *Let G be a discrete group of type I. Then*

- [3, 3.1.4, 4.1.11] *The dual space $\widehat{G}$ is a T_1-topological space.*
- [15] *Any irreducible representation of G is finite-dimensional.*

REMARK 5.2. In fact a discrete group G is of type I if and only if it has a normal, Abelian subgroup M of finite index. The dimension of any irreducible representation of G is at most $[G : M]$ [15].

Suppose $R = R(\phi) < \infty$, and let $F \subset L^\infty(G)$ be the R-dimensional space of all twisted-invariant functionals on $L^1(G)$. Let $K \subset L^1(G)$ be the intersection of kernels of functionals from F. Then K is a linear subspace of $L^1(G)$ of codimension R. For each $\rho \in \widehat{G}$ let us denote by K_ρ the image $\rho(K)$. This is a subspace of a (finite-dimensional) full matrix algebra. Let cd_ρ be its codimension.

Let us introduce the following set

$$\widehat{G}_F = \{\rho \in \widehat{G} \mid \mathrm{cd}_\rho \neq 0\}.$$

LEMMA 5.3. *One has* $\mathrm{cd}_\rho \neq 0$ *if and only if* ρ *is a fixed point of* $\widehat{\phi}$. *In this case* $\mathrm{cd}_\rho = 1$.

PROOF. Suppose, $\mathrm{cd}_\rho \neq 0$ and let us choose a functional φ_ρ on the (finite-dimensional full matrix) algebra $\rho(L^1(G))$ such that $K_\rho \subset \mathrm{Ker}\,\varphi_\rho$. Then for the corresponding functional $\varphi_\rho^* = \varphi_\rho \circ \rho$ on $L^1(G)$ one has $K \subset \mathrm{Ker}\,\varphi_\rho^*$. Hence, $\varphi_\rho^* \in F$ and is twisted-invariant, as well as φ_ρ. Then we argue as in the case of compact group (after Lemma 3.1).

Conversely, if ρ is a fixed point of $\widehat{\phi}$, it gives rise to a (unique up to scaling) non-trivial twisted-invariant functional φ_ρ. Let $x = \rho(a)$ be any element in $\rho(L^1(G))$ such that $\varphi_\rho(x) \neq 0$. Then $x \notin K_\rho$, because $\varphi_\rho^*(a) = \varphi_\rho(x) \neq 0$, while φ_ρ^* is a twisted-invariant functional on $L^1(G)$. So, $\mathrm{cd}_\rho \neq 0$.

The uniqueness (up to scaling) of the intertwining operator implies the uniqueness of the corresponding twisted-invariant functional. Hence, $\mathrm{cd}_\rho = 1$. □

Hence,

(3) $$\widehat{G}_F = \mathrm{Fix}(\widehat{\phi}).$$

From the property $\mathrm{cd}_\rho = 1$ one obtains for this (unique up to scaling) functional φ_ρ:

(4) $$\mathrm{Ker}\,\varphi_\rho = K_\rho.$$

LEMMA 5.4. $R = \#\widehat{G}_F$, *in particular, the set* $\widehat{G}_F$ *is finite*.

PROOF. First of all we remark that since G is finitely generated almost Abelian (cf. Remark 5.2) there is a normal Abelian subgroup H of finite index invariant under all ϕ. Hence we can apply Lemma 4.3 to G, H, ϕ. So there is a finite collection of irreducible representations of G such that any twisted-invariant functional is a linear combination of matrix elements of them, i.e. linear combination of functionals on them. If each of them gives a non-trivial contribution, it has to be a twisted-invariant functional on the corresponding matrix algebra. Hence, by the argument above, these representations belong to $\widehat{G}_F$, and the appropriate functional is unique up to scaling. Hence, $R \leq S$.

Then we use T_1-separation property. More precisely, suppose some points $\rho_1, \ldots, \rho_s$ belong to $\widehat{G}_F$. Let us choose some twisted-invariant functionals $\varphi_i = \varphi_{\rho_i}$ corresponding to these points as it was described (i.e. choose some scaling). Assume that $\|\varphi_i\| = 1$, $\varphi_i(x_i) = 1$, $x_i \in \rho_i(L^1(G))$. If we can find $a_i \in L^1(G)$ such that $\varphi_i(\rho_i(a_i)) = \varphi_i^*(a_i)$ is sufficiently large and $\rho_j(a_i)$, $i \neq j$, are sufficiently small (in fact it is sufficient $\rho_j(a_i)$ to be close enough to $K_j := K_{\rho_j}$), then $\varphi_j^*(a_i)$ are small for $i \neq j$, and φ_i^* are linear independent and hence, $s < R$. This would imply $S := \#\widehat{G}_F \leq R$ is finite. Hence, $R = S$.

So, the problem is reduced to the search of the above a_i. Let $d = \max\limits_{i=1,\ldots,s} \dim \rho_i$.

For each i let $c_i := \|b_i\|$, where x_i is the unitary equivalence of ρ_i and $\widehat{\phi}\rho_i$ and $x_i = \rho_i(b_i)$.

Let $c := \max\limits_{i=1,\ldots,s} c_i$ and $\varepsilon := \frac{1}{2 \cdot s^2 \cdot d \cdot c}$.

One can find a positive element $a_i' \in L^1(G)$ such that $\|\rho_i(a_i')\| \geq 1$ and $\|\rho_j(a_i')\| < \varepsilon$ for $j \neq i$. Indeed, since ρ_i can be separated from one point, and hence from the finite number of points: ρ_j, $j \neq i$. Hence, one can find an element v_i such that $\|\rho_i(v_i)\| > 1$, $\|\rho_j(v_i)\| < 1$ for $j \neq i$ [3, Lemma 3.3.3]. The same is true for the positive element $u_i = v_i^* v_i$. (Due to density we do not distinguish elements of L^1 and C^*). Now for a sufficiently large n the element $a_i' := (u_i)^n$ has the desired properties.

Let us take $a_i := a_i' b_i^*$. Then

(5) $\quad \varphi_i^*(a_i) = \mathrm{Tr}(x_i \rho_i(a_i)) = \mathrm{Tr}(x_i \rho_i(a_i')\rho_i(b_i)^*) = \mathrm{Tr}(x_i \rho_i(a_i') x_i^*) =$

$$= \mathrm{Tr}(x_i \rho_i(a_i')(x_i)^{-1}) = \mathrm{Tr}(\rho_i(a_i')) \geq \frac{1}{\dim \rho_i} \geq \frac{1}{d}.$$

For $j \neq i$ one has

(6) $\quad\quad\quad\quad\quad \|\varphi_j^*(a_i)\| = \|\varphi_j(\rho_j(a_i' b_i^*))\| \leq c_i \cdot \varepsilon.$

Then the $s \times s$ matrix $\Phi = \varphi_j^*(a_i)$ can be decomposed into the sum of the diagonal matrix Δ and off-diagonal Σ. By (5) one has $\Delta \geq \frac{1}{d}$. By (6) one has

$$\|\Sigma\| \leq s^2 \cdot c_i \cdot \varepsilon \leq s^2 \cdot c \cdot \frac{1}{2 \cdot s^2 \cdot d \cdot c} = \frac{1}{2d}.$$

Hence, Φ is non-degenerate and we are done. □

Lemma 5.4 together with (3) completes the proof of Theorem 1.4 for automorphisms.

We need the following additional observations for the proof of Theorem 1.4 for a general endomorphism (in which (3) is false for infinite-dimensional representations).

LEMMA 5.5.   (1) *If ϕ is an epimorphism, then $\widehat{G}$ is $\widehat{\phi}_R$-invariant.*

(2) *For any ϕ the set $\mathrm{Rep}(G) \setminus \widehat{G}$ is $\widehat{\phi}_R$-invariant.*

(3) *The dimension of the space of intertwining operators between $\rho \in \widehat{G}$ and $\widehat{\phi}_R(\rho)$ is equal to 1 if and only if $\rho \in \mathrm{Fix}(\widehat{\phi})$. Otherwise it is 0.*

PROOF. (1) and (2): This follows from the characterization of irreducible representation as that one for which the centralizer of $\rho(G)$ consists exactly of scalar operators.

(3) Let us decompose $\widehat{\phi}_R(\rho)$ into irreducible ones. Since $\dim H_\rho = \dim H_{\widehat{\phi}(\rho)}$ one has only 2 possibilities: ρ does not appear in $\widehat{\phi}(\rho)$ and the intertwining number is 0, otherwise $\widehat{\phi}_R(\rho)$ is equivalent to ρ. In this case $\rho \in \mathrm{Fix}(\widehat{\phi})$. □

The proof of Theorem 1.4 can be now repeated for the general endomorphism with the new definition of $\mathrm{Fix}(\widehat{\phi})$. The item (3) supplies us with the necessary property.

6. Examples and their discussion

The natural candidate for the dual object to be used instead of $\widehat{G}$ in the case when the different notions of the dual do not coincide (i.e. for groups more general than type I one groups) is the so-called quasi-dual $\widehat{\widehat{G}}$, i.e. the set of quasi-equivalence classes of factor-representations (see, e.g. [3]). This is a usual object when we need a sort of canonical decomposition for regular representation or group C^*-algebra. More precisely, one needs the support $\widehat{G}_p$ of the Plancherel measure.

Unfortunately the following example shows that this is not the case.

EXAMPLE 6.1. Let G be a non-elementary Gromov hyperbolic group. As it was shown by Fel'shtyn [5] with the help of geometrical methods, for any automorphism ϕ of G the Reidemeister number $R(\phi)$ is infinite. In particular this is true for free group in two generators F_2. But the support $(\widehat{F_2})_p$ consists of one point (i.e. regular representation is factorial).

The next hope was to exclude from this dual object the II_1-points assuming that they always give rise to an infinite number of twisted invariant functionals. But this is also wrong:

EXAMPLE 6.2. (an idea of Fel'shtyn realized in [10]) Let $G = (Z \oplus Z) \rtimes_\theta Z$ be the semi-direct product by a hyperbolic action $\theta(1) = \begin{pmatrix} 2 & 1 \\ 1 & 1 \end{pmatrix}$. Let ϕ be an automorphism of G whose restriction to Z is $-id$ and restriction to $Z \oplus Z$ is $\begin{pmatrix} 0 & 1 \\ -1 & 0 \end{pmatrix}$. Then $R(\phi) = 4$, while the space $\widehat{G}_p$ consists of a single II_1-point once again (cf. [2, p. 94]).

These examples show that powerful methods of the decomposition theory do not work for more general classes of groups.

On the other hand Example 6.2 disproves the old conjecture of Fel'shtyn and Hill [8] who supposed that the Reidemeister numbers of a injective endomorphism for groups of exponential growth are always infinite. More precisely, this group is amenable and of exponential growth. Together with some calculations for concrete groups which are too routine to be included in the present paper, this allow us to formulate the following question.

Question. Is the Reidemeister number $R(\phi)$ infinite for any automorphism ϕ of (non-amenable) finitely generated group G containing F_2 ?

In this relation the following example seems to be interesting.

EXAMPLE 6.3. [7] For amenable and non-amenable Baumslag-Solitar groups Reidemeister numbers are always infinite.

For Example 6.2 recently we have found 4 fixed points of $\widehat{\phi}$ being finite dimensional irreducible representations. They give rise to 4 linear independent twisted invariant functionals. These functionals can also be obtained from the regular factorial representation. There also exist fixed points (at least one) that are infinite dimensional irreducible representations. The corresponding functionals are evidently linear dependent with the first 4. This example will be presented in detail in a forthcoming paper.

7. Congruences for Reidemeister numbers of endomorphisms

LEMMA 7.1 ([12]). *For any endomorphism ϕ of a group G and any $x \in G$ one has $\phi(x) \in \{x\}_\phi$.*

PROOF. $\phi(x) = x^{-1}x\phi(x)$.   □

The following lemma is useful for calculating Reidemeister numbers.

LEMMA 7.2. *Let $\phi : G \to G$ be any endomorphism of any group G, and let H be a subgroup of G with the properties*

$$\phi(H) \subset H$$

$$\forall x \in G \; \exists n \in \mathbb{N} \text{ such that } \phi^n(x) \in H.$$

Then

$$R(\phi) = R(\phi_H),$$

where $\phi_H : H \to H$ is the restriction of ϕ to H.

PROOF. Let $x \in G$. Then there is n such that $\phi^n(x) \in H$. By Lemma 7.1 it is known that x is ϕ-conjugate to $\phi^n(x)$. This means that the ϕ-conjugacy class $\{x\}_\phi$ of x has non-empty intersection with H.

Now suppose that $x, y \in H$ are ϕ-conjugate, i.e. there is a $g \in G$ such that

$$gx = y\phi(g).$$

We shall show that x and y are ϕ_H-conjugate, i.e. we can find a $g \in H$ with the above property. First let n be large enough that $\phi^n(g) \in H$. Then applying ϕ^n to the above equation we obtain

$$\phi^n(g)\phi^n(x) = \phi^n(y)\phi^{n+1}(g).$$

This shows that $\phi^n(x)$ and $\phi^n(y)$ are ϕ_H-conjugate. On the other hand, one knows by Lemma 7.1 that x and $\phi^n(x)$ are ϕ_H-conjugate, and y and $\phi^n(y)$ are ϕ_H conjugate, so x and y must be ϕ_H-conjugate.

We have shown that the intersection with H of a ϕ-conjugacy class in G is a ϕ_H-conjugacy class in H. Therefore, we have a map

$$\begin{array}{rccc} Rest: & \mathcal{R}(\phi) & \to & \mathcal{R}(\phi_H) \\ & \{x\}_\phi & \mapsto & \{x\}_\phi \cap H \end{array}$$

It is evident that it has the two-sided inverse

$$\{x\}_{\phi_H} \mapsto \{x\}_\phi.$$

Therefore *Rest* is a bijection and $R(\phi) = R(\phi_H)$. □

COROLLARY 7.3. *Let $H = \phi^n(G)$. Then $R(\phi) = R(\phi_H)$.*

Now we pass to the proof of Theorem 1.5.

PROOF. From Theorems 1.4 and Lemma 7.2 it follows immediately that for every n

$$R(\phi^n) = \# \operatorname{Fix}\left[\widehat{\phi_H}^n : \widehat{H} \to \widehat{H}\right].$$

Let P_n denote the number of periodic points of $\widehat{\phi_H}$ of least period n. One obtains immediately

$$R(\phi^n) = \# \operatorname{Fix}\left[\widehat{\phi_H}^n\right] = \sum_{d|n} P_d.$$

Applying Möbius' inversion formula, we have,

$$P_n = \sum_{d|n} \mu(d) R(\phi^{\frac{n}{d}}).$$

On the other hand, we know that P_n is always divisible by n, because P_n is exactly n times the number of $\widehat{\phi_H}$-orbits in $\widehat{H}$ of cardinality n.

Now we pass to the proof of Theorem 1.5 for general endomorphisms.

From Theorem 1.4, Lemma 7.2 it follows immediately that for every n

$$R(\phi^n) = R(\phi_H^n) = \# \operatorname{Fix}(\widehat{(\phi_H^n)}_R|_{\widehat{H}})$$

Let P_n denote the number of periodic points of $(\widehat{\phi_H})_R|_{\widehat{H}}$ of least period n. One obtains by Lemma 5.5 (2)

$$R(\phi^n) = \# \operatorname{Fix}(\widehat{(\phi_H^n)}_R|_{\widehat{H}}) = \sum_{d|n} P_d.$$

The proof can be completed as in the case of automorphisms.   □

8. Congruences for Reidemeister numbers of a continuous map

Now we pass to the formulation of the topological counterpart of the main theorems. Let X be a connected, compact polyhedron and $f : X \to X$ be a continuous map. Let $p : \widetilde{X} \to X$ be the universal cover of X and $\widetilde{f} : \widetilde{X} \to \widetilde{X}$ a lifting of f, i.e. $p \circ \widetilde{f} = f \circ p$. Two liftings $\widetilde{f}$ and $\widetilde{f}'$ are called *conjugate* if there is an element γ in the deck transformation group $\Gamma \cong \pi_1(X)$ such that $\widetilde{f}' = \gamma \circ \widetilde{f} \circ \gamma^{-1}$. The subset $p(\operatorname{Fix}(\widetilde{f})) \subset \operatorname{Fix}(f)$ is called *the fixed point class of f determined by the lifting class* $[\widetilde{f}]$. Two fixed points x_0 and x_1 of f belong to the same fixed point class iff there is a path c from x_0 to x_1 such that $c \cong f \circ c$ (homotopy relative to endpoints). This fact can be considered as an equivalent definition of a non-empty fixed point class. Every map f has only finitely many non-empty fixed point classes, each a compact subset of X.

The number of lifting classes of f (and hence the number of fixed point classes, empty or not) is called the *Reidemeister number* of f, which is denoted by $R(f)$. This is a positive integer or infinity.

Let a specific lifting $\widetilde{f} : \widetilde{X} \to \widetilde{X}$ be fixed. Then every lifting of f can be written in a unique way as $\gamma \circ \widetilde{f}$, with $\gamma \in \Gamma$. So the elements of Γ serve as "coordinates" of liftings with respect to the fixed $\widetilde{f}$. Now, for every $\gamma \in \Gamma$, the composition $\widetilde{f} \circ \gamma$ is a lifting of f too; so there is a unique $\gamma' \in \Gamma$ such that $\gamma' \circ \widetilde{f} = \widetilde{f} \circ \gamma$. This correspondence $\gamma \to \gamma'$ is determined by the fixed $\widetilde{f}$, and is obviously a homomorphism.

DEFINITION 8.1. The endomorphism $\widetilde{f}_* : \Gamma \to \Gamma$ determined by the lifting $\widetilde{f}$ of f is defined by

$$\widetilde{f}_*(\gamma) \circ \widetilde{f} = \widetilde{f} \circ \gamma.$$

We shall identify $\pi = \pi_1(X, x_0)$ and Γ in the following way. Choose base points $x_0 \in X$ and $\widetilde{x}_0 \in p^{-1}(x_0) \subset \widetilde{X}$ once and for all. Now points of $\widetilde{X}$ are in 1-1 correspondence with homotopy classes of paths in X which start at x_0: for $\widetilde{x} \in \widetilde{X}$ take any path in $\widetilde{X}$ from $\widetilde{x}_0$ to $\widetilde{x}$ and project it onto X; conversely, for a path c starting at x_0, lift it to a path in $\widetilde{X}$ which starts at $\widetilde{x}_0$, and then take its endpoint. In this way, we identify a point of $\widetilde{X}$ with a path class $\langle c \rangle$ in X starting from x_0. Under this identification, $\widetilde{x}_0 = \langle e \rangle$ is the unit element in $\pi_1(X, x_0)$. The action of the loop class $\alpha = \langle a \rangle \in \pi_1(X, x_0)$ on $\widetilde{X}$ is then given by

$$\alpha = \langle a \rangle : \langle c \rangle \to \alpha \cdot c = \langle a \cdot c \rangle.$$

Now we have the following relationship between $\widetilde{f}_* : \pi \to \pi$ and

$$f_* : \pi_1(X, x_0) \longrightarrow \pi_1(X, f(x_0)).$$

LEMMA 8.2. *Suppose* $\widetilde{f}(\widetilde{x}_0) = \langle w \rangle$. *Then the following diagram commutes:*

$$
\begin{array}{ccc}
\pi_1(X, x_0) & \xrightarrow{\ f_*\ } & \pi_1(X, f(x_0)) \\
& \searrow{\scriptstyle \widetilde{f}_*} & \downarrow{\scriptstyle w_*} \\
& & \pi_1(X, x_0)
\end{array}
$$

where w_* *is the isomorphism induced by the path* w.

In other words, for every $\alpha = \langle a \rangle \in \pi_1(X, x_0)$, we have

$$
\widetilde{f}_*(\langle a \rangle) = \langle w(f \circ a)w^{-1} \rangle.
$$

REMARK 8.3. *In particular, if* $x_0 \in p(\mathrm{Fix}(\widetilde{f}))$ *and* $\widetilde{x}_0 \in \mathrm{Fix}(\widetilde{f})$, *then* $\widetilde{f}_* = f_*$.

LEMMA 8.4 (*see, e.g.* [12]). *Lifting classes of* f *(and hence fixed point classes, empty or not) are in 1-1 correspondence with* $\widetilde{f}_*$-*conjugacy classes in* π, *the lifting class* $[\gamma \circ \widetilde{f}]$ *corresponding to the* $\widetilde{f}_*$-*conjugacy class of* γ. *We therefore have* $R(f) = R(\widetilde{f}_*)$.

We shall say that the fixed point class $p(\mathrm{Fix}(\gamma \circ \widetilde{f}))$, which is labeled with the lifting class $[\gamma \circ \widetilde{f}]$, *corresponds* to the $\widetilde{f}_*$-conjugacy class of γ. Thus $\widetilde{f}_*$-conjugacy classes in π serve as "coordinates" for fixed point classes of f, once a fixed lifting $\widetilde{f}$ is chosen.

Using Lemma 8.4 we may apply the Theorem 1.5 to the Reidemeister numbers of continuous maps.

THEOREM 8.5. *Let* $f : X \to X$ *be a continuous map of a compact polyhedron* X *such that all numbers* $R(f^n)$ *are finite. Let* $f_* : \pi_1(X) \to \pi_1(X)$ *be an induced endomorphism of the group* $\pi_1(X)$ *and let* H *be a subgroup of* $\pi_1(X)$ *with the properties*

(1) $f_*(H) \subset H$,
(2) $\forall x \in \pi_1(X) \ \exists n \in \mathbb{N}$ *such that* $f_*^n(x) \in H$.

If the couple (H, f_*^n) *satisfies the conditions of Theorem 1.4 for any* $n \in \mathbb{N}$, *then one has for all* n,

$$
\sum_{d|n} \mu(d) \cdot R(f^{n/d}) \equiv 0 \mod n.
$$

Acknowledgment. The second author is partially supported by RFFI Grant 02-01-00574, Grant for the support of leading scientific schools HIII-619.203.1 and Grant "Universities of Russia."

References

[1] James Arthur and Laurent Clozel, *Simple algebras, base change, and the advanced theory of the trace formula*, Princeton University Press, Princeton, NJ, 1989. MR **90m:**22041

[2] Alain Connes, *Noncommutative geometry*, Academic Press Inc., San Diego, CA, 1994. MR **95j:**46063

[3] J. Dixmier, *C*-algebras*, North-Holland, Amsterdam, 1982.

[4] P. Eymard, *L'algèbre de Fourier d'un groupe localement compact*, Bull. Soc. math. France **92** (1964), 181–236.

[5] A. L. Fel'shtyn, *The Reidemeister number of any automorphism of a Gromov hyperbolic group is infinite*, Zap. Nauchn. Sem. S.-Peterburg. Otdel. Mat. Inst. Steklov. (POMI) **279** (2001), no. Geom. i Topol. 6, 229–240, 250. MR **2002e:**20081

[6] Alexander Fel'shtyn, *Dynamical zeta functions, Nielsen theory and Reidemeister torsion*, Mem. Amer. Math. Soc. **147** (2000), no. 699, xii+146. MR **2001a:**37031

[7] Alexander Fel'shtyn and Daciberg Gonçalves, *Twisted conjugacy classes of automorphisms of Baumslag-Solitar groups*, http://de.arxiv.org/abs/math.GR/0405590.

[8] Alexander Fel'shtyn and Richard Hill, *The Reidemeister zeta function with applications to Nielsen theory and a connection with Reidemeister torsion*, K-Theory **8** (1994), no. 4, 367–393. MR **95h:**57025

[9] _____, *Dynamical zeta functions, congruences in Nielsen theory and Reidemeister torsion*, Nielsen theory and Reidemeister torsion (Warsaw, 1996), Polish Acad. Sci., Warsaw, 1999, pp. 77–116. MR **2001h:**37047

[10] Daciberg Gonçalves and Peter Wong, *Twisted conjugacy classes in exponential growth groups*, Bull. London Math. Soc. **35** (2003), no. 2, 261–268. MR **2003j:**20054

[11] A. Grothendieck, *Formules de Nielsen-Wecken et de Lefschetz en géométrie algébrique*, Séminaire de Géométrie Algébrique du Bois-Marie 1965-66. SGA 5, Lecture Notes in Math., vol. 569, Springer-Verlag, Berlin, 1977, pp. 407–441.

[12] B. Jiang, *Lectures on Nielsen fixed point theory*, Contemp. Math., vol. 14, Amer. Math. Soc., Providence, RI, 1983.

[13] A. A. Kirillov, *Elements of the theory of representations*, Springer-Verlag, Berlin Heidelberg New York, 1976.

[14] Salahoddin Shokranian, *The Selberg-Arthur trace formula*, Springer-Verlag, Berlin, 1992, Based on lectures by James Arthur. MR **93j:**11029

[15] Elmar Thoma, *Über unitäre Darstellungen abzählbarer, diskreter Gruppen*, Math. Ann. **153** (1964), 111–138. MR 28 #3332

FACHBEREICH MATHEMATIK, EMMY-NOETHER-CAMPUS, UNIVERSITÄT SIEGEN, WALTER-FLEX-STR. 3, D-57068 SIEGEN, GERMANY AND INSTYTUT MATEMATYKI, UNIWERSYTET SZCZECINSKI, UL. WIELKOPOLSKA 15, 70-451 SZCZECIN, POLAND
   E-mail address: felshtyn@math.uni-siegen.de
   URL: http://www.math.uni-siegen.de/felshtyn

DEPT. OF MECH. AND MATH., MOSCOW STATE UNIVERSITY, 119992 GSP-2 MOSCOW, RUSSIA
   E-mail address: troitsky@mech.math.msu.su
   URL: http://mech.math.msu.su/~troitsky

Introduction to Hopf-Cyclic Cohomology

Masoud Khalkhali and Bahram Rangipour

ABSTRACT. We review the recent progress in the study of cyclic cohomology in the presence of Hopf symmetry.

1. Introduction

In their study of the index theory of transversally elliptic operators [9], Connes and Moscovici develope a cyclic cohomology theory for Hopf algebras which can be regarded, *post factum*, as the right noncommutative analogue of group homology and Lie algebra homology. One of the main reasons for introducing this theory was to obtain a *noncommutative characteristic map*

$$\chi_\tau : HC^*_{(\delta,\sigma)}(H) \longrightarrow HC^*(A),$$

for an action of a Hopf algebra H on an algebra A endowed with an "invariant trace" $\tau : A \to \mathbb{C}$. Here, the pair (δ, σ) consists of a grouplike element $\sigma \in H$ and a characater $\delta : H \to \mathbb{C}$ satisfying certain compatibility conditions explained in Section 2.3 below.

In [21] we found a new approach to this subject and extended it, among other things, to a cyclic cohomology theory for triples (C, H, M), where C is a coalgebra endowed with an action of a Hopf algebra H and M is an H-module and an H-comodule satisfying some extra compatibility conditions. It was observed that the theory of Connes and Moscovici corresponds to $C = H$ equipped with the regular action of H and M a one dimensional H-module with an extra structure.

One of the main ideas of [21] was to view the Hopf-cyclic cohomology as the cohomology of the *invariant* part of certain natural complexes attached to (C, H, M). This is remarkably similar to interpreting the cohomology of the Lie algebra of a Lie group as the invariant part of the de Rham cohomology of the Lie group. The second main idea was to introduce *coefficients* into the theory. This also explained the important role played by the so called modular pair (δ, σ) in [9].

Since the module M is a noncommutative analogue of coefficients for Lie algebra and group homology theories, it is of utmost importance to understand the most general type of coefficients allowable. In fact the periodicity condition $\tau_n^{n+1} = id$ for the cyclic operator and the fact that all simplicial and cyclic operators have to descend to the invrainat complexes, puts very stringent conditions on the type of the H-module M. This problem that remained unsettled in our paper [21] is completely solved in Hajac-Khalkhali-Rangipour-Sommerhäuser papers [14, 15] by introducing the class of *stable anti-Yetter-Drinfeld modules* over a Hopf algebra.

The category of anti-Yetter-Drinfeld modules over a Hopf algebra H is a twisting, or 'mirror image' of the category of Yetter-Drinfeld H-modules. Technically it is obtained from the latter by replacing the antipode S by S^{-1} although this connection is hardly illuminating.

In an effort to make this paper more accessible, we cover basic background material, with many examples, on Hopf algebras and the emerging role of *Hopf symmetry* in noncommutative geometry and its applications [**9, 10, 11, 12**]. This is justified since certain doses of the *"yoga of Hopf algebras"*, in the noncommutative and non-cocommutative case, is necessary to understand these works. Following these works, we emphasize the universal role played by the Connes-Moscovici Hopf algebra $\mathcal{H}_1$ and its cyclic cohomology in applications of noncommutative geometry to transverse geometry and number theory. See also Marcolli's article [**24**] as well as Connes-Marcolli articles [**6, 7**] and references therein for recent interactions between number theory and noncommutative geometry.

2. Preliminaries

In this section we recall some aspects of Hopf algebra theory that are most relevant to the current status of Hopf-cyclic cohomology theory.

2.1. Coalgebras, bialgebras, and Hopf algebras. We assume our Hopf algebras, coalgebras, and algebras are over a fixed field k of characteristic zero. Most of our definitions and constructions however continue to work over an arbitrary commutative ground ring k. Unadorned $\otimes$ and *Hom* will always be over k and I will always denote an identity map whose domain will be clear from the context. By a *coalgebra* over k we mean a k-linear space C endowed with k-linear maps

$$\Delta : C \longrightarrow C \otimes C, \quad \varepsilon : C \longrightarrow k,$$

called *comultiplication* and *counit* respectively, such that Δ is *coassociative* and ε is the counit of Δ. That is,

$$(\Delta \otimes I) \circ \Delta = (I \otimes \Delta) \circ \Delta : C \longrightarrow C \otimes C \otimes C,$$

$$(\varepsilon \otimes I) \circ \Delta = (I \otimes \varepsilon) \circ \Delta = I.$$

C is called *cocommutative* if $\tau\Delta = \Delta$, where $\tau : C \otimes C \to C \otimes C$ is the *fillip* $x \otimes y \mapsto y \otimes x$.

We use Sweedler-Heynemann's notation for comultiplication, with summation suppressed, and write

$$\Delta(c) = c^{(1)} \otimes c^{(2)}.$$

With this notation the axioms for a coalgebra read as

$$c^{(1)} \otimes c^{(2)(1)} \otimes c^{(2)(2)} = c^{(1)(1)} \otimes c^{(1)(2)} \otimes c^{(2)},$$

$$\varepsilon(c^{(1)})(c^{(2)}) = c = (c^{(1)})\varepsilon(c^{(2)}),$$

for all $c \in C$. We put

$$c^{(1)} \otimes c^{(2)} \otimes c^{(3)} := (\Delta \otimes I)\Delta(c).$$

Similarly, for *iterated comultiplication* maps

$$\Delta^n := (\Delta \otimes I) \circ \Delta^{n-1} : C \longrightarrow C^{\otimes(n+1)}, \quad \Delta^1 = \Delta,$$

we write

$$\Delta^n(c) = c^{(1)} \otimes \cdots \otimes c^{(n+1)},$$

where summation is understood. Many notions for algebras have their dual analogues for coalgebras. Thus, one can easily define such notions like, *subcoalgebra*, (left, right, two sided) *coideal*, *quotient coalgebra*, and *morphism of coalgebras* [26, 29].

A left *C-comodule* is a linear space M together with a linear map $\rho : M \to C \otimes M$ such that $(\rho \otimes 1)\rho = \Delta\rho$ and $(\varepsilon \otimes 1)\rho = \rho$. We write

$$\rho(m) = m^{(-1)} \otimes m^{(0)},$$

where summation is understood, to denote the coaction ρ. Similarly if M is a right C-comodule, we write

$$\rho(m) = m^{(0)} \otimes m^{(1)}$$

to denote its coaction $\rho : M \to M \otimes C$. Note that module elements are always assigned zero index. Let M and N be left C-comodules. A *C-colinear map* is a linear map $f : M \to N$ such that $\rho_N f = (1 \otimes f)\, \rho_M$. The category of left C-comodules is an abelian category; note that, unlike the situation for algebras, for this to be true, it is important that C be a flat k-module which will be the case if k is a field.

Let C be a coalgebra, A be a unital algebra, and $f, g : C \to A$ be k-linear maps. The *convolution product* of f and g, denoted by $f * g$, is defined as the composition

$$C \xrightarrow{\Delta} C \otimes C \xrightarrow{f \otimes g} A \otimes A,$$

or equivalently by

$$(f * g)(c) = f(c^{(1)})g(c^{(2)}).$$

It is easily checked that under the convolution product $Hom(C, A)$ is an associative unital algebra. Its unit is the map $e : C \to A$, $e(c) = \varepsilon(c)1_A$. In particular the linear dual of a coalgebra C, $C^* = Hom(C, k)$, is an algebra.

A *bialgebra* is a unital algebra endowed with a compatible coalgebra structure. This means that the coalgebra structure maps $\Delta : B \longrightarrow B \otimes B$, $\varepsilon : B \longrightarrow k$, are morphisms of unital algebras. This is equivalent to multiplication and unit maps of A being morphisms of coalgebras.

A *Hopf algebra* is a bialgebra endowed with an *antipode*. By definition, an antipode for a bialgebra H is a linear map $S : H \to H$ such that

$$S * I = I * S = \eta\varepsilon,$$

where $\eta : k \to H$ is the unit map of H. Equivalently,

$$S(h^{(1)})h^{(2)} = h^{(1)}S(h^{(2)}) = \varepsilon(h)1,$$

for all $h \in H$. Thus S is the inverse of the identity map $I : H \to H$ in the convolution algebra $Hom(H, H)$. This shows that the antipode is unique, if it exists at all. The following properties of the antipode are well known:

1. If H is commutative or cocommutative then $S^2 = I$. The converse need not be true.

2. S is an anti-algebra map and an anti-coalgebra map. The latter means

$$S(h^{(2)}) \otimes S(h^{(1)}) = S(h)^{(1)} \otimes S(h)^{(2)},$$

for all $h \in H$.

We give a few examples of Hopf algebras:

1. Commutative or cocommutative Hopf algebras are closely related to groups and Lie algebras. We give a few examples to indicate this connection

1.a. Let G be a discrete group (need not be finite) and $H = kG$ the group algebra of G over k. Let

$$\Delta(g) = g \otimes g, \quad S(g) = g^{-1}, \quad \text{and } \varepsilon(g) = 1,$$

for all $g \in G$ and linearly extend them to H. Then it is easy to check that $(H, \Delta, \varepsilon, S)$ is a cocommutative Hopf algebra. It is of course commutative if and only if G is commutative.

1.b. Let $\mathfrak{g}$ be a k-Lie algebra and $H = U(\mathfrak{g})$ be the universal enveloping algebra of $\mathfrak{g}$. Using the universal property of $U(\mathfrak{g})$ one checks that there are uniquely defined algebra homomorphisms $\Delta : U(\mathfrak{g}) \to U(\mathfrak{g}) \otimes U(\mathfrak{g})$, $\varepsilon : U(\mathfrak{g}) \to k$ and an anti-algebra map $S : U(\mathfrak{g}) \to U(\mathfrak{g})$, determined by

$$\Delta(X) = X \otimes 1 + 1 \otimes X, \quad \varepsilon(X) = 0, \quad \text{and } S(X) = -X,$$

for all $X \in \mathfrak{g}$. One then checks easily that $(U(\mathfrak{g}), \Delta, \varepsilon, S)$ is a cocommutative Hopf algebra. It is commutative if and only if $\mathfrak{g}$ is an abelian Lie algebra, in which case $U(\mathfrak{g}) = S(\mathfrak{g})$ is the symmetric algebra of $\mathfrak{g}$.

1.c. Let G be a compact topological group. A continuous function $f : G \to \mathbb{C}$ is called *representable* if the set of left translates of f by all elements of G forms a finite dimensional subspace of the space $C(G)$ of all continuous complex valued functions on G. It is then easily checked that the set of all representable functions, $H = Rep(G)$, is a subalgebra of the algebra of continuous functions on G. Let $m : G \times G \to G$ denote the multiplication of G and $m^* : C(G \times G) \to C(G)$, $m^* f(x,y) = f(xy)$, denote its dual map. It is easy to see that if f is representable, then $m^* f \in Rep(G) \otimes Rep(G) \subset C(G \times G)$. Let e denote the identity of G. One can easily check that the relations

$$\Delta f = m^* f, \quad \varepsilon f = f(e), \quad \text{and } (Sf)(g) = f(g^{-1}),$$

define a Hopf algebra structure on $Rep(G)$. Alternatively, one can describe $Rep(G)$ as the linear span of matrix coefficients of all finite dimensional complex representations of G. By *Peter-Weyl's Theorem*, $Rep(G)$ is a dense subalgebra of $C(G)$. This algebra is finitely generated (as an algebra) if and only if G is a Lie group.

1.d. The coordinate ring of an affine algebraic group $H = k[G]$ is a commutative Hopf algebra. The maps Δ, ε, and S are the duals of the multiplication, unit, and inversion maps of G, respectively. More generally, an *affine group scheme*, over a commutative ring k, is a commutative Hopf algebra over k. Given such a Hopf algebra H, it is easy to see that for any commutative k-algebra A, the set $Hom_{Alg}(H, A)$ is a group under convolution product and $A \mapsto Hom_{Alg}(H, A)$ is a functor from the category $ComAlg_k$ of commutative k-algebras to the category of groups. Conversely, any representable functor $ComAlg_k \to Groups$ is represented by a, unique up to isomorphism, commutative k-Hopf algebra. Thus the category of affine group schemes is equivalent to the category of representable functors $ComAlg_k \to Groups$.

2. Compact quantum groups and quantized enveloping algebras are examples of noncommutative and noncommutative Hopf algebras [26]. We won't recall these examples here. A very important example for noncommutative geometry and its applications to transverse geometry and number theory is the *Connes-Moscovici Hopf algebra* $\mathcal{H}_1$ [9, 11, 12] which we recall now. Let $\mathfrak{g}_{aff}$ be the Lie algebra of the group of affine transformations of the line with linear basis X and Y and the relation $[Y, X] = X$. Let $\mathfrak{g}$ be an abelian Lie algebra with basis $\{\delta_n; \, n = 1, 2, \cdots\}$. It is easily seen that $\mathfrak{g}_{aff}$ acts on $\mathfrak{g}$ via

$$[Y, \delta_n] = n\delta_n, \quad [X, \delta_n] = \delta_{n+1},$$

for all n. Let $\mathfrak{g}_{CM} := \mathfrak{g}_{aff} \ltimes \mathfrak{g}$ be the corresponding semidirect product Lie algebra. As an algebra, $\mathcal{H}_1$ coincides with the universal enveloping algebra of the Lie algebra $\mathfrak{g}_{CM}$. Thus $\mathcal{H}_1$ is the universal algebra generated by $\{X, Y, \delta_n; n = 1, 2, \cdots\}$ subject to relations

$$[Y, X] = X, \quad [Y, \delta_n] = n\delta_n, \quad [X, \delta_n] = \delta_{n+1}, \quad [\delta_k, \delta_l] = 0,$$

for $n, k, l = 1, 2, \cdots$. We let the counit of $\mathcal{H}_1$ coincide with the counit of $U(\mathfrak{g}_{CM})$. Its coproduct and antipode, however, will be certain deformations of the coproduct and antipode of $U(\mathfrak{g}_{CM})$ as follows. Using the universal property of $U(\mathfrak{g}_{CM})$, one checks that the relations

$$\Delta Y = Y \otimes 1 + 1 \otimes Y, \quad \Delta \delta_1 = \delta_1 \otimes 1 + 1 \otimes \delta_1,$$

$$\Delta X = X \otimes 1 + 1 \otimes X + \delta_1 \otimes Y,$$

determine a unique algebra map $\Delta : \mathcal{H}_1 \to \mathcal{H}_1 \otimes \mathcal{H}_1$. Note that Δ is not cocommutative and it differs from the corrodent of the enveloping algebra $U(\mathfrak{g}_{CM})$. Similarly, one checks that there is a unique antialgebra map $S : \mathcal{H}_1 \to \mathcal{H}_1$ determined by the relations

$$S(Y) = -Y, \quad S(X) = -X + \delta_1 Y, \quad S(\delta_1) = -\delta_1.$$

Again we note that this antipode also differs from the antipode of $U(\mathfrak{g}_{CM})$, and in particular $S^2 \neq I$. In fact $S^n \neq I$ for all n. In the next section we will show, following Connes-Moscovici [9], that $\mathcal{H}_1$ is a bicrossed product of Hopf algebras $U(\mathfrak{g}_{aff})$ and $U(\mathfrak{g})^*$, where $\mathfrak{g}$ is a pro-nilpotent Lie algebra to be described precisely in the next section.

Let H be a Hopf algebra. A *grouplike* element of H is a non-zero element $g \in H$ such that $\Delta g = g \otimes g$. We have, from the axioms for the antipode, $gS(g) = S(g)g = 1_H$ which shows that g is invertible. It is easily seen that grouplike elements of H form a subgroup of the multiplicative group of H. For example, for $H = kG$ the set of grouplike elements coincide with the group G itself. A *primitive element* is an element $x \in H$ such that $\Delta x = 1 \otimes x + x \otimes 1$. It is easily seen that the bracket $[x, y] := xy - yx$ of two primitive elements is again a primitive element. It follows that primitive elements form a Lie algebra. Using the *Poincare-Birkhoff-Witt* theorem, one shows that the set of primitive elements of $H = U(\mathfrak{g})$ coincide with the Lie algebra $\mathfrak{g}$.

A *character* of a Hopf algebra is a unital algebra map $\delta : H \to k$. For example the counit $\varepsilon : H \to k$ is a character. For a less trivial example, let G be a *non-unimodular* real Lie group and $H = U(\mathfrak{g})$ the universal enveloping algebra of the Lie algebra of G. The *modular function* of G, measuring the difference between the right and left Haar measure on G, is a smooth group homomorphism $\Delta : G \to \mathbb{R}^+$. Its derivative at identity defines a Lie algebra map $\delta : \mathfrak{g} \to \mathbb{R}$. We denote its natural extension by $\delta : U(\mathfrak{g}) \to \mathbb{R}$. It is obviously a character of $U(\mathfrak{g})$. For a concrete example, let $G = Aff(\mathbb{R})$ be the group of affine transformations of the real line. It is a non-unimodular group with modular homomorphism given by

$$\Delta \begin{pmatrix} a & b \\ 0 & 1 \end{pmatrix} = |a|.$$

The corresponding infinitesimal character on $\mathfrak{g}_{aff} = Lie(G)$ is given by

$$\delta(Y) = 1, \quad \delta(X) = 0,$$

where $Y = \begin{pmatrix} 1 & 0 \\ 0 & 0 \end{pmatrix}$ and $X = \begin{pmatrix} 0 & 1 \\ 0 & 0 \end{pmatrix}$ are a basis for $\mathfrak{g}_{aff}$. We will see that this character plays an important role in constructing a *twisted antipode* for the Connes-Moscovici Hopf algebra $\mathcal{H}_1$.

If H is a Hopf algebra, by a left H-module (resp. left H-comodule), we mean a left module (resp. left comodule) over the underlying algebra (resp. the underlying coalgebra) of H.

Recall that a *monoidal*, or *tensor*, category $(\mathcal{C}, \otimes, U, a, l, r)$ consists of a category $\mathcal{C}$, a functor $\otimes : \mathcal{C} \times \mathcal{C} \to \mathcal{C}$, an object $U \in \mathcal{C}$ (called the *unit object*), and natural isomorphisms

$$a = a_{A,B,C} : A \otimes (B \otimes C) \to (A \otimes B) \otimes C,$$

$$l = l_A : U \otimes A \to A, \qquad r = r_A : A \otimes U \to A,$$

(called the *associativity* and *unit constraints*, respectively) such that the so called *pentagon* and *triangle* diagrams commute:

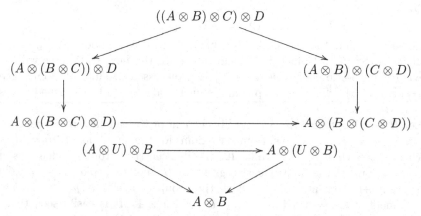

The coherence theorem of MacLane [25] guarantees that all diagrams composed from a, l, r by tensoring, substituting and composing, commute.

A *braided monoidal* category, is a monoidal category $\mathcal{C}$ endowed with a natural family of isomorphisms

$$c_{A,B} : A \otimes B \to B \otimes A,$$

called *braiding* such that the following diagram and the one obtained from it by replacing c by c^{-1} commute (*Hexagon axioms*):

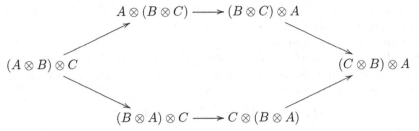

A braiding is called a *symmetry* if we have

$$c_{A,B} \circ c_{B,A} = I$$

for all A, B. A *symmetric monoidal category*, is a monoidal category endowed with a symmetry.

Let H be a bialgebra. Then thanks to the existence of a comultiplication on H, the category $H - Mod$ of left H-modules is a monoidal category: if M and N are left H-modules, their tensor product over k, $M \otimes N$, is again an H-module via

$$h(m \otimes n) = h^{(1)}m \otimes h^{(2)}n.$$

One can easily check that with associativity constraints defined as the natural isomorphism of H-modules $M \otimes (N \otimes P) \to (M \otimes N) \otimes P$, $m \otimes (n \otimes p) \mapsto (m \otimes n) \otimes p$, and with unit object $U = k$ with trivial H-action via the counit ε,

$$h(1) = \varepsilon(h)1,$$

$H - Mod$ is a monoidal category. If H is cocommutative, then one checks that the map

$$c_{M,N} : M \otimes N \to N \otimes M, \quad c_{M,N}(m \otimes n) = n \otimes m,$$

is a morphism of H-modules and is a symmetry operator on $H - Mod$, turning it into a symmetric monoidal category.

The category $H - Mod$ is not braided in general. For that to happen, one must either restrict the class of modules to what is called Yetter-Drinfeld modules, or restrict the class of Hopf algebras to quasisimilar Hopf algebras to obtain a braiding on $H - Mod$ [26]. We will discuss the first scenario in the next section and will see that, quite unexpectedly, this question is closely related to Hopf-cyclic cohomology.

Similarly, the category $H - Comod$ of left H-comodules is a monoidal category: if M and N are left H-comodules, their tensor product $M \otimes N$ is again an H-comodule via

$$\rho(m \otimes n) = m^{(-1)}n^{(-1)} \otimes m^{(0)} \otimes n^{(0)}.$$

Its unit object is $U = k$ endowed with the H-coaction $r \in k \mapsto 1_H \otimes r$. If H is commutative then $H - Comod$ is a symmetric monoidal category. More generally, when H is co-quasitriangular, $H - Comod$ can be endowed with a braiding [26].

2.2. Symmetry in noncommutative geometry.

The idea of *symmetry* in noncommutative geometry is encoded by the *action* or *coaction* of a Hopf algebra on an algebra or on a coalgebra. Thus there are four possibilities in general that will be referred to as (Hopf-) *module algebra, module coalgebra, comodule algebra,* and *comodule coalgebra* (for each type, of course, we still have a choice of left or right action or coaction). We call them *symmetries of type A, B, C* and *D*, respectively. We will see in the next sections that associated to each type of symmetry there is a corresponding cyclic cohomology theory with coefficients. These theories in a certain sense are generalizations of equivariant de Rham cohomology with coefficients in an equivariant local system.

Let H be a Hopf algebra. An algebra A is called a left H-*module algebra*, if it is a left H-module and the multiplication map $A \otimes A \to A$ and the unit map $k \to A$ are morphisms of H-modules. This means

$$h(ab) = h^{(1)}(a)h^{(2)}(b), \quad \text{and} \quad h(1) = \varepsilon(h)1,$$

for all $h \in H$ and $a, b \in A$ (summation is understood). Using the relations $\Delta g = g \otimes g$ and $\Delta x = 1 \otimes x + x \otimes 1$, it is easily seen that in an H-module algebra, grouplike elements act as unit preserving automorphisms while primitive elements act as derivations. In particular, for $H = kG$ the group algebra of a discrete group, an H-module algebra structure on A is simply an action of G by unit preserving automorphisms on A. Similarly, we have a 1-1 correspondence between $U(\mathfrak{g})$-module algebra structures on A and Lie actions of the Lie algebra $\mathfrak{g}$ by derivations on A.

An algebra B is called a left H-*comodule algebra*, if B is a left H-comodule and the multiplication and unit maps of B are H-comodule maps. That is

$$(ab)^{(-1)} \otimes (ab)^{(0)} = a^{(-1)}b^{(-1)} \otimes a^{(0)}b^{(0)}, \qquad (1_B)^{(-1)} \otimes (1_B)^{(0)} = 1_H \otimes 1_B,$$

for all a, b in B.

A left H-*module coalgebra* is a coalgebra C which is a left H-module such that the comultiplication map $\Delta : C \to C \otimes C$ and the counit map $\varepsilon : C \to k$ are H-module maps. That is

$$(hc)^{(1)} \otimes (hc)^{(2)} = h^{(1)}c^{(1)} \otimes h^{(2)}c^{(2)}, \qquad \varepsilon(hc) = \varepsilon(h)\varepsilon(c),$$

for all $h \in H$ and $c \in C$.

Finally, a coalgebra D is called a left H-*comodule coalgebra* when the comultiplication and counit maps of D are morphisms of H-comodules. That is

$$c^{(1)(-1)}c^{(2)(-1)} \otimes c^{(1)(0)} \otimes c^{(2)(0)} = c^{(-1)} \otimes c^{(0)(1)} \otimes c^{(0)(2)},$$

$$\varepsilon(c)1_H = c^{(-1)}\varepsilon(c^{(0)}),$$

for all $c \in C$.

We call the above four types of symmetries, *symmetries of type A, B, C,* and *D*, respectively. All four types of symmetries are present within an arbitrary Hopf algebra. For example, the coproduct $\Delta : H \to H \otimes H$ gives H the structure of a left (and right) H-comodule algebra while the product $H \otimes H \to H$ turns H into a left (and right) H-module coalgebra. These are noncommutative analogues of translation action of a group on itself. The *conjugation action* $H \otimes H \to H$, $g \otimes h \mapsto g^{(1)}hS(g^{(2)})$ gives H the structure of a left H-module algebra. The *co-conjugation* action $H \to H \otimes H$, $h \mapsto h^{(1)}S(h^{(3)}) \otimes h^{(2)}$ turns H into an H-comodule coalgebra.

For a different example, we turn to the Connes-Moscovici Hopf algebra $\mathcal{H}_1$. An important feature of $\mathcal{H}_1$, and in fact its *raison d'être*, is that it acts as quantum symmetries of various objects of interest in noncommutative geometry, like the 'space' of leaves of codimension one foliations or the 'space' of modular forms modulo the action of Hecke correspondences. Let M be a one dimensional manifold and $A = C_0^\infty(F^+M)$ denote the algebra of smooth functions with compact support on the bundle of positively oriented frames on M. Given a discrete group

$\Gamma \subset Diff^+(M)$ of orientation preserving diffeomorphisms of M, one has a natural prolongation of the action of Γ to $F^+(M)$ by

$$\varphi(y, y_1) = (\varphi(y), \varphi'(y)(y_1)).$$

Let $A_\Gamma = C_0^\infty(F^+M) \rtimes \Gamma$ denote the corresponding crossed product algebra. Thus the elements of A_Γ consist of finite linear combinations (over $\mathbb{C}$) of terms fU_φ^* with $f \in C_0^\infty(F^+M)$ and $\varphi \in \Gamma$. Its product is defined by

$$fU_\varphi^* \cdot gU_\psi^* = (f \cdot \varphi(g))U_{\psi\varphi}^*.$$

There is an action of $\mathcal{H}_1$ on A_Γ given by [**9, 12**]:

$$Y(fU_\varphi^*) = y_1\frac{\partial f}{\partial y_1}U_\varphi^*, \quad X(fU_\varphi^*) = y_1\frac{\partial f}{\partial y}U_\varphi^*,$$

$$\delta_n(fU_\varphi^*) = y_1^n\frac{d^n}{dy^n}(log\frac{d\varphi}{dy})fU_\varphi^*.$$

Once these formulas are given, it can be checked, by a long computation, that A_Γ is indeed an $\mathcal{H}_1$-module algebra. In the original application, M is a transversal for a codimension one foliation and thus $\mathcal{H}_1$ acts via transverse differential operators [**9**].

We recall, very briefly, the action of the Hopf algebra $\mathcal{H}_1$ on the so called *modular Hecke algebras*, discovered by Connes and Moscovici in [**12, 13**] where detailed discussions and a very intriguing dictionary comparing transverse geometry notions with modular forms notions can be found. For each $N \geq 1$, let

$$\Gamma(N) = \left\{\begin{pmatrix} a & b \\ c & d \end{pmatrix} \in SL(2,\mathbb{Z}); \begin{pmatrix} a & b \\ c & d \end{pmatrix} = \begin{pmatrix} 1 & 0 \\ 0 & 1 \end{pmatrix} \bmod N\right\}$$

denote the level N congruence subgroup of $\Gamma(1) = SL(2,\mathbb{Z})$. Let $\mathcal{M}_k(\Gamma(N))$ denote the space of modular forms of level N and weight k and

$$\mathcal{M}(\Gamma(N)) := \oplus_k\mathcal{M}_k(\Gamma(N))$$

be the graded algebra of modular forms of level N. Finally let

$$\mathcal{M} := \lim_{\overrightarrow{N}} \mathcal{M}(\Gamma(N))$$

denote the algebra of modular forms of all levels, where the inductive system is defined by divisibility. The group

$$G^+(\mathbb{Q}) := GL^+(2,\mathbb{Q}),$$

acts on $\mathcal{M}$ by its usual action on functions on the upper half plane (with corresponding weight):

$$(f, \alpha) \mapsto f|_k\alpha(z) = det(\alpha)^{k/2}(cz + d)^{-k}f(\alpha \cdot z),$$

$$\alpha = \begin{pmatrix} a & b \\ c & d \end{pmatrix}, \quad \alpha.z = \frac{az + b}{cz + d}.$$

The elements of the corresponding crossed-product algebra

$$\mathcal{A} = \mathcal{A}_{G^+(\mathbb{Q})} := \mathcal{M} \rtimes G^+(\mathbb{Q}),$$

are finite sums

$$\sum fU_\gamma^*, \quad f \in \mathcal{M}, \quad \gamma \in G^+(\mathbb{Q}),$$

with a product defined by

$$fU_\alpha^* \cdot gU_\beta^* = (f \cdot g|\alpha)U_{\beta\alpha}^*.$$

$\mathcal{A}$ can be thought of as the algebra of 'noncommutative coordinates' on the 'non-commutative quotient space' of modular forms modulo Hecke correspondences [12].

Consider the operator X of degree two on the space of modular forms defined by

$$X := \frac{1}{2\pi i}\frac{d}{dz} - \frac{1}{12\pi i}\frac{d}{dz}(log\Delta) \cdot Y,$$

where

$$\Delta(z) = (2\pi)^{12}\eta^{24}(z) = (2\pi)^{12}q\prod_{n=1}^\infty(1-q^n)^{24}, \quad q = e^{2\pi iz},$$

and Y denotes the grading operator

$$Y(f) = \frac{k}{2} \cdot f, \quad \text{for all } f \in \mathcal{M}_k.$$

The following proposition is proved in [12]. It shows that $\mathcal{A}_{G^+(\mathbb{Q})}$ is an $\mathcal{H}_1$-module algebra:

PROPOSITION 2.1. *There is a unique action of $\mathcal{H}_1$ on $\mathcal{A}_{G^+(\mathbb{Q})}$ determined by*

$$X(fU_\gamma^*) = X(f)U_\gamma^*, \quad Y(fU_\gamma^*) = Y(f)U_\gamma^*,$$

$$\delta_1(fU_\gamma^*) = \mu_\gamma \cdot f(U_\gamma^*),$$

where

$$\mu_\gamma(z) = \frac{1}{2\pi i}\frac{d}{dz}log\frac{\Delta|\gamma}{\Delta}.$$

More generally, for any congruence subgroup Γ an algebra $A(\Gamma)$ is constructed in [12] that contains as subalgebras both the algebra of Γ-modular forms and the Hecke ring at level Γ. There is also a corresponding action of $\mathcal{H}_1$ on $A(\Gamma)$.

For symmetries of type A, B, C, and D there is a corresponding *crossed product*, or *smash product*, construction that generalizes crossed products for actions of a group. We recall these constructions only for A and D symmetries, as well as a more elaborate version called *bicrossed product* construction. We shall see that Connes-Moscovici's Hopf algebra $\mathcal{H}_1$ is a bicrossed product of two, easy to describe, Hopf algebras.

Let A be a left H-module algebra. The underlying vector space of the crossed product algebra $A \rtimes H$ is the vector space $A \otimes H$ and its product is determined by

$$(a \otimes g)(b \otimes h) = a(g^{(1)}b) \otimes g^{(2)}h.$$

One checks that endowed with $1 \otimes 1$ as its unit, $A \rtimes H$ is an associative unital algebra. For example, for $H = kG$, the group algebra of a discrete group G acting by automorphisms on an algebra A, the algebra $A \rtimes \mathcal{H}$ is isomorphic to the crossed product algebra $A \rtimes G$. For a second simple example, let a Lie algebra $\mathfrak{g}$ act via derivations on a commutative algebra A. Then the crossed product algebra $A \rtimes U(\mathfrak{g})$ is a subalgebra of the algebra of differential operators on A generated by derivations from $\mathfrak{g}$ and multiplication operators by elements of A. The simples example is when $A = k[x]$ and $\mathfrak{g} = k$ acting via $\frac{d}{dx}$ on A. Then $A \rtimes U(\mathfrak{g})$ is the *weyl algebra* of differential operators on the line with polynomial coefficients.

Let D be a right H-comodule coalgebra via the coaction $d \mapsto d^{(0)} \otimes d^{(1)} \in D \otimes H$. The underlying linear space of the *crossed product coalgebra* $H \rtimes D$ is $H \otimes D$. It is a coalgebra whose coproduct and counit are defined by

$$\Delta(h \otimes d) = h^{(1)} \otimes (d^{(1)})^{(0)} \otimes h^{(2)}(d^{(1)})^{(1)} \otimes d^{(2)}, \quad \varepsilon(h \otimes d) = \varepsilon(d)\varepsilon(h).$$

The above two constructions deform multiplication or comultiplication, of algebras or coalgebras, respectively. Thus to obtain a simultaneous deformation of multiplication and comultiplication of a Hopf algebra it stands to reason to try to apply both constructions simultaneously. This idea, going back to G. I. Kac in 1960's in the context of Kac-von Neumann Hopf algebras, has now found its complete generalization in the notion of *bicrossed product* of *matched pairs* of Hopf algebras. See Majid's book [26] for extensive discussions and references. There are many variations of this construction of which the most relevant for the structure of the Connes-Moscovici Hopf algebra is the following.

Let U and F be two Hopf algebras. We assume that F is a left U-module algebra and U is a right F-comodule coalgebra via $\rho : U \to U \otimes F$. We say that (U, F) is a *matched pair* if the action and coaction satisfy the compatibility condition:

$$\epsilon(u(f)) = \epsilon(u)\epsilon(f), \quad \Delta(u(f)) = (u^{(1)})^{(0)}(f^{(1)}) \otimes (u^{(1)})^{(1)}(u^{(2)}(f^{(2)})),$$

$$\rho(1) = 1 \otimes 1, \quad \rho(uv) = (u^{(1)})^{(0)}v^{(0)} \otimes (u^{(1)})^{(1)}(u^{(2)}(v^{(1)})),$$

$$(u^{(2)})^{(0)} \otimes (u^{(1)}(f))(u^{(2)})^{(1)} = (u^{(1)})^{(0)} \otimes (u^{(1)})^{(1)}(u^{(2)}(f)).$$

Given a matched pair as above, we define its bicrossed product Hopf algebra $F \rtimes U$ to be $F \otimes U$ with crossed product algebra structure and crossed coproduct coalgebra structure. Its antipode S is defined by

$$S(f \otimes u) = (1 \otimes S(u^{(0)}))(S(fu^{(1)}) \otimes 1).$$

It is a remarkable fact that, thanks to the the above compatibility conditions, all the axioms of a Hopf algebra are satisfied for $F \rtimes U$.

The simplest and first example of this bicrossed product construction is as follows. Let $G = G_1 G_2$ be a *factorization* of a finite group G. This means that G_1 and G_2 are subgroups of G, $G_1 \cap G_2 = \{e\}$, and $G_1 G_2 = G$. We denote the factorization of g by $g = g_1 g_2$. The relation $g \cdot h := (gh)_2$ for $g \in G_1$ and $h \in G_2$ defines a left action of G_1 on G_2. Similarly $g \bullet h := (gh)_1$ defines a right action of G_2 on G_1. The first action turns $F = F(G_2)$ into a left $U = kG_1$-module algebra. The second action turns U into a right F-comodule coalgebra. The later coaction is simply the dual of the map $F(G_1) \otimes kG_2 \to F(G_1)$ induced by the right action of G_2 on G_1. Details of this example can be found in [26] and [9].

EXAMPLE 2.1. *1. By a theorem of Kostant [29], any cocommutative Hopf algebra H over an algebraically closed field k of characteristic zero is isomorphic (as a Hopf algebra) with a crossed product algebra $H = U(P(H)) \rtimes kG(H)$, where $P(H)$ is the Lie algebra of primitive elements of H and $G(H)$ is the group of all grouplike elements of H and $G(H)$ acts on $P(H)$ by inner automorphisms $(g, h) \mapsto ghg^{-1}$, for $g \in G(H)$ and $h \in P(H)$. The coalgebra structure of $H = U(P(H)) \rtimes kG(H)$ is simply the tensor product of the two coalgebras $U(P(H))$ and $kG(H)$.*

2. We show that the Connes-Moscovici Hopf algebra is a bicrossed Hopf algebra. Let $G = Diff(\mathbb{R})$ denote the group of diffeomorphisms of the real line. It has a *factorization of the form*

$$G = G_1 G_2,$$

where G_1 is the subgroup of diffeomorphisms that satisfy

$$\varphi(0) = 0, \quad \varphi'(0) = 1,$$

and G_2 is the $ax + b$- group of affine transformations. The first Hopf algebra, F, is formally speaking, the algebra of polynomial functions on the prounipotent group G_1. It can also be defined as the "continuous dual" of the enveloping algebra of the Lie algebra of G_1. It is a commutative Hopf algebra generated by functions δ_n, $n = 1, 2, \ldots$, defined by

$$\delta_n(\varphi) = \frac{d^n}{dt^n} (log(\varphi'(t))|_{t=0}.$$

The second Hopf algebra, U, is the universal enveloping algebra of the Lie algebra $\mathfrak{g}_2$ of the $ax + b$-group. It has generators X and Y and one relation $[X, Y] = X$.

The Hopf algebra F has a right U-module algebra structure defined by

$$\delta_n(X) = -\delta_{n+1}, \quad \text{and } \delta_n(Y) = -n\delta_n.$$

The Hopf algebra U is a left F-comodule coalgebra via

$$X \mapsto 1 \otimes X + \delta_1 \otimes X, \quad \text{and } Y \mapsto 1 \otimes Y.$$

One can check that they are a matched pair of Hopf algebras and the resulting bi-crossed product Hopf algebra is isomorphic to the Connes-Moscovici Hopf algebra $\mathcal{H}_1$. See [9] for a slightly different approach and fine points of the proof.

3. *Another important example of a bicrossed construction is the Drinfeld double $D(H)$ of a finite dimensional Hopf algebra H defined as a bicrossed product $H \bowtie H^*$* [26].

2.3. Modular pairs. Let H be a Hopf algebra, $\delta : H \to k$ a character and $\sigma \in H$ a grouplike element. Following [9, 10], we say that (δ, σ) is a *modular pair* if $\delta(\sigma) = 1$, and a *modular pair in involution* if in addition we have

$$\widetilde{S}_\delta^2 = Ad_\sigma, \quad \text{or,} \quad \widetilde{S}_\delta^2(h) = \sigma h \sigma^{-1},$$

for all h in H. Here the δ-*twisted antipode* $\widetilde{S}_\delta : H \to H$ is defined by $\widetilde{S}_\delta = \delta * S$, i.e.

$$\widetilde{S}_\delta(h) = \delta(h^{(1)}) S(h^{(2)}),$$

for all $h \in H$.

The notion of an invariant trace for actions of groups and Lie algebras can be extended to the Hopf setting. For applications to transverse geometry and number theory, it is important to formulate a notion of 'invariance' under the presence of a modular pair. Let A be an H-module algebra, δ a character of H, and $\sigma \in H$ a grouplike element. A k-linear map $\tau : A \to k$ is called δ-*invariant* if for all $h \in H$ and $a \in A$,

$$\tau(h(a)) = \delta(h)\tau(a).$$

τ is called a σ-*trace* if for all a, b in A,

$$\tau(ab) = \tau(b\sigma(a)).$$

For the following lemma from [11] the fact that A is unital is crucial. For $a, b \in A$, let

$$< a, b > := \tau(ab).$$

LEMMA 2.1. *(Integration by parts formula). Let τ be a σ-trace on A. Then τ is δ-invariant if and only if the integration by parts formula holds:*

$$< h(a), b >=< a, \widetilde{S}_\delta(h)(b) >,$$

for all $h \in H$ and $a, b \in A$.

Loosely speaking, the lemma says that the formal adjoint of the differential operator h is $\widetilde{S}_\delta(h)$.

EXAMPLE 2.2. *1. For any Hopf algebra H, the pair $(\varepsilon, 1)$ is modular. It is involutive if and only if $S^2 = id$. This happens, for example, when H is a commutative or cocommutative Hopf algebra.*

2. The original non-trivial example of a modular pair in involution is the pair $(\delta, 1)$ for Connes-Moscovici Hopf algebra $\mathcal{H}_1$. Let δ denote the unique extension of the modular character

$$\delta : \mathfrak{g}_{aff} \to \mathbb{R}, \quad \delta(X) = 1, \ \delta(Y) = 0,$$

to a character $\delta : U(\mathfrak{g}_{aff}) \to \mathbb{C}$. There is a unique extension of δ to a character, denoted by the same symbol $\delta : \mathcal{H}_1 \to \mathbb{C}$. Indeed the relations $[Y, \delta_n] = n\delta_n$ show that we must have $\delta(\delta_n) = 0$, for $n = 1, 2, \cdots$. One can then check that these relations are compatible with the algebra structure of $\mathcal{H}_1$. Recall the algebra $A_\Gamma = C_0^\infty(F^+(M)) \rtimes \Gamma$ from Section 1.2. It admits a δ-invariant trace $\tau : A_\Gamma \to \mathbb{C}$ under its canonical $\mathcal{H}_1$ action given by [9]:

$$\tau(fU_\varphi^*) = \int_{F^+(M)} f(y, y_1) \frac{dy dy_1}{y_1^2}, \quad if \varphi = 1,$$

and $\tau(fU_\varphi^*) = 0$, otherwise.

3. Let $H = A(SL_q(2, k))$ denote the Hopf algebra of functions on quantum $SL(2, k)$. As an algebra it is generated by symbols x, u, v, y, with the following relations:

$$ux = qxu, \quad vx = qxv, \quad yu = quy, \quad yv = qvy,$$

$$uv = vu, \quad xy - q^{-1}uv = yx - quv = 1.$$

The coproduct, counit and antipode of $\mathcal{H}$ are defined by

$$\Delta(x) = x \otimes x + u \otimes v, \quad \Delta(u) = x \otimes u + u \otimes y,$$

$$\Delta(v) = v \otimes x + y \otimes v, \quad \Delta(y) = v \otimes u + y \otimes y,$$

$$\epsilon(x) = \epsilon(y) = 1, \quad \epsilon(u) = \epsilon(v) = 0,$$

$$S(x) = y, \quad S(y) = x, \quad S(u) = -qu, \quad S(v) = -q^{-1}v.$$

Define a character $\delta : H \to k$ by:

$$\delta(x) = q, \quad \delta(u) = 0, \quad \delta(v) = 0, \quad \delta(y) = q^{-1}.$$

One checks that $\widetilde{S}_\delta^2 = id$. This shows that $(\delta, 1)$ is a modular pair for H. This example and its Hopf-cyclic cohomology is studied in [20].

More generally, it is shown in [10] that coribbon Hopf algebras and compact quantum groups are endowed with canonical modular pairs of the form $(\delta, 1)$ and, dually, ribbon Hopf algebras have canonical modular pairs of the type $(1, \sigma)$.

4. We will see in the next section that modular pairs in involution are in fact one dimensional cases of what we call stable anti-Yetter-Drinfeld modules, i.e. they are one dimensional noncommutative local systems that one can introduce into Hopf algebra equivariant cyclic cohomology.

3. Anti-Yetter-Drinfeld modules

An important question left open in our paper [21] was the issue of identifying the most general class of coefficients allowable in cyclic (co)homology of Hopf algebras and Hopf-cyclic cohomology in general. This problem is completely solved, among other things, in [14]. It is shown in this paper that the most general coefficients are the class of so called stable anti-Yetter-Drinfeld modules. In Section 3.2 we briefly report on this very recent development as well.

It is quite surprising that when the general formalism of cyclic cohomology theory, namely the theory of (co)cyclic modules [3], is applied to Hopf algebras, variations of such standard notions like Yetter-Drinfeld (YD) modules appear naturally. The so called anti-Yetter-Drinfeld modules introduced in [14] are twistings, by modular pairs in involution, of YD modules. This means that the category of anti-Yetter-Drinfeld modules is a "mirror image" of the category of YD modules. We mention that anti-Yetter-Drinfeld modules were considered independently also by C. Voigt in connection with his work on equivariant cyclic cohomology [31].

3.1. Yetter-Drinfeld modules. Yetter-Drinfeld modules were introduced by D. Yetter under the name *crossed bimodules* [32]. The present name was coined in [27]. One of the motivations was to define a *braiding* on the monoidal category $H - Mod$ of representations of a, not necessarily commutative or cocommutative, Hopf algebra H. To define such a braiding one should either restrict to special classes of Hopf algebras, or, to special classes of modules. Drinfeld showed that when H is a *quasitriangular* Hopf algebra, then $H - Mod$ is a braided monoidal category. See [26] for definitions and references. Dually, when H is *coquasitriangular*, the category $H - Comod$ of H-comodules is a braided monoidal category. In [32] Yetter shows that to obtain a braiding on a subcategory of $H - Mod$, for an arbitrary H, one has essentially one choice and that is restricting to the class of Yetter-Drinfeld modules as we explain now.

Let H be a Hopf algebra and M be a left H-module and left H-comodule simultaneously. We say that M is a left-left *Yetter-Drinfeld H-module* if the two structures on M are compatible in the sense that

$$\rho(hm) = h^{(1)}m^{(-1)}S(h^{(3)}) \otimes h^{(2)}m^{(0)},$$

for all $h \in H$ and $m \in M$ [26, 27, 32]. One can similarly define left-right, right-left and right-right YD modules. A morphism of YD modules $M \to N$ is an H-linear and H-colinear map $f : M \to N$. We denote the category of left-left YD modules over H by $^H_H\mathcal{YD}$.

This notion is closely related to the Drinfeld double of finite dimensional Hopf algebras. In fact if H is finite dimensional, then one can show that the category $^H_H\mathcal{YD}$ is isomorphic to the category of left modules over the Drinfeld double $D(H)$ of H [26].

The following facts about the category $^H_H\mathcal{YD}$ are well known [26, 27, 32]:

1. The tensor product $M \otimes N$ of two YD modules is a YD module. Its module and comodule structure are the standard ones:

$$h(m \otimes n) = h^{(1)}m \otimes h^{(1)}n, \quad (m \otimes n) \mapsto m^{(-1)}n^{(-1)} \otimes m^{(0)} \otimes n^{(0)}.$$

This shows that the category $^H_H\mathcal{YD}$ is a monoidal subcategory of the monoidal category $H - Mod$.

2. The category $^H_H\mathcal{YD}$ is a braided monoidal category under the braiding

$$c_{M,N} : M \otimes N \rightarrow N \otimes M, \quad m \otimes n \mapsto m^{(-1)} \cdot n \otimes m^{(0)}.$$

In fact Yetter proves a strong inverse to this statement as well [32]: for any small strict monoidal category $\mathcal{C}$ endowed with a monoidal functor $F : \mathcal{C} \rightarrow Vect_f$ to the category of finite dimensional vector spaces, there is a Hopf algebra H and a monoidal functor $\tilde{F} : \mathcal{C} \rightarrow^H_H \mathcal{YD}$ such that the following diagram comutes

$$
\begin{array}{ccc}
\mathcal{C} & \longrightarrow & ^H_H\mathcal{YD} \\
\downarrow & & \downarrow \\
Vect_f & \longrightarrow & Vect
\end{array}
$$

3. The category $^H_H\mathcal{YD}$ is the *center* of the monoidal category $H - Mod$. Recall that the (left) center $\mathcal{ZC}$ of a monoidal category [18] is a category whose objects are pairs $(X, \sigma_{X,-})$, where X is an object of $\mathcal{C}$ and $\sigma_{X,-} : X \otimes - \rightarrow - \otimes X$ is a natural isomorphism satisfying certain compatibility axioms with associativity and unit constraints. It can be shown that the center of a monoidal category is a braided monoidal category and $\mathcal{Z}(H - Mod) =^H_H \mathcal{YD}$ [18].

EXAMPLE 3.1. *1. Let $H = kG$ be the group algebra of a discrete group G. A left kG-comodule is simply a G-graded vector space*

$$M = \bigoplus_{g \in G} M_g$$

where the coaction is defined by

$$m \in M_g \mapsto g \otimes m.$$

An action of G on M defines a YD module structure iff for all $g, h \in G$,

$$hm \in M_{hgh^{-1}}.$$

This example can be explained as follows. Let $\mathcal{G}$ be a groupoid whose objects are G and its morphisms are defined by

$$Hom(g, h) = \{k \in G; \ kgk^{-1} = h\}.$$

Recall that an action of a groupoid $\mathcal{G}$ on the category $Vect$ of vector spaces is simply a functor $F : \mathcal{G} \rightarrow Vect$. Then it is easily seen that we have a one-one correspondence between YD modules for kG and actions of $\mathcal{G}$ on $Vect$. This example clearly shows that one can think of an YD module over kG as an 'equivariant sheaf' on G and of YD modules as noncommutative analogues of equivariant sheaves on a topological group.

2. If H is cocommutative then any left H-module M can be turned into a left-left YD module via the coaction $m \mapsto 1 \otimes m$. Similarly, when H is cocmmuntative then any left H-comodule M can be turned into a YD module via the H-action $h \cdot m := \varepsilon(h)m$.

3. Any Hopf algebra acts on itself via conjugation action $g \cdot h := g^{(1)} h S(g^{(2)})$ and coacts via translation coaction $h \mapsto h^{(1)} \otimes h^{(2)}$. It can be checked that this endows $M = H$ with a YD module structure.

3.2. Stable anti-Yetter-Drinfeld modules. This class of modules for Hopf algebras were introduced for the first time in [**14**]. Unlike Yetter-Drinfeld modules, its definition, however, was entirely motivated and dictated by cyclic cohomology theory: the anti-Yetter-Drinfeld condition guarantees that the simplicial and cyclic operators are well defined on invariant complexes and the stability condition implies that the crucial periodicity condition for cyclic modules are satisfied.

DEFINITION 3.1. *A left-left anti-Yetter-Drinfeld H-module is a left H-module and left H-comodule such that*

$$\rho(hm) = h^{(1)} m^{(-1)} S(h^{(3)}) \otimes h^{(2)} m^{(0)},$$

for all $h \in H$ and $m \in M$. We say that M is stable if in addition we have

$$m^{(-1)} m^{(0)} = m,$$

for all $m \in M$.

There are of course analogous definitions for left-right, right-left and right-right stable anti-Yetter-Drinfeld (SAYD) modules. We note that by changing S to S^{-1} in the above equation, we obtain the compatibility condition for a Yetter-Drinfeld module from the previous subsection.

The following lemma from [**14**] shows that 1-dimensional SAYD modules correspond to Connes-Moscovici's modular pairs in involution:

LEMMA 3.1. *There is a one-one correspondence between modular pairs in involution (δ, σ) on H and SAYD module structure on $M = k$, defined by*

$$h.r = \delta(h) r, \quad r \mapsto \sigma \otimes r,$$

for all $h \in H$ and $r \in k$. We denote this module by $M = {}^{\sigma}k_{\delta}$.

Let ${}^{H}_{H}\mathcal{AYD}$ denote the category of left-left anti-Yetter-Drinfeld H-modules, where morphisms are H-linear and H-colinear maps. Unlike YD modules, anti-Yetter-Drinfeld modules do not form a monoidal category under the standard tensor product. This can be checked easily on 1-dimensional modules given by modular pairs in involution. The following result of [**14**], however, shows that the tensor product of an anti-Yetter-Drinfeld module with a Yetter-Drinfeld module is again anti-Yetter-Drinfeld.

LEMMA 3.2. *Let M be a Yetter-Drinfeld module and N be an anti-Yetter-Drinfeld module (both left-left). Then $M \otimes N$ is an anti-Yetter-Drinfeld module under the diagonal action and coaction:*

$$h(m \otimes n) = h^{(1)} m \otimes h^{(1)} n, \quad (m \otimes n) \mapsto m^{(-1)} n^{(-1)} \otimes m^{(0)} \otimes n^{(0)}.$$

In particular, using a modular pair in involution (δ, σ), we obtain a functor

$$ {}^{H}_{H}\mathcal{YD} \to {}^{H}_{H}\mathcal{AYD}, \quad M \mapsto \bar{M} = {}^{\sigma}k_{\delta} \otimes M.$$

This result clearly shows that AYD modules can be regarded as the *twisted analogue* or *mirror image* of YD modules, with twistings provided by modular pairs in involution. This result was later strengthened by the following result, pointed out to us by M. Staic [28]. It shows that if the Hopf algebra has a modular pair in involution then the category of YD modules is equivalent to the category of AYD modules:

PROPOSITION 3.1. *Let H be a Hopf algebra, (δ, σ) a modular pair in involution and M an anti-Yetter-Drinfeld module. If we define $m \cdot h = mh^{(1)}\delta(S(h^{(2)}))$ and $\rho(m) = \sigma^{-1}m^{(-1)} \otimes m^{(0)}$, then $(M, \cdot, \rho)$ is an Yetter-Drinfeld module. Moreover in this way we get an isomorphism between the categories of AYD modules and YD mpdules.*

It follows that tensoring with $^{\sigma^{-1}}k_{\delta \circ S}$ turns the anti-Yetter-Drinfeld modules to Yetter-Drinfeld modules and this is the inverse for the operation of tensoring with $^1k_\delta$.

EXAMPLE 3.2. 1. *For Hopf algebras with $S^2 = I$, e.g. commutative or co-commutative Hopf algebras, there is no distinction between YD and AYD modules. This applies in particular to $H = kG$ and Example 2.1.1. The stability condition $m^{(-1)}m^{(0)} = m$ is equivalent to*

$$g \cdot m = m, \quad \textit{for all } g \in G, \, m \in M_g.$$

2. *Hopf-Galois extensions are noncommutative analogues of principal bundles in (affine) algebraic geometry. Following [14] we show that they give rise to large classes of examples of SAYD modules. Let P be a right H-comodule algebra, and let*

$$B := P^H = \{p \in P; \, \rho(p) = p \otimes 1\}$$

be the space of coinvariants of P. It is easy to see that B is a subalgebra of P. The extension $B \subset P$ is called a Hopf-Galois extension if the map

$$can : P \otimes_B P \to B \otimes H, \quad p \otimes p' \mapsto p\rho(p'),$$

is bijective. (Note that in the commutative case this corresponds to the condition that the action of the structure group on fibres is free). The bijectivity assumption allows us to define the translation map $T : H \to P \otimes_B P$,

$$T(h) = can^{-1}(1 \otimes h) = h^{(\bar{1})} \otimes h^{(\bar{2})}.$$

It can be checked that the centralizer $Z_B(P) = \{p \mid bp = pb \; \forall b \in B\}$ of B in P is a subcomodule of P. There is an action of H on $Z_B(P)$ defined by $ph = h^{(\bar{1})}ph^{(\bar{2})}$ called the Miyashita-Ulbrich action. It is shown that this action and coaction satisfy the Yetter-Drinfeld compatibility condition. On the other hand if B is central, then by defining the new action $ph = (S^{-1}(h))^{(\bar{2})}p(S^{-1}(h))^{(\bar{1})}$ and the right coaction of P we have a SAYD module. This example was the starting point of [16] where the relative cyclic homology of Hopf-Galois extensions is related to a variant of Hopf-cyclic cohomology with coefficients in stable anti-Yetter-Drinfeld modules. In [23] we showed that the theory introduced in [16] is isomorphic to one of the theories introduced in [15].

3. *Let $M = H$. Then with conjugation action $g \cdot h = g^{(1)}hS(g^{(2)})$ and comultiplication $h \mapsto h^{(1)} \otimes h^{(2)}$ as coaction, M is an SAYD module.*

4. Hopf-cyclic cohomology

In this section we first recall the approach by Connes and Moscovici towards the definition of their cyclic cohomology theory for Hopf algebras. The characteristic map χ_τ palys an imporant role here. Then we switch to the point of view adopted in [21] based on invariant complexes, culminating in our joint work [14, 15]. The resulting Hopf-cyclic cohomology theories of type A, B, and C, and their corresponding cyclic modules, contain all known examples of cyclic theory discovered so far. We note that very recently A. Kaygun has extended the Hopf-cyclic cohomology to a cohomology for bialgebras with coefficients in stable modules. For Hopf algebras it reduces to Hopf-cyclic cohomology [19].

4.1. Connes-Moscovici's breakthrough.
Without going into details, in this section we formulate one of the problems that was faced and solved by Connes and Moscovici in the course of their study of an index problem on foliated manifolds [9]. See also [11] for a survey. As we shall see, this led them to a new cohomology theory for Hopf algebras that is now an important example of Hopf-cyclic cohomology.

The local index formula of Connes and Moscovici [8] gives the Chern character $Ch(A, h, D)$ of a regular spectral triple (A, h, D) as a cyclic cocycle in the (b, B)-bicomplex of the algebra A. For spectral triples of interest in transverse geometry [9], this cocycle is *differentiable* in the sense that it is in the image of the Connes-Moscovici characteristic map χ_τ defined below, with $H = \mathcal{H}_1$ and $A = \mathcal{A}_\Gamma$. To identify this class in terms of characteristic classes of foliations, it would be extremely helpful to show that it is the image of a cocycle for a cohomology theory for Hopf algebras. This is rather similar to the situation for classical characteristic classes which are pull backs of group cohomology classes.

We can formulate this problem abstractly as follows: Let H be a Hopf algebra endowed with a modular pair in involution (δ, σ), and A be an H-module algebra. Let $\tau : A \to k$ be a δ- invariant σ-trace on A as defined in Section 1.3. Consider the Connes-Moscovici *characteristic map*

$$\chi_\tau : H^{\otimes n} \longrightarrow Hom(A^{\otimes(n+1)}, k),$$

$$\chi_\tau(h_1 \otimes \cdots \otimes h_n)(a_0 \otimes \cdots \otimes a_n) = \tau(a_0 h_1(a_1) \cdots h_n(a_n)).$$

Now the burning question is: can we promote the collection of spaces $\{H^{\otimes n}\}_{n \geq 0}$ to a *cocyclic module* such that the characteristic map χ_τ turns into a morphism of cocyclic modules? We recall that the face, degeneracy, and cyclic operators for $Hom(A^{\otimes(n+1)}, k)$ are defined by [3]:

$$\delta_i^n \varphi(a_0, \cdots, a_{n+1}) = \varphi(a_0, \cdots, a_i a_{i+1}, \cdots, a_{n+1}), \quad i = 0, \cdots, n,$$
$$\delta_{n+1}^n \varphi(a_0, \cdots, a_{n+1}) = \varphi(a_{n+1} a_0, a_1, \cdots, a_n),$$
$$\sigma_i^n \varphi(a_0, \cdots, a_n) = \varphi(a_0, \cdots, a_i, 1, \cdots, a_n), \quad i = 0, \cdots, n,$$
$$\tau_n \varphi(a_0, \cdots, a_n) = \varphi(a_n, a_0, \cdots, a_{n-1}).$$

The relation

$$h(ab) = h^{(1)}(a) h^{(2)}(b)$$

shows that in order for χ_τ to be compatible with face operators, the face operators on $H^{\otimes n}$ must involve the coproduct of H. In fact if we define, for $0 \leq i \leq n$,

$\delta_i^n : H^{\otimes n} \to H^{\otimes(n+1)}$, by

$$\delta_0^n(h_1 \otimes \cdots \otimes h_n) = 1 \otimes h_1 \otimes \cdots \otimes h_n,$$
$$\delta_i^n(h_1 \otimes \cdots \otimes h_n) = h_1 \otimes \cdots \otimes h_i^{(1)} \otimes h_i^{(2)} \otimes \cdots \otimes h_n,$$
$$\delta_{n+1}^n(h_1 \otimes \cdots \otimes h_n) = h_1 \otimes \cdots \otimes h_n \otimes \sigma,$$

then we have, for all n and i,

$$\chi_\tau \delta_i^n = \delta_i^n \chi_\tau.$$

Similarly, the relation $h(1_A) = \varepsilon(h)1_A$, shows that the degeneracy operators on $H^{\otimes n}$ should involve the counit of H. We thus define

$$\sigma_i^n(h_1 \otimes \cdots \otimes h_n) = h_1 \otimes \cdots \otimes \varepsilon(h_i) \otimes \cdots \otimes h_n.$$

The most difficult part in this regard is to guess the form of the *cyclic operator* $\tau_n : H^{\otimes n} \to H^{\otimes n}$. Compatibility with χ_τ demands that

$$\tau(a_0 \tau_n(h_1 \otimes \cdots \otimes h_n)(a_1 \otimes \cdots \otimes a_n)) = \tau(a_n h_1(a_0)h_2(a_1) \cdots h_n(a_{n-1})),$$

for all a_i's and h_i's. Now integration by parts formula in Lemma 2.1, combined with the σ-trace property of τ, gives us:

$$\tau(a_1 h(a_0)) = \tau(h(a_0)\sigma(a_1)) = \tau(a_0 \tilde{S}_\delta(h)(\sigma(a_1))).$$

This suggests that we should define $\tau_1 : H \to H$ by

$$\tau_1(h) = \tilde{S}_\delta(h)\sigma.$$

Note that the condition $\tau_1^2 = I$ is equivalent to the involutive condition $\tilde{S}_\delta^2 = Ad_\sigma$.

For any n, integration by parts formula together with the σ-trace property shows that:

$$\tau(a_n h_1(a_0) \cdots h_n(a_{n-1})) = \tau(h_1(a_0) \cdots h_n(a_{n-1})\sigma(a_n))$$
$$= \tau(a_0 \tilde{S}_\delta(h_1)(h_2(a_1) \cdots h_n(a_{n-1})\sigma(a_n)))$$
$$= \tau(a_0 \tilde{S}_\delta(h_1) \cdot (h_2 \otimes \cdots \otimes h_n \otimes \sigma)(a_1 \otimes \cdots \otimes a_n)).$$

This suggests that the *Hopf-cyclic operator* $\tau_n : H^{\otimes n} \to H^{\otimes n}$ should be defined as

$$\tau_n(h_1 \otimes \cdots \otimes h_n) = \tilde{S}_\delta(h_1) \cdot (h_2 \otimes \cdots \otimes h_n \otimes \sigma),$$

where $\cdot$ denotes the diagonal action defined by

$$h \cdot (h_1 \otimes \cdots \otimes h_n) := h^{(1)}h_1 \otimes h^{(2)}h_2 \otimes \cdots \otimes h^{(n)}h_n.$$

We let $\tau_0 = I : H^{\otimes 0} = k \to H^{\otimes 0}$, be the identity map. The remarkable fact, proved by Connes and Moscovici [9, 10], is that endowed with the above face, degeneracy, and cyclic operators, $\{H^{\otimes n}\}_{n \geq 0}$ is a cocyclic module. The resulting cyclic cohomology groups are denoted by $\bar{H}C_{(\delta,\sigma)}^n(H)$, $n = 0, 1, \cdots$ and we obtain the desired characteristic map

$$\chi_\tau : HC_{(\delta,\sigma)}^n(H) \to HC^n(A).$$

As with any cocyclic module, cyclic cohomology can also be defined in terms of cyclic cocycles. In this case a cyclic n-cocycle is an element $x \in H^{\otimes n}$ satisfying the conditions

$$bx = 0, \quad (1 - \lambda)x = 0,$$

where $b : H^{\otimes n} \to H^{\otimes (n+1)}$ and $\lambda : H^{\otimes n} \to H^{\otimes n}$ are defined by

$$
\begin{aligned}
b(h^1 \otimes \cdots \otimes h^n) &= 1 \otimes h_1 \otimes \cdots \otimes h_n \\
&+ \sum_{i=1}^{n} (-1)^i h_1 \otimes \cdots \otimes h_i^{(1)} \otimes h_i^{(2)} \otimes \cdots \otimes h_n \\
&+ (-1)^{n+1} h_1 \otimes \cdots \otimes h_n \otimes \sigma, \\
\lambda(h_1 \otimes \cdots \otimes h_n) &= (-1)^n \tilde{S}_\delta(h_1) \cdot (h_2 \otimes \cdots \otimes h_n \otimes \sigma).
\end{aligned}
$$

The cyclic cohomology groups $HC^n_{(\delta,\sigma)}(H)$ are computed in several cases in [9]. Of particular interest for applications to transverse index theory and number theory is the the (periodic) cyclic cohomology of the Connes-Moscovici Hopf algebra $\mathcal{H}_1$. It is shown in [9] that the periodic groups $HP^n_{(\delta,1)}(\mathcal{H}_1)$ are canonically isomorphic to the Gelfand-Fuchs cohomology of the Lie algebra of formal vector fields on the line:

$$
H^*(\mathfrak{a}_1, \mathbb{C}) = HP^*_{(\delta,1)}(\mathcal{H}_1).
$$

Calculation of the unstable groups is an interesting open problem. The following interesting elements are however already been identified. It can be directly checked that the elements $\delta'_2 := \delta_2 - \frac{1}{2}\delta_1^2$ and δ_1 are cyclic 1-cocycles on $\mathcal{H}_1$, and

$$
F := X \otimes Y - Y \otimes X - \delta_1 Y \otimes Y
$$

is a cyclic 2-cocycle. See [12] for detailed calculations and relations between these cocycles and the Schwarzian derivative, Godbillon-Vey cocycle, and the transverse fundamental class of Connes [5], respectively.

4.2. Hopf-cyclic cohomology: type A, B, and C theories. We recall the definitions of the three cyclic cohomology theories that were defined in [15]. We call them A, B and C theories. In the first case the algebra A is endowed with an action of a Hopf algebra; in the second case the algebra B is equipped with a coaction of a Hopf algebra; and finally in theories of type C, we have a coalgebra endowed with an action of a Hopf algebra. In all three theories the module of coefficients is a stable anti-Yetter-Drinfeld (SAYD) module over the Hopf algebra and we attach a cocyclic module in the sense of Connes [3] to the given data. Along the same lines one can define a Hopf-cyclic cohomology theory for comodule coalgebras as well (type D theory). Since so far we have found no applications of such a theory we won't give its definition here. We also show that all known examples of cyclic cohomology theories that are introduced so far such as: ordinary cyclic cohomology for algebras, Connes-Moscovici's cyclic cohomology for Hopf algebras [8], and equivariant cyclic cohomology [1, 2], are special cases of these theories.

Let A be a left H-module algebra and M be a left-right SAYD H-module. Then the spaces $M \otimes A^{\otimes (n+1)}$ are right H-modules via the diagonal action

$$
(m \otimes \tilde{a})h := mh^{(1)} \otimes S(h^{(2)})\tilde{a},
$$

where the left H-action on $\tilde{a} \in A^{\otimes (n+1)}$ is via the left diagonal action of H.

We define the space of *equivariant cochains on A with coefficients in M* by

$$
\mathcal{C}^n_H(A, M) := Hom_H(M \otimes A^{\otimes (n+1)}, k).
$$

More explicitly, $f : M \otimes A^{\otimes (n+1)} \to k$ is in $\mathcal{C}^n_H(A, M)$, if and only if

$$
f((m \otimes a_0 \otimes \cdots \otimes a_n)h) = \varepsilon(h)f(m \otimes a_0 \otimes \cdots \otimes a_n),
$$

for all $h \in H, m \in M$, and $a_i \in A$. It is shown in [15] that the following operators define a cocyclic module structure on $\{C_H^n(A, M)\}_{n \in \mathbb{N}}$:

$$(\delta_i f)(m \otimes a_0 \otimes \cdots \otimes a_n) = f(m \otimes a_0 \otimes \cdots \otimes a_i a_{i+1} \otimes \cdots \otimes a_n), \; 0 \le i < n,$$

$$(\delta_n f)(m \otimes a_0 \otimes \cdots \otimes a_n) = f(m^{(0)} \otimes (S^{-1}(m^{(-1)})a_n)a_0 \otimes a_1 \otimes \cdots \otimes a_{n-1}),$$

$$(\sigma_i f)(m \otimes a_0 \otimes \cdots \otimes a_n) = f(m \otimes a_0 \otimes \cdots \otimes a_i \otimes 1 \otimes \cdots \otimes a_n), \; 0 \le i \le n,$$

$$(\tau_n f)(m \otimes a_0 \otimes \cdots \otimes a_n) = f(m^{(0)} \otimes S^{-1}(m^{(-1)})a_n \otimes a_0 \otimes \cdots \otimes a_{n-1}).$$

We denote the resulting cyclic cohomology theory by $HC_H^n(A, M), n = 0, 1, \cdots$.

EXAMPLE 4.1. *1. For $H = k = M$ we obviously recover the standard cocyclic module of the algebra A. The resulting cyclic cohomology theory is the ordinary cyclic cohomology of algebras.*

2. For $M = H$ and H acting on M by conjugation and coacting via coproduct (Example 2.2.3.), we obtain the equivariant cyclic cohomology theory of Akbarpour and Khalkhali For H-module algebras [1, 2].

3. For $H = k[\sigma, \sigma^{-1}]$ the Hopf algebra of Laurent polynomials, where σ acts by automorphisms on an algebra A, and $M = k$ is a trivial module, we obtain the so called twisted cyclic cohomology of A with respect to σ. A twisted cyclic n-cocycle is a linear map $f : A^{\otimes(n+1)} \to k$ satisfying:

$$f(\sigma a_n, a_0, \cdots, a_{n-1}) = (-1)^n f(a_0, \cdots, a_n), \quad b_\sigma f = 0,$$

where b_σ is the twisted Hochschild boundary defined by

$$b_\sigma f(a_0, \cdots, a_{n+1}) = \sum_{i=0}^{n} (-1)^i f(a_0, \cdots, a_i a_{i+1}, \cdots, a_{n+1})$$
$$+ \; (-1)^{n+1} f(\sigma(a_{n+1})a_0, a_1, \cdots, a_n).$$

4. It is easy to see that for $M = {}^\sigma k_\delta$, the SAYD module attached to a modular pair in involution (δ, σ), $HC_H^0(A, M)$ is the space of δ-invariant σ-traces on A in the sense of Connes-Moscovici [9, 10] (cf. Section 1.3.).

Next, let B be a right H-comodule algebra and M be a right-right SAYD H-module. Let

$$C^{n,H}(B, M) := Hom^H(B^{\otimes(n+1)}, M),$$

denote the space of right H-colinear $(n + 1)$-linear functionals on B with values in M. Here $B^{\otimes(n+1)}$ is considered a right H-comodule via the diagonal coaction of H:

$$b_0 \otimes \cdots \otimes b_n \mapsto (b_0^{(0)} \otimes \cdots \otimes b_n^{(0)}) \otimes (b_0^{(1)} b_1^{(1)} \cdots b_n^{(1)}).$$

It is shown in [15] that, thanks to the invariance property imposed on our cochains and the SAYD condition on M, the following maps define a cocyclic module structure on $\{C^{n,H}(B, M)\}_{n \in \mathbb{N}}$:

$$(\delta_i f)(b_0, \cdots, b_{n+1}) = f(b_0, \cdots, b_i b_{i+1}, \cdots, b_{n+1}), \; 0 \le i < n,$$

$$(\delta_n f)(b_0, \cdots, b_{n+1}) = f(b_{n+1}^{(0)} b_0, b_1, \cdots, b_n) b_{n+1}^{(1)},$$

$$(\sigma_i f)(b_0, \cdots, b_{n-1}) = f(b_0, \cdots, b_i, 1, \cdots b_{n-1}), \; 0 \le i < n - 1,$$

$$(\tau_n f)(b_0, \cdots, b_n) = f(b_n^{(0)}, b_0, \cdots, b_{n-1}) b_n^{(1)}.$$

We denote the resulting cyclic cohomology groups by $HC^{n,H}(B, M)$, $n = 0, 1, \cdots$.

EXAMPLE 4.2. *1. For $B = H$, equipped with comultiplication as coaction, and $M = {}^\sigma k_\delta$ associated to a modular pair in involution, we obtain the Hopf-cyclic cohomology of Hopf algebras as defined in [20]. This theory is different from Connes-Moscovici's theory for Hopf algebras. It is dual, in the sense of Hopf algebras and not Hom dual, to Connes-Moscovici's theory [22]. It is computed in the following cases [20]: $H = kG$, $H = U(\mathfrak{g})$, where it is isomorphic to group cohomology and Lie algebra cohomology, respectively; $H = SL_2(q, k)$, and $H = U_q(sl_2)$.*

2. For $H = k$, and $M = k$ a trivial module, we obviously recover the cyclic cohomology of the algebra B.

Finally we describe theories of type C and their main examples. As we shall see, Connes-Moscovici's original example of Hopf-cyclic cohomology belong to this class of theories. Let C be a left H-module coalgebra, and M be a right-left SAYD H-module. Let

$$\mathcal{C}^n(C, M) := M \otimes_H C^{\otimes(n+1)} \quad n \in \mathbb{N}.$$

It can be checked that, thanks to the SAYD condition on M, the following operators are well defined and define a cocyclic module, denoted $\{\mathcal{C}^n(C, M)\}_{n \in \mathbb{N}}$. In particular the crucial periodicity conditions $\tau_n^{n+1} = id$, $n = 0, 1, 2 \cdots$, are satisfied [15]:

$$\delta_i(m \otimes c_0 \otimes \cdots \otimes c_{n-1}) = m \otimes c_0 \otimes \cdots \otimes c_i^{(1)} \otimes c_i^{(2)} \otimes c_{n-1}, 0 \leq i < n,$$

$$\delta_n(m \otimes c_0 \otimes \cdots \otimes c_{n-1}) = m^{(0)} \otimes c_0^{(2)} \otimes c_1 \otimes \cdots \otimes c_{n-1} \otimes m^{(-1)} c_0^{(1)},$$

$$\sigma_i(m \otimes c_0 \otimes \cdots \otimes c_{n+1}) = m \otimes c_0 \otimes \cdots \otimes \varepsilon(c_{i+1}) \otimes \cdots \otimes c_{n+1}, 0 \leq i \leq n,$$

$$\tau_n(m \otimes c_0 \otimes \cdots \otimes c_n) = m^{(0)} \otimes c_1 \otimes \cdots \otimes c_n \otimes m^{(-1)} c_0.$$

EXAMPLE 4.3. *1. For $H = k = M$, we recover the cocyclic module of a coalgebra which defines its cyclic cohomology.*

2. For $C = H$ and $M = {}^\sigma k_\delta$, the cocyclic module $\{\mathcal{C}_H^n(C, M)\}_{n \in \mathbb{N}}$ is isomorphic to the cocyclic module of Connes-Moscovici [9], attached to a Hopf algebra endowed with a modular pair in involution. This example is truly fundamental and started the whole theory.

Acknowledgment. It is a great pleasure to thank Alain Connes and Henri Moscovici for their support and many illuminating discussions, and Matilde Marcolli for organizing the MPI, Bonn, conferences on noncommutative geometry and number theory and for her support. We are also much obliged to our collaborators Piotr M. Hajac and Yorck Sommerhäuser.

References

[1] R. Akbarpour, and M. Khalkhali, Hopf algebra equivariant cyclic homology and cyclic homology of crossed product algebras. J. Reine Angew. Math., **559** (2003), 137-152.

[2] R. Akbarpour, and M. Khalkhali, Equivariant cyclic cohomology of H-algebras. K-Theory **29** (2003), no. 4, 231–252.

[3] A. Connes, Cohomologie cyclique et foncteurs Ext^n. (French) (Cyclic cohomology and functors Ext^n) C. R. Acad. Sci. Paris Sr. I Math. **296** (1983), no. 23, 953–958.

[4] A. Connes, Noncommutative differential geometry. Inst. Hautes tudes Sci. Publ. Math. No. **62** (1985), 257–360.

[5] A. Connes, Cyclic cohomology and the transverse fundamental class of a foliation. Geometric methods in operator algebras (Kyoto, 1983), 52–144, Pitman Res. Notes Math. Ser., 123, Longman Sci. Tech., Harlow, 198.

[6] A. Connes, and M. Marcolli, From physics to number theory via noncommutative geometry. Part I: Quantum statistical mechanics of Q-lattices. math.NT/0404128.

[7] A. Connes, and M. Marcolli, From physics to number theory via noncommutative geometry, Part II: Renormalization, the Riemann-Hilbert correspondence, and motivic Galois theory. hep-th/0411114.

[8] A. Connes and H. Moscovici, Local index formula in noncommutative geometry, Geom. Funct. Anal. **5** (1995), no. 2, 174-243.

[9] A. Connes and H. Moscovici, Hopf algebras, Cyclic Cohomology and the transverse index theorem, Comm. Math. Phys. **198** (1998), no. 1, 199–246.

[10] A. Connes and H. Moscovici, Cyclic cohomology and Hopf algebras. Lett. Math. Phys. **48** (1999), no. 1, 97–108.

[11] A. Connes, and H. Moscovici, Cyclic cohomology and Hopf algebra symmetry. Conference Mosh Flato 1999 (Dijon). Lett. Math. Phys. **52** (2000), no. 1, 1–28.

[12] A. Connes and H. Moscovici, Modular Hecke algebras and their Hopf symmetry, to appear in the Moscow Mathematical Journal (volume dedicated to Pierre Cartier).

[13] A. Connes and H. Moscovici, Rankin-Cohen brackets and the Hopf algebra of transverse geometry, to appear in the Moscow Mathematical Journal (volume dedicated to Pierre Cartier).

[14] P. M. Hajac, M. Khalkhali, B. Rangipour, and Y. Sommerhäuser, Stable anti-Yetter-Drinfeld modules. C. R. Math. Acad. Sci. Paris **338** (2004), no. 8, 587–590.

[15] P. M. Hajac, M. Khalkhali, B. Rangipour, and Y. Sommerhäuser, Hopf-cyclic homology and cohomology with coefficients. C. R. Math. Acad. Sci. Paris **338** (2004), no. 9, 667–672.

[16] P. Jara, D. Stefan, Cyclic homology of Hopf Galois extensions and Hopf algebras. Preprint math/0307099.

[17] A. Joyal, and R. Street, The geometry of tensor calculus. I. Adv. Math. **88** (1991), no. 1, 55–112.

[18] Ch. Kassel, Quantum groups. Graduate Texts in Mathematics, 155. Springer-Verlag, New York, 1995.

[19] A. Kaygun, Bialgebra cyclic homology with coefficients. Preprint math/0408094.

[20] M. Khalkhali, and B. Rangipour, A new cyclic module for Hopf algebras. *K*-Theory **27** (2) (2002), 111-131.

[21] M. Khalkhali, and B. Rangipour, Invariant cyclic homology. *K*-Theory **28** (2) (2003), 183-205.

[22] M. Khalkhali, and B. Rangipour, A note on cyclic duality and Hopf algebras. Comm. in Algebra **33** (2005) 763-773.

[23] M. Khalkhali, and B. Rangipour, Cup products in Hopf-cyclic cohomology. C.R. Acad. Sci. Paris Ser. I **340** (2005), 9-14.

[24] M. Marcolli, Lectures on arithmetic noncommutative geometry. math.QA/0409520.

[25] S. Mac Lane, categories for the working mathematician. Second edition. Graduate Texts in Mathematics, 5. Springer-Verlag, New York, 1998.

[26] S. Majid, Foundations of quantum group theory. Cambridge University Press, Cambridge, 1995.

[27] D. Radford, and J. Towber, Yetter-Drinfeld categories associated to an arbitrary bialgebra. J. Pure Appl. Algebra **87** (1993), 259-279.

[28] M. Staic, personal communication.

[29] M. Sweedler, Hopf algebras. Mathematics Lecture Note Series W. A. Benjamin, Inc., New York 1969.

[30] R. Taillefer, Cyclic Homology of Hopf Algebras. *K*-Theory **24** (2001), no. 1, 69–85.

[31] C. Voigt, personal communication.

[32] D. Yetter, Quantum groups and representations of monoidal categories. Math. Proc. Cambridge Philos. Soc. **108** (1990), no. 2, 261–290.

DEPARTMENT OF MATHEMATICS, UNIVERSITY OF WESTERN ONTARIO, LONDON ONTARIO, CANADA

Current address: Department of Mathematics, University of Western Ontario, London Ontario, Canada

E-mail address: masoud@uwo.ca

DEPARTMENT OF MATHEMATICS AND STATISTICS, UNIVERSITY OF VICTORIA, VICTORIA BC, CANADA

E-mail address: bahram@uvic.ca

The non-abelian (or non-linear) method of Chabauty

Minhyong Kim

ABSTRACT. This article is a brief introduction to the ideas surrounding the *non-linear Albanese map* that provides an approach to Diophantine finiteness theorems in the spirit of the method of Chabauty.

I do not claim to have a clear understanding of the full relationship between the two adjectives occurring in my title. However, one particular connection goes as follows. Let us be given a group object G in a certain category. Then the consideration of *torsors* for G becomes unavoidable and even useful in certain contexts. The central example for us will be the homotopy classes of paths with fixed end-points on a topological space which, therefore, is a torsor for a fundamental group. An example more familiar in arithmetic consists of representatives for the Tate-Shafarevich group of an elliptic curve.

Recall that a torsor refers to an object P in our category endowed with an action of G

$$P \times G \to P$$

with the property that on points, that is, maps from some other object T, the action

$$P(T) \times G(T) \to P(T)$$

is transitive. Typically, our category will have a final object $*$, and the torsor P may or may not have a $*$ point. If $P(*) \neq \phi$, then we say the torsor is trivial. In all natural examples, the classification of torsors will give rise to some cohomology set $H^1(G)$ possibly endowed in a natural fashion with extra structure. Now there is the convenient fact that when G is abelian, torsors themselves can be added: we send P_1 and P_2 to the torsor $P_1 \times_G P_2 := (P_1 \times P_2)/G$ where the action is diagonal: $(p_1, p_2)g = (p_1 g, p_2 g)$. This is of course the 'associated bundle' construction, and it yields another torsor when G is abelian. However, there is no natural way to make G act on $P_1 \times_G P_2$ if it is non-abelian.

Now, the objects that are most naturally linear are those that form an abelian category. From this perspective, non-abelian groups, for example, should not be regarded as linear. However, since we are used to studying wide classes of non-abelian groups through linear representations, this point may seem debatable (as was pointed out to me by Arthur Ogus). Nevertheless, one way in which non-linearity arises unavoidably from non-commutativity is via the torsor set $H^1(G)$, which cannot be given a group structure in any way when G is non-abelian. In fact, our main interest is in a situation where the set has the natural structure of a (non-linear) pro-algebraic variety. It is perhaps surprising that one obtains from such a construction concrete and substantive applications to number theory.

To lead up to it we start from a seemingly different direction, namely, the classical abelian method of Chabauty [2]. Here, we are given a smooth curve C over a number field F and assume that we also have a smooth integral model over some ring $R = \mathcal{O}_F[1/S]$ of S-integers of F, where S is a finite set of primes of $\mathcal{O}_F$, so that it makes sense to speak of the R-integral points of C. We also assume that the model has a smooth compactification over R in such a way that the compactification divisor is itself smooth over R. Denote by J the (generalized) Jacobian of C. Let us assume that we have an R-point $x \in C(R)$, giving us an Albanese map $i_x : C \to J$. Let v be a prime of R and consider the completion R_v of R at v. Denote by C_v and J_v the base changes of C and J to the completion F_v. Thus, we are also given R_v-models for these schemes with respect to which we can discuss integral points. Then we have a log map on the R_v-integral points of J_v (depending in general on the choice of a log map for the multiplicative group):

$$\log : J(R_v) \to T_e J_v$$

Composing the Albanese map with the log map, we get a map

$$j_1 : C(R_v) \to T_e J_v$$

which, thereby, 'linearizes' the set of points of C. To get results on the global points, Chabauty makes the important hypothesis

$$\mathrm{rank} J(R) < \dim J \quad \text{(Rank Hypothesis)}$$

and studies its effect on the image $j_1(C(R))$ of the global points of C. The assumption implies that $j_1(C(R))$ lies inside a proper linear subspace of $T_e J_v$, and hence, that there is a non-zero linear form α on $T_e J_v$ that vanishes on $j_1(C(R))$. A finiteness result then follows from the properties of the function $\alpha \circ j_1$ on $C(R_v)$. In fact, since α can be interpreted as a differential form on C_v, it is convenient to express the function in modern language as being

$$\alpha \circ j_1(y) = \int_x^y \alpha$$

where the integral is Coleman integration [1]. So locally, the function is an anti-derivative of the form α, and hence, has a non-zero convergent power-series expansion on any given residue disc of $C(R_v)$. Thus, its zero-set is discrete, and hence, by compactness, finite. Since we had already argued that $C(R)$ is in this zero set, we get finiteness.

This method was made effective by Coleman [3]. In fact, one can work out an effective bound on the number of points by explicitly bounding the number of zeroes on each residue disc. We note that the method of Coleman and Chabauty includes the effective finiteness of rational points in the compact case. Because of this, it has been used as a powerful tool for computing completely the set of rational solutions to an equation in a number of interesting cases.

What can be done if the rank hypothesis is violated? This will happen for example if $C = \mathbf{P}^1 \setminus \{0, 1, \infty\}$, $F = \mathbb{Q}$, and $|S| \geq 1$. Of course, we know the theorems of Siegel and of Faltings that give us finiteness in any case, whenever it is to be expected. However, the known methods of proof are never as direct as Chabauty's, involving rather heavy use of Archimedean machinery stemming from a study of heights. Also, there is hope that a sufficiently powerful extension of

Chabauty's method will give us efficient effective bounds on the number of points whereas the method of Faltings' is not effective in a useful way.

We would like to give a brief outline of our strategy [8] for generalizing Chabauty's method to situations where the abelian method fails, by using unipotent fundamental groups and torsors for them. The way fundamental groups come in is by interpreting $T_e J_v$ as $H_1^{dr}(C_v)/F^0$, where the De Rham homology H_1^{dr} refers to the dual of the first De Rham cohomology of C_v endowed with the dual Hodge filtration which is such that

$$F^{-1} = H_1^{dr} \supset F^0 H_1^{dr} \supset F^1 = 0.$$

But $H_1^{dr}(C_v)$ is just the abelianization of the De Rham fundamental group $\pi_{1,DR}(C_v)$ [4], which we now go on to discuss. Denote by $\mathrm{Un}_n(C_v)$ the category of unipotent vector bundles with connection having index of unipotence $\leq n$. So the objects are bundles with connections (V, ∇) for which there is a filtration

$$0 \subset V_n \subset V_{n-1} \subset \cdots \subset V_1 \subset V_0 = V$$

by sub-bundles, stabilized by the connection, such that

$$(V_i/V_{i+1}, \nabla) \simeq (\mathcal{O}_{C_v}^{r_i}, d),$$

where the last object is the rank r_i trivial bundle with trivial connection. We obviously have an inclusion $\mathrm{Un}_n(C_v) \hookrightarrow \mathrm{Un}_{n+1}(C_v)$ and we denote by $\mathrm{Un}(C_v)$ the union of these categories. The point x defines a fiber functor $e_x : \mathrm{Un}(C_v) \to \mathrm{Vect}_{F_v}$ such that $(V, \nabla) \mapsto V_x$. Now, we have $\pi_{1,DR}(C_v, x) := Aut(e_x)$ as functors of F_v algebras in the obvious sense, which is then represented by a pro-algebraic group by the usual Tannakian formalism. Let Z^i be the descending central series of $\pi_{1,DR}(C_v, x)$ defined by $Z^1 = \pi_{1,DR}(C_v, x)$ and $Z^{n+1} = [\pi_{1,DR}(C_v, x), Z^n]$. Then $Z^{n+1} \backslash \pi_{1,DR}(C_v, x) = Aut(e_x^n)$, where e_x^n is the fiber functor restricted to the subcategory $\mathrm{Un}_n(C_v)$. This turns out to be a unipotent algebraic group (of finite type) [4]. It will be convenient to use the notation $U^{dr} = \pi_{1,DR}(C_v, x)$ and $U_n^{dr} = Z^{n+1} \backslash U^{dr}$.

The point is that using the fundamental group, we can 'lift' the logarithmic Albanese map j_1 to a sequence of 'non-linear' (or 'polylogarithmic,' or 'higher,' or 'unipotent',...) Albanese maps $j_2, j_3, j_4, \ldots$,

$$j_i : C(R_v) \to U_i^{dr}/F^0$$

that fit into a stack of arrows as

$$
\begin{array}{ccc}
\vdots & & \vdots \\
C(R_v) & \xrightarrow{j_3} & U_3^{dr}/F^0 \\
\| & & \downarrow \\
C(R_v) & \xrightarrow{j_2} & U_2^{dr}/F^0 \\
\| & & \downarrow \\
C(R_v) & \xrightarrow{j_1} & U_1^{dr}/F^0
\end{array}
$$

The dimensions of the target varieties in the tower grow with n, and we will attempt to 'resolve' the image of the global points using them.

To describe briefly the definition of the maps, we now bring in the torsor of paths for U^{dr}. By definition $P^{dr}(y) = \pi_{1,DR}(C_v, x, y) := Isom(e_x, e_y)$ and $P_n^{dr}(y) := Isom(e_x^n, e_y^n)$, so that they are right torsors for U^{dr} and U_n^{dr}.

Now, the $P_n^{dr}(y)$ are equipped with Hodge filtrations ([10], [6])

$$P_n^{dr} \supset \cdots F^{-i}P_n^{dr} \supset F^{-i+1}P_n^{dr} \supset \cdots \supset F^{-1}P_n^{dr} \supset F^0P_n^{dr} \supset 0$$

as in the complex situation. In fact, one can show that $F^0P_n^{dr}$ is a torsor for $F^0U_n^{dr}$ and that P_n^{dr} is just the push-out with respect to the inclusion $F^0U_n^{dr}\hookrightarrow U_n^{dr}$. Furthermore, comparison with a crystalline fundamental group equips U_n^{dr} with a Frobenius map ϕ, which is a homomorphism of algebraic groups. The torsor structure is compatible with both the Hodge filtration and the Frobenius. But in fact, one can trivialize the torsor with respect to both structures. This is achieved by choosing a 'Hodge path' $p^h(y) \in F^0P^{dr}$ and then showing that there is a unique Frobenius invariant path $p^c(y)$ [1]. Comparing the two gives rise to a class mod F^0

$$g(y) := [(p^c(y))^{-1} \circ p^h(y)] \in U^{dr}/F^0$$

that is independent of our choices. Considering this at finite levels gives rise to $g_n(y) \in U_n^{dr}/F^0$. In fact, U_n^{dr}/F^0 can be interpreted as a classifying space of U_n^{dr}-torsors compatible with the two structures, and $g_n(y)$ just represents the isomorphism class of $P_n^{dr}(y)$. In any case, we get thereby a compatible collection of maps

$$j_n : C(R_v) \to U_n^{dr}/F^0$$

which is just j_1 for $n = 1$. This is a p-adic analytic map which is very non-degenerate in that its image is Zariski-dense, at least when $F_v = \mathbb{Q}_p$. We stress once more that the targets of our maps are not naturally linear varieties as soon $n > 1$.

What plays the role of the rank hypothesis in this setting? For this we need to introduce a classifying space for global étale torsors, which will replace the Mordell-Weil group of the Jacobian and control the global points. That is, we have a commutative diagram:

$$
\begin{array}{ccc}
C(R) & \to & C(R_v) \\
\downarrow & & \downarrow \\
H_f^1(\Gamma_T, U_n^{et})(\mathbb{Q}_p) & \to & H_f^1(G, U_n^{et})(\mathbb{Q}_p)
\end{array}
$$

We will give just brief descriptions of the ingredients, referring the reader to [8] for a detailed exposition. Here $H_f^1(\Gamma_T, \pi_{1,\text{ét}})$ classifies torsors for the $\mathbb{Q}_p$ pro-unipotent étale fundamental group $U^{et} = \pi_{1,\text{ét}}(\bar{C}, x)$ with an action of the global Galois group Γ_T which classifies extensions of F that are unramified outside of T. The local classifying space $H_f^1(G, U^{et})$ classifies torsors for U^{et} endowed with an action of the local Galois group $G = \text{Gal}(\bar{F}_v/F_v)$. The subscript n refers to the levels of the descending central series as before. The lower horizontal map just restricts the Γ_T-action to a G-action. The subscript 'f' refers to a 'Selmer condition' on the G-action, that requires that the torsor trivializes (locally) when the base is changed to B_{DR}, Fontaine's ring of p-adic periods. The vertical maps send a point y to the class of the torsor of paths $\pi_{1,\text{ét}}(C; x, y)$. The notation and terminology indicates the fact that both $H_f^1(\Gamma_T, U_n^{et})$ and $H_f^1(G, U_n^{et})$ are endowed with the structure of varieties over $\mathbb{Q}_p$ in a natural way and that the maps are going to the $\mathbb{Q}_p$-points of the varieties. The localization map from the global to local cohomology becomes an algebraic map with respect to this structure. We have another commutative diagram

$$C(R_v) \qquad \rightarrow \quad U_n^{dr}/F^0$$
$$\downarrow \qquad \nearrow$$
$$H_f^1(G, U_n^{et})(\mathbb{Q}_p)$$

where the diagonal arrow is now a p-adic comparison map. It takes a torsor $P = \mathrm{Spec}(\mathcal{P})$ for the étale fundamental group and sends it to $\mathrm{Spec}((\mathcal{P} \otimes B_{DR})^G)$ which ends up being a torsor for U_n^{dr}. The commutativity of the diagram follows from a π_1 version of the p-adic comparison isomorphism proved by Shiho, Vologodsky, Tsuji, and Olsson. It is an algebraic map as well when interpreted as going from $H_f(G, U_n^{et})$ to $\mathrm{Res}_{\mathbb{Q}_p}^{F_v}(U_n^{dr}/F^0)$, the Weil restriction of scalars. The utility of this construction is that the image of the global points under j_n is contained in the image of $H_f^1(\Gamma_T, U_n^{et})$ which we can attempt to control in a Galois-theoretic way. This is very much in the spirit of the weak Mordell-Weil theorem. However, we wish to prove a strong enough version that the second step in proving finiteness, that is, the descent, is eliminated. As an intermediate step, we introduce the hypothesis

$$\dim H_f(G, U_n^{et}) < \dim \mathrm{Res}_{\mathbb{Q}_p}^{F_v}(U_n^{dr}/F^0) \quad \text{(Dimension Hypothesis)}$$

and then, the

CONJECTURE 1. *Suppose C is hyperbolic and $F = \mathbb{Q}$. Then the dimension hypothesis holds for n sufficiently large.*

This conjecture is proved in [8] when C has genus zero. Notice that the truth of this conjecture immediately implies the theorems of Faltings and Siegel over $\mathbb{Q}$. This is because there would be an algebraic function on U_n^{dr} that restricts (via j_n) non-trivially to $C(R_v)$ but is zero on $C(R)$. The map can in fact be locally expressed via p-adic iterated integrals, and hence, such a function is locally analytic and non-zero on all residue discs of $C(R_v)$. As in the linear case, we deduce from this the finiteness of the zero set, and hence, of $C(R)$. Expressed differently, we see that $C(R)$ is contained inside

$$Im(C(R_v)) \cap Im(H_f^1(\Gamma_T, U_n^{et}))$$

which is finite. So in some sense, this proof (modulo the conjecture) can be viewed as realizing the old idea of Lang, whereby one attempts to prove finiteness of global points by showing that the intersection of a curve and a finitely generated subgroup of the Jacobian is finite. The only difference is that the Jacobian is replace by a homogeneous space for the De Rham fundamental group, while the finitely generated subgroup is replaced by a finite-dimensional algebraic subvariety. That is to say, homology is replaced by homotopy and a finitely generated group by a non-linear finite-dimensional variety.

Besides being a vindication of the non-linear/non-abelian philosophy, perhaps the main interest of this program is the link it provides between two rather distinct strands of investigation in Diophantine geometry. To explain this remark, we recall briefly (one half of) a conjecture of Jannsen's [7].

Let X be a smooth projective variety over a number field F and consider the geometric étale cohomology of X, $H^n(\bar{X}, \mathbb{Q}_p)$ with $\mathbb{Q}_p$-coefficients. The action of the Galois group of $\mathbb{Q}$ on this space factors through Γ_T where T is the set of primes including the Archimedean ones, p, and those of bad reduction for X. Jannsen's conjectures says

$$H^2(\Gamma_T, H^n(\bar{X}, \mathbb{Q}_p(r))) = 0$$

for $r \geq n + 2$. In the case of a curve C with Jacobian J, $V = T_p J \otimes \mathbb{Q}_p$, the p-adic Tate module of J, is isomorphic to $H^1(\bar{C}, \mathbb{Q}_p)(1)$. Therefore,

$$V^{\otimes n} \simeq H^1(\bar{C}, \mathbb{Q}_p)^{\otimes n}(n)$$

and the last group is a direct summand of $H^n(\bar{C}^n)(n)$. Thus, Jannsen's conjecture implies that $H^2(\Gamma_T, V^{\otimes n}(r)) = 0$ for $r \geq 2$. Elaborating on this idea, one gets

THEOREM 2. *Jannsen's conjecture implies conjecture 1 for affine curves.*

The details will be contained in a forthcoming manuscript, but the basic idea of the proof goes as follows. The fundamental groups sit inside exact sequences

$$0 \to Z^n/Z^{n+1} \to U_n \to U_{n-1} \to 0$$

and Z^n/Z^{n+1} can be expressed as a quotient of $V^{\otimes n}$. By analyzing the decomposition of $V^{\otimes n}$ into irreducibles, one sees that enough of it arises as twists from $V^{\otimes(n-4)}$ to give moderate growth for $H^2(\Gamma_T, V^{\otimes n})$. This, in turn, gives us control of $H^2(\Gamma_T, Z^n/Z^{n+1})$, and hence, of $H^1(\Gamma_T, Z^n/Z^{n+1})$ via the Euler characteristic formula:

$$\dim H^1(\Gamma_T, Z^n/Z^{n+1}) - \dim H^2(\Gamma_T, Z^n/Z^{n+1}) = \dim(Z^n/Z^{n+1}) - \dim(Z^n/Z^{n+1})^c$$

where the superscript c refers to the part fixed by complex conjugation. Thus, we get control of the dimension of $H^1(\Gamma_T, U_n^{et})$ as n grows. When the calculations are done, the upshot is that in this circumstance, $\dim H^1(\Gamma_T, U_n^{et}) < \dim U_n^{dr}/F^0$ for large n, giving us the conjecture. [8] contains this argument in the simple case of genus zero curves.

As shown in [7], Jannsen's conjecture is a consequence of Beilinson's conjecture on the bijectivity of the regulator map together with a bijectivity conjecture concerning the Chern class maps to Galois cohomology. As such, it belongs squarely to the realm of the 'structure theory of motives'. It is thus rather surprising that a result like Siegel's theorem, that does not usually fit into the motivic philosophy in an obvious way, can be deduced from a 'standard conjecture on motives' as outlined above. That is to say, the usual theory of motives relates to Diophantine geometry via L-functions which are constructed from homology. Thus, in this approach it can carry only *linearized* Diophantine information, such as might be contained in Chow groups. Therefore, it is rather surprising that applying the same theory to a mildly non-linear homotopy group yields non-linear information.

In view of the topic of this workshop, perhaps it is not entirely out of order to conclude with a few remarks on relations to non-commutative geometry. Given a representation like $H^n(\bar{X}, \mathbb{Q}_p) = H^n(\bar{X}, \mathbb{Z}_p) \otimes \mathbb{Q}_p$ of Γ_T, one has the image L of G_T inside $Aut(H^n(\bar{X}, \mathbb{Z}_p))$ and the fixed field K of the kernel of this action. Let K' be the maximal abelian pro-p extension of K unramified outside all primes dividing primes of T. Let $H = Gal(K'/F)$. Thus, H sits inside an exact sequence

$$0 \to M \to H \to L \to 0,$$

where $M = Gal(K'/K)$. The analysis of Galois cohomology relies crucially (e.g., [9]) on the structure of the completed group algebra $A := \mathbb{Z}_p[[H]] = \varprojlim \mathbb{Z}_p[H/N]$, where N runs through the finite-index normal subgroups of H, and especially its augmentation ideal $I := Ker(\mathbb{Z}_p[[H]] \to \mathbb{Z}_p)$. In theory of group algebras, I is identified with the module of differentials $\Omega_{A/\mathbb{Z}_p}$, the target for the universal derivation. Here, derivations are taken in the sense of Fox's differential calculus [5]. It is then tempting to ask if the techniques of non-commutative geometry could be made to bear upon questions of Galois cohomology like Jannsen's conjecture. In this connection, there is a way, perhaps artificial, of viewing group algebras of Galois groups as analogous to the 'quantization' of a field. This analogy is most tight in the case of a local field like $F = \mathbb{Q}_p$. Is this case, G^{ab}, the Galois group of the maximal abelian extension of F, is the profinite completion of F^*. Hence, at the level of completed group algebras, $\hat{\mathbb{Z}}[[G^{ab}]] = \hat{\mathbb{Z}}[[F^*]]$. Note that $\hat{\mathbb{Z}}[[F^*]]$, which is a completed enveloping algera, can already be interpreted as a sort of completed quantization of the field F, where the multiplicative relations are retained but the additive ones removed. Furthermore, if G denotes the full Galois group of F, $\hat{\mathbb{Z}}[[G]]$ is a natural non-commutative algebra with quotient $\hat{\mathbb{Z}}[[F^*]]$. Hence, 'Spec($\hat{\mathbb{Z}}[[G]]$)' can be thought of as a quantization of a completion of (maybe a quotient space of) Spec(F).

Acknowledgment. I am grateful to Katia Consani and Matilde Marcolli for inviting me to participate in the stimulating workshop NCGNT. I thank the referee for a thorough reading of the first version of this note. The overall philosophy of non-linearity was impressed upon me by Shinichi Mochizuki, whose influence was also critical in [**8**]. For this I am deeply grateful.

References

[1] Besser, Amnon *Coleman integration using the Tannakian formalism.* Math. Ann. **322** (2002), no. 1, 19–48.

[2] Chabauty, Claude, *Sur les points rationnels des courbes algébriques de genre supérieur l'unité.* C. R. Acad. Sci. Paris **212**, (1941). 882–885.

[3] Coleman, Robert F. *Effective Chabauty.* Duke Math. J. **52** (1985), no. 3, 765–770.

[4] Deligne, Pierre *Le groupe fondamental de la droite projective moins trois points.* Galois groups over $\mathbb{Q}$ (Berkeley, CA, 1987), 79–297, Math. Sci. Res. Inst. Publ., **16**, Springer, New York, 1989.

[5] Fox, Ralph H. *Free differential calculus. I. Derivation in the free group ring.* Ann. of Math. (2) **57**, (1953). 547–560.

[6] Hain, Richard M. *Higher Albanese manifolds.* Hodge theory (Sant Cugat, 1985), 84–91, Lecture Notes in Math., **1246**, Springer, Berlin, 1987.

[7] Jannsen, Uwe *On the l-adic cohomology of varieties over number fields and its Galois cohomology.* Galois groups over $\mathbb{Q}$ (Berkeley, CA, 1987), 315–360, Math. Sci. Res. Inst. Publ., **16**, Springer, New York, 1989.

[8] Kim, Minhyong *The motivic fundamental group of* $\mathbf{P}^1 \setminus \{0, 1, \infty\}$ *and the theorem of Siegel.* Inventiones Mathematicae (to be published).

[9] Ochi, Yoshihiro; Venjakob, Otmar *On the ranks of Iwasawa modules over p-adic Lie extensions.* Math. Proc. Cambridge Philos. Soc. **135** (2003), no. 1, 25–43.

[10] Vologodsky, Vadim *Hodge structure on the fundamental group and its application to p-adic integration.* Mosc. Math. J. **3** (2003), no. 1, 205–247, 260.

DEPARTMENT OF MATHEMATICS, UNIVERSITY OF ARIZONA, TUCSON, AZ 85721
E-mail address: kim@math.arizona.edu

The residues of quantum field theory - numbers we should know

Dirk Kreimer

ABSTRACT. We discuss in an introductory manner structural similarities between the polylogarithm and Green functions in quantum field theory.

1. Introduction: ambiguities in the choice of either a branch or a finite part

It is a pleasure to report here on a connection between mathematics and physics through the study of Dyson–Schwinger equations (DSE) which has been left mostly unexplored so far. While a thorough study of these quantum equations of motions for four-dimensional renormalizable gauge field theories is to be presented in [1], here we have a much more limited goal: to introduce this connection in simple examples and use it as a pedagogical device to explain how the Hopf algebraic structure of a perturbative expansion in quantum field theory (QFT), those non-perturbative quantum equations of motion, renormalization and (breaking of) scaling behaviour fit together.

1.1. The polylog.

We will start our exploration in the rather distinguished world of polylogarithms and mixed Tate Hodge structures to have examples for such phenomena. We emphasize right away though that non-trivial algebraic geometry considerations are beyond our scope. If the remarks below familiarize the reader with this very basic connection between the structure of quantum field theory and such objects they have fulfilled their goal.

Consider the following matrix $M^{(N)}$ borrowed from Spencer Bloch's function theory of the polylogarithm [2]:

$$
(1) \quad
\begin{matrix}
\alpha^0 \\
\alpha^1 \\
\alpha^2 \\
\alpha^3 \\
\cdots
\end{matrix}
\left(
\begin{array}{c|c|c|c|c}
+1 & 0 & 0 & 0 & \cdots \\
-\mathrm{Li}_1(z) & 2\pi i & 0 & 0 & \cdots \\
-\mathrm{Li}_2(z) & 2\pi i \ln z & [2\pi i]^2 & 0 & \cdots \\
-\mathrm{Li}_3(z) & 2\pi i \frac{\ln^2 z}{2!} & [2\pi i]^2 \ln z & [2\pi i]^3 & \cdots \\
\cdots & \cdots & \cdots & \cdots & \cdots
\end{array}
\right)
$$

It is not really meant to be a four by four matrix, but a generic N by N matrix, here spelled out for $N = 4$. Note that we assign an order in a small parameter α to each row, counting rows $0, 1, \ldots$ from top to bottom, similarly we count columns

$0, 1, \ldots$ from left to right. We use the polylog defined by

$$(2) \qquad \mathrm{Li}_n(z) = \sum_{k=1}^{\infty} \frac{z^k}{k^n}$$

inside the unit circle and analytically continued with a branch cut along the real axis from one to plus infinity, say.

The matrix above is highly structured in that the ambiguity reflected by the branch cut, for any entry $M_{i,j}$, is nicely stored in the same row i at $i, j + 1$. Furthermore in each column from top (disregarding the trivial uppermost row $1, 0, 0, \ldots$) to bottom each transcendental function $\mathrm{Li}_n(z)$ or $\ln^m(z)/m!$ has the same coefficient: -1 in the first column, $2\pi i$ in the second, and so forth. This structure allows for the construction of unambiguous univalent polylogs [2] assembled from real and imaginary parts of those rows, for example the univalent dilog is $\Im(\mathrm{Li}_2(z)) + \ln|z|\arg(1 - z)$.

1.2. DSE for the polylog. With this motivic object thrown at us, we can familiarize ourselves with it by considering the following Dyson–Schwinger equation, where the use of this name is justified from the basic observation that it can be written using the Hochschild cohomology of a Hopf algebra of the underlying perturbative expansion [3, 4] as exemplified below. Consider, for suitable z off the cut,

$$(3) \qquad F(\alpha, z) = 1 - \frac{1}{1 - z} + \alpha \int_0^z \frac{F(\alpha, x)}{x} dx,$$

where we continue to name-drop as follows: We call $F(\alpha, z)$ a renormalized Green function, α the coupling (a small parameter, $0 < \alpha < 1$) and consider the perturbative expansion

$$(4) \qquad F(\alpha, z) = 1 - \frac{1}{1 - z} + \sum_{k=1}^{\infty} \alpha^k f_k(z),$$

where we distinguished the tree-level term $f_0(z) = z/(z - 1)$ at order α^0 which equals $-\mathrm{Li}_0(z)$, as it should. We immediately find

$$(5) \qquad f_1(z) = \ln(1 - z) = -\mathrm{Li}_1(z)$$

and if we remind ourselves that the log is a multivalued function with ambiguity an integer multiple of $2\pi i$, we reproduce the second row in the above. Identifying row numbers with powers of α increasing from top to bottom the above matrix is then nothing but the solution of the DSE so constructed:

$$(6) \qquad \sum_{j \leq k} M_{k,j} = f_k(a), \ k > 0.$$

We now utilize the Hopf algebra H of non-planar undecorated rooted trees [5]. It has a Hochschild one-cocycle $B_+ : H \to H$, such that it determines the coproduct Δ via the closedness of this cocycle,

$$(7) \qquad bB_+ = 0 \Leftrightarrow \Delta B_+ = B_+ \otimes 1 + [\mathrm{id} \otimes B_+]\Delta$$

and $\Delta(1) = 1 \otimes 1$.

There is a sub Hopf algebra of "ladder trees"

(8)
$$t_n := \underbrace{B_+(B_+(\cdots(B_+(1))\cdots)}_{n-\text{times}}.$$

For them, we have

(9)
$$\Delta(t_n) = \sum_{j=0}^{n} t_j \otimes t_{n-j},$$

which is cocommutative and we identify $t_0 = 1_H$. For these ladder trees t_n we also introduce an extra dedicated commutative product $t_n \cdot t_m = \frac{(n+m)!}{n!m!} t_{n+m}$. In general, the commutative product in the Hopf algebra H is the disjoint union of trees into forests [5].

We now define Feynman rules as characters on the Hopf algebra. It thus suffices to give them on the generators t_n. Also, as H decomposes as $H = 1_H \mathbb{C} \oplus H_{\text{aug}}$, each $h \in H$ decomposes as $h = h_1 + h_{\text{aug}}$. We now define our Feynman rules by

(10)
$$\phi(B_+(h))(z, z_0) = \int_{z_0}^{z} \frac{\phi(h_{\text{aug}})(x)}{x} dx + \int_{z_0}^{z} \frac{\phi(h_1)(x)}{x - 1} dx, \ \forall h \in H,$$

while we set $\phi(1_H)(z, z_0) = 1$, and $\phi(h_1 h_2) = \phi(h_1)\phi(h_2)$, as they are elements of the character group of the Hopf algebra. Note that for $X = t_{n+1} = B_+(t_n)$, this gives iterated integrals.

Next, we introduce the series

(11)
$$H[[\alpha]] \ni X \equiv c_1 + \sum_{k=1}^{\infty} x_k \alpha^k = c_1 1_H + \alpha B_+ \left(\frac{1}{c_1} \bar{e}(X) + P(X) \right),$$

$c_1 = 1 - \frac{1}{1-z}$ fixing the inhomogenous part. Solving this fix-point equation determines

(12)
$$X = c_1 + \alpha B_+(1) + \alpha^2 B_+(B_+(1)) + \cdots,$$

hence $x_k = t_k$, $k > 0$. We then have $\forall k \geq 0$

(13)
$$f_k(z) = \phi(t_k)(z, 0),$$

and the Hochschild closed one-cocycle B_+ maps to an integral operator $\phi(B_+) \to \int dx/x$, as it should.

Now, let Li and L be the characters on the Hopf algebra defined by

(14)
$$-\phi(t_n)(z, 0) \equiv Li(t_n) = \text{Li}_n(z), \ L(t_n) = \frac{\ln^n(z)}{n!}.$$

From [2] we know that the elimination of all ambiguities due to a choice of branch lies in the construction of functions $a_p(z) = (2\pi i)^{-p} \tilde{a}_p(z)$ where

(15)
$$\tilde{a}_p(z) := \left[\text{Li}_p(z) - \cdots + (-1)^j \text{Li}_{p-j}(z) \frac{\ln^j(z)}{j!} + \cdots + (-1)^{p-1} \text{Li}_1(z) \frac{\ln^{p-1}(z)}{(p-1)!} \right].$$

We have

PROPOSITION 1.

(16)
$$\tilde{a}_p(z) = m \circ ((L^{-1} \otimes Li) \circ (\text{id} \otimes P) \circ \Delta(t_p),$$

where $L^{-1} = L \circ S$, with S the antipode in H and P the projection into the augmentation ideal.

Proof: elementary combinatorics confirming that $L \circ S(t_n/n!) = (-\ln(z))^n/n!$.

There is a strong analogy here to the Bogoliubov R operation in renormalization theory [3, 6], thanks to the fact that Li and L have matching asymptotic behaviour for $|z| \to \infty$. Indeed, if we let R be defined by $R(Li) = L$ and P the projector into the augmentation ideal of H, then

(17) $$L \circ S = S_R^{Li} = -R[m \circ (S_R^{Li} \otimes Li)(\mathrm{id} \otimes P)\Delta] \equiv -R\left[\overline{Li}\right],$$

for example

(18) $$S_R^{Li}(t_2) = -R[Li(t_2) + S_R^{Li}(t_1)Li(t_1)] = -L(t_2) + L(t_1)L(t_1) = \frac{+\ln^2(z)}{2!},$$

where $\overline{Li}(t_2) = Li(t_2) - L(t_1)Li(t_1)$. Thus, a_p is the result of the Bogoliubov map $\overline{Li}$ acting on t_n. The notions of quantum field theory and polylogs are close indeed.

Let us us now reconsider the above function $F(\alpha, z)$ as a function of the lower boundary as well:

(19) $$F(\alpha, z) \equiv F(\alpha, z, 0)$$

and let us return to a generic lower boundary ($\neq 1$, say) z_0, with corresponding DSE

(20) $$F(\alpha, z, z_0) = 1 + \frac{1}{1-z} + \alpha \int_{z_0}^z \frac{F(\alpha, x, z_0)}{x}dx,$$

and returning to Feynman characters (for $h \in H_{\mathrm{aug}}$)

(21) $$\phi(B_+(h))(z, z_0) = \int_{z_0}^z \frac{\phi(h)(x, z_0)}{x}dx.$$

How can we express $F(\alpha, z, \tilde{z}_0)$ in terms of characters $\phi(z, z_0)$ and $\phi(z_0, \tilde{z}_0)$?

The answer is given by reminding ourselves that along with the Hopf algebra structure comes the convolution

(22) $$\phi(z, z_0) = m \circ (\phi(\tilde{z}_0, z_0) \otimes \phi(z, \tilde{z}_0)) \circ (S \otimes \mathrm{id}) \circ \Delta,$$

which answers this question. This is a first example of renormalization, aimed at a reparametrization in the DSE.

2. Renormalization vs polylogs

Having made first contact with renormalization as a modification of a boundary condition in a DSE, we now investigate its greatest strength: the definition of locality and the absorption of short-distance singularities. To do so, we start with examples which are even simpler than the polylog. So let us now introduce a first toy model for renormalization still in analogy with the previous section.

2.1. The simplest model: $F(\alpha, z) = z^{-\mathrm{Res}(\wp)\alpha}$. To make close contact with the situation in perturbative quantum field theory we introduce a regulator ε, which is a complex parameter with small positive real part. For fixed $0 < \alpha < 1$ we then consider the following equation:

(23) $$F_Z(\alpha, z; \varepsilon) = Z + \alpha \int_z^\infty dx \frac{F(\alpha, x; \varepsilon)}{x^{1+\varepsilon}}.$$

Here,

$$(24) \qquad Z = 1 + \sum_{k=1}^{\infty} \alpha^k p_k(\varepsilon)$$

is assumed to be a series in α with coefficients which are Laurent series in the regulator ε with poles of finite order and we thus set $p_k(\varepsilon) = \sum_{j=-k}^{\infty} p_{k,j} \varepsilon^j$ for some real numbers $p_{k,j}$. A glance at Eq.(23) shows that the integrals involved in solving it as a fixpoint equation in α are all logarithmically divergent at the upper boundary for $\varepsilon = 0$. All these integrals will indeed give Laurent series in ε with poles of finite order. Hence we attempt to choose the $p_k(\varepsilon)$ such that the limit $\varepsilon \to 0$ exists in Eq.(23). We want to understand the remaining ambiguity in that choice.

Let us first define the residue of our DSE as the pole at $\varepsilon = 0$ associated to the integral operator $\wp$ involved in it:

$$(25) \qquad \mathrm{Res}(\wp) = \lim_{\varepsilon \to 0} \varepsilon \int_z^{\infty} \frac{1}{x^{1+\varepsilon}} dx.$$

Equally well Res can be defined as the coefficient of the logarithmic growth at plus infinity of the integral operator underlying our DSE:

$$(26) \qquad \mathrm{Res}(\wp) = - \lim_{\Lambda \to \infty} \frac{\alpha \int_z^{\Lambda} \frac{1}{x} dx}{\ln(\Lambda)}.$$

So by residue we mean the coefficient of $\ln(z)$ in this integral, and hence it is closely related to the anomalous dimension $\gamma(\alpha)$, defined as the coefficient of logarithmic growth on a dimensionfull variable.

This is in accordance with the operator-theoretic residue to which this generalizes in the case of Feynman graphs considering the primitive elements of their corresponding Hopf algebra. In the models in subsequent sections below we will see that in general the function $\gamma(\alpha)$ is not merely given by the residue at the primitive element t_1 as will be the case in this section, though the residue continues to play the most crucial role in the determination of an anomalous dimension. Here, for our DSE above, $\mathrm{Res}(\wp) = 1$.

Regard (23) as a fixpoint equation for F_Z and set

$$(27) \qquad F_Z = Z + \sum_{k=1}^{\infty} \alpha^k c_k^Z(z; \varepsilon).$$

The notation emphasizes the dependence on the "counterterm" Z. Let us first set $Z = 1$ in (23), ie. $p_k(\varepsilon) = 0 \; \forall k$. We regard Eq.(23) as an unrenormalized DSE for the Hopf algebra of ladder trees, with Feynman rules exemplified shortly.

We find, plugging (27) in (23),

$$(28) \qquad c_1^{Z=1}(z; \varepsilon) = \int_z^{\infty} dx \frac{x^{-\varepsilon}}{x} = z^{-\varepsilon} \frac{1}{\varepsilon},$$

$$(29) \qquad c_2^{Z=1}(z; \varepsilon) = \frac{z^{-2\varepsilon}}{2! \varepsilon^2},$$

and in general

$$(30) \qquad c_k^{Z=1}(z; \varepsilon) = z^{-k\varepsilon} \frac{1}{k! \varepsilon^k}.$$

Let us set

(31)
$$c_k^{Z=1}(z; \varepsilon) = \sum_{j=0}^{k} c_{k,j}^{Z=1}(\varepsilon) \ln^j(z),$$

upon expanding $z^{-\varepsilon}$ (discarding terms $\ln(z)^j$ with $j > k$ as they will always drop out ultimately when $\varepsilon \to 0$ as the powers of $\ln(z)$ are always bounded by the augmentation degree), heading towards the two gradings in α and $\ln(z)$. The coefficients $c_{k,j}^{Z=1}$ are Laurent series in ε with poles of finite order as promised. Actually, we see that they are extremely simple in this first example. This will change soon enough, and certainly does in full QFT.

Before we solve our DSE exactly, let us set up the perturbative approach in analogy to perturbative quantum field theory. We use the ladder trees t_n as elements of the Hopf algebra H and with multiplication $t_n \cdot t_m$, so that

(32)
$$\Delta(t_n \cdot t_m) = \Delta(t_n) \cdot \Delta(t_m),$$

where $(h_1 \otimes h_2) \cdot (h_3 \otimes h_4) = h_1 \cdot h_3 \otimes h_2 \cdot h_4$.

Again, define Feynman rules ϕ this time by

(33)
$$\phi[B_+(h)](z; \varepsilon) = \int_z^\infty dx \frac{\phi[h](x; \varepsilon)}{x},$$

and $\phi[1](z; \varepsilon) = 1 \; \forall z, \varepsilon$. With such Feynman rules we immediately have

PROPOSITION 2.

(34)
$$c_k^{Z=1}(z; \varepsilon) = \phi(t_k)(z; \varepsilon).$$

This allows to regard Eq.(23) as the image under those Feynman rules ϕ of the already familiar combinatorial fix-point equation

(35)
$$X = 1 + \alpha B_+(X).$$

As a side remark, we note that
(36)
$$\phi(t_n \cdot t_m)(z; \varepsilon) = \frac{(n+m)!}{n!m!} \phi(t_{n+m})(z; \varepsilon) = \frac{z^{-(n+m)\varepsilon}}{n!m!z^{n+m}} = \phi(t_n)(z; \varepsilon)\phi(t_m)(z; \varepsilon).$$

This factorization of the Feynman rules even on a perturbative level is a property of the simplicity of this first model. It holds in general in any renormalizable quantum field theory for the leading pole term, as can be easily shown in any complex regularization like dimensional regularization or analytic regularization, for that manner [7, 8].

Note that we have two different expansion parameters in our DSE. There is α, but for each coefficient $c_k(z; \varepsilon)$ we can expand this coefficient in terms of powers of $\ln(z)$. As we are interested in the limit $\varepsilon \to 0$, it is consistent to maintain only coefficients which have a pole or finite part in ε as we did above. This gives a second grading which, in accord with quantum field theory [3], is provided by the augmentation degree [3, 4]. Note that this is consistent with what we did in the previous section, upon noticing that $\mathrm{Li}_k(z) \sim \ln(z)^k/k! = \mathrm{L}_k(z)$ for $|z| \to \infty$, so that indeed all rows had decreasing degree in $\ln(z)$ from right to left.

Hence we should feel tempted to organize the perturbative solution to our unrenormalized DSE in a manner using again a lower triangular matrix. This does not look very encouraging for the unrenormalized solution though: let us set

$$M_{i,j}^{(N)} = c_{i,i-j}^{Z=1}(\varepsilon)\ln^{i-j}(z), \tag{37}$$

making use of both gradings. Looking at this matrix for say $N = 4$, we find

$$M^{(4)} = \begin{pmatrix} 1 & | & 0 & | & 0 & | & 0 \\ -L_1(z) & | & \frac{1}{\varepsilon} & | & 0 & | & 0 \\ 2L_2(z) & | & -\frac{1}{\varepsilon}L_1(z) & | & +\frac{1}{2!\varepsilon^2} & | & 0 \\ -\frac{9}{2}L_3(z) & | & +\frac{3}{2\varepsilon}L_2(z) & | & -\frac{1}{2!\varepsilon^2}L_1(z) & | & +\frac{1}{3!\varepsilon^3} \end{pmatrix} \tag{38}$$

where again orders in α increase top to bottom and orders in $\ln(z)$ from right to left. This matrix M is an unrenormalized matrix, its evaluation at $\varepsilon = 0$ is impossible. Worse, it does not reveal much structure similar to what we had previously. But so far, this matrix is completely meaningless, being unrenormalized. Thus, being good physicist, our first instinct should be to renormalize it by local counterterms. This will lead us, as we will see, just back to the desired structural properties.

To renormalize it, we have to choose $Z \neq 1$ such that the poles in ε disappear, by choosing appropriate $p_k(\varepsilon) = \sum_{j=-k}^{\infty} p_{k,j}\varepsilon^j$. To understand the possible choices let us go back to the simple case $N = 2$ (i.e. calculating to order α merely) for which we obtain for a generic $Z = 1 + \alpha p_1(\varepsilon)$

$$\begin{pmatrix} 1 & | & 0 \\ -\ln z & | & \frac{1}{\varepsilon} + p_1(\varepsilon) \end{pmatrix} \tag{39}$$

As we require that $M^{(2)}(z;\varepsilon)$ exists at $\varepsilon = 0$, this fixes $p_{1,-1}$:

$$\left\langle \frac{1}{\varepsilon} + p_1(\varepsilon) \right\rangle = 0 \rightarrow p_{1,-1} = -\frac{1}{\varepsilon}, \tag{40}$$

where $\langle \ldots \rangle$ means projection onto the pole part. All higher coefficients $p_{1,j}$, for $j = 0, 1, \ldots$, are left undetermined. To understand better the full freedom in that choice of a renormalized $M^{(N)}$, let us reconsider perturbative renormalization for $M^{(N)}$. It is indeed clear that we are confronted with a choice here: we absorb singularities located at $\varepsilon = 0$ and hence there is a freedom to choose the remaining finite part. In physicists parlance this corresponds to the choice of a renormalization scheme. But such maps can not be chosen completely arbitrarily: they must be in accord with the group structure of the character group of the Hopf algebra, and they must leave the short-distance singularities untouched. Both requirements are easily formulated. For the first, we introduce a Rota–Baxter map R [7, 6],

$$R[ab] + R[a]R[b] = R[R[a]b] + R[aR[b]]. \tag{41}$$

For the second we demand that it is chosen such that

$$R\left[c_k^{Z=1}(z_0;\varepsilon)\right] - c_k^{Z=1}(z_0;\varepsilon) \tag{42}$$

exists at $\varepsilon = 0$ for all k: at a given reference point z_0, usually called the renormalization point, we require that the Rota–Baxter map leaves the short-distance singularities reflected in the poles in ε unaltered. From now on we shall set the renormalization point to $z_0 = 1$ for simplicity.

Define the Bogoliubov map with respect to R, $\overline{\phi}_R$, by

$$\overline{\phi}_R(t_n) = m(S_R^\phi \otimes \phi)(\mathrm{id} \otimes P)\Delta(t_n), \tag{43}$$

with P still the projector into the augmentation ideal. Note that indeed we had this equation before in (16).

Now we have a Birkhoff decomposition of the Feynman character ϕ with respect to R

(44) $$\phi_+ = [\mathrm{id} - R](\overline{\phi}_R), \quad \phi_- = -R(\overline{\phi}_R),$$

for any Rota–Baxter map as above [7], into the renormalized character ϕ_+ and the counterterm ϕ_-, thanks to the existence of a double construction which brings renormalization close to integrable systems for any renormalization scheme R [6]. The crucial fact here is that the pole parts which are still present in the Bogoliubov map are free of $\ln(z)$, which makes sure that ϕ_- provides local counterterms:

THEOREM 3. $\lim_{\varepsilon \to 0} \frac{\partial}{\partial \ln(z)} \overline{\phi}_R(t_n)$ *exists for all* n.

Proof: The theorem has been proven much more generally [3, 9]. A proof follows immediately from induction over the augmentation degree, using that

(45) $$S_R^\phi(B_+(t_n)) = -R[m \circ (S_R^\phi \otimes \phi) \circ (\mathrm{id} \otimes B_+)\Delta(t_n)],$$

using the Hochschild closedness $bB_+ = 0$ and the fact that each element in the perturbation series is in the image of such a closed one-cocycle. This connection between Hochschild closedness and locality is universal in quantum field theory [4, 9], and will be discussed in detail in [1].

Let us look at an example.

(46) $$\begin{aligned} \lim_{\varepsilon \to 0} \frac{\partial}{\partial \ln(z)} \overline{\phi}_R(t_2) &= \lim_{\varepsilon \to 0} \frac{\partial}{\partial \ln(z)} \left(\frac{1}{2!\varepsilon^2} z^{-2\varepsilon} - \left(\frac{1}{\varepsilon} + p_{1,0} \right) \frac{1}{\varepsilon} z^{-\varepsilon} \right) \\ &= (1 + p_{1,0}) \ln(z), \end{aligned}$$

where $p_{1,0}$ depends on the chosen renormalization scheme R.

So this theorem tells us that the pole terms in $\overline{\phi}$ are local, independent of $\ln(z)$. Now, every choice of R as above determines a possible Z in the DSE by setting

(47) $$Z = 1 + \sum_{n=1}^{\infty} \alpha^n S_R^\phi(t_n).$$

We can hence introduce the renormalized matrix $M_{i,j}^{N,R}(z, \varepsilon)$ for any such R. In particular, we can consider this matrix for the renormalized character $[\mathrm{id} - R](\overline{\phi}_R) \equiv S_R^\phi \star \phi(X)$. The above proposition then guarantees that the corresponding matrix exists at $\varepsilon = 0$, by the choice of $\ln(z)$-independent $p_k(\varepsilon)$.

Renormalization has achieved our goal. Now the renormalized matrix $M^{N,R}(z, 0)$ has the same structure as before: columnwise, the coefficient of a power of $\ln(z)$ is inherited from the row above. Let us look at $M^{4,R}$ chosing a renormalized character ϕ_+ with R chosen to be evaluation at $z = 1$, which in this simple model agrees with the projection onto the pole part so that subtraction at the renormalization point is a minimal subtraction (MS) scheme (as $\phi(t_n) \sim \frac{z^{-n\varepsilon}}{n!\varepsilon^n}$ only has poles and no finite parts in ε).

(48) $$\begin{pmatrix} 1 & | & 0 & | & 0 & | & 0 \\ -L_1(z) & | & 0 & | & 0 & | & 0 \\ +L_2(z) & | & 0 & | & 0 & | & 0 \\ -L_3(z) & | & 0 & | & 0 & | & 0 \end{pmatrix}$$

which is so simple for this choice of R that almost no structure remains. We nevertheless urge the reader to work $S_R^\phi(t_n)$ out for several n as in (49)

$$S_R^\phi(t_2) = -R[\phi(t_2) + S_R^\phi(t_1)\phi(t_1)] = -R\left[\frac{1}{2!\varepsilon^2}z^{-2\varepsilon}\right] + R\left[R\left[\frac{1}{\varepsilon}z^{-\varepsilon}\right]\frac{1}{\varepsilon}z^{-\varepsilon}\right] = \frac{1}{2\varepsilon^2}.$$

Due to the simplicity of this DSE we can now show that its perturbative solution in this MS scheme agrees with the non-perturbative (NP) solution for the same renormalization point: at $z = 1$, we require $F(\alpha, z) = 1$. We immediately find that this leads to a Dyson–Schwinger equation

$$(50) \qquad F^{\mathrm{NP}}(\alpha, a) = 1 + \alpha\left\{\int_z^\infty dx \frac{F^{\mathrm{NP}}(\alpha, x)}{x} - \int_1^\infty dx \frac{F^{\mathrm{NP}}(\alpha, x)}{x}\right\}.$$

This reproduces the result Eq.(48) above. This agreement between the Taylor expansion of the non-perturbative solution and the renormalized solution in the MS scheme is a degeneracy of this simple model.

We obviously have

$$(51) \qquad \phi^{\mathrm{NP}}(\alpha, z) = z^{-\alpha}$$

and

$$(52) \qquad \phi^{\mathrm{NP}}(t_{m+n}) = \phi^{\mathrm{NP}}(t_m)\phi^{\mathrm{NP}}(t_n),$$

a hallmark of a non-perturbative approach not available for a perturbative scheme, in particular not for a MS scheme.

Note that we obtain scaling behaviour: $F(\alpha; z) = z^{-\alpha}$, thanks to the basic fact that the DSE was linear. Indeed, the *Ansatz* $F(\alpha, z) = z^{-\gamma(\alpha)}$ solves the DSE above immediately as

$$(53) \qquad z^{-\gamma(\alpha)} = 1 + \alpha\frac{1}{\gamma(\alpha)}\left\{z^{-\gamma(\alpha)} - 1\right\} \Leftrightarrow 1 = \frac{\alpha\mathrm{Res}(\wp)}{\gamma(\alpha)},$$

delivering $\gamma(\alpha) = \alpha$, as $\mathrm{Res}(\wp) = 1$. Note that

$$(54) \qquad \mathrm{Res}(\wp) = \mathrm{Res}_{\varepsilon=0}(\phi(t_1)),$$

the residue of the primitive element of the Hopf algebra, evaluated under the Feynman rules. This holds in general: at a conformal point (a non-trivial fixpoint of the renormalization group) of a QFT one is to find scaling in a DSE and the anomalous dimension is just the sum of the residues of the primitive elements of the Hopf algebra underlying the DSE.

It is high time to come back to the question about the freedom in chosing R. The simple Rota–Baxter map R considered above led to Laurent polynomials $p_k(\varepsilon)$ which were extremely simple, in particular, $p_{k,j}$ was zero for $j \geq k$. Assume you make other choices, such that the requirements on R are still fulfilled. In general, for such a generic R, we find here a solution

$$(55) \qquad F^R(\alpha, z) = (\tilde{z})^{-\alpha},$$

where

$$(56) \qquad \tilde{z} = z\exp\{\Gamma(\alpha)\} \equiv z\exp\left\{\sum_{j=0}^\infty \frac{\gamma_j\alpha^j}{(j+1)!}\right\},$$

for coefficients γ_j recursively determined by the choice of R (or $p_{j,k}$, respectively), and for example $M^{4,R}$ looks like

(57)
$$\begin{pmatrix} 1 & \mid & 0 & \mid & 0 & \mid & 0 \\ -L_1(z) & \mid & -\gamma_0 & \mid & 0 & \mid & 0 \\ +L_2(z) & \mid & +\gamma_0 L_1(z) & \mid & +\frac{1}{2}(\gamma_0^2 - \gamma_1) & \mid & 0 \\ -L_3(z) & \mid & -\gamma_0 L_2(z) & \mid & -\frac{1}{2}(\gamma_0^2 - \gamma_1) L_1(z) & \mid & -\frac{1}{3!}(\gamma_0^3 + 3\gamma_0\gamma_1 - \gamma_2) \end{pmatrix}$$

Note that the associated DSE has the form

(58)
$$F_R(\alpha, z) = F_R(\alpha, 1) - \alpha \int_z^1 \frac{F_R(\alpha, x)}{x},$$

where $F_R(\alpha, 1) = \exp\left\{\sum_{j=0}^\infty \gamma_j \alpha^j\right\}$.

So finally, the ambiguities in the choice of a finite part in renormalization and in the choice of a branch for the log are closely related, a fact which is similarly familiar in quantum field theory in the disguise of the optical theorem connecting real and imaginary parts of quantum field theory amplitudes, as will be discussed elsewhere. Finally, we note that the solution to our DSE fulfills

(59)
$$\frac{\partial \ln F_R(\alpha, z)}{\partial \ln(z)} = -\alpha$$

for all R, confirming the renormalization scheme independence of the anomalous dimension $\gamma_F(\alpha) = -\alpha \mathrm{Res}(\wp)$ of the Green function $F_R(\alpha, z)$, a fact which generally holds when dealing with a DSE which is linear. This last equation actually is a remnant of the propagator coupling duality in quantum field theory, first explored in [10].

2.2. Another toy: $F(\alpha, z) = z^{\arcsin[\alpha\pi\mathrm{Res}(\wp)]/\pi}$. Next, let us study yet another DSE, which is slightly more interesting in so far as that the anomalous dimension is not just given by the residue of the integral operator on the rhs of the equation. Consider the DSE

(60)
$$F(\alpha, z; \varepsilon) = Z + \alpha \int_0^\infty \frac{F(\alpha, x; \varepsilon)}{x + z} dx.$$

First note that again $\mathrm{Res}(\wp) = 1$. Continuing, we find

(61)
$$c_1^{Z=1}(z; \varepsilon) = \int_0^\infty dx \frac{x^{-\varepsilon}}{x + z} = z^{-\varepsilon} \frac{1}{\varepsilon} B(1 - \varepsilon, 1 + \varepsilon),$$

(62)
$$c_2^{Z=1}(z; \varepsilon) = \frac{z^{-2\varepsilon}}{2!\varepsilon^2} B_1 B_2,$$

and in general

(63)
$$c_k^{Z=1}(z; \varepsilon) = z^{-k\varepsilon} \frac{1}{k!\varepsilon^k} B_1 \ldots B_k,$$

where $B_k := B(1 - k\varepsilon, 1 + k\varepsilon)$.

As before, let us set

(64)
$$c_k^{Z=1}(z; \varepsilon) = \sum_{j=0}^k c_{k,j}^{Z=1}(\varepsilon) \ln^j(z),$$

upon expanding $z^{-\varepsilon}$. The coefficients $c_{k,j}^{Z=1}$ are again Laurent series in ε with poles of finite order. In this example we can indeed distinguish between the perturbative solution in the MS scheme and the non-perturbative solution of the DSE, as $c_k(1, \varepsilon)$ is a Laurent series in ε which has non-vanishing finite and higher order parts. It has some merit to study both the MS and the NP case. In the MS scheme we define R to evaluate at the renormalization point $z = 1$ and to project onto the proper pole part. This defines indeed a Rota–Baxter map [7, 6], and as an example, let us calculate

$$\phi(t_2)_{\text{MS}} = \frac{1}{2! \varepsilon^2} B_1 B_2 z^{-2\varepsilon} - \frac{1}{\varepsilon^2} B_1 z^{-\varepsilon},$$

$$(65) \qquad\qquad = -\frac{1}{2\varepsilon^2} - \frac{3}{2}\zeta(2) + L_2(z),$$

disregarding terms which vanish at $\varepsilon = 0$. In accordance with our theorem, no pole terms involve powers of $\ln(z)$. The counterterm $S^\phi_{\text{MS}}(t_2)$ subtracts these pole terms only, leaving $\phi_+(t_2) = L_2(z) - \frac{3}{2}\text{Li}_2(1)$, where $\zeta(2) = \text{Li}_1(1)$.

For the MS scheme we hence find a renormalized matrix

$$(66) \qquad M^{4,\text{MS}}(z, 0) = \begin{pmatrix} 1 & | & 0 & | & 0 & | & 0 \\ -L_1(z) & | & 0 & | & 0 & | & 0 \\ +L_2(z) & | & 0 & | & -\frac{3}{2}\text{Li}_2(1) & | & 0 \\ -L_3(z) & | & 0 & | & +\frac{3}{2}L_1(z)\text{Li}_2(1) & | & 0 \end{pmatrix}$$

Note that this still has non-zero entries along the diagonal, so that $F_{\text{MS}}(\alpha, 1) = 1 + \mathcal{O}(\alpha)$.

Non-perturbatively, we find a solution by imposing the side constraint $F(\alpha, 1) = 1$ as

$$(67) \qquad\qquad F_{\text{NP}}(\alpha, z) = z^{\arcsin[\alpha\pi\text{Res}(\wp)]/\pi}.$$

Note that now the corresponding entries in the Matrix $M^{\text{NP}}_{i,j}$ are not only located in the leftmost column, but are given by the double Taylor expansion

$$(68) \qquad\qquad M_{i,j} = \frac{\partial^i_\alpha \partial^j_{\ln(z)}}{i! j!} \exp\left\{ \frac{\arcsin(\pi\alpha)}{\pi} \ln(z) \right\}_{\alpha=0, \ln(z)=0}.$$

The residue here is still one: $\text{Res}(\wp) = 1$. Again, we can find the above solution with the Ansatz (scaling)

$$(69) \qquad\qquad F(\alpha, z) = z^{-\gamma(\alpha)}$$

where we assume $\gamma(\alpha)$ to vanish at $\alpha = 0$. With this Ansatz we immediately transform the DSE into

$$(70) \qquad z^{-\gamma(\alpha)} - 1 = \alpha \frac{\text{Res}(\wp)}{\gamma(\alpha)} B(1 - \gamma(\alpha), 1 + \gamma(\alpha))[z^{-\gamma(\alpha)} - 1]$$

from which we conclude

$$(71) \qquad\qquad \gamma(\alpha) = \frac{\arcsin \alpha\pi}{\pi} = \alpha + \zeta(2)\alpha^3 + \cdots.$$

Note that this solutions has branch cuts outside the perturbative regime $|\alpha| < 1$. Furthermore, note that the same solution is obtained for the DSE

$$(72) \qquad\qquad F(\alpha, z) = \alpha \int \frac{F(\alpha, x)}{x + z} dx,$$

as the inhomogenous term is an artefact of the perturbative expansion which is absorbed in the scaling behaviour. Finally we note that the appearance of scaling is again a consequence of the linearity of this DSE, and if we were to consider a DSE like

$$(73) \qquad F(\alpha, z) = \int_0^\infty \frac{\mathcal{F}(F(\alpha, x))}{x + z} dx,$$

say, for $\mathcal{F}$ some non-linear polynomial or series, then indeed we would not find scaling behaviour. An Ansatz of the form

$$(74) \qquad F(\alpha, z) = z^{-\gamma(\alpha)} \sum_{k=0}^\infty c_k(\alpha) \ln^k(z),$$

is still feasible though, and leads back to the propagator-coupling dualities [10] explored elsewhere.

Having determined the non-perturbative solution here by the boundary condition $F_{NP}(\alpha, 1) = 1$, other renormalization schemes can be expressed through this solution as before introducing $\tilde{z} = z \exp(\Gamma(\alpha))$ for a suitable series $\Gamma(\alpha)$. The difficulty with perturbative schemes like MS is simply that we do not know off-hand the corresponding boundary condition for a non-perturbative solution of such a scheme as the series $\Gamma(\alpha)$ has to be calculated itself perturbatively, and often is in itself highly divergent as an asymptotic series in α. Even if one were able to resum the perturbative coefficients of a minimal subtraction scheme, the solution so obtained will solve the DSE only with rather meaningless boundary conditions which reflect the presence not only of instanton singularities but, worse, renormalon singularities in the initial asymptotic series. While it is fascinating to import quantum field theory methods into number theory, which suggest to resum perturbation theory amplitudes of a MS scheme making use of the Birkhoff decomposition of [12] combined with progress thanks to Ramis and others in resumation of asymptotic series, as suggested recently [13], the problem is unfortunately much harder still for a renormalizable quantum field theory. We indeed have almost no handle outside perturbation theory on such schemes, while on the other hand the NP solution of DSE with physical side-constraints like $F(\alpha, 1) = 1$ is amazingly straightforward and resums perturbation theory naturally once one has recognized the role of the Hochschild closed one-cocycles [3, 10, 4]. Such an approach will be exhibited in detail in [1]. The idea then to reversely import number-theoretic methods into quantum field theory is to my mind very fruitful and needed to make progress at a level beyond perturbation theory. It is here where in my mind the structures briefly summarized in section three shall ultimately be helpful to overcome these difficulties and make the kinship between numbers and quantum fields even closer.

2.3. More like QFT: $F(\alpha, q^2/\mu^2) = [q^2/\mu^2]^{(1-\sqrt{5-4\sqrt{1-2\alpha\text{Res}(\varphi)}})}$. Let us finish this section with one simple DSE originating in QFT, say a massless scalar field theory with cubic coupling in six dimensions, φ_6^3, with its well-known Feynman rules [5]. We consider the vertex function at zero-momentum transfer which obeys the following DSE (in a NP scheme such that $F(\alpha, 1)) = 1$)

$$(75) \qquad F\left(\alpha, \frac{q^2}{\mu^2}\right) = 1 + \alpha \int d^6 k \, \frac{F(\alpha, \frac{k^2}{\mu^2})}{[k^2]^2[(k+q)^2]}.$$

The scaling Ansatz

(76)
$$F\left(\alpha, \frac{q^2}{\mu^2}\right) = \left(\frac{q^2}{\mu^2}\right)^{-\gamma(\alpha)}$$

still works which is rather typical [10, 11] and delivers

(77) $$1 = \frac{\alpha \text{Res}(\wp)}{\gamma(\alpha)(1-\gamma(\alpha))(1+\gamma(\alpha))(1-\gamma(\alpha)/2)} \Rightarrow \gamma(\alpha) = (1 - \sqrt{5 - 4\sqrt{1-\alpha}})$$

using that the residue is still very simple:

(78)
$$\text{Res}(\wp) = \frac{1}{2}.$$

We used that

(79) $$\int d^6k \frac{\left(\frac{k^2}{\mu^2}\right)^{-x}}{[k^2]^2(k+q)^2} = \text{Res}(\wp) \left[\frac{q^2}{\mu^2}\right]^{-x} \frac{1}{x(1-x)(1+x)(1-x/2)},$$

which is elementary. Note the invariance under the transformation $\gamma(\alpha) \to 1 - \gamma(\alpha)$ in the denominator polynomial of Eq.(77) which reflects the invariance of the primitive $\text{Res}(\wp) = \phi(t_1)$ under the conformal transformation in momentum space $k_\nu \to k_\nu/k^2$ at the renormalization point $q^2 = \mu^2$.

3. The real thing

And how does this fare in the real world of local interactions, mediated by quantum fields which asymptotically approximate free fields specified by covariant wave equations and a Fourier decomposition into raising and lowering operators acting on a suitable state space? The following discussion was essentially given already in [4] and is repeated here with special emphasis on the analogies pointed out in the previous two sections.

Considering DSEs in QFT, one usually obtains them as the quantum equations of motion of some Lagrangian field theory using some generating functional technology in the path integral. DSEs for 1PI Green functions can all be written in the form

(80)
$$\Gamma^{\underline{n}} = 1 + \sum_{\substack{\gamma \in H_L^{[1]} \\ \text{res}(\gamma) = \underline{n}}} \frac{\alpha^{|\gamma|}}{\text{Sym}(\gamma)} B_+^\gamma(X_R^\gamma),$$

where the B_+^γ are Hochschild closed one-cocycles of the Hopf algebra of Feynman graphs indexed by Hopf algebra primitives γ with external legs $\underline{n}$, and X_R^γ is a monomial in superficially divergent Green functions which dress the internal vertex and edges of γ [4, 1]. This allows to obtain the quantum equations of motion, the DSEs for 1PI Green functions, without any reference to actions, Lagrangians or path integrals, but merely from the representation theory of the Poincaré group for free fields.

Hence we were justified in this paper to call any equation of the form (and we only considered the linear case $k = 1$ in some detail)

(81)
$$X = 1 + \alpha B_+(X^k),$$

with B_+ a closed Hochschild one-cocycle, a Dyson Schwinger equation. In general, this motivates an approach to quantum field theory which is utterly based on the Hopf and Lie algebra structures of graphs [3].

3.1. Determination of H. The first step aims at finding the Hopf algebra suitable for the description of a chosen renormalizable QFT. For such a QFT, identify the one-particle irreducible (1PI) diagrams. Identify all edges and propagators in them and define a pre-Lie product on 1PI graphs by using the possibility to replace a local vertex by a vertex correction graph, or, for internal edges, by replacing a free propagator by a self-energy. For any local QFT this defines a pre-Lie algebra of graph insertions [**3**]. For a renormalizable theory, the corresponding Lie algebra will be non-trivial for only a finite number of types of 1PI graphs (self-energies, vertex-corrections) corresponding to the superficially divergent graphs, while the superficially convergent ones provide a semi-direct product with a trivial abelian factor [**12**].

The combinatorial graded pre-Lie algebra so obtained provides not only a Lie-algebra $\mathcal{L}$, but a commutative graded Hopf algebra H as the dual of its universal enveloping algebra $\mathcal{U}(L)$, which is not cocommutative if $\mathcal{L}$ was non-abelian. Dually one hence obtains a commutative but non-cocommutative Hopf algebra H which underlies the forest formula of renormalization. This generalizes the examples discussed in the previous sections as they were all cocommutative. The main structure, and the interplay between the gradings in α and $\ln(z)$ are maintained though, as a glance at [**4**] easily confirms.

3.2. Character of H. For such a Hopf algebra $H = H(m, E, \bar{e}, \Delta, S)$, a Hopf algebra with multiplication m, unit e with unit map $E : \mathbb{Q} \to H$, $q \to qe$, with counit $\bar{e}$, coproduct Δ and antipode S, $S^2 = e$, we immediately have at our disposal the group of characters $G = G(H)$ which are multiplicative maps from G to some target ring V. This group contains a distinguished element: the Feynman rules φ are indeed a very special character in G. They will typically suffer from short-distance singularities, and the character φ will correspondingly reflect these singularities. We will here typically take V to be the ring of Laurent polynomials in some indeterminate ε with poles of finite orders for each finite Hopf algebra element, and design Feynman rules so as to reproduce all salient features of QFT. The Feynman rules of the previous sections were indeed a faithful model for such behaviour.

As $\varphi : H \to V$, with V a ring, with multiplication m_V, we can introduce the group law

$$(82) \qquad \varphi * \psi = m_V \circ (\varphi \otimes \psi) \circ \Delta \,,$$

and use it to define a new character

$$(83) \qquad S_R^\phi * \phi \in G \,,$$

where $S_R^\phi \in G$ twists $\phi \circ S$ and furnishes the counterterm of $\phi(\Gamma)$, $\forall \Gamma \in H$, while $S_R^\phi * \phi(\Gamma)$ corresponds to the renormalized contribution of Γ. S_R^ϕ depends on the Feynman rules $\phi : H \to V$ and the chosen renormalization scheme $R : V \to V$. It is given by

$$(84) \qquad S_R^\phi = -R \left[m_V \circ (S_R^\phi \otimes \phi) \circ (\mathrm{id}_H \otimes P) \circ \Delta \right] \,,$$

where R is supposed to be a Rota-Baxter operator in V, and the projector into the augmentation ideal $P : H \to H$ is given by $P = \mathrm{id} - E \circ \bar{e}$.

The $\bar{R}$ operation of Bogoliubov is then given by

$$(85) \qquad \bar{\phi} := \left[m_V \circ (S_R^\phi \otimes \phi) \circ (\mathrm{id}_H \otimes P) \circ \Delta \right] ,$$

and

$$(86) \qquad S_R^\phi \star \phi \equiv m_V \circ (S_R^\phi \otimes \phi) \circ \Delta = \bar{\phi} + S_R^\phi = (\mathrm{id}_H - R)(\bar{\phi})$$

is the renormalized contribution. Again, this is in complete analogy with the study in the previous sections.

3.3. Locality from H.

The next step aims to show that locality of counterterms is utterly determined by the Hochschild cohomology of Hopf algebras [3, 9]. Again, one can dispense of the existence of an underlying Lagrangian and derive this crucial feature from the Hochschild cohomology of H. What we are considering are spaces $\mathcal{H}^{(n)}$ of maps from the Hopf algebra into its own n-fold tensor product,

$$(87) \qquad \mathcal{H}^{(n)} \ni \psi \Leftrightarrow \psi : H \to H^{\otimes n}$$

and an operator

$$(88) \qquad b : \mathcal{H}^{(n)} \to \mathcal{H}^{(n+1)}$$

which squares to zero: $b^2 = 0$. We have for $\psi \in \mathcal{H}^{(1)}$

$$(89) \qquad (b\psi)(a) = \psi(a) \otimes e - \Delta(\psi(a)) + (\mathrm{id}_H \otimes \psi)\Delta(a)$$

and in general

$$(90) \qquad (b\psi)(a) = (-1)^{n+1}\psi(a) \otimes e + \sum_{j=1}^{n}(-1)^j \Delta_{(j)}(\psi(a)) + (\mathrm{id} \otimes \psi)\Delta(a),$$

where $\Delta_{(i)} : H^{\otimes n} \to H^{\otimes n+1}$ applies the coproduct in the j-th slot of $\psi(a) \in H^{\otimes n}$.

Locality of counterterms and finiteness of renormalized quantities follow indeed from the Hochschild properties of H: the Feynman graph is in the image of a closed Hochschild one-cocycle B_+^γ, $b\,B_+^\gamma = 0$, i.e.

$$(91) \qquad \Delta \circ B_+^\gamma(X) = B_+^\gamma(X) \otimes e + (\mathrm{id} \otimes B_+^\gamma) \circ \Delta(X) ,$$

and this equation suffices to prove the above properties by a recursion over the augmentation degree of H, again in analogy to the study in the previous section.

3.4. Combinatorial DSEs from Hochschild cohomology.

Having understood the mechanism which achieves locality step by step in the perturbative expansion, one realizes that this mechanism delivers the quantum equations of motion, our DSEs. Once more, they typically are of the form

$$(92) \qquad \underline{\Gamma^n} = 1 + \sum_{\substack{\gamma \in H_L^{[1]} \\ \mathrm{res}(\gamma)=\underline{n}}} \frac{\alpha^{|\gamma|}}{\mathrm{Sym}(\gamma)} B_+^\gamma(X_{\mathcal{R}}^\gamma) = 1 + \sum_{\substack{\Gamma \in H_L \\ \mathrm{res}\,(\Gamma)=\underline{n}}} \frac{\alpha^{|\Gamma|}\Gamma}{\mathrm{Sym}(\Gamma)} ,$$

where the first sum is over a finite (or countable) set of Hopf algebra primitives γ,

$$(93) \qquad \Delta(\gamma) = \gamma \otimes e + e \otimes \gamma,$$

indexing the closed Hochschild one-cocycles B_+^γ above, while the second sum is over all one-particle irreducible graphs contributing to the desired Green function, all weighted by their symmetry factors. Here, $\Gamma^{\underline{n}}$ is to be regarded as a formal series

$$(94) \qquad \Gamma^{\underline{n}} = 1 + \sum_{k \geq 1} c_k^{\underline{n}} \alpha^k, \quad c_k^{\underline{n}} \in H.$$

These coefficients of the perturbative expansion deliver form sub Hopf algebras in their own right [4].

There is a very powerful structure behind the above decomposition into Hopf algebra primitives - the fact that the sum over all Green functions G^r is indeed the sum over all 1PI graphs, and this sum, the effective action, gets a very nice structure: $\prod \frac{1}{1-\gamma}$, a product over "prime" graphs - graphs which are primitive elements of the Hopf algebra and which index the closed Hochschild one-ccocycles, in complete factorization of the action. A single such Euler factor with its corresponding DSE and Feynman rules was evaluated in [10], a calculation which was entirely based on a generalization of our study: an understanding of the weight of contributions $\sim \ln(z)$ from a knowledge of the weight of such contributions of lesser degree in α, dubbed propagator-coupling duality in [10]. Altogether, this allows to summarize the structure in QFT as a vast generalization of the introductory study in the previous sections. It turns out that even the quantum structure of gauge theories can be understood along these lines [4]. A full discussion is upcoming [1].

Let us finish this paper by a discussion of the role of matrices $M^{(\gamma)}$ which one can set up for any Hochschild closed one-cocycle B_+^γ in the Hopf algebra. The above factorization indeed allows to gain a great deal of insight into QFT from studying these matrices separately, disentangling DSEs into one equation for each of them, of the form

$$(95) \qquad F^{(\gamma)}(\alpha, z) = 1 + \alpha^{|\gamma|} \int \mathcal{D}(\gamma, F^{(\gamma)}(\alpha, k)),$$

where $\mathcal{D}(\gamma, F^{(\gamma)}(\alpha, k))$ is the integrand for the primitive, which determines a residue which typically and fascinatingly is not a boring number $1, 1/2, \ldots$ as in our previous examples, but a multiple zeta value in its own right [3]. Those are the numbers we should know and understand for the benefit of quantum field theory - know them as motives and understand the contribution of their DSE to the full non-perturbative theory.

The above gives a linear DSE whose solution can be obtained by a scaling *Ansatz* as before. This determines an equation

$$(96) \qquad 1 = \alpha^{|\gamma|} J_\gamma(\mathrm{anom}_\gamma(\alpha))$$

leading to a dedicated anomalous dimension $\mathrm{anom}_\gamma(\alpha)$, just as we did before, with J_γ an algebraic or transcendental function as to yet only known in very few examples. Realizing that the breaking of scaling is parametrized by insertions of logs into the integrand $\mathcal{D}$ with weights prescribed by the β-function of the theory one indeed finds a vast but fascinating generalization of the considerations before.

In particular, matrices $M^{(\gamma)}$ can be obtained from a systematic study of the action of operators $S \star Y^k$, where Y is the grading wrt to the augmentation degree, which faithfully project onto the coefficients of $\ln^k(z)$ apparent in the expansion of $\ln F^{(\gamma)}(\alpha, z)$, as in [10]. In our previous examples this was simply reflected by the fact that $S \star Y^k(t_m) = 0$ for $m > k$, so that for example the coefficient of $\ln(z)$

was only given by the residue of $\phi(t_1)$, which upon exponentiation delivers the subdiagonal entries $M_{j+1,j}$, and similarly $S \star Y^k$ delivers the subdiagonals $M_{j+k,j}$. To work these matrices out for primitives γ beyond one loop (essentially, [10] did it for one-loop) is a highly non-trivial exercise in QFT, with great potential though for progress in understanding of those renormalizable theories. Apart from the perturbative results well-published already, and the introductory remarks here and in [4], a detailed study of DSEs in QFT will be given in [1].

Acknowledgments

Let me first thank Katia Consani, Yuri Manin and Matilde Marcolli for making the workshop possible. Whilst in my talk I reviewed mainly the Hopf algebraic approach to renormalization in more detail than section three above provides, the somewhat polylogarithmically enthused view presented here is a result of stimulating discussions with Spencer Bloch and Herbert Gangl, which I very gratefully acknowledge.

References

[1] D. Kreimer: Locality, Hochschild cohomology and Dyson–Schwinger equations. In preparation.

[2] S. Bloch: Function Theory of Polylogarithms. In: Structural properties of polylogarithms, 275–285, Math. Surveys Monogr., **37**, Amer. Math. Soc., Providence, RI, 1991.

[3] D. Kreimer: New mathematical structures in renormalizable quantum field theories. Annals Phys. **303** (2003) 179 [Erratum-ibid. **305** (2003) 79] [arXiv:hep-th/0211136].

[4] D. Kreimer: Factorization in quantum field theory: an exercise in Hopf algebras and local singularities. Preprint IHES/P/03/38, hep-th/0306020.

[5] A. Connes and D. Kreimer: Hopf algebras, renormalization and non-commutative geometry. Commun. Math. Phys. **199** (1998), 203-242; hep-th/9808042

[6] K. Ebrahimi-Fard, D. Kreimer: Integrable Renormalization II: The General Case. IHES/P/04/08, hep-th/0403118.

[7] D. Kreimer: Chen's iterated integral represents the operator product expansion. Adv. Theor. Math. Phys. **3** (2000) 3 [Adv. Theor. Math. Phys. **3** (1999) 627] [arXiv:hep-th/9901099].

[8] R. Delbourgo, D. Kreimer: Using the Hopf algebra structure of QFT in calculations. Phys. Rev. D **60** (1999) 105025 [arXiv:hep-th/9903249].

[9] C. Bergbauer and D. Kreimer: The Hopf algebra of rooted trees in Epstein-Glaser renormalization. Preprint IHES/P/04/12, hep-th/0403207.

[10] D.J. Broadhurst, D. Kreimer: Exact solutions of Dyson-Schwinger equations for iterated one-loop integrals and propagator-coupling duality. Nucl. Phys. B **600** (2001) 403 [arXiv:hep-th/0012146].

[11] R. Delbourgo: Self-consistent nonperturbative anomalous dimensions. J. Phys. A **36** (2003) 11697 [arXiv:hep-th/0309047].

[12] A. Connes and D. Kreimer: Renormalization in quantum field theory and the Riemann-Hilbert problem I: the Hopf algebra structure of graphs and the main theorem. Commun. Math. Phys. **210** (2000) 249-273; hep-th/9912092

[13] A. Connes and M. Marcolli: From Physics to Number Theory via Noncommutative Geometry I: Quantum Statistical Mechanics of Q-lattices. Preprint math.nt/0404128.

INSTITUT DES HAUTES ÉTUDES SCIENTIFIQUES, 35 RTE. DE CHARTRES, F-91440 BURES-SUR-YVETTE   AND CENTER FOR MATHEMATICAL PHYSICS, BOSTON UNIVERSITY, BOSTON MA 02215.

E-mail address: kreimer@ihes.fr

Phase transitions with spontaneous symmetry breaking on Hecke C*-algebras from number fields

Marcelo Laca and Machiel van Frankenhuijsen

When one attempts to generalize the results of Bost and Connes [**BC**] to algebraic number fields, one has to face sooner or later the fact that in a number field there is no unique factorization in terms of primes. As is well known, this failure is twofold: the ring of integers has nontrivial units, and even if one considers integers modulo units, (equivalently the principal integral ideals), it turns out that factorization in terms of these can fail too, essentially because 'irreducible' does not mean 'prime'. The first difficulty, with the units, already arises in the situation of [**BC**, Remark 33.b], but is easily dealt with by considering elements fixed by a symmetry corresponding to complex conjugation. In the existing generalizations the lack of unique factorization has been dealt with in various ways. It has been eliminated, through replacing the integers by a principal ring that generates the same field [**HLe**], it has been sidestepped, by basing the construction of the dynamical system on the additive integral adeles with multiplication by (a section of) the integral ideles [**Coh**], and it has been ignored, by considering an almost normal subgroup that makes no reference to multiplication [**ALR**]. These simplifications make the construction and analysis of interesting dynamical systems possible, but they come at a price. Indeed, the noncanonical choices introduced in [**HLe**] and [**Coh**] lead to phase transitions with groups of symmetries that are not obviously isomorphic to actual Galois groups of maximal abelian extensions, and have slightly perturbed zeta functions in the case of [**HLe**], while the units not included in the almost normal subgroup in [**ALR**], reappear as a (possibly infinite) group of symmetries under which KMS states have to be invariant, which causes severe difficulties in their computation.

Here we study the phase transition on the Hecke algebra of the almost normal inclusion of the full '$ax + b$' group of algebraic integers of a number field in that of the field. This inclusion is a canonical one, arising from the ring inclusion of the integers in the field. A compactification of the group of units arises naturally and enables us to give, in Theorem 2, a full description of the associated Hecke algebra for all algebraic number fields and, in Theorem 6, a full description of the phase transition for class number one. We also state in the final section some preliminary

---

This note consists of a summary of the results in our paper [**LvF**] and an announcement of results for higher class numbers that will appear in a forthcoming paper. The contents correspond roughly to the talks given by the first author at the MPI workshops on Noncommutative Geometry and Number Theory in 2003 and 2004. We emphasize the results and discuss their relevance but we offer little in terms of details and proofs, for which we refer to [**LvF**].

results on the phase transition for higher class numbers, Theorem 15. For fields of class number one we obtain a group of symmetries isomorphic to a Galois group, of the maximal abelian extension for purely imaginary fields, Proposition 9, and of the maximal abelian extension modulo complex conjugations for fields that have real embeddings, Proposition 10. However, these symmetry groups fail to act as one would wish, as Galois symmetries via the Artin map, on the values that KMS_∞ states take on an arithmetic Hecke algebra defined with coefficients in the field. Essentially the reason for this is that the arithmetic Hecke algebra is based on torsion alone, so only the real subfield of the maximal cyclotomic extension plays a role, Theorem 11. Thus our construction fails to generalize what is perhaps the most intriguing feature of the Bost-Connes system. On a positive note, recent work of Connes and Marcolli [**CM**] seems to provide the right context for such a generalization, at least for imaginary quadratic number fields.

The Hecke C*-algebra. We use the following notation: $\mathcal{K}$ will denote an algebraic number field with ring of integers $\mathcal{O}$. The invertible elements of $\mathcal{K}$ form a multiplicative group, which will be denoted by $\mathcal{K}^*$; since $\mathcal{K}$ is a field, these are simply the nonzero elements of $\mathcal{K}$. Similarly, the group of invertible elements (units) of $\mathcal{O}$ will be denoted by $\mathcal{O}^*$; notice that $\mathcal{O}^*$ is strictly smaller than the multiplicative semigroup $\mathcal{O}^\times$ of nonzero integers.

Following [**BC**], but taking now into account the nontrivial group of units $\mathcal{O}^*$, we associate to the ring inclusion of $\mathcal{O}$ in $\mathcal{K}$ the inclusion of full '$ax + b$ groups',

$$(1) \qquad P_\mathcal{O} := \begin{pmatrix} 1 & \mathcal{O} \\ 0 & \mathcal{O}^* \end{pmatrix} \subset P_\mathcal{K} := \begin{pmatrix} 1 & \mathcal{K} \\ 0 & \mathcal{K}^* \end{pmatrix}.$$

Strictly speaking, this is a not a generalization of the inclusion considered in [**BC**] but of the inclusion $\begin{pmatrix} 1 & \mathbb{Z} \\ 0 & \pm 1 \end{pmatrix} \subset \begin{pmatrix} 1 & \mathbb{Q} \\ 0 & \mathbb{Q}^* \end{pmatrix}$ instead, cf. [**BC**, Remark 33.b].

The pair of groups in (1) is a *Hecke pair* (or an *almost normal inclusion*), which, by definition, means that every double coset contains finitely many right cosets, i.e $R(\gamma) := |P_\mathcal{O} \backslash P_\mathcal{O} \gamma P_\mathcal{O}| < \infty$. Indeed, for $\gamma = \begin{pmatrix} 1 & y \\ 0 & x \end{pmatrix} \in P_\mathcal{K}$, a direct computation yields

$$(2) \qquad R(\gamma) = [\mathcal{O}^* : \mathcal{O}^*_{y(\mathcal{O}+x\mathcal{O})^{-1}}] \cdot [\mathcal{O} : (\mathcal{O} \cap x\mathcal{O})],$$

where $\mathcal{O}^*_{y(\mathcal{O}+x\mathcal{O})^{-1}}$ denotes the subgroup of $\mathcal{O}^*$ that fixes the elements of the fractional ideal $y(\mathcal{O} + x\mathcal{O})^{-1}$ of $\mathcal{K}$ modulo $\mathcal{O}$, i.e. $\mathcal{O}^*_{y(\mathcal{O}+x\mathcal{O})^{-1}} := \{u \in \mathcal{O}^* : ur = r \bmod \mathcal{O}$ for each $r \in y(\mathcal{O} + x\mathcal{O})^{-1}\}$, see [**LvF**, Lemma 1.2]. It is also possible to prove that (1) is a Hecke pair, without computing (2), e.g. by first guessing the associated topological pair and then using [**Tz**, Proposition 4.1], or by verifying directly the conditions of [**LLar1**, Proposition 1.7].

The *Hecke algebra* $\mathcal{H}(P_\mathcal{K}, P_\mathcal{O})$ is, by definition, the convolution *-algebra of complex–valued bi-invariant functions on $P_\mathcal{K}$ that are supported on finitely many double cosets. The convolution product, denoted here simply by juxtaposition, is defined by

$$(3) \qquad fg(\gamma) := \sum_{\gamma_1 \in P_\mathcal{O} \backslash P_\mathcal{K}} f(\gamma \gamma_1^{-1}) g(\gamma_1),$$

which is a finite sum because each double coset contains finitely many right cosets. The involution is given by $f^*(\gamma) = \overline{f(\gamma^{-1})}$, and the identity is the characteristic function of $P_{\mathcal{O}}$. The same convolution formula, with g replaced by a square integrable function ξ on the space $P_{\mathcal{O}}\backslash P_{\mathcal{K}}$ defines a *-preserving, representation λ, the *Hecke representation* of $\mathcal{H}(P_{\mathcal{K}}, P_{\mathcal{O}})$ on the Hilbert space $\ell^2(P_{\mathcal{O}}\backslash P_{\mathcal{K}})$:

$$(4) \qquad \lambda(f)\xi := \sum_{\gamma_1 \in P_{\mathcal{O}}\backslash P_{\mathcal{K}}} f(\gamma\gamma_1^{-1})\xi(\gamma_1) \qquad \text{for } f \in \mathcal{H} \text{ and } \xi \in \ell^2(P_{\mathcal{O}}\backslash P_{\mathcal{K}}).$$

The *Hecke C*-algebra* $C_r^*(P_{\mathcal{K}}, P_{\mathcal{O}})$ is, by definition, the norm closure of $\lambda(\mathcal{H}(P_{\mathcal{K}}, P_{\mathcal{O}}))$. It contains a faithful copy of the Hecke algebra because $\lambda(f)[P_{\mathcal{O}}] = f$ for every $f \in \mathcal{H}(P_{\mathcal{K}}, P_{\mathcal{O}})$ implies that λ is injective. Both the Hecke algebra and C*-algebra come with a canonical *time evolution* by automorphisms $\{\sigma_t : t \in \mathbb{R}\}$, given by

$$(5) \qquad \sigma_t(f)(\gamma) = (R(\gamma)/L(\gamma))^{it} f(\gamma), \qquad \gamma \in P_{\mathcal{K}}, \, t \in \mathbb{R};$$

where $L(\gamma)$ denotes the number of left cosets in the double coset of γ, and is equal to $R(\gamma^{-1})$. A routine calculation shows that this action is spatially implemented on $\ell^2(P_{\mathcal{O}}\backslash P_{\mathcal{K}})$, by the one-parameter unitary group defined by $(U_t\xi)(\gamma) = (R(\gamma)/L(\gamma))^{it}\xi(\gamma)$ for $\xi \in \ell^2(P_{\mathcal{O}}\backslash P_{\mathcal{K}})$. See [**K, Bi, BC**] for details.

To describe our Hecke algebra we shall follow [**BC**] and consider certain special elements determined by the range of two maps μ and θ into $\mathcal{H}(P_{\mathcal{K}}, P_{\mathcal{O}})$, which we define as follows. Let $N_a = |\mathcal{O}/a\mathcal{O}|$ be the absolute norm of $a \in \mathcal{O}^\times$, and, using square brackets to indicate the characteristic function of a subset of $P_{\mathcal{K}}$, let

$$(6) \qquad \mu_a := \frac{1}{\sqrt{N_a}}\left[P_{\mathcal{O}}\begin{pmatrix} 1 & 0 \\ 0 & a \end{pmatrix}P_{\mathcal{O}}\right].$$

Also, for $r \in \mathcal{K}$, let

$$(7) \qquad \theta_r := \frac{1}{R(r)}\left[P_{\mathcal{O}}\begin{pmatrix} 1 & r \\ 0 & 1 \end{pmatrix}P_{\mathcal{O}}\right],$$

where $R(r) := [\mathcal{O}^* : \mathcal{O}_r^*]$ coincides with the number of right cosets in the double coset of $\begin{pmatrix} 1 & r \\ 0 & 1 \end{pmatrix}$, by (2).

The μ_a satisfy the relations

$$(8) \qquad \mu_w = 1, \qquad w \in \mathcal{O}^*,$$

$$(9) \qquad \mu_a^*\mu_a = 1 \qquad a \in \mathcal{O}^\times,$$

$$(10) \qquad \mu_a\mu_b = \mu_{ab} \qquad a, b \in \mathcal{O}^\times,$$

from which it is clear that μ_a depends only on the class of a in the semigroup $\mathcal{O}^\times/\mathcal{O}^*$, which is (canonically isomorphic to) the semigroup of principal integral ideals in $\mathcal{O}$. The relations say that μ is a representation of $\mathcal{O}^\times/\mathcal{O}^*$ by isometries.

The θ_r satisfy the relations

$$(11) \qquad \theta_0 = 1$$

$$(12) \qquad \theta_{wr+b} = \theta_r = \theta_r^*, \qquad r \in \mathcal{K}/\mathcal{O}, \, w \in \mathcal{O}^*, \, b \in \mathcal{O}$$

$$(13) \qquad \theta_r\theta_s = \frac{1}{R(r)}\frac{1}{R(s)}\sum_{u\in\mathcal{O}^*/\mathcal{O}_r^*}\sum_{v\in\mathcal{O}^*/\mathcal{O}_s^*}\theta_{ur+vs}, \qquad r, s \in \mathcal{K}/\mathcal{O}.$$

By (12), θ_r is self-adjoint and depends only on the orbit of $r + \mathcal{O} \in \mathcal{K}/\mathcal{O}$ under the action of $\mathcal{O}^*$, and by (13), their linear span is a commutative, unital *-subalgebra of $\mathcal{H}(P_{\mathcal{K}}, P_{\mathcal{O}})$, which we denote by $\mathfrak{A}$. This algebra is universal for the relations (11) – (13) in the category of *-algebras, and its closure inside $C_r^*(P_{\mathcal{K}}, P_{\mathcal{O}})$

is a commutative, unital C*-algebra, denoted A_θ, which is also universal for the same relations, in the category of C*-algebras, see [**LvF**, Prop. 1.6].

The elements μ_a do not commute with the θ_r, instead, they satisfy the relation:

$$(14) \qquad \mu_a \theta_r \mu_a^* = \frac{1}{N_a} \sum_{b \in \mathcal{O}/a\mathcal{O}} \theta_{\frac{r+b}{a}}, \qquad a \in \mathcal{O}^\times,\ r \in \mathcal{K}/\mathcal{O},$$

[**LvF**, Prop. 1.7]. The right hand side of this relation can be interpreted as defining an action α of the semigroup $\mathcal{O}^\times/\mathcal{O}^*$ by endomorphisms of the *-algebra $\mathfrak{A}$ and of the C*-algebra A_θ, given by

$$\alpha_a(\theta_r) := \frac{1}{N_a} \sum_{b \in \mathcal{O}/a\mathcal{O}} \theta_{\frac{r+b}{a}}.$$

That this defines indeed such an action follows easily from the left-hand side of (14) using the relations (8) – (10).

We pause briefly to review the notion of semigroup crossed product we will be using. When S is a semigroup that acts by endomorphisms α_s of the unital C*-algebra A, we say that (A, S, α) is a semigroup dynamical system. A *covariant pair* for such a system is a pair (π, v) consisting of a unital *-homomorphism π of A into a C*-algebra and a representation v of S by isometries in the same C*-algebra such that the covariance condition $\pi(\alpha_s(a)) = v_s\pi(a)v_s^*$ is satisfied for every $a \in A$ and $s \in S$. There is no loss of generality in taking the target C*-algebra to consist of operators on Hilbert space. The semigroup crossed product associated to (A, S, α) is the C*-algebra $A \rtimes_\alpha S$ generated by a universal covariant pair, that is, a pair through which each covariant pair factors. $A \rtimes_\alpha S$, taken together with its generating universal covariant pair is unique up to canonical isomorphism. When the endomorphisms α_s are injective, the component π of the universal covariant pair is injective and it is customary to drop it from the notation and to think of $A \rtimes_\alpha S$ as being generated by a copy of A and a semigroup of isometries $\{v_s : s \in S\}$ that are universal for the covariance condition. See [**LR1**] for more details on semigroup crossed products. The specific endomorphisms defined above are *injective corner endomorphisms*, this means that each α_s is an isomorphism of A_θ onto the corner $p_s A_\theta p_s$ determined by the projection $p_s = \alpha_s(1)$. See [**L2**] for general facts about crossed products by semigroups of corner endomorphisms.

It turns out that the Hecke C*-algebra $C_r^*(P_\mathcal{K}, P_\mathcal{O})$ is universal for the relations (8) – (10), and this allows us to describe it as a semigroup crossed product of the type discussed above. This description, or rather a slight modification thereof, will be crucial in our analysis of the KMS states. See [**LvF**, Prop. 1.7], [**LvF**, Prop. 1.8].

THEOREM 1. *The Hecke C*-algebra $C_r^*(P_\mathcal{K}, P_\mathcal{O})$ is canonically isomorphic to the C*-algebra with presentation (8) – (14) and to the semigroup crossed product $A_\theta \rtimes_\alpha (\mathcal{O}^\times/\mathcal{O}^*)$. Similarly, the Hecke algebra is canonically isomorphic to the *-algebra with presentation (8) – (14) and to the 'algebraic' semigroup crossed product $\mathfrak{A} \rtimes_\alpha (\mathcal{O}^\times/\mathcal{O}^*)$, which embeds in $A_\theta \rtimes_\alpha (\mathcal{O}^\times/\mathcal{O}^*)$ as the *-subalgebra spanned by the monomials $\mu_a^* \theta_r \mu_b$.*

There still remains the task of understanding the C*-algebra A_θ, since its nature is not at all obvious from the defining relations (11) – (13). It may thus come as a bit of a surprise that A_θ is itself a Hecke algebra that can also be viewed as a subalgebra of the group C*-algebra $C^*(\mathcal{K}/\mathcal{O})$. To see this, we observe first that the subgroup $N := \begin{pmatrix} 1 & \mathcal{K} \\ 0 & \mathcal{O}^* \end{pmatrix}$ of $P_\mathcal{K}$ is normal and that the quotient $P_\mathcal{K}/N$ is isomorphic to $\mathcal{K}^*/\mathcal{O}^*$, the group of principal fractional ideals of $\mathcal{K}$. Moreover, the inclusions $P_\mathcal{O} \subset N \subset P_\mathcal{K}$ are in the conditions of [**LLar1**, Theorem 1.9]. This is not entirely obvious, since in particular, it requires the existence of a multiplicative cross section for $\mathcal{K}^* \to \mathcal{K}^*/\mathcal{O}^*$, shown to exist in [**LvF**, Lemma 1.11]. We also notice that the action of $\mathcal{O}^*$ on $\mathcal{K}$ by multiplication induces an action on $\mathbb{C}[\mathcal{K}/\mathcal{O}]$ that has finite orbits, and thus has a fixed point algebra $\mathbb{C}[\mathcal{K}/\mathcal{O}]^{\mathcal{O}^*}$. At the level of C*-algebras, the analogous fixed point algebra $C^*(\mathcal{K}/\mathcal{O})^{\mathcal{O}^*}$ also exists, but this is more subtle since it depends on the existence of a compactification $\overline{\mathcal{O}^*}$ of $\mathcal{O}^*$, obtained either by taking the inverse limit of finite groups $\mathcal{O}^*/\mathcal{O}^*_{1/a}$, as a tends to infinity multiplicatively in $\mathcal{O}^\times/\mathcal{O}^*$, or by closing the diagonally embedded copy of $\mathcal{O}^*$ in the finite integral ideles, acting on the dual of $\mathcal{K}/\mathcal{O}$, see [**LvF**, Lemma 2.3] and the ensuing discussion.

PROPOSITION 2. *The C*-algebra A_θ, with presentation* (11) – (13), *is canonically isomorphic to the Hecke C*-algebra $C^*_r(N, P_\mathcal{O})$ of the 'intermediate' Hecke inclusion $P_\mathcal{O} \subset N$, and to the fixed point algebra $C^*(\mathcal{K}/\mathcal{O})^{\mathcal{O}^*} = C^*(\mathcal{K}/\mathcal{O})^{\overline{\mathcal{O}^*}}$. Similarly, the *-algebra $\mathfrak{A}$, with the same presentation, is canonically isomorphic to the Hecke algebra $H(N, P_\mathcal{O})$, and to the fixed point algebra $\mathbb{C}[\mathcal{K}/\mathcal{O}]^{\mathcal{O}^*}$. This gives canonical isomorphisms at the level of C*-algebras:*

$$(15) \qquad C^*_r(P_\mathcal{K}, P_\mathcal{O}) \cong C^*_r(N, P_\mathcal{O}) \rtimes (\mathcal{O}^\times/\mathcal{O}^*) \cong C^*_r(\mathcal{K}/\mathcal{O})^{\mathcal{O}^*} \rtimes (\mathcal{O}^\times/\mathcal{O}^*)$$

*and of *-algebras*

$$(16) \qquad \mathcal{H}(P_\mathcal{K}, P_\mathcal{O}) \cong \mathcal{H}(N, P_\mathcal{O}) \rtimes (\mathcal{O}^\times/\mathcal{O}^*) \cong \mathbb{C}[\mathcal{K}/\mathcal{O}]^{\mathcal{O}^*} \rtimes (\mathcal{O}^\times/\mathcal{O}^*).$$

See [**LvF**, Theorem 1.10] and [**LvF**, Theorem 2.5] for the proofs of these assertions. We shall need yet another (dual, adelic) version of the crossed product (15), but before going into such considerations, we digress momentarily to discuss the C*-algebra generated by the isometries μ_a. Using (14) and the multiplication rules of the θ_r's, one shows [**LvF**, Proposition 1.8] that if a and b have a least common multiple $[a, b] \in \mathcal{O}^\times/\mathcal{O}^*$, then

$$(17) \qquad \mu_a \mu_a^* \mu_b \mu_b^* = \mu_{[a,b]} \mu_{[a,b]}^*.$$

In the particular case of a field $\mathcal{K}$ of class number one, least upper bounds always exist and are defined up to units, so (17) says that the range projections $\mu_a \mu_a^*$ respect the lattice structure of the semigroup $\mathcal{O}^\times/\mathcal{O}^*$, in the sense that the product of projections corresponds to the operation of taking least common multiples. Thus, when $\mathcal{K}$ is of class number one, $C^*(\mu_a : a \in \mathcal{O}^\times/\mathcal{O}^*)$ is universal for the relations (8) – (10) and (17), that is, for Nica–covariant isometric representations [**Ni**], and is isomorphic to the Toeplitz C*-algebra of $\mathcal{O}^\times/\mathcal{O}^*$, by an easy application of [**LR1**, Theorem 3.7].

The adelic semigroup crossed product. Being unital and commutative, the C*-algebra $A_\theta \cong C^*(\mathcal{K}/\mathcal{O})^{\mathcal{O}^*}$ is isomorphic to the continuous functions on a compact Hausdorff topological space, and clearly this space ought to be better understood in terms of the dual of $\mathcal{K}/\mathcal{O}$. With this purpose in mind, we begin by recalling that the *adeles* $\mathbb{A}(\mathcal{K})$ of $\mathcal{K}$ are the restricted product

$$\mathbb{A}(\mathcal{K}) := \prod_{v \in M^0_\mathcal{K}} (\mathcal{K}_v; \mathcal{O}_v) \times \prod_{v \in M^\infty_\mathcal{K}} \mathcal{K}_v,$$

where $\mathcal{K}_v$ denotes the completion of $\mathcal{K}$ with respect to either the nonarchimedean absolute values $v \in M_{\mathcal{K}}^0$ (the finite places) or the archimedean ones $v \in M_{\mathcal{K}}^\infty$ (the infinite places), and $\mathcal{O}_v$ denotes the v-completion of $\mathcal{O}$, which is the maximal compact open subring of $\mathcal{K}_v$. The compact ring of *finite integral adeles* is $\mathcal{R} := \prod_{v \in M_{\mathcal{K}}^0} \mathcal{O}_v$, and its group of units, the *finite integral ideles*, is $W := \prod_{v \in M_{\mathcal{K}}^0} \mathcal{O}_v^*$.

As in [**Ta**], a self-dual Haar measure on $\mathbb{A}(\mathcal{K})$ gives a duality pairing of $\mathbb{A}(\mathcal{K})$ to itself,

$$\langle a, b \rangle := \chi_{\mathbb{A}(\mathbb{Q})}(\mathrm{Tr}_{\mathbb{A}(\mathcal{K})/\mathbb{A}(\mathbb{Q})}(ab)) = \prod_{p=\infty,2,3,\cdots} \exp\left(2\pi i \lambda_p(\textstyle\sum_{v|p} \mathrm{Tr}_{\mathcal{K}_v/\mathbb{Q}_p} a_v b_v)\right)$$

where for each finite prime p and $z \in \mathbb{Q}_p$ we choose a rational number $\lambda_p(z)$ such that $z - \lambda_p(z) \in \mathbb{Z}_p$, and for $p = \infty$, we set $\lambda_p(z) = -z$; the exponentials are well defined because λ_p is determined modulo $\mathbb{Z}$. See the beginning of [**LvF**, Section 2] for more details. By passing to subgroups and quotients, this pairing induces an isomorphism of compact groups

$$\widehat{\mathcal{K}/\mathcal{O}} \cong \mathcal{D}^{-1} \times \prod_{v \in M_{\mathcal{K}}^\infty} \{0\},$$

where $\mathcal{D}^{-1}$ is the product, over the finite places, of the local inverse differents

$$\mathfrak{D}_v^{-1} := \{b \in \mathcal{K}_v : \mathrm{Tr}_{\mathcal{K}_v/\mathbb{Q}_p}(ab) \in \mathbb{Z}_p \text{ for all } a \in \mathcal{O}_v\},$$

see [**LvF**, Proposition 2.1].

Since $\overline{\mathcal{O}^*}$ is a subgroup of W, and W leaves the set $\mathcal{D}^{-1} \subset \mathcal{R}$ invariant under multiplication, there is a restricted action of $\overline{\mathcal{O}^*}$ on $\mathcal{D}^{-1}$ which obviously has compact orbits. Taking the quotient with respect to this action gives a compact, Hausdorff orbit space $\Omega := \mathcal{D}^{-1}/\overline{\mathcal{O}^*}$.

THEOREM 3. *The duality pairing of $\mathcal{K}/\mathcal{O}$ with $\mathcal{D}^{-1}$ induces a homeomorphism of the orbit space Ω to the maximal ideal space of A_θ. The transposition to $C(\Omega)$ of the action α of $\mathcal{O}^\times/\mathcal{O}^*$ by endomorphisms of A_θ via this homeomorphism is given by*

$$(18) \qquad \alpha_a(f)(x) := \begin{cases} f(a^{-1}x) & \text{if } x \in a\Omega \\ 0 & \text{if } x \notin a\Omega, \end{cases}$$

for $a\mathcal{O}^* \in \mathcal{O}^\times/\mathcal{O}^*$ and $f \in C(\Omega)$. Moreover, a representation of the crossed product $C(\Omega) \rtimes_\alpha (\mathcal{O}^\times/\mathcal{O}^*)$ is faithful if and only if it is faithful on $C(\Omega)$, and, in particular, $C(\Omega) \rtimes_\alpha (\mathcal{O}^\times/\mathcal{O}^*)$ is canonically isomorphic to $C_r^*(P_{\mathcal{K}}, P_{\mathcal{O}})$.

See [**LvF**, Theorem 2.5] for a more detailed statement and the proof.

REMARK 4. When $\mathcal{K} = \mathbb{Q}$, the unit group is $\mathcal{O}^* = \{\pm 1\}$ and [**BC**, Remark 33.b] shows that the Hecke C*-algebra of $P_{\mathcal{O}} \subset P_{\mathcal{K}}$ is the subalgebra of the Bost-Connes C*-algebra consisting of fixed points under conjugation by $\begin{pmatrix} 1 & 0 \\ 0 & -1 \end{pmatrix}$. It is also interesting to compare our Hecke C*-algebra to that of the almost normal inclusion

$$\Gamma_{\mathcal{O}} := \begin{pmatrix} 1 & \mathcal{O} \\ 0 & 1 \end{pmatrix} \subset \Gamma_{\mathcal{K}} := \begin{pmatrix} 1 & \mathcal{K} \\ 0 & \mathcal{K}^* \end{pmatrix},$$

considered in [**ALR**]. Using a cross section $s : \mathcal{O}^\times/\mathcal{O}^* \to \mathcal{O}^\times$, as constructed in [**LvF**, Lemma 1.11], it is possible to embed $C_r^*(P_{\mathcal{K}}, P_{\mathcal{O}})$ as a C*-subalgebra H_s of $C_r^*(\Gamma_{\mathcal{K}}, \Gamma_{\mathcal{O}})$, in such a way that

$$C_r^*(\Gamma_K, \Gamma_{\mathcal{O}})^{\overline{\mathcal{O}^*}} \cong H_s \otimes C^*(\mathcal{O}^*).$$

See [**LvF**, Proposition 2.7].

A phase transition theorem. The adelic semigroup crossed product $C(\Omega) \rtimes_\alpha (\mathcal{O}^\times / \mathcal{O}^*)$ gives a complementary version of the C*-dynamical system $(C_r^*(P_K, P_{\mathcal{O}}), \sigma)$ on which the time evolution is given by

(19) $$\sigma_t(v_a^* f v_b) = (N_b/N_a)^{it} v_a^* f v_b \qquad t \in \mathbb{R}$$

for $a, b \in \mathcal{O}^\times$ and $f \in C(\Omega)$. There is a natural strongly continuous action γ of the group $W/\overline{\mathcal{O}^*}$ as symmetries. For $\chi \in W/\overline{\mathcal{O}^*}$ let $f_\chi(x) := f(\chi x)$ for $x \in \Omega$, and let

$$\gamma_\chi(v_a^* f v_b) = v_a^* f_\chi v_b.$$

Since γ obviously commutes with σ, these are symmetries of the C*-dynamical system $(C(\Omega) \rtimes (\mathcal{O}^\times / \mathcal{O}^*), \sigma)$.

Back at the level of the Hecke C*-algebra, where the time evolution is given by (5), the symmetries are given by

$$\gamma_\chi(\theta_r) = \theta_{\chi r},$$

where the product of $\chi \in W/\overline{\mathcal{O}^*}$ by $(r + \mathcal{O})/\mathcal{O}^* \in (K/\mathcal{O})^{\mathcal{O}^*}$ is a well defined class in $(K/\mathcal{O})^{\mathcal{O}^*}$ because, in the adelic picture,

$$\gamma_\chi(\hat{\theta}_r)(z) = \frac{1}{R(r)} \sum_{u \in \mathcal{O}^*/\mathcal{O}_r^*} \hat{\delta}_{ur}(\chi z) = \frac{1}{R(r)} \sum_{u \in \mathcal{O}^*/\mathcal{O}_r^*} \chi_{\mathbb{A}(\mathbb{Q})}(\mathrm{Tr}_{\mathbb{A}(K)/\mathbb{A}(\mathbb{Q})}(ru\chi z)).$$

We recall that in the crossed product picture, σ is induced from the dual action $\hat{\alpha}$ of $\widehat{K^*/\mathcal{O}^*}$ via the norm cf. [**L1**]. We denote also by $\hat{\alpha}$ the corresponding action at the level of $C_r^*(P_K, P_{\mathcal{O}})$, and notice that its fixed point algebra is A_θ. This becomes relevant in the following theorem because the equilibrium condition forces full $\hat{\alpha}$–invariance on the KMS states. This is due to the stability of equilibrium and is a strictly stronger property than just σ–invariance, since for instance, when the norm is not injective on principal integral ideals, $\overline{\sigma_\mathbb{R}}$ is strictly smaller than $\hat{\alpha}_{\widehat{K^*/\mathcal{O}^*}}$.

We need to consider a subset of the orbit space to parametrize the extreme KMS states.

DEFINITION 5. An *extreme point* of the inverse different (or an *extreme inverse different*) is, by definition, an element χ of $\mathcal{D}^{-1}$ such that $\chi \mathcal{R} = \mathcal{D}^{-1}$; equivalently, χ has v-component of maximal absolute value (minimal exponential valuation v) in $\mathfrak{D}_v^{-1}$ at every finite place v. The set of extreme points will be denoted by $\partial \mathcal{D}^{-1}$, and the corresponding set of orbits by $\partial \Omega$.

THEOREM 6. *Suppose K is an algebraic number field with class number $h_K = 1$. Let $(C_r^*(P_K, P_{\mathcal{O}}), \sigma)$ be the Hecke C*-dynamical system associated to the almost normal inclusion $P_{\mathcal{O}} \subset P_K$, and let γ be the action of $W/\overline{\mathcal{O}^*}$ as symmetries of this system. Then all KMS$_\beta$ states are $\hat{\alpha}$-invariant, and hence determined by their values on the generators θ_r of $A_\theta \subset C_r^*(P_K, P_{\mathcal{O}})$. Moreover,*

(i) *for each $\beta \in [0, \infty]$ there exists a unique $W/\overline{\mathcal{O}^*}$-invariant KMS$_\beta$ state ϕ_β of σ, given by*

$$\phi_\beta(\theta_r) = N_b^{-\beta} \prod_{p \mid b} \left(\frac{1 - N_p^{\beta-1}}{1 - N_p^{-1}} \right),$$

with $r = a/b$, for $a, b \in \mathcal{O}^\times$, in reduced form.

(ii) *For $\beta \in [0, 1]$ the state ϕ_β is the unique KMS_β state for σ; it is a type III hyperfinite factor state when $\beta \neq 0$.*

(iii) *For $\beta \in (1, \infty]$, the extreme KMS_β states are indexed by $\partial\Omega$, the $\overline{\mathcal{O}^*}$-orbits of extreme points in the inverse different:*

- *The extreme ground states are pure and faithful, and are given by*

$$\phi_{\chi,\infty}(\theta_r) = \frac{1}{R(r)} \sum_{u \in \mathcal{O}^*/\mathcal{O}_r^*} \langle r, u\chi \rangle$$

where $R(r) = |\mathcal{O}^*/\mathcal{O}_r^*|$, the element $\chi \in \partial\mathcal{D}^{-1}$ is a representative of an $\overline{\mathcal{O}^*}$-orbit, and $\langle \cdot, \cdot \rangle$ indicates the duality pairing between $\mathcal{K}/\mathcal{O}$ and $\mathcal{D}^{-1}$; ground states corresponding to different $\overline{\mathcal{O}^*}$-orbits are mutually inequivalent.

- *For $\beta \in (1, \infty)$ the extreme KMS_β states are given by*

$$\phi_{\chi,\beta}(\theta_r) = \frac{1}{\zeta_\mathcal{K}(\beta)} \sum_{a \in \mathcal{O}^\times/\mathcal{O}^*} N_a^{-\beta} \Big(\frac{1}{R(r)} \sum_{u \in \mathcal{O}^*/\mathcal{O}_r^*} \langle ar, u\chi \rangle \Big),$$

where $\zeta_\mathcal{K}$ denotes the Dedekind zeta function of $\mathcal{K}$.

The state $\phi_{\chi,\beta}$ is quasiequivalent to $\phi_{\chi,\infty}$ and the map $T_\beta : \phi_{\chi,\infty} \mapsto \phi_{\chi,\beta}$ extends to an affine isomorphism of the simplex of KMS_∞ states onto the KMS_β states. The action of the symmetry group $W/\overline{\mathcal{O}^*}$ on the extreme KMS_β states, given by $\gamma_w(\phi_{\chi,\beta}) = \phi_{\chi,\beta} \circ \gamma_w = \phi_{\chi w,\beta}$, is free and transitive.

(iv) *The eigenvalue list of the Hamiltonian H_χ associated to $\phi_{\chi,\infty}$ is $\{\log N_a : a \in \mathcal{O}^\times/\mathcal{O}^*\}$ for every χ, so the 'represented partition function' $\mathrm{Tr}\, e^{\beta H_\chi}$ is independent of χ and equals the Dedekind zeta function $\zeta_\mathcal{K}(\beta)$ of $\mathcal{K}$.*

For $\beta > 1$ the proof is a straightforward application of [**L1**, Theorem], while for $\beta \in (0, 1)$ we first need to compute the minimal automorphic extension of the renormalization semigroup.

PROPOSITION 7. *Denote the finite adeles over $\mathcal{K}$ by $\mathbb{A}^0(\mathcal{K})$. The semigroup crossed product $C(\Omega) \rtimes (\mathcal{O}^\times/\mathcal{O}^*)$ is canonically isomorphic to the full corner of $C_0(\mathbb{A}^0(\mathcal{K})^{\overline{\mathcal{O}^*}}) \rtimes (\mathcal{K}^*/\mathcal{O}^*)$ determined by the projection $1_\Omega \in C_0(\mathbb{A}^0(\mathcal{K})^{\overline{\mathcal{O}^*}})$.*

PROOF. By uniqueness of the minimal automorphic extension of the semigroup action α, it suffices to check that $\{q_* f \in C_0(\mathbb{A}^0(\mathcal{K})^{\overline{\mathcal{O}^*}}) : q \in \mathcal{K}^*/\mathcal{O}^*, f \in C(\Omega)\}$ is dense in $C_0(\mathbb{A}^0(\mathcal{K})^{\overline{\mathcal{O}^*}})$, and this is easy to see from the density of $\bigcup_q q^{-1}\Omega$ in $\mathbb{A}^0(\mathcal{K})^{\overline{\mathcal{O}^*}}$. □

The proof of uniqueness in the case $\beta \in (0, 1)$ is by a modified version of Neshveyev's ergodicity proof [**N**]. See Propositions 3.2, 3.3, and 3.4 of [**LvF**].

PROPOSITION 8. *Let μ on $\mathbb{A}^0(\mathcal{K})^{\overline{\mathcal{O}^*}}$ be a positive measure, which we view, indistinctly, also as a positive linear functional on $C_0(\mathbb{A}^0(\mathcal{K})^{\overline{\mathcal{O}^*}})$. If μ satisfies*

$$(20) \qquad \mu(\Omega) = 1 \qquad \text{and} \qquad q_*\mu = N_q^\beta \mu, \quad \text{where} \quad q_*\mu(X) = \mu(q^{-1}X),$$

then the state ϕ_μ of $C_r^*(P_\mathcal{K}, P_\mathcal{O})$ obtained by restricting $\mu \circ E_\gamma$ is a KMS_β state. Conversely, every KMS state arises this way, in fact, the map $\mu \mapsto \phi_\mu$ is an affine isomorphism of simplices. For each $\beta \in (0, 1]$ and any measure μ satisfying (20), the action of $\mathcal{K}^*/\mathcal{O}^*$ on $\mathbb{A}^0(\mathcal{K})^{\overline{\mathcal{O}^*}}$ is ergodic.

Connection with Classfield theory. Using the Artin map it is possible to show that the symmetry groups parametrizing the phase transition at low temperature are isomorphic to Galois groups.

PROPOSITION 9. *If $\mathcal{K}$ is a number field of class number one with no real embeddings, then $W/\overline{\mathcal{O}^*}$ is isomorphic to $G(\mathcal{K}^{ab} : \mathcal{K})$. Via this isomorphism, the extreme KMS_β states of the system $(C_r^*(P_\mathcal{K}, P_\mathcal{O}), \sigma)$ for $\beta > 1$ are indexed by the complex embeddings of the maximal abelian extension $\mathcal{K}^{ab}$ of $\mathcal{K}$.*

PROPOSITION 10. *If the class number of $\mathcal{K}$ is still one, but there are nontrivial unramified extensions (at the finite places), then extreme KMS_β states for $\beta > 1$ are indexed by the complex embeddings of the maximal abelian extension, modulo the complex conjugations over each real embedding of $\mathcal{K}$.*

It is also possible to define another action of $W/\overline{\mathcal{O}^*}$, coming from 'arithmetic' symmetries, on extreme KMS_∞ states. First define an *arithmetic Hecke algebra* $\mathcal{K}(P_\mathcal{K}, P_\mathcal{O})$ to be the algebra over $\mathcal{K}$ generated by the θ_r and the μ_a. Evaluation of an extreme KMS_∞ state on θ_r corresponds to evaluation of $\hat{\theta}_r$ on (the $W/\overline{\mathcal{O}^*}$–orbit of) a point in the extreme inverse different. Since $\hat{\theta}_r$ is a $\mathbb{Q}$–linear combination of characters, it follows that the image of $\mathcal{K}(P_\mathcal{K}, P_\mathcal{O})$ under an extreme KMS_∞ state is contained in the maximal *cyclotomic* extension of $\mathcal{K}$, and hence there is an action of $G(\mathcal{K}^{ab} : \mathcal{K})$ on the values of extreme KMS_∞ states. The arithmetic action of $W/\overline{\mathcal{O}^*}$ is obtained by pulling this back via the isomorphism of Proposition 9. That these two actions coincide for $\mathcal{K} = \mathbb{Q}$ [**BC**, Proposition 20] is an important feature of the Bost–Connes system. But this feature does not carry over to our Hecke algebras: one can compute both actions and see how they differ when $\mathcal{K} \neq \mathbb{Q}$.

THEOREM 11. *The geometric (free, transitive) action of $W/\overline{\mathcal{O}^*}$ on extreme KMS_∞ states from Theorem 6 coincides with the arithmetic action of $W/\overline{\mathcal{O}^*}$ obtained via the Artin map of class field theory, using Proposition 9 and the Galois action on their values, $\phi_{\chi,\infty}(\theta_r)$, if and only if $\mathcal{K} = \mathbb{Q}$.*

PROOF. Let $\phi_{\chi,\infty}$ be an extreme KMS_∞ state, and let j be an idele. The action of j, viewed as a symmetry of $\Omega = \mathcal{D}^{-1}/\overline{\mathcal{O}^*}$, on $\phi_{\chi,\infty}$ is given by

$$\phi_{\chi,\infty}(\theta_r) \mapsto \phi_{j\chi,\infty}(\theta_r) = \frac{1}{R(r)} \sum_{u \in \mathcal{O}^*/\mathcal{O}_r^*} \chi_{\mathbb{A}(\mathbb{Q})}(\mathrm{Tr}_{\mathbb{A}(\mathcal{K})/\mathbb{A}(\mathbb{Q})}(jru\chi)).$$

Next we compute the action of the idele j, viewed now as a Galois element acting on the values of the extreme KMS states via the Artin map from class field theory using Proposition 9. Note first that the complex number $\phi_{\chi,\infty}(\theta_r)$ is a linear combination, with rational coefficients, of character values of the profinite group $\mathcal{K}/\mathcal{O}$, and thus is in the maximal cyclotomic extension of $\mathcal{K}$, in fact it is on the real subfield thereof, because the θ_r are self-adjoint. By class field theory, the Galois action of an idele $j \in W$, when restricted to $\mathbb{Q}^{cycl}$, coincides with the action of $N(j)$, where $N(j)$ is the norm of j to the rational ideles (see [**We**, Corollary 1, p. 246]). Thus the Galois action of j on values of KMS_∞ states is given by

$$\phi_{\chi,\infty}(\theta_r) \mapsto \frac{1}{R(r)} \sum_{u \in \mathcal{O}^*/\mathcal{O}_r^*} \chi_{\mathbb{A}(\mathbb{Q})}(\mathrm{Tr}_{\mathbb{A}(\mathcal{K})/\mathbb{A}(\mathbb{Q})}(N(j)ru\chi)),$$

where we have replaced $N(j) \mathrm{Tr}_{\mathbb{A}(\mathcal{K})/\mathbb{A}(\mathbb{Q})}(ru\chi)$ by $\mathrm{Tr}_{\mathbb{A}(\mathcal{K})/\mathbb{A}(\mathbb{Q})}(N(j)ru\chi)$ using the $\mathbb{A}(\mathbb{Q})$-linearity of the trace. If we now take j to be a rational idele, then $N(j) = j^d$, where $d = \deg(\mathcal{K}/\mathbb{Q})$, and it follows that the two actions are different unless $d = 1$, i.e. unless $\mathcal{K} = \mathbb{Q}$. □

REMARK 12. The particular case of Corollary 9 corresponding to the nine quadratic imaginary fields of class number one, i.e. $\mathcal{K} = \mathbb{Q}[\sqrt{-d}]$ for $d = 1$, 2, 3, 7, 11, 19, 43, 67, 163, is already implicit in [**HLe**]. As indicated by Harari and Leichtnam, the argument of [**BC**, Remark 33.b)] can be used to show that $C_r^*(P_\mathcal{K}, P_\mathcal{O})$ is the fixed point algebra of their Hecke C*-algebra $C^*(\Gamma_S, \Gamma_\mathcal{O})$ under the action of $\mathcal{O}^*$, see also [**LLar1**, Example 2.9]. That the extreme KMS_β states of $C_r^*(P_\mathcal{K}, P_\mathcal{O})$ for $\beta > 1$ are indexed by $\mathcal{G}(\mathcal{K}^{ab} : \mathcal{K})$ then follows from [**HLe**, Theorem 0.2 and Proposition 8.5].

This observation and the final remarks [**HLe**, pp. 241–242] concerning the Artin map, were a strong motivation for the present work, by reinforcing our belief that the almost normal inclusion of full "$ax + b$" groups considered here might lead to groups of symmetries isomorphic to the right Galois groups, and by suggesting that an appropriate compactification of $\mathcal{O}^*$, like the one given in Lemma 2.3 of [**LvF**], was the key to understand the corresponding Hecke C*-algebra.

Higher class numbers.[1] When the class number $h_\mathcal{K}$ of $\mathcal{K}$ is larger than 1, the 'renormalization semigroup' $\mathcal{O}^\times/\mathcal{O}^*$ appearing in the crossed product version of the Hecke algebra is not lattice-ordered, so [**L1**, Theorem 12] cannot be used to study the phase transition. However, $\mathcal{O}^\times/\mathcal{O}^*$ is isomorphic to the semigroup of principal integral ideals, which embeds in the lattice-ordered (free, abelian) semigroup $\mathcal{I}^+$ of all integral ideals. This lattice has a counterpart in the Hecke algebra, given by the characteristic functions of the clopen sets

$$\Omega_\mathfrak{a} := \bigcup_{a \in \mathfrak{a}} a\mathcal{D}^{-1}/\overline{\mathcal{O}^*} = \{\omega \in \Omega : v(\omega) \geq v(\mathfrak{a}) + v(\mathfrak{D}_v^{-1}) \text{ for all } v \in M_\mathcal{K}^0\} \qquad \mathfrak{a} \in \mathcal{I}^+.$$

Let $P_\mathfrak{a}$ denote the associated projection. Because of the isomorphism of Theorem 3 one may view $P_\mathfrak{a}$ indistinctly as function on Ω and as an element of A_θ. The projections $P_\mathfrak{a}$ form a lattice that extends the family of range projections of the isometries μ_a. Indeed, if $\mathfrak{a}$ is principal, say, $\mathfrak{a} = (a)$, then $P_{(a)} = \alpha_a(1) = \mu_a \mu_a^*$, see [**ALR**, Proposition 3.4]. This lattice is the key to the following extension of [**L1**, Theorem 12], and thus to the computation of the KMS states.

THEOREM 13. *Assume $0 < \beta < \infty$, and let ϕ be a state of $C_r^*(P_\mathcal{K}, P_\mathcal{O})$. Then the following are equivalent:*

(i) *ϕ is a KMS_β state*

(ii) *$\phi = \phi \circ E_{\hat{\mathfrak{a}}}$ and $\phi \circ \alpha_a = N_a^{-\beta} \phi$ for every $a \in \mathcal{O}^\times/\mathcal{O}^*$.*

Characterizing ground states is more involved, and we have only been able to do it under the extra assumption of $\hat{\alpha}$-invariance. Even then, there is an unexpected feature: in contrast with the case $h_\mathcal{K} = 1$, when we view a ground state as a measure on Ω, we find that its support is not restricted to the set of orbits of points in $\mathcal{D}^{-1}$ with maximal absolute value at every place.

PROPOSITION 14. *The following are equivalent for an $\hat{\alpha}$-invariant state ϕ of $C_r^*(P_\mathcal{K}, P_\mathcal{O}) \cong C(\Omega) \rtimes \mathcal{O}^\times/\mathcal{O}^*$:*

(i) *ϕ is a ground state.*

---

[1]This section is a preliminary announcement of results to appear in a forthcoming paper.

(ii) $\phi(\mu_b^*\mu_a\mu_a^*\mu_b) = 0$ *whenever* $N_a > N_b$.

(iii) ϕ *is supported on the set* $\Omega_0 := \bigcap_{N_x>1}(\Omega_{[x,1]})^c$

It follows that the $\hat\alpha$-invariant ground states form a Choquet simplex affinely isomorphic to the Borel probability measures on Ω_0.

For example, to obtain the equivalence between (ii) and (iii), one uses the endomorphism β_b, right inverse to α_b such that $\alpha_b \circ \beta_b$ is multiplication by $\alpha_b(1)$ on A_θ. By the covariance condition (14) and its consequences, $\mu_b^*\mu_a\mu_a^*\mu_b = \beta_b \circ \alpha_a(1) = P_{a/(a,b)}$. This projection, in turn, equals $P_{[x,1]}$, with $x = a/b$, and $N_x > 1$ iff $N_a > N_b$.

But not every ground state is a limit of KMS$_\beta$ states. Indeed, from the characterization of KMS$_\beta$ states at finite temperature it is easy to deduce that they satisfy

$$N_b^{-\beta}\phi_n(b^*aa^*b) = \phi(bb^*aa^*bb^*) = \phi(bb^*aa^*) = \phi(aa^*bb^*aa^*) = N_a^{-\beta}\phi(a^*bb^*a)$$

so, assuming $N_a = N_b$ and taking limits as $\beta \to \infty$, we get that KMS$_\infty$ states satisfy

$$(21) \qquad \phi(\beta_b \circ \alpha_a(1)) = \phi(\beta_a \circ \alpha_b(1)) \qquad \text{when } N_a = N_b.$$

Define an equivalence relation on Ω_0 by $\omega \sim \omega'$ if and only if there exists a and b in $\mathcal{O}^\times$ with $N_a = N_b$ such that $a\omega = b\omega'$. The states of Ω induced from states of the quotient space $\Omega_0/\sim$ automatically satisfy (21). This gives the KMS$_\beta$ states for $\beta > 1$, with extreme points parametrized by $\Omega_0/\sim$, and highlights the distinction between ground states and KMS$_\infty$ states.

THEOREM 15. *For $1 < \beta \le \infty$, the extreme KMS$_\beta$ states are parametrized by $\Omega_0/\sim$, and they have a (geometric) free transitive action of a Galois group.*

Denote by $\partial\mathcal{D}^{-1}$ the set of finite adeles in the extreme inverse different, that is, elements in $\mathcal{D}^{-1}$ with maximal absolute value at every finite place. Since $[1, x]$ is an integral ideal for every $x \in \mathcal{K}$, we have $\partial\mathcal{D}^{-1}/\overline{\mathcal{O}^*} \subset \Omega_0$, but the inclusion may be strict.

PROPOSITION 16.

$$\Omega_0 = \bigcap\{(\Omega_b)^c \colon \mathfrak{b} \text{ is an ideal in } \mathcal{O} \text{ such that } \exists \mathfrak{a} \sim \mathfrak{b} \text{ with } N_\mathfrak{a} < N_\mathfrak{b}\}$$

$$= \{y \in \Omega \colon y \notin x\Omega \text{ for every } x \in \mathcal{K} \text{ with } N_x > 1\}$$

$$= \bigcup_{c \in \mathcal{Cl}_\mathcal{K}} \bigcup \{\mathfrak{a}\partial\Omega \colon \mathfrak{a} \text{ integral and of minimal norm in } c\}.$$

REMARK 17. It seems possible to give a better description of the classes in $\Omega_0/\sim$, by noticing that two $\overline{\mathcal{O}^*}$-orbits in Ω_0 are in the same class modulo $\sim$ if any two adeles representing them are in the same $\mathcal{K}^*$-orbit, modulo multiplication by $\overline{\mathcal{O}^*}$, so for $\beta \in (1,\infty]$, the extreme points of the phase transition are in one to one correspondence with by the elements of

$$\bigcup_{c \in \mathcal{Cl}_\mathcal{K}} \bigcup \{\mathfrak{a}\partial\mathcal{D}^{-1} \colon \mathfrak{a} \text{ integral and of minimal norm in } c\}.$$

taken modulo $\mathcal{K}\overline{\mathcal{O}^*}$ acting by multiplication.

Acknowledgment: The work discussed here was carried through several visits, of M.L to I.H.E.S and Utah Valley State University, and of M.v.F. to the University of Newcastle, Australia and to the SFB 478 at the Westfälische Wilhelms-Universität, Münster. The authors acknowledge here with pleasure the support and the hospitality received from these institutions. M.L. also acknowledges support from the National Sciences and Engineering Research Council of Canada.

References

[ALR]   J. Arledge, M. Laca, I. Raeburn, *Semigroup crossed products and Hecke algebras arising from number fields*, Doc. Math. **2** (1997) 115–138.

[Bi]   M.W. Binder, *Induced factor representations of discrete groups and their types*, J. Functional Analysis **115** (1993), 294–312.

[BC]   J.-B. Bost and A. Connes, *Hecke algebras, Type III factors and phase transitions with spontaneous symmetry breaking in number theory*, Sel. Math. (New Series) **1** (1995), 411–457.

[Coh]   P. B. Cohen, *A C*-dynamical system with Dedekind zeta partition function and spontaneous symmetry breaking*, Journées Arithmétiques de Limoges, 1997.

[CM]   A. Connes and M. Marcolli, *Quantum statistical mechanics of Q-lattices*, preprint (2004).

[HLe]   D. Harari and E. Leichtnam, *Extension du phénomène de brisure spontanée de symétrie de Bost-Connes au cas de corps globaux quelconques*, Sel. Math. (New Series), **3** (1997), 205–243.

[K]   A. Krieg, *Hecke Algebras*, Mem. Amer. Math. Soc. **87** (1990), No. 435.

[L1]   M. Laca, *Semigroups of *-endomorphisms, Dirichlet series and phase transitions*, J. Funct. Anal. **152** (1998), 330–378.

[L2]   M. Laca, *From endomorphisms to automorphisms and back: dilations and full corners*, J. London Math. Soc. **61** (2000), 893–904.

[LLar1] M. Laca and N. S. Larsen, *Hecke algebras of semidirect products*, Proc. Amer. Math. Soc., **131** (2003), 2189–2199.

[LR1]   M. Laca and I. Raeburn, *Semigroup crossed products and the Toeplitz algebras of non-abelian groups*, J. Functional Analysis, **139** (1996), 415–440.

[LR2]   M. Laca and I. Raeburn, *A semigroup crossed product arising in number theory*, J. London Math. Soc.,(2) **59** (1999), 330–344.

[LvF]   M. Laca and M. van Frankenhuijsen, *Phase transitions on Hecke C*-algebras and class-field theory over Q*, preprint, 2004.

[N]   S. Neshveyev, *Ergodicity of the action of the positive rationals on the group of finite adeles and the Bost-Connes phase transition theorem*, Proc. Amer. Math. Soc. **130** (2002), 2999–3003.

[Ni]   A. Nica, *C*-algebras generated by isometries and Wiener-Hopf operators*, J. Operator Theory **27** (1992), 17-52.

[Ta]   J. T. Tate, *Fourier Analysis in Number Fields and Hecke's Zeta-Functions*, Ph.D. Dissertation, Princeton University, Princeton, N. J., 1950. (Reprinted in: *Algebraic Number Theory*, J. W. S. Cassels and A. Fröhlich (eds.), Academic Press, New York, 1967, pp. 305–347.)

[Tz]   K. Tzanev, *Hecke C*-algebras and amenability*, J. Operator Theory **50** (2003), 169–178.

[We]   A. Weil, *Basic Number Theory*, 3rd edition, Springer-Verlag, New York, 1974.

DEPARTMENT OF MATHEMATICS AND STATISTICS, UNIVERSITY OF VICTORIA, VICTORIA, BC, V8W 3P4, CANADA
E-mail address: laca@math.uvic.ca

DEPARTMENT OF MATHEMATICS, UTAH VALLEY STATE COLLEGE, OREM, UT 84058-5999, USA
E-mail address: vanframa@uvsc.edu

On harmonic maps in noncommutative geometry

Giovanni Landi

This paper is dedicated to Marta.

ABSTRACT. We report on some recent work on harmonic maps and non-linear σ-models in noncommutative geometry. After a general discussion we concentrate mainly on models on noncommutative tori.

1. Introduction

Recent applications of noncommutative geometry started with the papers [**CDS**] and [**SW**] on matrix and string theories. These developments involve gauge theories defined on noncommutative spaces, a first model of which was constructed almost twenty years ago [**CR**]. One has a vast formalism for gauge fields in noncommutative geometry which allows one to define connections, their curvatures, the associated Yang-Mills action and most of the classical concepts [**C2**]. Methods of differential topology are also available within the realm of cyclic and Hochschild homology and cohomology and this leads, via the coupling of the former with K-theory, to quantities that are stable under deformation and that generalize topological invariants like, for instance, winding numbers and topological charges. A notably results is the fact that one can prove a topological bound for the Yang-Mills action in four dimension [**C2**]. Also, one can construct a Chern-Simons type theory and interpret its behavior under large gauge transformations as a coupling between cyclic cohomology and K-theory [**T**].

In [**DKL1**] and [**DKL2**] we have constructed noncommutative analogues of two dimensional non-linear σ-models. In a more mathematical parlance, these are examples of noncommutative harmonic maps. We have proposed three different models: a continuous analogue of the Ising model which admits instantonic solutions, the analogue of the principal chiral model together with its infinite number of conserved currents and a noncommutative Wess-Zumino-Witten model together with its modified conformal invariance. In particular, we constructed instantonic solutions carrying a nontrivial topological charge q and satisfying a Belavin-Polyakov bound [**BP**], an example of the use of cyclic and Hochschild homology and cohomology mentioned above. The moduli space of these instantons can be identified with an ordinary torus endowed with a complex structure times a projective space $\mathbb{CP}^{q-1}$. This work have relations with recent work on quantized theta-functions

[M1, M2, M3, M4, S-A] as well as work on complex and holomorphic structures on noncommutative spaces, notably noncommutative tori [C2, C4, DS, PS, S-M].

2. Noncommutative harmonic maps

Ordinary non-linear σ-models are field theories whose configuration space consists of maps X from the *source* space, a Riemannian manifold (Σ, g) which we assume to be compact and orientable, to a *target* space, an other Riemannian manifold $(\mathcal{M}, G)$. From a more mathematical point of view they are examples of a theory of harmonic maps (see [Z] and [E] for reviews).

In local coordinates one has an action functional given by

$$(2.1) \qquad S[X] = \frac{1}{2\pi} \int_\Sigma \sqrt{g}\, g^{\mu\nu}\, G_{ij}(X) \partial_\mu X^i\, \partial_\nu X^j,$$

where $g = \det(g_{\mu\nu})$ and $g^{\mu\nu}$ is the inverse of $g_{\mu\nu}$; moreover $\mu, \nu = 1, \ldots, \dim \Sigma$, and $i, j = 1, \ldots, \dim \mathcal{M}$. Here and in the following we use the convention of summing over repeated indexes, and indices are lowered and raised by means of a metric.

The stationary points of the function (2.1) are harmonic maps from Σ to $\mathcal{M}$ and describe minimal surfaces embedded in $\mathcal{M}$.

Different choices of the source and target spaces lead to different field theories, some of them playing a major role in physics. Especially interesting are their applications to conformal field theory and statistical field theory (see for instance [G]). Furthermore, in supersymmetric generalizations, they are the basic building blocks of superstring theory (see for instance [GSW] and [P-J]).

When Σ is two dimensional the action S is conformally invariant, namely it is left invariant by any rescaling of the metric $g \to ge^\sigma$, where σ is any map from Σ to $\mathbb{R}$. As a consequence, the action only depends on the conformal class of the metric and may be rewritten using a complex structure on Σ as

$$(2.2) \qquad S[X] = \frac{i}{\pi} \int_\Sigma G_{ij}(X)\, \partial X^i \wedge \overline{\partial} X^j,$$

where $\partial = \partial_z dz$ and $\overline{\partial} = \partial_{\bar{z}} d\bar{z}$, z being a suitable local complex coordinate.

To construct a noncommutative generalization of the previous construction, we must first dualize the picture and reformulate it in terms of the $*$-algebras $\mathcal{A}$ and $\mathcal{B}$ of complex valued smooth functions defined respectively on Σ and $\mathcal{M}$. Embeddings X of Σ into $\mathcal{M}$ are then in one to one correspondence with $*$-algebra morphisms π from $\mathcal{B}$ to $\mathcal{A}$, the correspondence being simply given by pullback, $\pi_X(f) = f \circ X$.

All this makes perfectly sense for (fixed) not necessarily commutative algebras $\mathcal{A}$ and $\mathcal{B}$; both algebras are over $\mathbb{C}$ and for simplicity we take them to be unital. Then, we take as configuration space the space of all $*$-algebra morphisms from $\mathcal{B}$ (the target algebra) to $\mathcal{A}$ (the source algebra). To define an action functional we need noncommutative generalizations of the conformal and Riemannian geometries. Following an idea of Connes [C2, C4] conformal structures can be understood within the framework of positive Hochschild cohomology. In the commutative case the tri-linear map $\Phi : \mathcal{A} \otimes \mathcal{A} \otimes \mathcal{A} \to \mathbb{R}$ defined by

$$(2.3) \qquad \Phi(f_0, f_1, f_2) = \frac{i}{\pi} \int_\Sigma f_0 \partial f_1 \wedge \overline{\partial} f_2$$

is an extremal element of the space of positive Hochschild cocycles that belongs to the Hochschild cohomology class of the cyclic cocycle Ψ given by

$$(2.4) \qquad \Psi(f_0, f_1, f_2) = \frac{i}{2\pi} \int_\Sigma f_0 df_1 \wedge df_2.$$

Clearly, both (2.3) and (2.4) still make perfectly sense for a general noncommutative algebra $\mathcal{A}$. One can say that Ψ allows to integrate 2-forms in dimension 2,

$$(2.5) \qquad \frac{i}{2\pi} \int a_0 da_1 da_2 = \Psi(a_0, a_1, a_2)$$

so that it is a metric independent object, whereas Φ defines a suitable scalar product

$$(2.6) \qquad \langle a_0 da_1, b_0 db_1 \rangle = \Phi(b_0^* a_0, a_1, b_1^*)$$

on the space of 1-forms and thus depends on the conformal class of the metric. Furthermore, this scalar product is positive and invariant with respect to the action of the unitary elements of $\mathcal{A}$ on 1-forms, and its relation to the cyclic cocycle Ψ allows to prove various inequalities involving topological quantities. In particular we shall get a topological bound for the action which is a two dimensional analogue of the inequality in four dimensional Yang-Mills theory. In analogy with this theory, the configurations giving (absolute) minima will be called σ-model instantons.

Having such a cocycle Φ we can compose it with a morphism $\pi : \mathcal{B} \to \mathcal{A}$ to obtain a positive cocycle on $\mathcal{B}$ defined by

$$(2.7) \qquad \Phi_\pi = \Phi \circ (\pi \otimes \pi \otimes \pi).$$

Since our goal is to build an action functional, which assigns a number to any morphism π, we have to evaluate the cocycle Φ_π on a suitably chosen element of $\mathcal{B} \otimes \mathcal{B} \otimes \mathcal{B}$. Such an element is provided by the noncommutative analogue of the metric on the target, which we take simply as a positive element

$$(2.8) \qquad G = \sum_i b_0^i \delta b_1^i \delta b_2^i$$

of the space of universal 2-forms $\Omega^2(\mathcal{B})$. Thus, the quantity

$$(2.9) \qquad S[\pi] = \Phi_\pi(G)$$

is well defined and positive. We take it as a noncommutative analogue of the action functional (2.2) for non-linear σ-models.

Clearly, we consider π as the dynamical variable (the embedding) whereas Φ (the conformal structure on the source) and G (the metric on the target) are background structures that have been fixed. Alternatively, one could take only the metric G on the target as a background field and use the morphism $\pi : \mathcal{B} \to \mathcal{A}$ to define the induced metric $\pi_* G$ on the source as

$$(2.10) \qquad \pi_* G = \sum_i \pi(b_0^i) \delta \pi(b_1^i) \delta \pi(b_2^i),$$

which is obviously a positive universal 2-form on $\mathcal{A}$. To such an object one can associate, by means of a variational problem (see [**C2**] and [**C4**]), a positive Hochschild cocycle that stands for the conformal class of the induced metric.

The critical points of the σ-model corresponding to the action functional (2.9) are noncommutative generalization of harmonic maps and describe 'minimally embedded surfaces' in the noncommutative space associated with $\mathcal{B}$.

3. Models on the noncommutative torus

In this section we shall work out explicitly the previous construction for the noncommutative torus as the source space and a two point space as a target. The cocycle Φ and the metric G will be replaced by their explicit simple expressions.

3.1. Two points as a target space. The simplest example of a target space one can think of is that of a finite space made of two points $\mathcal{M} = \{1,2\}$, like in the Ising model. Of course, any continuous map from a connected surface to a discrete space is constant and the resulting (commutative) theory would be trivial. However, this is no longer true if the source is a noncommutative space and one has, in general, lots of such maps (i.e. algebra morphisms). Now, the algebra of functions over $\mathcal{M} = \{1,2\}$ is just $\mathcal{B} = \mathbb{C}^2$ and any element $f \in \mathcal{M}$ is a couple of complex numbers (f_1, f_2) with $f_a = f(a)$, the value of f at the point a. As a vector space $\mathcal{B}$ is generated by the function e defined by $e(1) = 1$, $e(2) = 0$. Clearly e is a hermitian idempotent (a projection) , $e^2 = e^* = e$, and $\mathcal{B}$ can be thought of as the unital *-algebra generated by such a projection e. As a consequence, any *-algebra morphism π from $\mathcal{B}$ to $\mathcal{A}$ is given by a projection $p = \pi(e)$ in $\mathcal{A}$.

Choosing the metric $G = \delta e \delta e$ on the space $\mathcal{M}$ of two points, the action functional (2.9) simply becomes

$$(3.1) \qquad\qquad S[p] = \Phi(1, p, p),$$

where Φ is a given Hochschild cocycle standing for the conformal structure.

As we have already mentioned, from general consideration of positivity in Hochschild cohomology this action is bounded by a topological term [**C2**]. In the following we shall explicitly prove this fact when taking the noncommutative torus as source space.

3.2. The noncommutative torus. We recall the very basic aspects of the noncommutative torus that we shall need in the following and refer the reader to [**R4**] for a survey.

Consider an ordinary square two-torus $\mathbb{T}^2$ with coordinate functions $U_1 = \mathrm{e}^{2\pi i x}$ and $U_2 = \mathrm{e}^{2\pi i y}$, where $x, y \in [0,1]$. By Fourier expansion the algebra $C^\infty(\mathbb{T}^2)$ of complex-valued smooth functions on the torus is made up of all power series of the form

$$(3.2) \qquad\qquad a = \sum_{(m,n)\in\mathbb{Z}^2} a_{m,n}\, U_1^m\, U_2^n ,$$

with $\{a_{m,n}\} \in S(\mathbb{Z}^2)$ a complex-valued Schwartz function on $\mathbb{Z}^2$. This means that the sequence of complex numbers $\{a_{m,n} \in \mathbb{C} \mid (m,n) \in \mathbb{Z}^2\}$ decreases rapidly at 'infinity', i.e. for any $k \in \mathbb{N}_0$ one has bounded semi-norms

$$(3.3) \qquad\qquad \|a\|_k = \sup_{(m,n)\in\mathbb{Z}^2} |a_{m,r}| \left(1 + |m| + |r|\right)^k < \infty .$$

Let us now fix a real number θ. The algebra $\mathcal{A}_\theta = C^\infty(\mathbb{T}^2_\theta)$ of smooth functions on the noncommutative torus is the associative algebra made up of all elements of the form (3.2), but now the two generators U_1 and U_2 satisfy

$$(3.4) \qquad\qquad U_2\, U_1 = \mathrm{e}^{2\pi i\theta}\, U_1\, U_2 .$$

The algebra $\mathcal{A}_\theta$ can be made into a *-algebra by defining a involution by

$$(3.5) \qquad\qquad U_1^* := U_1^{-1} , \quad U_2^* := U_2^{-1} .$$

From (3.3) with $k = 0$ one gets a C^*-norm and the corresponding closure of $\mathcal{A}_\theta$ in this norm is the universal C^*-algebra A_θ generated by two unitaries with the relation (3.4); the algebra $\mathcal{A}_\theta$ is dense in A_θ and is thus a pre-C^*-algebra.

There is a one-to-one correspondence between elements of the noncommutative torus algebra $\mathcal{A}_\theta$ and the commutative torus algebra $C^\infty(\mathbb{T}^2)$ given by the Weyl map Ω and its inverse, the Wigner map. As is usual for a Weyl map there are operator ordering ambiguities, and so we will take the prescription

$$(3.6) \qquad \Omega\left(\sum_{(m,r)\in\mathbb{Z}^2} f_{m,r}\, e^{2\pi i(m\,x + r\,y)}\right) := \sum_{(m,r)\in\mathbb{Z}^2} f_{m,r}\, e^{\pi i m\, r\,\theta}\, U_1^m\, U_2^r \ .$$

This choice (called Weyl or symmetric ordering) maps real-valued functions on $\mathbb{T}^2$ into hermitian elements of $\mathcal{A}_\theta$. The inverse map is given by

$$(3.7) \qquad \Omega^{-1}\left(\sum_{(m,r)\in\mathbb{Z}^2} a_{m,r}\, U_1^m\, U_2^r\right) = \sum_{(m,r)\in\mathbb{Z}^2} a_{m,r}\, e^{-\pi i m\, r\,\theta}\, e^{2\pi i(m\,x + r\,y)} \ .$$

Clearly, the map $\Omega : C^\infty(\mathbb{T}^2) \to \mathcal{A}_\theta$ is not an algebra homomorphism. It can be used to deform the commutative product on the algebra $C^\infty(\mathbb{T}^2)$ into a noncommutative star-product by defining

$$(3.8) \qquad f \star g := \Omega^{-1}\big(\Omega(f)\,\Omega(g)\big) \ , \quad f,g \in C^\infty\left(\mathbb{T}^2\right) \ .$$

A straightforward computation gives

$$(3.9) \qquad f \star g = \sum_{(r_1,r_2)\in\mathbb{Z}^2} (f \star g)_{r_1,r_2}\, e^{2\pi i(r_1 x + r_2 y)} \ ,$$

with the coefficients of the expansion of the star-product given by a twisted convolution

$$(3.10) \qquad (f \star g)_{r_1,r_2} = \sum_{(s_1,s_2)\in\mathbb{Z}^2} f_{s_1,s_2}\, g_{r_1-s_1,r_2-s_2}\, e^{\pi i(r_1 s_2 - r_2 s_1)\,\theta}$$

which reduces to the usual Fourier convolution product in the limit $\theta = 0$. Up to isomorphism, the deformed product depends only on the cohomology class in the group cohomology $\mathrm{H}^2(\mathbb{Z}^2, U(1))$ of the $U(1)$-valued two-cocycle on $\mathbb{Z}^2$ given by

$$(3.11) \qquad \lambda(r,s) := e^{\pi i(r_1 s_2 - r_2 s_1)\,\theta}$$

with $r = (r_1, r_2), s = (s_1, s_2) \in \mathbb{Z}^2$.

Heuristically, the noncommutative structure (3.4) of the torus is the exponential of the Heisenberg commutation relation $[y, x] = i\theta/2\pi$. Acting on functions of x alone, the operator U_1 is represented as multiplication by $e^{2\pi i x}$ while conjugation by U_2 yields the shift $x \mapsto x + \theta$,

$$\Omega^{-1}\big(U_1\,\Omega(f(x))\big) = e^{2\pi i x}\, f(x) \ ,$$

$$(3.12) \qquad \Omega^{-1}\big(U_2\,\Omega(f(x))\, U_2^{-1}\big) = f(x + \theta) \ .$$

Analogously, on functions of y alone we have

$$\Omega^{-1}\big(U_1\,\Omega(g(y))\, U_1^{-1}\big) = g(y - \theta) \ ,$$

$$(3.13) \qquad \Omega^{-1}\big(U_2\,\Omega(g(y))\big) = e^{2\pi i y}\, g(y) \ .$$

From (3.4) it follows that $\mathcal{A}_\theta$ is commutative if and only if θ is an integer, and one identifies $\mathcal{A}_0$ with the algebra $C^\infty(\mathbb{T}^2)$. Also, for any $n \in \mathbb{Z}$ there is an isomorphism $\mathcal{A}_\theta \cong \mathcal{A}_{\theta+n}$ since (3.4) does not change under integer shifts $\theta \mapsto \theta+n$. Thus we may restrict the noncommutativity parameter to the interval $0 \le \theta < 1$. Furthermore, since $U_1 U_2 = \mathrm{e}^{-2\pi i\theta} U_2 U_1 = \mathrm{e}^{2\pi i(1-\theta)} U_2 U_1$, the correspondence $U_2 \mapsto U_1, U_1 \mapsto U_2$ yields an isomorphism $\mathcal{A}_\theta \cong \mathcal{A}_{1-\theta}$. These are the only possible isomorphisms and the interval $\theta \in [0, \frac{1}{2}]$ parametrizes a family of non-isomorphic algebras.

When the deformation parameter θ is a rational number the corresponding algebra is related to the commutative torus algebra $C^\infty(\mathbb{T}^2)$, i.e. $\mathcal{A}_\theta$ is Morita equivalent to it in this case [**R1**]. Let $\theta = M/N$, with M and N integers which we take to be relatively prime with $N > 0$. Then $\mathcal{A}_{M/N}$ is isomorphic to the algebra of all smooth sections of an algebra bundle $\mathcal{B}_{M/N} \to \mathbb{T}^2$ whose typical fiber is the algebra $\mathbb{M}_N(\mathbb{C})$ of $N \times N$ complex matrices. Moreover, there is a smooth vector bundle $E_{M/N} \to \mathbb{T}^2$ with typical fiber $\mathbb{C}^N$ such that $\mathcal{B}_{M/N}$ is the endomorphism bundle $\mathrm{End}(E_{M/N})$.

With $\omega = \mathrm{e}^{2\pi i M/N}$, one introduces the $N \times N$ clock and shift matrices

$$(3.14) \quad \mathcal{C}_N = \begin{pmatrix} 1 & & & & \\ & \omega & & & \\ & & \omega^2 & & \\ & & & \ddots & \\ & & & & \omega^{N-1} \end{pmatrix}, \quad \mathcal{S}_N = \begin{pmatrix} 0 & 1 & & & 0 \\ & 0 & 1 & & \\ & & \ddots & \ddots & \\ & & & \ddots & 1 \\ 1 & & & & 0 \end{pmatrix},$$

which are unitary and traceless (since $\sum_{k=0}^{N-1} \omega^k = 0$), satisfy

$$(3.15) \qquad\qquad (\mathcal{C}_N)^N = (\mathcal{S}_N)^N = \mathbb{I}_N \ ,$$

and obey the commutation relation

$$(3.16) \qquad\qquad \mathcal{S}_N \mathcal{C}_N = \omega \, \mathcal{C}_N \mathcal{S}_N \ .$$

Since M and N are relatively prime the matrices (3.14) generate the finite dimensional algebra $\mathbb{M}_N(\mathbb{C})$: they generate a C^*-subalgebra which commutes only with multiples of the identity matrix $\mathbb{I}_N$ which thus has to be the full matrix algebra. Were M and N not coprime the generated algebra would be a proper subalgebra of $\mathbb{M}_N(\mathbb{C})$.

The algebra $\mathcal{A}_{M/N}$ has a 'huge' center $\mathcal{C}(\mathcal{A}_{M/N})$ which is generated by the elements U_1^q and U_2^q, and one identifies $\mathcal{C}(\mathcal{A}_{M/N})$ with the commutative algebra $C^\infty(\mathbb{T}^2)$ of smooth functions on an ordinary torus $\mathbb{T}^2$ which is 'wrapped' N times onto itself. There is a surjective algebra homomorphism

$$(3.17) \qquad\qquad \pi_N : \mathcal{A}_{M/N} \longrightarrow \mathbb{M}_N(\mathbb{C})$$

given by

$$(3.18) \qquad \pi_N \left(\sum_{(m,r)\in\mathbb{Z}^2} a_{m,r} \, U_1^m U_2^r \right) = \sum_{(m,r)\in\mathbb{Z}^2} a_{m,r} \, (\mathcal{C}_N)^m \, (\mathcal{S}_N)^r \ .$$

Under this homomorphism the whole center $\mathcal{C}(\mathcal{A}_{M/N})$ is mapped to $\mathbb{C}$.

From now on we will assume that θ is an irrational number unless otherwise explicitly stated. On $\mathcal{A}_\theta$ there is a unique normalized positive definite trace which we shall denote by the symbol $f : \mathcal{A}_\theta \to \mathbb{C}$ and which is given by

$$\fint \sum_{(m,r)\in\mathbb{Z}^2} a_{m,r}\, U_1^m U_2^r \quad := \quad a_{0,0}$$

$$(3.19) \qquad\qquad = \int_{\mathbb{T}^2} dx\, dy\, \Omega^{-1} \left(\sum_{(m,r)\in\mathbb{Z}^2} a_{m,r}\, U_1^m U_2^r \right)(x,y)\,.$$

Then, for any $a, b \in \mathcal{A}_\theta$, one readily checks the properties

$$(3.20) \qquad\qquad \fint a\,b = \fint b\,a\,, \quad \fint \mathbb{I} = 1\,, \quad \fint a^* a > 0\,, \quad a \neq 0\,,$$

with $\fint a^* a = 0$ if and only if $a = 0$ (i.e. the trace is faithful). This trace is invariant under the natural action of the commutative torus $\mathbb{T}^2$ on $\mathcal{A}_\theta$ whose infinitesimal form is by two commuting derivations ∂_1, ∂_2 acting by

$$(3.21) \qquad\qquad \partial_\mu(U_\nu) = 2\pi i\, \delta_\mu^\nu U_\nu\,, \quad \mu,\nu = 1,2\,.$$

Invariance is just the statement that $\fint \partial_\mu(a) = 0$, $\mu = 1,2$, for any $a \in \mathcal{A}_\theta$.

The cyclic 2-cocycle allowing the integration of two forms is simply given by

$$(3.22) \qquad\qquad \Psi(a_0, a_1, a_2) = \frac{i}{2\pi} \fint a_0\, (\partial_1 a_1 \partial_2 a_2 - \partial_2 a_1 \partial_1 a_2)\,.$$

Its normalization ensures that for any projection $p \in \mathcal{A}_\theta$ the quantity $\Psi(p, p, p)$ is an integer: it is indeed the index of a Fredholm operator.

In two dimensions, the conformal class of a general constant metric is parametrized by a complex number $\tau \in \mathbb{C}$, $\Im\tau > 0$. Up to a conformal factor, the metric is given by

$$(3.23) \qquad\qquad g = (g_{\mu\nu}) = \begin{pmatrix} 1 & \Re\tau \\ \Re\tau & |\tau|^2 \end{pmatrix}\,.$$

Clearly $\sqrt{\det g} = \Im\tau$, and the inverse metric is found to be

$$(3.24) \qquad\qquad g^{-1} = (g^{\mu\nu}) = \frac{1}{(\Im\tau)^2} \begin{pmatrix} |\tau|^2 & -\Re\tau \\ -\Re\tau & 1 \end{pmatrix}\,.$$

By using the two derivations ∂_1, ∂_2 defined in (3.21) we may think of 'the complex torus' $\mathbb{T}^2$ as acting on the noncommutative torus $\mathcal{A}_\theta$ and construct two associated derivations of $\mathcal{A}_\theta$ given by

$$(3.25) \qquad \partial_{(\tau)} = \frac{1}{(\tau - \bar\tau)}\, (-\bar\tau \partial_1 + \partial_2)\,, \quad \bar\partial_{(\tau)} = \frac{1}{(\tau - \bar\tau)}\, (\tau \partial_1 - \partial_2)\,.$$

In the terminology of [S-A, DS] we are endowing the noncommutative torus $\mathcal{A}_\theta$ with a complex structure. Then, with the two complex derivatives (3.25) one easily finds that

$$(3.26) \qquad\qquad \partial_{(\tau)}\bar\partial_{(\tau)} = \bar\partial_{(\tau)}\partial_{(\tau)} = \frac{1}{4} g^{\mu\nu} \partial_\mu \partial_\nu = \frac{1}{4}\Delta\,,$$

the operator $\Delta = g^{\mu\nu}\partial_\mu\partial_\nu$ being just the Laplacian of the metric (3.23).

By working with the metric (3.23) the positive Hochschild cocycle Φ associated with the cyclic one (3.22) will be given by

$$(3.27) \qquad \Phi(a_0, a_1, a_2) = \frac{2}{\pi} \oint \sqrt{\det g}\ a_0 \partial_{(\tau)} a_1 \overline{\partial}_{(\tau)} a_2\ .$$

A construction of the cocycle (3.27) as the conformal class of a general constant metric on the torus can be found in [**C2**] and [**C4**].

Before we procede, we briefly mention how to construct a spectral geometry in the sense of [**C2, C3**] for the noncommutative torus.

The algebra $\mathcal{A}_\theta$ can be represented faithfully as operators acting on a separable Hilbert space $\mathcal{H}_0$, the GNS representation space $\mathcal{H}_0 = L^2(\mathcal{A}_\theta, \oint)$, defined as the completion of $\mathcal{A}_\theta$ itself in the Hilbert norm

$$(3.28) \qquad \|a\|_{\mathrm{GNS}} := \left(\oint a^* a \right)^{1/2},$$

with $a \in \mathcal{A}_\theta$. Since the trace is faithful the map $\mathcal{A}_\theta \ni a \mapsto \widehat{a} \in \mathcal{H}_0$ is injective and the faithful GNS representation $\pi : \mathcal{A}_\theta \to \mathcal{B}(\mathcal{H})$ is simply given by

$$(3.29) \qquad \pi(a)\widehat{b} = \widehat{a\,b},$$

for any $a, b \in \mathcal{A}_\theta$. The vector $1 = \widehat{\mathbb{I}}$ of $\mathcal{H}_0$ is cyclic (i.e. $\pi(\mathcal{A}_\theta)1$ is dense in $\mathcal{H}_0$) and separating (i.e. $\pi(a)1 = 0$ implies $a = 0$) so that the Tomita involution is just

$$(3.30) \qquad J(\widehat{a}) = \widehat{a^*}, \qquad \forall \widehat{a} \in \mathcal{H}_0\ .$$

It is worth mentioning that the C^*-algebra norm on $\mathcal{A}_\theta$ given in (3.3) with $k = 0$ coincides with the operator norm in (3.28) when $\mathcal{A}_\theta$ is represented on the Hilbert space $\mathcal{H}_0$, and also with the L^∞-norm in the Wigner representation. For ease of notation, in what follows we will not distinguish between elements of the algebra $\mathcal{A}_\theta$ and their corresponding operators in the GNS representation.

A two dimensional noncommutative geometry $(\mathcal{A}_\theta, \mathcal{H}, D, \gamma, J)$ for the torus $\mathcal{A}_\theta$ is constructed as follows. The Hilbert space $\mathcal{H}$ is just $\mathcal{H} = \mathcal{H}_0 \oplus \mathcal{H}_0$ on which $\mathcal{A}_\theta$ acts diagonally with two copies of the representation π in (3.29). Moreover,

$$(3.31) \qquad \gamma = \begin{pmatrix} 1 & 0 \\ 0 & -1 \end{pmatrix}, \qquad J = \begin{pmatrix} 0 & -J_0 \\ J_0 & 0 \end{pmatrix}.$$

The Dirac operator D is

$$(3.32) \qquad D = \begin{pmatrix} 0 & \overline{\partial}_{(\tau)} \\ \partial_{(\tau)} & 0 \end{pmatrix}.$$

For the particular choice $\tau = i$ this reduces to $D = \partial_1 \sigma_1 + \partial_2 \sigma_2$ with σ_1, σ_2, two of the Pauli matrices.

The Hochschild 2-cycle c giving the orientation and volume form is

$$(3.33) \qquad c = \frac{1}{(2i\pi)^2} \frac{1}{(\tau - \overline{\tau})} (U_2^{-1} U_1^{-1} \otimes U_1 \otimes U_2 - U_1^{-1} U_2^{-1} \otimes U_2 \otimes U_1)\ .$$

Detail on this geometry can be found in [**C3**] (see also [**V**]) where one shows that all properties of a real spectral geometry are satisfied. Here we only mention that the 'area' $\oint D^{-2}$ of the torus depends on the complex parameter τ, as it should be.

Indeed, with Φ the cyclic 2-cocycle in (3.22), after some computations one finds that

$$(3.34) \qquad \int D^{-2} = \langle \Phi, c \rangle = \frac{1}{2\pi \, \Im \tau} .$$

Here $\langle \cdot, \cdot \rangle$ indicates the pairing between cocycles and cycles.

3.3. The action and the field equations. With $\mathcal{P}_\theta = Proj(\mathcal{A}_\theta)$ denoting the collection of all projections in the algebra $\mathcal{A}_\theta$ we construct an action functional $S : \mathcal{P}_\theta \to \mathbb{R}^+$ by

$$(3.35) \qquad S_{(\tau)}(p) = \Phi(1, p, p) = \frac{2}{\pi} \int \sqrt{detg} \; \partial_{(\tau)} p \bar{\partial}_{(\tau)} p .$$

By using (3.25) this action functional can also be written as

$$(3.36) \qquad S_{(\tau)}(p) = \frac{1}{2\pi} \int \sqrt{detg} \; g^{\mu\nu} \partial_\mu p \partial_\nu p = \frac{1}{\pi} \int \sqrt{detg} \; g^{\mu\nu} p \partial_\mu p \partial_\nu p .$$

Here the two derivations ∂_μ are the ones defined in (3.21) while the metric g is the one in (3.23)-(3.24) which carries also the dependence on the complex parameter τ. The equality follows from the constraint $p^2 = p$ and the use of Leibniz rule. That the value of the action is a positive real number follows from the properties of the trace f.

Then, we shall look for critical points of the action functional (3.36) in a given connected component of the space $\mathcal{P}_\theta$. It is well known [**R1, PV**] that these are parametrized by two integers $r, q \in \mathbb{Z}$ such that $r + \theta q \geq 0$. When θ is irrational the corresponding projections are of trace $r + \theta q$ and of K_0-class equal to (r, q).

In order to derive field equations from the action functional (3.36), we need to have a look at the tangent space to $\mathcal{P}_\theta$ at any of its 'point' p. An element $\delta p \in T_p(\mathcal{P}_\theta)$ is not arbitrary but needs to fulfill two requirements. First of all it must be hermitian, $(\delta p)^* = \delta p$, and this implies that it is written as

$$(3.37) \qquad \delta p = p x p + (1 - p) y (1 - p) + (1 - p) z p + p z^* (1 - p) ,$$

with $x = x^*, y = y^*, z$ three arbitrary elements in $\mathcal{A}_\theta$. Furthermore, it must be such that $(p + \delta p)^2 = p + \delta p + O(\delta p)$ which implies that $(1 - p)\delta p = \delta pp$. When using (3.37) we get the most general tangent vector as

$$(3.38) \qquad \delta p = (1 - p) z p + p z^* (1 - p) ,$$

with z an arbitrary elements in $\mathcal{A}_\theta$.

As usual, the equation of motion are obtained by equating to zero the first variation of the action functional (3.36),

$$
\begin{aligned}
0 &= \delta S_{(\tau)}(p) \\
&= -\frac{1}{2\pi} \int \sqrt{detg} \; g^{\mu\nu} (\partial_\mu \partial_\nu p) \, \delta p \\
&= -\frac{1}{2\pi} \int \sqrt{detg} \; g^{\mu\nu} (\partial_\mu \partial_\nu p) \big[(1 - p) z p + p z^* (1 - p) \big] \\
(3.39) \quad &= -\frac{1}{2\pi} \int \Big[\sqrt{detg} \; p \, \Delta(p) \, (1 - p) \Big] z + \Big[\sqrt{detg} \; (1 - p) \, \Delta(p) \, p \Big] z^* .
\end{aligned}
$$

With $\Delta = g^{\mu\nu} \partial_\mu \partial_\nu$ the laplacian of the metric g. We have 'integrated by part' and used the invariance of the integral to get rid of the 'boundary terms', and we have used ciclicity of the integral. Since z is arbitrary we get the field equations

$$(3.40) \qquad p\,\Delta(p)\,(1-p) = 0 \ , \text{ and } (1-p)\,\Delta(p)\,p = 0 \ ,$$

or, equivalently

$$(3.41) \qquad p\,\Delta(p) \ - \ \Delta(p)\,p = 0 \ .$$

The previous equations are non-linear equations of the second order and it is rather difficult to explicit their solutions in closed form. Presently, we shall show that the absolute minima of (3.36) in a given connected component of $\mathcal{P}_\theta$ actually fulfill first order equations which are easier to solve.

3.4. Topological Charges and Self-Duality Equations. Given a projection $p \in \mathcal{A}_\theta$, its 'topological charge' (the first Chern number) is computed by using the cyclic 2-cocycle in (3.22),

$$(3.42) \qquad \Psi(p) := \frac{i}{2\pi} \int p\Big[\partial_1(p)\partial_2(p) - \partial_2(p)\partial_1(p)\Big] \ \in \ \mathbb{Z}$$

(we refer to [**C1**] for a detailed discussion). Then we have the following

PROPOSITION 3.1. *For any $p \in \mathcal{P}_\theta$ there is the inequality*

$$(3.43) \qquad S_{(\tau)}(p) \geq 2\|\Psi(p)\| \ .$$

PROOF. Due to positivity of the trace $\int$ and its cyclic properties, we have that

$$0 \ \leq \ \int \sqrt{detg}\ g^{\mu\nu}\Big[\partial_\mu(p)\,p \pm i\epsilon_\mu{}^\alpha\partial_\alpha(p)\,p\Big]^*\Big[\partial_\nu(p)\,p \pm i\epsilon_\nu{}^\theta\partial_\theta(p)\,p\Big]$$

$$\leq \ \int \sqrt{detg}\ \Big[g^{\mu\nu}p\partial_\mu(p)\partial_\nu(p) + g^{\mu\nu}\epsilon_\mu{}^\alpha\epsilon_\nu{}^\theta p\partial_\alpha(p)\partial_\theta(p)\Big]$$

$$\pm i\int \sqrt{detg}\ \Big[\epsilon^{\mu\theta}p\partial_\mu(p)\partial_\theta(p) - \epsilon^{\nu\alpha}p\partial_\alpha(p)\partial_\nu(p)\Big]$$

$$(3.44) \qquad \leq \ 2\int \sqrt{detg}\ g^{\mu\nu}p\partial_\mu(p)\partial_\nu(p) \pm 2i\int \epsilon^{\mu\nu}_{flat}\ p\partial_\mu(p)\partial_\nu(p) \ .$$

By comparing (3.44) with the definitions (3.36) and (3.42) we get the inequality (3.43). □

In the derivation of (3.44) we have used the following equalities which are valid for any metric g,

$$(3.45) \qquad g^{\mu\nu}\epsilon_\mu{}^\alpha\epsilon_\nu{}^\theta = g^{\alpha\theta} \ , \quad \epsilon^{\mu\nu} = \frac{1}{\sqrt{detg}}\ \epsilon^{\mu\nu}_{flat} \ ,$$

with $\epsilon^{\mu\nu}_{flat}$ the antisymmetric two-tensor determined by $\epsilon^{1,1}_{flat} = 1$.

From (3.44) it is clear that the equality in (3.43) occurs when the projection p satisfies the following *self-duality* or *anti-self duality* equations

$$(3.46) \qquad \Big[\partial_\mu p \pm i\epsilon_\mu{}^\alpha\partial_\alpha p\Big]\,p = 0 \text{ and/or } p\Big[\partial_\mu p \mp i\epsilon_\mu{}^\alpha\partial_\alpha p\Big] = 0.$$

By using the equality

$$(3.47) \qquad \epsilon_\mu{}^\alpha = \sqrt{detg}\ g^{\alpha\nu}\epsilon^{flat}_{\mu\nu}$$

and the derivations $\partial_{(\tau)}$ and $\overline{\partial}_{(\tau)}$ in (3.25), the self-duality equations (3.46) reduce to

$$(3.48) \qquad \overline{\partial}_{(\tau)}(p) \, p = 0 \text{ and/or } p \, \partial_{(\tau)}(p) = 0 \; ,$$

while the anti-self-duality one is

$$(3.49) \qquad \partial_{(\tau)}(p) \, p = 0 \text{ and/or } p \, \overline{\partial}_{(\tau)}(p) = 0 \; .$$

Simple manipulations show directly that each of the equations (3.48) and (3.49) implies the field equations (3.41) as it should be.

4. Noncommutative σ-model instantons

From now on we shall concentrate on the self-dual equations (3.48), a solution of which we shall call σ-model instanton. A similar analysis would be possible for the anti self-dual equations (3.49), whose solutions would be σ-model anti-instantons .

As we have mentioned already, the connected components of $\mathcal{P}_\theta$ are parametrized by two integers r and q and when θ is irrational, the corresponding projections are exactly the projections of trace $r+q\theta$ and the topological charge $\Psi(p)$ appearing in (3.42) is just given by q. Thus we have to find projections that belongs to the previous homotopy classes and satisfy the self-duality equation $\overline{\partial}_{(\tau)}(p) \, p = 0$ or, equivalently, $p\partial_{(\tau)}p = 0$. Although these equations look very simple, they are far from being easy to solve because of their non-linear nature; the next step will be to reduce them to a linear problem. The key point is to identify the algebra $\mathcal{A}_\theta$ as the endomorphism algebra of a suitable bundle and to think of any projection in it as an operator on such a bundle. The bundle in question will be a finite projective module on a different copy $\mathcal{A}_\alpha$ of the noncommutative torus, the two algebras $\mathcal{A}_\theta$ and $\mathcal{A}_\alpha$ being related by the fact that they are Morita equivalent to each other.

4.1. The modules. Let us then consider another copy $\mathcal{A}_\alpha$ of the noncommutative torus with generators Z_1, Z_2 obeying the relation

$$(4.1) \qquad Z_2 \, Z_1 = e^{2\pi i \alpha} Z_1 \, Z_2 \; .$$

When α is not rational, every finitely generated projective module over the algebra $\mathcal{A}_\alpha$ which is not free is isomorphic to a Heisenberg module [**C1, R2**]. Any such a module $\mathcal{E}_{r,q}$ is characterized by two integers r, q. If $q = 0$, $\mathcal{E}_{r,q} = \mathcal{A}_\alpha^{|r|}$. Otherwise, they can be taken to be relatively coprime with $q > 0$ (a similar construction being possible for $q < 0$), or $r = 0$ and $q = 1$. We shall briefly describe them. As a vector space

$$(4.2) \qquad \mathcal{E}_{r,q} = \mathcal{S}(\mathbb{R} \times \mathbb{Z}_q) \simeq \mathcal{S}(\mathbb{R}) \otimes \mathbb{C}^q \; ,$$

the space of Schwartz functions of one continuous variable $s \in \mathbb{R}$ and a discrete one $k \in \mathbb{Z}_q$ (we shall implicitly understand that such a variable is defined modulo q). By introducing the shorthand

$$(4.3) \qquad \varepsilon = r/q - \alpha \; ,$$

the space $\mathcal{E}_{r,q}$ is made a *right* module over $\mathcal{A}_\alpha$ by defining

$$(4.4) \qquad \begin{aligned} (\xi \, Z_1)(s, k) &:= \xi(s - \varepsilon, k - r) \; , \\ (\xi \, Z_2)(s, k) &:= e^{2\pi i (s - k/q)} \xi(s, k) \; , \end{aligned}$$

for any $\xi \in \mathcal{E}_{r,q}$, the relations (4.1) being easily verified.

On the module $\mathcal{E}_{r,q}$ one defines an $\mathcal{A}_\alpha$-valued hermitian structure, namely a sesquilinear form $\langle \cdot, \cdot \rangle_\alpha : \mathcal{E}_{r,q} \times \mathcal{E}_{r,q} \to \mathcal{A}_\alpha$, which is antilinear in the first variable and such that

$$\langle \xi, \eta a \rangle_\alpha = \langle \xi, \eta \rangle_\alpha a \ ,$$
$$(\langle \xi, \eta \rangle_\alpha)^* = \langle \eta, \xi \rangle_\alpha \ ,$$
(4.5) $$\langle \xi, \xi \rangle_\alpha \geq 0 \ , \ \langle \xi, \xi \rangle_\alpha = 0 \iff \xi = 0 \ ,$$

for all $\xi, \eta \in \mathcal{E}_{r,q} \ , a \in \mathcal{A}_\alpha$.
This hermitian structure is explicitly given by

(4.6) $$\langle \xi, \eta \rangle_\alpha := \sum_{m,n} \sum_{k=0}^{q-1} \int_{-\infty}^{+\infty} ds \, \overline{\xi(s - m\varepsilon, k - mp)} \, \eta(s,k) e^{-2\pi i n(s-k/q)} \, Z_1^m \, Z_2^n \ .$$

It is proven in [**CR**] that the endomorphism algebra $End_{\mathcal{A}_\alpha}(\mathcal{E}_{r,q})$, which acts on the left on $\mathcal{E}_{r,q}$, can be identified with another copy of the noncommutative torus $\mathcal{A}_\theta$ where the parameter θ is 'uniquely' determined by α in the following way. Since r and q are coprime, there exist integer numbers $a, b \in \mathbb{Z}$ such that $ar + bq = 1$. Then, the transformed parameter is given by

(4.7) $$\theta = \frac{a\alpha + b}{-q\alpha + r} \ .$$

Given any two other integers $a', b' \in \mathbb{Z}$ such that $a'r + b'q = 1$, one would find that $\theta' - \theta \in \mathbb{Z}$ so that $\mathcal{A}_{\theta'} \simeq \mathcal{A}_\theta$. Thus we are saying that the algebra $End_{\mathcal{A}_\alpha}(\mathcal{E}_{r,q})$ is generated by two operators U_1, U_2 acting on the *left* on $\mathcal{E}_{r,q}$ by

$$(U_1 \xi)(s, k) := \xi(s - 1/q, k - 1) \ ,$$
(4.8) $$(U_2 \xi)(s, k) := e^{2\pi i \, (s/\varepsilon - ak)/q} \, \xi(s, k) \ ,$$

and one easily verify that

(4.9) $$U_2 \, U_1 = e^{2\pi i\theta} U_1 \, U_2 \ ,$$

namely the defining relations of the algebra $\mathcal{A}_\theta$.

Furthermore, the $\mathcal{A}_\theta$-$\mathcal{A}_\alpha$-bimodule $\mathcal{E}_{r,q}$ is a *Morita equivalence* between the two algebras $\mathcal{A}_\theta$ and $\mathcal{A}_\alpha$. That is, there exists also a $\mathcal{A}_\theta$-valued hermitian structure on $\mathcal{E}_{r,q}$,

(4.10) $$\langle \cdot, \cdot \rangle_\theta : \mathcal{E}_{r,q} \times \mathcal{E}_{r,q} \to \mathcal{A}_\theta \ ,$$

which is compatible with the $\mathcal{A}_\alpha$-valued one $\langle \cdot, \cdot \rangle_\alpha$,

(4.11) $$\langle \xi, \eta \rangle_\theta \, \zeta = \xi \, \langle \eta, \zeta \rangle_\alpha \ ,$$

for all $\xi, \eta, \zeta \in \mathcal{E}_{r,q}$.
The second hermitian structure is explicitly given by

$$\langle \xi, \eta \rangle_\theta \quad := \quad \frac{1}{|q\varepsilon|} \sum_{m,n} \sum_{k=0}^{q-1} \int_{-\infty}^{+\infty} ds \, \xi(s, k) \, \overline{\eta(s - m/q, k - m)} \, \times$$
(4.12) $$\times \, e^{-2\pi i n/q[(s-m/q)/\varepsilon - ak]} \, Z_1^m \, Z_2^n \ .$$

Notice that now the antilinearity is in the second variable. The hermitian structure $\langle \cdot, \cdot \rangle_\theta$ will obey properties analogous to the ones in (4.5) but now there is a left linearity, $\langle a\xi, \eta \rangle_\theta = a \langle \xi, \eta \rangle_\theta$, with $a \in \mathcal{A}_\theta$.

The proof of the compatibility condition (4.11) goes as follows. By using the explicit formula (4.6) for the right hermitian structure, the right hand side of (4.11) is

$$(\xi \langle \eta, \zeta \rangle_\alpha)(s, k) = \sum_{m,n} \sum_{l=0}^{q-1} \int_{-\infty}^{+\infty} dt \, \xi(s - m\varepsilon, k - mp) \, \overline{\eta}(t - m\varepsilon, l - mp) \zeta(t, l) \times$$

$$\times \, e^{2\pi i n[s - t - 1/q(k-l)]}$$

$$= \sum_{m,n} \sum_{l=0}^{q-1} \xi(s - m\varepsilon, k - mp) \, \overline{\eta}(s + \frac{l-k}{q} + n - m\varepsilon, l - mp) \zeta(s + \frac{l-k}{q} + n, l) \, ,$$

where we have used a Poisson resummation formula for the periodic delta function, $\sum_{n \in \mathbb{Z}} e^{2\pi i n x} = \sum_{n \in \mathbb{Z}} \delta(x + n)$, and we have integrated over the variable t. On the other hand, by using similar techniques, we find that the left hand side of (4.11) is

$$((\langle \xi, \eta \rangle_\theta \, \zeta)(s, k) = \sum_{x,y} \sum_{d=0}^{q-1} \xi(s + \varepsilon[a(d-k) + qy], d) \times$$

$$\times \, \overline{\eta}(s + \varepsilon[a(d-k) + qy] - \frac{x}{q}, d - x) \zeta(s - \frac{x}{q}, k - x) \, ,$$

By comparing these two expressions we see that they coincide provide the indices on which we sum are related by $m = qy - a(d-k)$, $n = (k - l - x)/q$, $l = k - x \bmod q$; notice that the last relation assures that n is an integer.

By using the previous construction one can construct projections in the algebra $\mathcal{A}_\theta$ by picking suitable vectors $\xi \in \mathcal{E}_{p,q}$ with $\langle \xi, \xi \rangle_\alpha = \mathbb{I}$. Then, the bimodule property (4.11) implies that $p = \langle \xi, \xi \rangle_\theta$ is a non-trivial projection in $\mathcal{A}_\theta$,

$$(4.13) \qquad p^2 = \langle \xi, \xi \rangle_\theta \langle \xi, \xi \rangle_\theta = \langle \langle \xi, \xi \rangle_\theta \xi, \xi \rangle_\theta = \langle \xi \langle \xi, \xi \rangle_\alpha, \xi \rangle_\theta = \langle \xi, \xi \rangle_\theta = p \, .$$

By using the identification $\mathcal{A}_\theta \simeq \mathrm{End}_{\mathcal{A}_\alpha}(\mathcal{E}_{p,q})$, any such a projection may be equivalently written in the more suggestive form

$$(4.14) \qquad p_\psi = \left\langle \psi \left(\langle \psi, \psi \rangle_\alpha\right)^{-1/2}, \, \psi \left(\langle \psi, \psi \rangle_\alpha\right)^{-1/2} \right\rangle_\theta = |\psi\rangle \frac{1}{\langle \psi, \psi \rangle_\alpha} \langle \psi| \, ,$$

where for each vector $|\psi\rangle \in \mathcal{E}_{p,q}$ the corresponding dual vector $\langle \psi| \in (\mathcal{E}_{p,q})^*$ is defined by means of the $\mathcal{A}_\alpha$-valued hermitian structure as $\langle \psi| (\eta) := \langle \psi, \eta \rangle_\alpha \in \mathcal{A}_\alpha$. We are only assuming that $\langle \psi, \psi \rangle_\alpha$ is an invertible element of the algebra $\mathcal{A}_\alpha$.

In order to translate the self-duality equations (3.48) for p_ψ to equations for ψ, we need to introduce a gauge connection on the right $\mathcal{A}_\alpha$-module $\mathcal{E}_{p,q}$.

4.2. The constant curvature connections. Again the theory of gauge connection on noncommutative tori is worked out in [**CR**]. A gauge connection on the right $\mathcal{A}_\alpha$-module $\mathcal{E}_{r,q}$ is given by two covariant derivatives

$$(4.15) \qquad \nabla_\mu : \mathcal{E}_{r,q} \to \mathcal{E}_{r,q} \, , \quad \mu = 1, 2 \, ,$$

which satisfy a right Leibniz rule

$$(4.16) \qquad \nabla_\mu(\xi \, a) = (\nabla_\mu \xi) a + \xi(\partial_\mu a) \, , \quad \mu = 1, 2 \, ,$$

for any $\xi \in \mathcal{E}_{r,q}$ and any $a \in \mathcal{A}_\alpha$. One also requires compatibility with the $\mathcal{A}_\alpha$-valued hermitian structure,

$$(4.17) \qquad \partial_\mu(\langle \xi, \eta \rangle_\alpha) = \langle \nabla_\mu \xi, \eta \rangle_\alpha + \langle \xi, \nabla_\mu \eta \rangle_\alpha \, , \quad \mu = 1, 2 \, .$$

It is not difficult to see that any other compatible connection $\widetilde{\nabla}_\mu$ must be of the form $\widetilde{\nabla}_\mu = \nabla_\nu + \eta_\mu$ with the η_μ's (skew-adjoint) elements in $End_{\mathcal{A}_\alpha}(\mathcal{E}_{r,q})$.

Given any compatible connection, the operators ∇_μ can be used to define derivations on the endomorphism algebra $End_{\mathcal{A}_\alpha}(\mathcal{E}_{r,q})$ by

$$(4.18) \qquad \hat{\delta}_\mu(T) := \nabla_\mu \circ T - T \circ \nabla_\mu , \ \mu = 1, 2,$$

for any $T \in End_{\mathcal{A}_\alpha}(\mathcal{E}_{r,q})$. By using the compatibility (4.17) and the right Leibniz rule (4.16), one easily prove that the connection ∇_μ is compatible for the derivations $\hat{\delta}_\mu$ and the hermitian structure $\langle \cdot, \cdot \rangle_\theta$, that is,

$$(4.19) \qquad \hat{\delta}_\mu(\langle \xi, \eta \rangle_\theta) = \langle \nabla_\mu \xi, \eta \rangle_\theta + \langle \xi, \nabla_\mu \eta \rangle_\theta , \ \mu = 1, 2 .$$

Moreover, there is also a left Leibniz rule,

$$(4.20) \qquad \nabla_\mu(T\xi) = T(\nabla_\mu \xi) + (\hat{\delta}_\mu T)\xi , \ \mu = 1, 2 ,$$

for any $\xi \in \mathcal{E}_{r,q}$ and any $T \in End_{\mathcal{A}_\alpha}(\mathcal{E}_{r,q})$.

A particular connection on the right $\mathcal{A}_\alpha$-module $\mathcal{E}_{r,q}$ is given by the operators

$$(4.21) \qquad (\nabla_1 \xi)(s, k) := \frac{2\pi i}{\varepsilon} s \, \xi(s, k) , \ (\nabla_2 \xi)(s, k) := \frac{d\xi}{ds}(s, k) .$$

Notice that the discrete index k is not touched. The previous connection is of constant curvature,

$$(4.22) \qquad F_{1,2} := [\nabla_1, \nabla_2] - \nabla_{[\partial_1, \partial_2]} = -\frac{2\pi i}{\varepsilon} \mathbb{I}_{\mathcal{E}_{r,q}} ,$$

where $\mathbb{I}_{\mathcal{E}_{r,q}}$ is the identity operator on $\mathcal{E}_{r,q}$.

Finally, by remembering that $End_{\mathcal{A}_\alpha}(\mathcal{E}_{r,q}) \simeq \mathcal{A}_\theta$, one also proves that the derivations $\hat{\delta}_\mu$ on $End_{\mathcal{A}_\alpha}(\mathcal{E}_{r,q})$ determined by the constant curvature connection (4.21) are proportional to the generators of the infinitesimal action of the commutative torus $\mathbb{T}^2$ on $\mathcal{A}_\theta$, that is,

$$(4.23) \qquad \hat{\delta}_\mu(U_\nu) = \frac{2\pi i}{q\varepsilon} \delta_\mu^\nu U_\nu , \ \mu, \nu = 1, 2 .$$

That the induced derivations are the canonical derivations on $\mathcal{A}_\theta$ is crucial for our analysis in the following (see the proof of Prop. 4.1). Any other connection with this same property will have the form $\widetilde{\nabla}_\mu = \nabla_\nu + \lambda_\mu$ where λ_1 and λ_2 are (pure-imaginary) constants. Then, the corresponding curvature $\widetilde{F}_{1,2}$ will be constant and in fact will coincide with the constant curvature $F_{1,2}$ in (4.22). It is known [**CR**] that constant curvature connections are exactly the connections that minimize a suitable Yang-Mills functional and that all this minimizing connections must have the same curvature. Moreover, the moduli spaces of these connections is just (in the simplest possible case) an ordinary two-torus $\mathbb{T}^2$.

In the following we shall also need the covariant derivatives $\nabla_{(\tau)}$ and $\overline{\nabla}_{(\tau)}$ which lift the complex derivations $\partial_{(\tau)}$ and $\overline{\partial}_{(\tau)}$ in (3.25). By using (4.21) they are given by

$$(4.24) \qquad \nabla_{(\tau)} = \frac{1}{(\tau - \overline{\tau})} (-\overline{\tau} \nabla_1 + \nabla_2) , \ \overline{\nabla}_{(\tau)} = \frac{1}{(\tau - \overline{\tau})} (\tau \nabla_1 - \nabla_2)$$

These connections will endow the module $\mathcal{E}_{r,q}$ with a complex (or better holomorphic) structure (see also [DS, S-A, PS, S-M]).

4.3. The Instantons. We are now ready to look for solutions of the self-dual equations (3.48) of the form

$$(4.25) \qquad p_\psi := |\psi\rangle \, (\langle\psi,\psi\rangle_\alpha)^{-1} \, \langle\psi| \, ,$$

with the vector $|\psi\rangle \in \mathcal{E}_{p,q}$ assumed to be such that $\langle\psi,\psi\rangle_\alpha$ is invertible in $\mathcal{A}_\alpha$.

Then, one can prove the following

PROPOSITION 4.1.
Let the vector $|\psi\rangle \in \mathcal{E}_{r,q}$ be such that $\langle\psi,\psi\rangle_\alpha \in \mathcal{A}_\theta$ is invertible. And consider the projection $p_\psi := |\psi\rangle \, (\langle\psi,\psi\rangle_\alpha)^{-1} \, \langle\psi|$.
Then,
1.) If there exists an element $\lambda \in \mathcal{A}_\alpha$ such that $\psi = |\psi\rangle$ is a solution of the equations

$$(4.26) \qquad \overline{\nabla}\psi - \psi\lambda = 0 \, ,$$

where $\overline{\nabla}$ is the anti-holomorphic connection (4.24), then the projection p_ψ is a solution of the self-duality equations,

$$(4.27) \qquad \overline{\partial}_{(\tau)}(p_\psi) \, p_\psi = 0 \, .$$

2.) Conversely, if p_ψ is a solution of the equations (4.27), then $\psi = |\psi\rangle$ obeys the equations (4.26) with

$$(4.28) \qquad \lambda = (\langle\psi,\psi\rangle_\alpha)^{-1} \, \langle\psi,\overline{\nabla}\psi\rangle_\alpha \, .$$

PROOF. A straightforward computation using the properties of the connection, notably the fact that the induced derivations on the endomorphism algebra are the canonical derivations on $\mathcal{A}_\theta$, and left and right Leibniz rules and metric compatibilities. □

For a generic element $\lambda \in \mathcal{A}_\alpha$, equations (4.26) still are horribly complicated. A very important class of examples is obtained by considering constant parameters $\lambda \in \mathbb{C}$.

PROPOSITION 4.2. *Let $\lambda \in \mathbb{C}$. Then the Gaussian*

$$(4.29) \qquad \psi_\lambda(s,k) = A_k \, e^{i\tau\pi s^2/\varepsilon + \lambda(\overline{\tau}-\tau)s} \quad , \quad k = 1,\ldots,q \, ,$$

is a solution of the equation (4.26) such that the element $\langle\psi_\lambda,\psi_\lambda\rangle_\alpha \in \mathcal{A}_\alpha$ is invertible. The vector $A = (A_1,\ldots,A_q) \in \mathbb{C}^q$ can be taken to lie in the complex projective space $\mathbb{CP}^{q-1}$ by removing an inessential normalization.

PROOF. Recall first that we are taking $\Im\tau > 0$, hence the generalized gaussian $\psi_\lambda(s,k)$ in (4.29) is an element of $\mathcal{E}_{r,q}$. That it is also a solution of the equations (4.26) (with the constant curvature connection (4.21)) is a straightforward computation. As for invertibility of the corresponding element $\langle\psi_\lambda,\psi_\lambda\rangle_\alpha \in \mathcal{A}_\alpha$ this is more difficult. For the lowest values of the parameters, i.e. $q = 1$, $r = 0$, (and with $\tau = i$, $\lambda = 0$), the invertibility of

$$(4.30) \qquad \langle\psi_{\lambda=0},\psi_{\lambda=0}\rangle_{-1/\theta} = \frac{1}{\sqrt{2\theta}} \sum_{(m,n)\in\mathbb{Z}^2} e^{i\pi mn/\theta - \pi(m^2+n^2)/2\theta} \, Z_1^m \, Z_2^n$$

(now $\alpha = -1/\theta$) was proved in [**B**] for a restricted range of the deformation parameter θ. The invertibility was extended in [**W**] for all values $0 < \theta < 1$. As for the general case, a careful extension of the techniques of [**B**] and [**W**] provides invertibility. $\qquad\square$

Furthermore, a lengthy computation gives that the projection $p_\lambda \in \mathcal{A}_\theta$ corresponding to the Gaussian (4.29) has dimension (= rank) $r + q\theta$: $f(p_\lambda) = r + q\theta$, and topological charge q: $\Psi(p_\lambda) = q$.

Before we pass to the general case of nonconstant 'parameter' λ in the equations (4.26) we need to introduce gauge transformations.

4.4. Gauge Transformations. Having lifted our duality equations to equations on the bundle $\mathcal{E}_{r,q}$ has introduced what we could call 'gauge degrees of freedom'. Then, there are analogous of (complex) gauge transformations. These are given by invertible elements in $\mathcal{A}_\alpha$ acting on the right on $\mathcal{E}_{r,q}$,

$$(4.31) \qquad \mathcal{E}_{r,q} \ni |\psi\rangle \rightarrow |\psi^g\rangle = |\psi\rangle\, g \in \mathcal{E}_{r,q}\ ,\ \forall\, g \in \mathrm{GL}(\mathcal{A}_\alpha)\ .$$

Notice that we do not require g to be unitary. Then, it is straightforward to prove that projections of the form (4.25) are invariant under gauge transformations. Indeed,

$$
\begin{aligned}
p_\psi \rightarrow p_\psi^g &= |\psi^g\rangle\, (\langle\psi^g,\psi^g\rangle_\alpha)^{-1}\, \langle\psi^g| = |\psi\rangle\, g(g^*\,\langle\psi,\psi\rangle_\alpha\, g)^{-1} g^*\,\langle\psi| \\
&= |\psi\rangle\, gg^{-1}(\langle\psi,\psi\rangle_\alpha)^{-1}(g^*)^{-1} g^*\,\langle\psi| = |\psi\rangle\,(\langle\psi,\psi\rangle_\alpha)^{-1}\,\langle\psi| \\
&= p_\psi.
\end{aligned}
$$

Equally straightforward is to find the gauge transformed of the self duality equations (4.26). Indeed, let $|\psi\rangle \in \mathcal{E}_{r,q}$ be a solution of the equations (4.26): $\overline{\nabla}\psi - \psi\lambda = 0$; and let $g \in \mathrm{GL}(\mathcal{A}_\alpha)$. Then by using the Leibniz rule for the connection, one finds that the gauge transformed vector $|\psi^g\rangle$ will be a solution of an equation of the same form: $\overline{\nabla}\psi^g - \psi^g\lambda_g = 0$ with λ_g given by

$$(4.32) \qquad \lambda_g = g^{-1}\lambda g + g^{-1}\overline{\partial}_{(\tau)}g\ .$$

Let us then analyze how the gaussian projections transform under the action of the gauge transformations (4.31). We have the following

PROPOSITION 4.3. *Let $|\psi\rangle$ be a solution of equation (4.26), $\lambda \in \mathbb{C}$; and let $g \in \mathrm{GL}(\mathcal{A}_\alpha)$. Then the transformed λ_g will again be constant if an only if there exists a couple of integer $(m,n) \in \mathbb{Z}^2$ such that*

$$(4.33) \qquad g = g_{mn}U^m V^n\ \text{no sum.}$$

Furthermore,

$$(4.34) \qquad \lambda_g - \lambda = \frac{2\pi i\tau}{\tau - \overline{\tau}}\,(m - \frac{1}{\tau}n).$$

PROOF. The first statement in the direction right to left of the statement is obvious. On the other hand, assume that $\lambda_g \in \mathbb{C}$. Any $g \in \mathrm{GL}(\mathcal{A}_\alpha)$ is written as $g = \sum_{(m,n)\in\mathbb{Z}^2} g_{mn}U^m V^n$. When substituting in (4.32) one gets an equation which requires that only one term in the sum does not vanish. The relation (4.34) is straightforwardly worked out. $\qquad\square$

One could also act directly on gaussian functions, which are elements of the module $\mathcal{E}_{r,q}$ on which $\mathrm{GL}(\mathcal{A}_\alpha)$ acts on the left. Indeed, let the solution corresponding to the parameter $\lambda \in \mathbb{C}$ be the Gaussian ψ_λ as in (4.29). When acting on ψ_λ with the invertible element $g = g_{mn}U^m V^n \in \mathrm{GL}(\mathcal{A}_\alpha)$ one produces a new Gaussian ψ_{λ_g} with the constant parameter λ_g given exactly as in (4.34). By putting together these results with Proposition 4.2, we have the following result which parallel an analogous one for connections minimizing a Yang-Mill functional (for which the connections are forced to be of constant curvature) [**CR**].

COROLLARY 4.4. *The moduli space of 'gaussian projection' is given by an ordinary complex torus times an ordinary projective space*

$$(4.35) \qquad\qquad \mathbb{C}/(\tau\mathbb{Z} + \mathbb{Z}) \times \mathbb{CP}^{q-1} .$$

As for the generic case of the equations (4.26) for a nonconstant element $\lambda \in \mathcal{A}_\alpha$, at the moment we are unable to state a general result. In [**DKL1, DKL2**] it was suggested that any solution of the these self-duality equations could be gauged away to a gaussian solution. This would be equivalent to the statement that given any element $\rho \in \mathcal{A}_\alpha$, there exists an element $g \in \mathrm{GL}(\mathcal{A}_\alpha)$ such that $\rho = \lambda + g^{-1}\bar{\partial}_{(\tau)}g$, with $\lambda \in \mathbb{C}$. That this fact is not true follows from recent work in [**P-A**].

Acknowledgments.

This is a review of work done with Thomas Krajewski and Ludwik Dąbrowski most of which published in [**DKL1, DKL2**]; I am most grateful to them for the collaboration. It is based on a seminar delivered during the Workshop on 'Noncommutative Geometry and Number Theory' held at the MPIM, Bonn, Germany, August 18-22, 2003; I thank the organizers for the kind invitation and all the participants for the nice time in Bonn. Finally, I thank Florin Boca for drawing my attention to reference [**W**] and Hanfeng Li for reference [**P-A**].

References

[BP] A.A. Belavin, A.M. Polyakov, *Metastable states of two-dimensional isotropic ferromagnets*, JETP Lett. **22** (1975) 245-247.

[B] F.-P. Boca, *Projections in Rotation Algebras and Theta Functions*, Commun. Math. Phys. **202** (1999) 325-357.

[C1] A. Connes, *C*-algèbres et géométrie différentielle*, C.R. Acad. Sci. Paris Sér. A **290** (1980) 599-604.

[C2] A. Connes, *Noncommutative Geometry*, Academic Press, 1994.

[C3] A. Connes, *Gravity coupled with matter and the foundation of noncommutative geometry*, Commun. Math. Phys. **182** (1996) 155-176.

[C4] A. Connes, *A short survey of noncommutative geometry*, J. Math. Phys. **41** (2000) 3832-3866.

[CDS] A. Connes, M. Douglas, A. Schwarz, *Matrix theory compactification on tori*, J. High Energy Phys. **02** (1998) 003.

[CR] A. Connes, M. Rieffel, *Yang-Mills for Non-commutative Two-Tori*, in *Operator Algebras and Mathematical Physics*, Contemp. Math. **62** (1987) 237-266.

[DKL1] L. Dabrowski, T. Krajewski, G. Landi, *Some Properties of Non-linear σ-models in Noncommutative Geometry*, Int. J. Mod. Phys. **B14** (2000) 2367-2382.

[DKL2] L. Dabrowski, T. Krajewski, G. Landi, *Non-linear σ-models in noncommutative geometry: fields with values in finite spaces*, Mod. Phys. Lett. **A18** (2003) 2371-2380.

[DS] M. Dieng, A. Schwarz, *Differential and complex geometry of two-dimensional noncommutative tori*, Lett. Math. Phys. **61** (2002) 263-270.

[T] T. Krajewski, *Gauge invariance of the Chern-Simons action in noncommutative geometry*, ISI GUCCIA Conference 'Quantum Groups, Noncommutative Geometry and Fundamental Physical Interactions', Palermo December 1997, math-ph/9810015.

[E] J. Eells, *Harmonic maps: Selected papers of James Eells and collaborators*, World Scientific, 1992.

[G] K. Gawedzki, *Lectures on conformal field theory*, in 'Quantum Fields and Strings: a Course for Mathematicians', P. Deligne et al. editors, American Mathematical Society 1999; pp 727-805.

[GSW] M. Green, J.H. Schwarz, E. Witten, *Superstring theory*, Cambridge University Press, 1987.

[M1] Yu. I. Manin, *Quantized theta-function*, Prog. Theor. Phys. Suppl. **102** (1990) 219-228.

[M2] Yu. I. Manin, *Theta functions, quantum tori and Heisenberg group*, Lett. Math. Phys. **56** (2001) 295-320.

[M3] Yu. I. Manin, *Real multiplication and noncommutative geometry*, math.QA/0202109.

[M4] Yu. I. Manin. *Functional equations for quantum theta functions*, math.QA/0307393.

[PV] M. Pimsner, D. Voiculescu, *Exact Sequences for K-Groups and Ext-Groups of Certain Cross-Product C*-Algebras*, J. Oper. Theory **4** (1980) 93-118.

[P-J] J. Polchinski, *String theory*, Cambridge University Press, 1998.

[P-A] A. Polishchuk, *Analogues of the exponential map associated with complex structures on noncommutative two-tori*, math.QA/0404056.

[PS] A. Polishchuk, A. Schwarz, *Categories of holomorphic vector bundles on noncommutative two-tori*, Commun. Math. Phys. **236** (2003) 135-159.

[R1] M. Rieffel, *C*-algebras associated with irrational rotations*, Pacific J. Math. **93** (1981) 415-429.

[R2] M. Rieffel, *The Cancellation Theorem for projective Modules over irrational C*-algebras*, Proc. London Math. Soc. **47** (1983) 1285-302.

[R3] M. Rieffel, *Projective modules over higher-dimensional noncommutative tori*, Can. J. Math. **40** (1988) 257-338.

[R4] M. Rieffel, *Non-commutative Tori -A case study of Non-commutative Differentiable Manifolds* Contemp. Math. **105** (1991) 191-211.

[SW] N. Seiberg, E. Witten, *String Theory and Noncommutative Geometry*, JHEP **09** (1999) 032.

[S-M] M. Spera, *Yang-Mills theory in non commutative differential geometry*, Rend. Sem. Fac. Scienze Univ. Cagliari, Suppl. **58** (1988) 409-421.

[S-A] A. Schwarz, *Theta-functions on noncommutative tori*, Lett. Math. Phys. **58** (2001) 81-90.

[V] J.C. Varilly, *An Introduction to Noncommutative Geometry*, Lectures at EMS Summer School on NCG and Applications, Sept 1997, physics/9709045.

[W] S. Walters, *The AF Structure of Noncommutative Toroidal Z/4Z Orbifolds*, J. Reine Angew. Math. **568** (2004) 139-196.

[Z] W.J. Zakrzewski, *Low dimensional sigma models*, Adam Hilger, Bristol 1989.

DIPARTIMENTO DI MATEMATICA E INFORMATICA, UNIVERSITÀ DI TRIESTE, VIA A. VALERIO 12/B, I-34127, TRIESTE, ITALIA

E-mail address: landi@univ.trieste.it

Towards the fractional quantum Hall effect: a noncommutative geometry perspective

Matilde Marcolli and Varghese Mathai

ABSTRACT. In this paper we give a survey of some models of the integer and fractional quantum Hall effect based on noncommutative geometry. We begin by recalling some classical geometry of electrons in solids and the passage to noncommutative geometry produced by the presence of a magnetic field. We recall how one can obtain this way a single electron model of the integer quantum Hall effect. While in the case of the integer quantum Hall effect the underlying geometry is Euclidean, we then discuss a model of the fractional quantum Hall effect, which is based on hyperbolic geometry simulating the multi-electron interactions. We derive the fractional values of the Hall conductance as integer multiples of orbifold Euler characteristics. We compare the results with experimental data.

1. Electrons in solids – Bloch theory and algebraic geometry

We first recall some general facts about the mathematical theory of electrons in solids. In particular, after reviewing some basic facts about Bloch theory, we recall an approach pioneered by Gieseker at al. [16] [17], which uses algebraic geometry to treat the inverse problem of determining the pseudopotential from the data of the electric and optical properties of the solid.

Crystals. The Bravais lattice of a crystal is a lattice $\Gamma \subset \mathbb{R}^d$ (where we assume $d = 2, 3$), which describes the symmetries of the crystal.

The electron–ions interaction is described by a periodic potential

$$(1.1) \qquad U(x) = \sum_{\gamma \in \Gamma} u(x - \gamma),$$

namely, U is invariant under the translations in Γ,

$$(1.2) \qquad T_\gamma U = U, \quad \forall \gamma \in \Gamma.$$

When one takes into account the mutual interaction of electrons, one obtains an N-particles Hamiltonian of the form

$$(1.3) \qquad \sum_{i=1}^{N} (-\Delta_{x_i} + U(x_i)) + \frac{1}{2} \sum_{i \neq j} W(x_i - x_j).$$

This can be treated in the *independent electron approximation*, namely by introducing a modification V of the single electron potential

$$(1.4) \qquad \sum_{i=1}^{N} (-\Delta_{x_i} + V(x_i)).$$

It is remarkable that, even though the original potential U is unbounded, a reasonable independent electron approximation can be obtained with V a bounded function.

The wave function for the N-particle problem (1.4) is then of the form

$$\psi(x_1, \dots, x_N) = \det(\psi_i(x_j)),$$

for $(-\Delta + V)\psi_i = E_i \psi_i$ so that $\sum(-\Delta_{x_i} + V(x_i))\psi = E\psi$ with $E = \sum E_i$. This reduces a multi-electron problem to the single particle case.

However, in this approximation, usually the single electron potential V is not known explicitly, hence the focus shifts on the *inverse problem* of determining V.

Bloch electrons. Let T_γ denote the unitary operator on $\mathcal{H} = L^2(\mathbb{R}^d)$ implementing the translation by $\gamma \in \Gamma$, as in (1.2). We have, for $H = -\Delta + V$,

$$(1.5) \qquad T_\gamma H T_{\gamma^{-1}} = H, \quad \forall \gamma \in \Gamma.$$

Thus, we can simultaneously diagonalize these operators. This can be done via characters of Γ, or equivalently, via its Pontrjagin dual $\hat{\Gamma}$. In fact, the eigenvalue equation is of the form $T_\gamma \psi = c(\gamma)\psi$. Since $T_{\gamma_1 \gamma_2} = T_{\gamma_1} T_{\gamma_2}$, and the operators are unitaries, we have $c : \Gamma \to U(1)$, of the form

$$c(\gamma) = e^{i\langle k, \gamma \rangle}, \quad k \in \hat{\Gamma}.$$

The Pontrjagin dual $\hat{\Gamma}$ of the abelian group $\Gamma \cong \mathbb{Z}^d$ is a compact group isomorphic to T^d, obtained by taking the dual of $\mathbb{R}^d$ modulo the reciprocal lattice

$$(1.6) \qquad \Gamma^\sharp = \{k \in \mathbb{R}^d : \langle k, \gamma \rangle \in 2\pi\mathbb{Z}, \forall \gamma \in \Gamma\}.$$

Brillouin zones. By definition, the Brillouin zones of a crystals are fundamental domains for the reciprocal lattice $\Gamma^\sharp$ obtained via the following inductive procedure. The *Bragg hyperplanes* of a crystal are the hyperplanes along which a pattern of diffraction of maximal intensity is observed when a beam of radiation (X-rays for instance) is shone at the crystal. The N-th Brillouin zone consists of all the points in (the dual) $\mathbb{R}^d$ such that the line from that point to the origin crosses exactly $(n-1)$ Bragg hyperplanes of the crystal.

Band structure. One obtains this way (*cf.* [16]) a family self-adjoint elliptic boundary value problems, parameterized by the lattice momenta $k \in \mathbb{R}^d$,

$$(1.7) \qquad D_k = \begin{cases} (-\Delta + V)\psi = E\psi \\ \psi(x + \gamma) = e^{i\langle k, \gamma \rangle}\psi(x) \end{cases}$$

For each value of the momentum k, one has eigenvalues $\{E_1(k), E_2(k), E_3(k), \dots\}$. As functions of k, these satisfy the periodicity

$$E(k) = E(k + u) \quad \forall u \in \Gamma^\sharp.$$

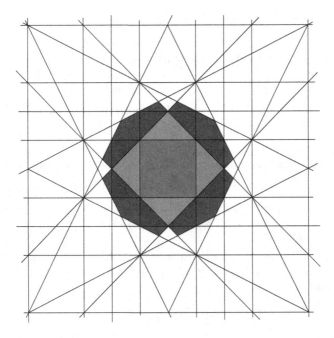

FIGURE 1. *Brillouin zones in a 2D crystal*

Chromium   Vanadium   Yttrium

FIGURE 2. *Examples of Fermi surfaces*

It is customary therefore to plot the eigenvalue $E_n(k)$ over the n-the Brillouin zone and obtain this way a map

$$k \mapsto E(k) \quad k \in \mathbb{R}^d$$

called the *energy–crystal momentum dispersion relation.*

Fermi surfaces and complex geometry. Many electric and optical properties of the solid can be read off the geometry of the Fermi surface. This is a hypersurface F in the space of crystal momenta k,

(1.8) $$F_\lambda(\mathbb{R}) = \{k \in \mathbb{R}^d : E(k) = \lambda\}.$$

A comprehensive archive of Fermi surfaces for various chemical elements can be found in [9], or online at http://www.phys.ufl.edu/fermisurface. We reproduce in Figure 2 an example of the complicated geometry of Fermi surfaces.

The approach to the theory of electrons in solids proposed by [16] [17] is based on the idea that the geometry of the Fermi surfaces can be better understood by passing to complex geometry and realizing (1.8) as a cycle on a complex hypersurface. One considers first the *complex Bloch variety* defined by the condition

$$
(1.9) \qquad B(V) = \left\{ (k, \lambda) \in \mathbb{C}^{d+1} : \begin{array}{c} \exists \psi \text{ nontrivial solution of} \\ (-\Delta + V)\psi = \lambda \psi \\ \psi(x + \gamma) = e^{i\langle k, \gamma \rangle} \psi(x) \end{array} \right\}
$$

Then the complex Fermi surfaces are given by the fibers of the projection to $\lambda \in \mathbb{C}$,

$$
(1.10) \qquad F_\lambda(\mathbb{C}) = \pi^{-1}(\lambda) \subset B(V).
$$

This is a complex hypersurface in $\mathbb{C}^d$. To apply the tools of projective algebraic geometry, one works in fact with a singular projective hypersurface (a compactification of $B(V)$, *cf.* [16]).

One can then realize the original Fermi surface (1.8) as a cycle $F_\lambda \cap \mathbb{R}^d = F_\lambda(\mathbb{R})$ representing a class in the homology $H_{d-1}(F_\lambda(\mathbb{C}), \mathbb{Z})$. A result of [16] is that the *integrated density of states*

$$
(1.11) \qquad \rho(\lambda) = \lim_{\ell \to \infty} \frac{1}{\ell} \#\{\text{eigenv of } H \leq \lambda\},
$$

for $H = -\Delta + V$ on $L^2(\mathbb{R}^d / \ell\Gamma)$, is obtained from a *period*

$$
(1.12) \qquad \frac{d\rho}{d\lambda} = \int_{F_\lambda(\mathbb{R})} \omega_\lambda,
$$

where ω_λ is a holomorphic differential on $F_\lambda(\mathbb{C})$.

Discretization. It is often convenient to treat problems like (1.7) by passing to a discretized model. On $\ell^2(\mathbb{Z}^d)$, one considers the *random walk* operator

$$
(1.13) \qquad \begin{aligned} \mathcal{R}\psi(n_1, \ldots, n_d) = \ & +\sum_{i=1}^d \psi(n_1, \ldots, n_i + 1, \ldots, n_d) \\ & +\sum_{i=1}^d \psi(n_1, \ldots, n_i - 1, \ldots, n_d). \end{aligned}
$$

This is related to the *discretized Laplacian* by

$$
(1.14) \qquad \Delta\psi(n_1, \ldots, n_d) = (2d - \mathcal{R}) \, \psi(n_1, \ldots, n_d).
$$

In this discretization, the complex Bloch variety is described by a polynomial equation in z_i, z_i^{-1} (*cf.* [16])

$$
B(V) = \left\{ (z_1 \ldots, z_d, \lambda) : \begin{array}{c} \exists \psi \in \ell^2(\Gamma) \text{ nontriv sol of} \\ (\mathcal{R} + V)\psi = (\lambda + 2d) \, \psi \\ R_{\gamma_i} \psi = z_i \, \psi \end{array} \right\},
$$

where $R_{\gamma_i}\psi(n_1, \ldots, n_d) = \psi(n_1, \ldots, n_i + a_i, \ldots, n_d)$.

It will be very useful for us in the following to also consider a more general random walk for a discrete group Γ, as an operator on $\mathcal{H} = \ell^2(\Gamma)$. Let γ_i be a symmetric set of generators of Γ, *i.e.* a set of generators and their inverses. The random walk operator is defined as

$$
(1.15) \qquad \mathcal{R}\,\psi(\gamma) = \sum_{i=1}^r R_{\gamma_i}\psi\,(\gamma) = \sum_{i=1}^r \psi(\gamma\gamma_i)
$$

and the corresponding discretized Laplacian is $\Delta = r - \mathcal{R}$.

The breakdown of classical Bloch theory. The approach to the study of electrons in solids via Bloch theory breaks down when either a magnetic field is present, or when the periodicity of the lattice is replaced by an aperiodic configuration, such as those arising in quasi–crystals. What is common to both cases is that the commutation relation $T_\gamma H = HT_\gamma$ fails.

Both cases can be studied by replacing ordinary geometry by *noncommutative geometry* (*cf.* [12]). Ordinary Bloch theory is replaced by noncommutative Bloch theory [18]. A good introduction to the treatment via noncommutative geometry of the case of aperiodic solids can be found in [4].

For our purposes, we are mostly interested in the other case, namely the presence of magnetic field, as that is the source of the Hall effects. Bellissard pioneered an approach to the quantum Hall effect via noncommutative geometry and derived a complete and detailed mathematical model for the Integer Quantum Hall Effect within this framework, [3].

2. Quantum Hall effect

We describe the main aspects of the classical and quantum (integer and fractional) Hall effects, and some of the current approaches used to produce a mathematical model. Our introduction will not be exhaustive. In fact, for reasons of space, we will not discuss many interesting mathematical results on the quantum Hall effect and will concentrate mostly on the direction leading to the use of noncommutative geometry.

Classical Hall effect. The classical Hall effect was first observed in the XIX century [19]. A thin metal sample is immersed in a constant uniform strong orthogonal magnetic field, and a constant current $\mathbf{j}$ flows through the sample, say, in the x-direction. By Flemming's rule, an electric field is created in the y-direction, as the flow of charge carriers in the metal is subject to a Lorentz force perpendicular to the current and the magnetic field. This is called the *Hall current*.

The equation for the equilibrium of forces in the sample

$$(2.1) \qquad N e\mathbf{E} + \mathbf{j} \wedge \mathbf{B} = 0,$$

defines a linear relation. The ratio of the intensity of the Hall current to the intensity of the electric field is the Hall conductance,

$$(2.2) \qquad \sigma_H = \frac{Ne\delta}{B}.$$

In the stationary state, σ_H is proportional to the dimensionless *filling factor* $\nu = \frac{\rho h}{eB}$, where ρ is the 2-dimensional density of charge carriers, h is the Planck constant, and e is the electron charge. More precisely, we have

$$(2.3) \qquad \sigma_H = \frac{\nu}{R_H},$$

where $R_H = h/e^2$ denotes the Hall resistance, which is a universal constant. This measures the fraction of Landau level filled by conducting electrons in the sample.

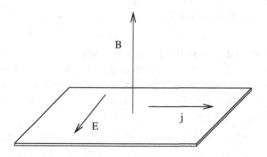

FIGURE 3. *Hall effect*

Integer quantum Hall effect. In 1980, von Klitzing's experiment showed that, lowering the temperature below 1 K, quantum effects dominate, and the relation of Hall conductance to filling factor shows plateaux at integer values, [**21**]. The effect is measured with very high precision (of the order of 10^{-8}) and allows for a very accurate measurement of the fine structure constant $e^2/\hbar c$.

Under the above conditions, one can effectively use the independent elector approximation discussed in the previous section and reduce the problem to a single particle case.

The main physical properties of the integer quantum Hall effect are the following:

- σ_H, as a function of ν, has plateaux at integer multiples of e^2/h.
- At values of ν corresponding to the plateaux, the conductivity along the current density axis (direct conductivity) vanishes.

Laughlin first suggested that IQHE should have a geometric explanation [**23**]. More precisely, the fact that the quantization of the Hall conductance appears as a very robust phenomenon, insensitive to changes in the sample and its geometry, or to the presence of impurities, suggests the fact that the effect should have the same qualities of the *index theorem*, which assigns an integer invariant to an elliptic differential operator, in a way that is topological and independent of perturbations. The prototype of such index theorems is the Gauss–Bonnet theorem, which extracts from an infinitesimally variable quantity, the curvature of a closed surface, a robust topological invariant, its Euler characteristic. The idea of modelling the integer quantum Hall effect on an index theorem started fairly early after the discovery of the effect. Laughlin's formulation can already be seen as a form of Gauss–Bonnet, while this was formalized more precisely in such terms shortly afterwards by Thouless et al. (1982) and by Avron, Seiler, and Simon (1983) (*cf.* [**31**] [**1**]).

One of the early successes of Connes' noncommutative geometry [**12**] was a rigorous mathematical model of the integer quantum Hall effect, developed by Bellissard, van Elst, and Schulz-Baldes, [**3**]. Unlike the previous models, this accounts for all the aspects of the phenomenon: integer quantization, localization, insensitivity to the presence of disorder, and vanishing of direct conductivity at plateaux levels. Again the integer quantization is reduced to an index theorem, albeit of a more sophisticated nature, involving the Connes–Chern character, the K-theory of C^*-algebras and cyclic cohomology (*cf.* [**11**]).

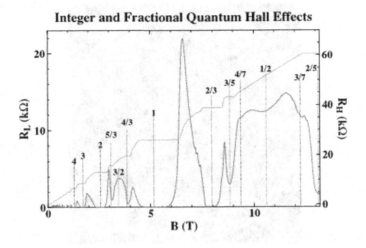

FIGURE 4. *Fractional quantum Hall effect*

Fractional quantum Hall effect. The fractional QHE was discovered by Stormer and Tsui in 1982. The setup is as in the quantum Hall effect: in a high quality semi-conductor interface, which will be modelled by an infinite 2-dimensional surface, with low carrier concentration and extremely low temperatures $\sim 10mK$, in the presence of a very strong magnetic field, the experiment shows that the same graph of $\frac{h}{e^2}\sigma_H$ against the filling factor ν exhibits plateaux at certain fractional values (Figure 4).

Under the conditions of the experiments, the independent electron approximation that reduces the problem to a single electron is no longer viable and one has to incorporate the Coulomb interaction between the electrons in a many-electron theory. For this reason, many of the proposed mathematical models of the fractional quantum Hall effect resort to quantum field theory and, in particular, Chern–Simons theory (*cf. e.g.* [28]).

In this survey we will only discuss a proposed model [24] [25], which is based on extending the validity of the Bellissard approach to the setting of hyperbolic geometry as in [5], where passing to a negatively curved geometry is used as a device to simulate the many-electrons Coulomb interaction while remaining within a single electron model.

What is expected of any proposed mathematical model? Primarily three things: to account for the strong electron interactions, to exhibit the observed fractions and predict new fractions, and to account for the varying width of the observed plateaux. We will discuss these various aspects in the rest of the paper.

3. Noncommutative geometry models

In the theory of the quantum Hall effect noncommutativity arises from the presence of the magnetic field, which has the effect of turning the Brillouin zone into a noncommutative space. In Bellissard's model of the integer quantum Hall effect [3] the noncommutative space obtained this way is the noncommutative torus

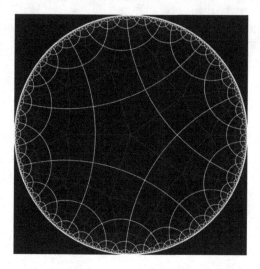

FIGURE 5. *Tiling of the hyperbolic plane*

and the integer values of the Hall conductance are obtained from the corresponding Connes–Chern character. We will consider a larger class of noncommutative spaces, associated to the action of a Fuchsian group of the first kind without parabolic elements on the hyperbolic plane. The idea is that, by effect of the strong interaction with the other electrons, a single electron "sees" the surrounding geometry as curved, while the lattice sites appear to the moving electron, in a sort of multiple image effect, as sites in a lattice in the hyperbolic plane. This model will recover the integer values but will also produce fractional values of the Hall conductance.

3.1. Hyperbolic geometry. Let $\mathbb{H}$ denote the hyperbolic plane. Its geometry is described as follows. Consider the pseudosphere $\{x^2 + y^2 + z^2 - t^2 = 1\}$ in 4-dimensional Minkowski space-time M. The $z = 0$ slice of the pseudosphere realizes an isometric embedding of the hyperbolic plane $\mathbb{H}$ in M. In this geometry, a periodic lattice on the resulting surface is determined by a Fuchsian group Γ of isometries of $\mathbb{H}$ of signature $(g; \nu_1, \ldots, \nu_n)$,

$$(3.1) \qquad\qquad \Gamma = \Gamma(g; \nu_1, \ldots, \nu_n).$$

This is a discrete cocompact subgroup $\Gamma \subset \mathrm{PSL}(2, \mathbb{R})$ with generators a_i, b_i, c_j, with $i = 1, \ldots, g$ and $j = 1, \ldots, n$ and a presentation of the form

$$(3.2) \qquad \Gamma(g; \nu_1, \ldots, \nu_n) = \langle a_i, b_i, c_j \ \Big| \ \prod_{i=1}^{g} [a_i, b_i] c_1 \cdots c_n = 1, \ c_j^{\nu_j} = 1 \rangle.$$

The quotient of the action of Γ by isometries on $\mathbb{H}$,

$$(3.3) \qquad\qquad \Sigma(g; \nu_1, \ldots, \nu_n) := \Gamma \backslash \mathbb{H},$$

is a hyperbolic orbifold, namely a compact Riemann surface of genus g with n cone points $\{x_1, \ldots, x_n\}$, which are the image of points in $\mathbb{H}$ with non-trivial stabilizer of the action of Γ. In the torsion free case, where we only have generators a_i and b_i, we obtain smooth compact Riemann surfaces of genus g.

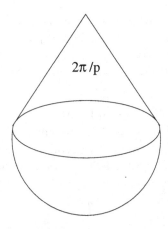

FIGURE 6. *Thurston's teardrop orbifold*

Orbifolds. The space $\Sigma(g; \nu_1, \ldots, \nu_n)$ of (3.3) is a special case of *good orbifolds.* These are orbifolds that are orbifold–covered by a smooth manifold. In dimension two, in the oriented compact case, the only exceptions (bad orbifolds) are the Thurston teardrop (Figure 6) with a single cone point of angle $2\pi/p$, and the double teardrop.

In particular, all the hyperbolic orbifolds $\Sigma(g; \nu_1, \ldots, \nu_n)$ are good orbifolds and they are orbifold–covered by a smooth compact Riemann surface,

$$(3.4) \qquad \Sigma_{g'} \xrightarrow{G} \Sigma(g; \nu_1, \ldots, \nu_n) = \Gamma\backslash\mathbb{H},$$

where the genus g' satisfies the Riemann–Hurwitz formula for branched covers

$$(3.5) \qquad g' = 1 + \frac{\#G}{2}(2(g-1) + (n - \nu)),$$

for $\nu = \sum_{j=1}^{n} \nu_j^{-1}$.

Notice moreover that the orbifolds $\Sigma = \Sigma(g; \nu_1, \ldots, \nu_n)$ are an example of classifying space for proper actions in the sense of Baum–Connes, namely they are of the form

$$(3.6) \qquad \Sigma = \underline{B}\Gamma = \Gamma\backslash\underline{E}\Gamma.$$

An important invariant of orbifold geometry, which will play a crucial role in our model of the fractional quantum Hall effect, is the *orbifold Euler characteristic.* This is an analog of the usual topological Euler characteristic, but it takes values in rational numbers, $\chi_{orb}(\Sigma) \in \mathbb{Q}$. It is multiplicative over orbifold covers, it agrees with the usual topological Euler characteristic χ for smooth manifolds, and it satisfies the inclusion–exclusion principle

$$(3.7) \qquad \begin{aligned} \chi_{orb}(\Sigma_1 \cup \cdots \cup \Sigma_r) = &\quad \sum_i \chi_{orb}(\Sigma_i) - \sum_{i,j} \chi_{orb}(\Sigma_i \cap \Sigma_j) \\ \cdots &\quad + (-1)^{r+1}\chi_{orb}(\Sigma_1 \cap \cdots \cap \Sigma_r). \end{aligned}$$

In the case of the hyperbolic orbifolds $\Sigma(g; \nu_1, \ldots, \nu_n)$, the orbifold Euler characteristic is given by the formula

$$(3.8) \qquad \chi_{orb}(\Sigma(g; \nu_1, \ldots, \nu_n)) = 2 - 2g + \nu - n.$$

Magnetic field and symmetries. The magnetic field can be described by a 2-form $\omega = d\eta$, where ω and η are the field and potential, respectively, subject to the customary relation $\mathbf{B} = \mathrm{curl}\mathbf{A}$.

One then considers the magnetic Schrödinger operator

$$(3.9) \qquad\qquad \Delta^\eta + V,$$

where the magnetic Laplacian is given by $\Delta^\eta := (d - i\eta)^* \, (d - i\eta)$ and V is the electric potential of the independent electron approximation.

The 2-form ω satisfies the periodicity condition $\gamma^*\omega = \omega$, for all $\gamma \in \Gamma = \mathbb{Z}^d$ (e.g. one might assume that the magnetic field is a constant field B perpendicular to the sample). Thus, we have the relation $0 = \omega - \gamma^*\omega = d(\eta - \gamma^*\eta)$, which implies

$$(3.10) \qquad\qquad \eta - \gamma^*\eta = d\phi_\gamma.$$

Due to the fact that η itself need not be periodic, but only subject to condition (3.10), the magnetic Laplacian no longer commutes with the Γ action, unlike the ordinary Laplacian. This is, in a nutshell, how turning on a magnetic field brings about noncommutativity.

What are then the symmetries of the magnetic Laplacian? These are given by the *magnetic translations*. Namely, after writing (3.10) in the form $\phi_\gamma(x) = \int_{x_0}^x (\eta - \gamma^*\eta)$, we consider the unitary operators

$$(3.11) \qquad\qquad T_\gamma^\phi \psi := \exp(i\phi_\gamma) \, T_\gamma \psi.$$

It is easy to check that these satisfy the desired commutativity $(d - i\eta)T_\gamma^\phi = T_\gamma^\phi(d - i\eta)$. However, commutativity is still lost in another way, namely, magnetic translations, unlike the ordinary translations by elements $\gamma \in \Gamma = \mathbb{Z}^d$, no longer commute among themselves (except in the case of integer flux). We have instead

$$(3.12) \qquad\qquad T_\gamma^\phi T_{\gamma'}^\phi = \sigma(\gamma, \gamma')T_{\gamma\gamma'}^\phi.$$

Instead of obtaining a representation of Γ, the magnetic translations give rise to a *projective representation*, with the cocycle

$$(3.13) \qquad\qquad \sigma(\gamma, \gamma') = \exp(-i\phi_\gamma(\gamma'x_0)),$$

where $\phi_\gamma(x) + \phi_{\gamma'}(\gamma x) - \phi_{\gamma'\gamma}(x)$ is independent of x.

Algebra of observables. The C^*-algebra of observables should be minimal, yet large enough to contain all of the spectral projections onto gaps in the spectrum of the magnetic Schrödinger operators $\Delta^\eta + V$ for any periodic potential V. Now let $\mathcal{U}$ denote the set of all bounded operators on $L^2(\mathbb{H})$ that commute with the magnetic translations. By a theorem of von Neumann, $\mathcal{U}$ is a von Neumann algebra. By Lemma 1.1, [**22**], any element $Q \in \mathcal{U}$ can be represented uniquely as

$$Q = \sum_{\gamma \in \Gamma} T_\gamma^{-\phi} \otimes Q(\gamma),$$

where $Q(\gamma)$ is a bounded operator on the Hilbert space $L^2(\mathbb{H}/\Gamma)$. Let $\mathcal{L}^1$ denote the subset of $\mathcal{U}$ consisting of all bounded operators on Q on $L^2(\mathbb{H})$ that commute with the magnetic translations and such that $\sum_{\gamma \in \Gamma} \|Q(\gamma)\| < \infty$. The norm closure of $\mathcal{L}^1$ is a C^*-algebra denoted by $\mathcal{C}^*$, that is taken to be the algebra of observables. Using the Riesz representation for projections onto spectral gaps cf.(3.26), one can show as in [22] that C^* contains all of the projections onto the spectral gaps of the magnetic Schrödinger operators. In fact, it can be shown that C^* is Morita equivalent to the reduced twisted group C^* algebra $C_r^*(\Gamma, \bar{\sigma})$, explained later in the text, showing that in both the continuous and the discrete models for the quantum Hall effect, the algebra of observables are Morita equivalent, so they describe the same physics. Hence we will mainly discuss the discrete model in this paper.

Semiclassical properties, as the electro-magnetic coupling constant goes to zero. Recall the magnetic Schrödinger operator

$$(3.14) \qquad\qquad \Delta^\eta + \mu^{-2}V,$$

where Δ^η is the magnetic Laplacian, V is the electric potential and μ is the electro-magnetic coupling constant. When V is a Morse type potential, i.e. for all $x \in \mathbb{H}$, $V(x) \geq 0$. Moreover, if $V(x_0) = 0$ for some x_0 in $\mathcal{M}$, then there is a positive constant c such that $V(x) \geq c|x - x_0|^2 I$ for all x in a neighborhood of x_0. Also assume that V has at least one zero point. Observe that all functions $V = |df|^2$, where $|df|$ denotes the pointwise norm of the differential of a Γ-invariant Morse function f on $\mathbb{H}$, are examples of Morse type potentials.

Under these assumptions, the semiclassical properties of the spectrum of the magnetic Schrödinger operator, and the Hall conductance were studied by Kordyukov, Mathai and Shubin in [22], as the electro-magnetic coupling constant μ goes to zero. One result obtained is that there exists an arbitrarily large number of gaps in the spectrum of the magnetic Schrödinger operator for all μ sufficiently small. Another result obtained in [22] is that the low energy bands do not contribute to the Hall conductance, again for all μ sufficiently small.

Extending Pontrjagin duality. An advantage of noncommutative geometry is that it provides a natural generalization of Pontrjagin duality. Namely, the duals of discrete groups are noncommutative spaces.

In fact, first recall that, if Γ is a discrete abelian group, then its Pontrjagin dual $\hat{\Gamma}$, which is the group of characters of Γ is a compact abelian group. The duality is given by Fourier transform $e^{i\langle k, \gamma \rangle}$, for $\gamma \in \Gamma$ and $k \in \hat{\Gamma}$.

In particular, this shows that the algebra of functions on $\hat{\Gamma}$ can be identified with the (reduced) C^*-algebra of the group Γ,

$$(3.15) \qquad\qquad C(\hat{\Gamma}) \cong C_r^*(\Gamma),$$

where the reduced C^*-algebra $C_r^*(\Gamma)$ is the C^*-algebra generated by Γ in the regular representation on $\ell^2(\Gamma)$.

When Γ is non-abelian, although Pontrjagin duality no longer works in the classical sense, the left hand side of (3.15) still makes perfect sense and behaves "like" the algebra of functions on the dual group. In other words, we can say that, for a non-abelian group, the Pontrjagin dual $\hat{\Gamma}$ still exists as a noncommutative space whose algebra of functions is $C_r^*(\Gamma)$.

This point of view can be adopted to work with the theory of electrons in solids whenever classical Bloch theory breaks down. In the case of aperiodicity, the dual $\hat{\Gamma}$ (which is identified with the Brillouin zone) is replaced by a noncommutative C^*-algebra. This is the case, similarly, for the presence of magnetic field in the quantum

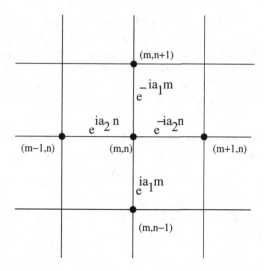

FIGURE 7. *The Harper operator on the square lattice*

Hall effect. The magnetic field deforms the Brillouin zone to a noncommutative space, given by the (noncommutative) algebra of the magnetic translation.

Harper operators. It is again convenient to discretize the problem. The discretized magnetic Laplacian is given in terms of the Harper operator, which is an analog of the random walk operator seen in the previous section, but defined using the magnetic translations. For $\Gamma = \mathbb{Z}^2$, the Harper operator is of the form

$$
\begin{aligned}
H_{\alpha_1,\alpha_2}\psi(m,n) = \ & e^{-i\alpha_1 n} \quad \psi(m+1,n) \\
+ \ & e^{i\alpha_1 n} \quad \psi(m-1,n) \\
+ \ & e^{-i\alpha_2 m} \quad \psi(m,n+1) \\
+ \ & e^{i\alpha_2 m} \quad \psi(m,n-1).
\end{aligned}
$$

(3.16)

Here the 2-cocycle $\sigma : \Gamma \times \Gamma \to U(1)$ is given by

(3.17) $$\sigma((m',n'),(m,n)) = \exp(-i(\alpha_1 m'n + \alpha_2 mn')).$$

The magnetic translations are generated by $U = R^\sigma_{(0,1)}$ and $V = R^\sigma_{(1,0)}$ of the form

(3.18) $$U\psi(m,n) = \psi(m,n+1)e^{-i\alpha_2 m} \quad V\psi(m,n) = \psi(m+1,n)e^{-i\alpha_1 n}.$$

These satisfy the commutation relations of the *noncommutative torus* $\mathcal{A}_\theta$, with $\theta = \alpha_2 - \alpha_1$, namely

(3.19) $$UV = e^{i\theta}VU.$$

The Harper operator (3.16) is in fact more simply written as $H_\sigma = U + U^* + V + V^*$. This shows that, on a 2-dimensional lattice, the effect of the magnetic field is to deform the usual Brillouin zone (which is an ordinary torus T^2) to a noncommutative torus, where the parameter θ depends on the magnetic flux through a cell of the lattice.

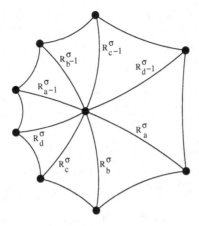

FIGURE 8. *Harper operator on a lattice in the hyperbolic plane*

As in the case of the discretization of the ordinary Laplacian, for the magnetic Laplacian we can also consider the corresponding Harper operator on a more general (possibly non-abelian) discrete group Γ. This will be useful later, in our model of the fractional quantum Hall effect, but we introduce it here for convenience. For the general setup for finitely generated discrete groups recalled here below, we follow [**5**].

Suppose given a finitely generated discrete group Γ and a *multiplier* $\sigma : \Gamma \times \Gamma \to U(1)$ (a 2-cocycle)

$$\sigma(\gamma_1, \gamma_2)\sigma(\gamma_1\gamma_2, \gamma_3) = \sigma(\gamma_1, \gamma_2\gamma_3)\sigma(\gamma_2, \gamma_3),$$

$\sigma(\gamma, 1) = \sigma(1, \gamma) = 1$.

On the Hilbert space $\ell^2(\Gamma)$, consider the *left/right σ-regular representations*

$$(3.20) \qquad L_\gamma^\sigma \psi(\gamma') = \psi(\gamma^{-1}\gamma')\sigma(\gamma, \gamma^{-1}\gamma') \qquad R_\gamma^\sigma \psi(\gamma') = \psi(\gamma'\gamma)\sigma(\gamma', \gamma).$$

These satisfy

$$(3.21) \qquad L_\gamma^\sigma L_{\gamma'}^\sigma = \sigma(\gamma, \gamma')L_{\gamma\gamma'}^\sigma \qquad R_\gamma^\sigma R_{\gamma'}^\sigma = \sigma(\gamma, \gamma')R_{\gamma\gamma'}^\sigma.$$

The cocycle identity can be used to show that the left σ-regular representation commutes with the right $\bar{\sigma}$-regular representation, where $\bar{\sigma}$ denotes the conjugate cocycle. Also the left $\bar{\sigma}$-regular representation commutes with the right σ-regular representation.

Let $\{\gamma_i\}_{i=1}^r$ be a symmetric set of generators of Γ. The Harper operator is given by

$$(3.22) \qquad \mathcal{R}_\sigma = \sum_{i=1}^r R_{\gamma_i}^\sigma.$$

The operator $r - \mathcal{R}_\sigma$ is the discrete analog of the magnetic Laplacian (*cf.* [**30**]).

Algebra of observables (discrete model). We continue in the same generality as above. The special case of interest for the integer quantum Hall effect will be for $\Gamma = \mathbb{Z}^2$, but we adopt a more general setting in view of applications to the fractional case.

For Γ a finitely generated discrete group, let $\mathbb{C}(\Gamma, \sigma)$ be the algebra generated by the magnetic translations represented as operators in $\mathcal{B}(\ell^2(\Gamma))$ through the right σ-regular representation R_γ^σ. Equivalently, the algebra $\mathbb{C}(\Gamma, \sigma)$ consists of functions

$$f : \Gamma \to \mathbb{C}$$

with the convolution product

$$f_1 * f_2(\gamma) = \sum_{\gamma_1 \gamma_2 = \gamma} f_1(\gamma_1) f_2(\gamma_2) \sigma(\gamma_1, \gamma_2),$$

acting on the Hilbert space $\ell^2(\Gamma)$.

By taking the weak closure of $\mathbb{C}(\Gamma, \sigma)$ one obtains the *twisted group von Neumann algebra* $\mathcal{U}(\Gamma, \sigma)$. This is equivalently (by the commutant theorem of von Neumann) described as

$$\mathcal{U}(\Gamma, \sigma) = \left\{ A \in B(\ell^2(\Gamma)) : [L_\gamma^{\bar\sigma}, A] = 0 \; \forall \gamma \in \Gamma \right\}.$$

That is, $\mathcal{U}(\Gamma, \sigma)$ is the commutant of the left $\bar\sigma$-regular representation. When taking the norm closure of $\mathbb{C}(\Gamma, \sigma)$ one obtains the twisted (reduced) group C^*-algebra $C_r^*(\Gamma, \sigma)$, which is the algebra of observables in the discrete model.

The key properties of these algebras are summarized as follows. $\mathcal{U}(\Gamma, \sigma)$ is generated by its projections and it is also closed under the *measurable* functional calculus, *i.e.* if $a \in \mathcal{U}(\Gamma, \sigma)$ and $a = a^*$, $a > 0$, then $f(a) \in \mathcal{U}(\Gamma, \sigma)$ for all *essentially bounded measurable* functions f on $\mathbb{R}$. On the other hand, $C_r^*(\Gamma, \sigma)$ has only at most countably many projections and is only closed under the *continuous* functional calculus.

In the case when $\sigma = 1$ (integer flux), with the group $\Gamma = \mathbb{Z}^2$, we simply have $\mathcal{U}(\Gamma, 1) \cong L^\infty(T^2)$ and $C^*(\Gamma, 1) \cong C(T^2)$, *i.e.* functions on the classical Brillouin zone. Here the ordinary torus T^2 is identified with the group $\widehat{\Gamma}$ of characters of the abelian group $\Gamma = \mathbb{Z}^2$.

In Bellissard's model of the integer quantum Hall effect, where $\Gamma = \mathbb{Z}^2$, with σ the nontrivial cocycle described in (3.17) and $\theta = \alpha_2 - \alpha_1$, the twisted (reduced) group C^*-algebra is the irrational rotation algebra of the noncommutative torus,

(3.23) $$C_r^*(\Gamma, \sigma) \cong A_\theta.$$

We will not describe in detail the derivation of the quantization of the Hall conductance in this model of the integer quantum Hall effect. In fact, we will concentrate mostly on a model for the fractional quantum Hall effect and we will show how to recover the integer quantization within that model, using the results of [5].

Spectral theory. For Γ a finitely generated discrete group and $\{g_i\}_{i=1}^r$ a symmetric set of generators, the *Cayley graph* $\mathcal{G} = \mathcal{G}(\Gamma, g_i)$ has as set of vertices the elements of Γ and as set of edges emanating from a given vertex $h \in \Gamma$ the set of translates $g_i h$.

The random walk operator (1.15) for Γ is then an average on nearest neighbors in the Cayley graph. The discrete analog of the Schrödinger equation is of the form

(3.24) $$i\frac{\partial}{\partial t}\psi = \mathcal{R}_\sigma \psi + V\psi,$$

where all physical constants have been set equal to 1. It describes the quantum mechanics of a single electron confined to move along the Cayley graph of Γ, subject to the periodic magnetic field. Here $\mathcal{R}_\sigma$ is the Harper operator encoding the magnetic

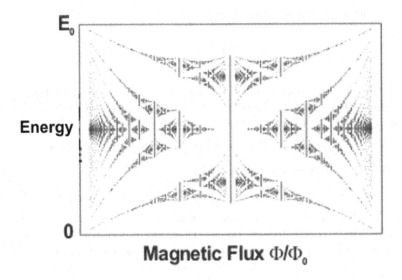

FIGURE 9. *Hofstadter butterfly*

field and V is the electric potential of the independent electron approximation. The latter can be taken to be an operator in the twisted group algebra, $V \in \mathbb{C}(\Gamma, \sigma)$. More precisely, equation (3.24) should be formulated with the discrete magnetic laplacian $\delta_\sigma = r - \mathcal{R}_\sigma$ in place of $\mathcal{R}_\sigma$, with r the cardinality of a symmetric set of generators for Γ. This does not really matter as far as the spectral properties are concerned, as the spectrum of one determines the spectrum of the other.

As in the case of the theory of electrons in solids without magnetic field recalled in the first section, an important problem is understanding the energy levels of the Hamiltonian $H_{\sigma,V} = \mathcal{R}_\sigma + V$, and the band structure (gaps in the spectrum).

The Harper operator $\mathcal{R}_\sigma$ is a bounded *self-adjoint* operator on $\ell^2(\Gamma)$, since it is defined in terms of a *symmetric* set of generators of Γ. Thus, the spectrum $\mathrm{Spec}(\mathcal{R}_\sigma)$ is a closed and bounded subset of $\mathbb{R}$. It follows that the complement $\mathbb{R}\backslash\mathrm{Spec}(\mathcal{R}_\sigma)$ is an open subset of $\mathbb{R}$, hence a countable union of disjoint open intervals. Each such interval is called a *gap in the spectrum.*

There are two very different situations. When the complement of the spectrum consists of a finite collection of intervals then the operator has a band structure, while if the complement consists of an infinite collection of intervals then the spectrum is a Cantor set. In the case of the group $\Gamma = \mathbb{Z}^2$, one or the other possibility occurs depending on the rationality or irrationality of the flux

$$\theta = \langle [\sigma], [\Gamma] \rangle.$$

This gives rise to a diagram known as the Hofstadter butterfly (Figure 9).

Range of the trace. In our model of the fractional quantum Hall effect, Γ is a cocompact Fuchsian group of signature $(g, \nu_1, \ldots, \nu_n)$. In this case (*cf.* [25]), if $[\sigma]$ is *rational*, then there is only a finite number of gaps in the spectrum of $H_\sigma + V$.

In fact, if $\theta = p/q$ then the number of gaps is at most

(3.25)
$$(q+1)\prod_{j=1}^{n}(\nu_j + 1).$$

In terms of the algebra of observables, the question of how many gaps there are in the spectrum of $H_{\sigma,V}$ can be reduced to studying the *number of projections* in the C^*-algebra $C_r^*(\Gamma, \sigma)$ (up to equivalence). In fact, we have

$$H_{\sigma,V} \in \mathbb{C}(\Gamma, \sigma) \subset C_r^*(\Gamma, \sigma) \subset \mathcal{U}(\Gamma, \sigma).$$

In particular, H_σ and its spectral projections

$$P_E = \chi_{(-\infty, E]}(H_{\sigma,V})$$

belong to the algebra $\mathcal{U}(\Gamma, \sigma)$. Moreover, when $E \notin \mathrm{Spec}(H_{\sigma,V})$, the spectral projection P_E is in $C_r^*(\Gamma, \sigma)$. In fact, suppose that the spectrum of $H_{\sigma,V}$ is contained in a closed interval, and that the open interval (a, b) is a spectral gap of $H_{\sigma,V}$. Suppose that $E \in (a, b)$, i.e. $E \notin \mathrm{Spec}(H_{\sigma,V})$. Then there is a holomorphic function ϕ on a neighborhood of $\mathrm{spec}(H_{\sigma,V})$ such that

(3.26)
$$P_E = \phi(H_{\sigma,V}) = \int_C \frac{d\lambda}{\lambda - H_{\sigma,V}}$$

where C is a closed contour enclosing the spectrum of $H_{\sigma,V}$ to the left of E. Since $C_r^*(\Gamma, \sigma)$ is closed under the holomorphic functional calculus, it follows that $P_E \in C_r^*(\Gamma, \sigma)$.

The equivalence relation we need to consider on projections, so that the counting will provide the counting of spectral gaps, is described as follows. Let $\mathrm{Proj}(C_r^*(\Gamma, \sigma) \otimes \mathcal{K})$ denote the projections in $C_r^*(\Gamma, \sigma) \otimes \mathcal{K}$, where $\mathcal{K}$ the C^* algebra of compact operators. Two projections $P, Q \in \mathrm{Proj}(C_r^*(\Gamma, \sigma) \otimes \mathcal{K})$ are said to be *Murray-von Neumann equivalent* if there is an element $V \in C_r^*(\Gamma, \sigma) \otimes \mathcal{K}$ such that $P = V^*V$ and $Q = VV^*$, and we write $P \sim Q$. It can be shown that $\mathrm{Proj}(C_r^*(\Gamma, \sigma) \otimes \mathcal{K})/\sim$ is an abelian semi-group under direct sums, and the Grothendieck group $K_0(C_r^*(\Gamma, \sigma)))$ is defined as the associated abelian group.

Now the estimate on the number of equivalence classes of projections is achieved by computing the range of a trace. The von Neumann algebra $\mathcal{U}(\Gamma, \sigma)$ and C^*-algebra $C_r^*(\Gamma, \sigma)$ have a canonical faithful finite trace τ, where

$$\tau(a) = \langle a\delta_1, \delta_1\rangle_{\ell^2(\Gamma)},$$

where δ_γ is the basis of $\ell^2(\Gamma)$. If Tr denotes the standard trace on bounded operators in an ∞-dimensional separable Hilbert space $\mathcal{H}$, then we obtain a trace

$$\mathrm{tr} = \tau \otimes \mathrm{Tr} : \mathrm{Proj}(C_r^*(\Gamma, \sigma) \otimes \mathcal{K}) \to \mathbb{R}.$$

This induces a trace on the K-group

$$[\mathrm{tr}] : K_0(C_r^*(\Gamma, \sigma))) \to \mathbb{R}$$

with

$$\mathrm{tr}(\mathrm{Proj}(C_r^*(\Gamma, \sigma))) = [\mathrm{tr}](K_0(C_r^*(\Gamma, \sigma))) \cap [0, 1].$$

The result quoted above in (3.25), counting the energy gaps in our hyperbolic model, can then be derived from the following result proved in [**25**].

THEOREM 3.1. *Let Γ be a cocompact Fuchsian group of signature $(g : \nu_1, \ldots, \nu_n)$ and σ be a multiplier on Γ with flux θ. Then the range of the trace is,*

$$(3.27) \qquad [\mathrm{tr}](K_0(C_r^*(\Gamma, \sigma))) = \mathbb{Z} + \theta\mathbb{Z} + \sum_j \frac{1}{\nu_j}\mathbb{Z}.$$

Here the flux is again given by the pairing $\theta = \langle [\sigma], [\Gamma] \rangle$, where $[\Gamma] = \frac{[\Sigma_{g'}]}{\#G}$ is the fundamental class of the group $\Gamma = \Gamma(g; \nu_1, \ldots, \nu_n)$ and g' is given by the formula (3.5).

The Baum–Connes conjecture holds for the Fuchsian groups $\Gamma = \Gamma(g; \nu_1, \ldots, \nu_n)$, and one can compute the K-theory of the C^*-algebra $C_r^*(\Gamma, \sigma)$ in terms of the orbifold K-theory of $\Sigma(g; \nu_1, \ldots, \nu_n)$. This uses a Morita equivalence $(A \otimes C_0(G)) \rtimes \Gamma \simeq C_0(\Gamma \backslash G, \mathcal{E})$, where $\mathcal{E} = A \times_\Gamma G \to \Gamma \backslash G$, in the case where $G = \mathrm{PSL}(2, \mathbb{R})$. Without the twisting by σ, one can identify

$$K_\bullet(C_r^*(\Gamma)) \cong K^\bullet_{SO(2)}(P(g; \nu_1, \ldots, \nu_n)),$$

where $P(g; \nu_1, \ldots, \nu_n)$ is the frame bundle $\Gamma \backslash \mathrm{PSL}(2, \mathbb{R})$. The result can be identified with the orbifold K-theory

$$K^\bullet_{orb}(\Sigma(g; \nu_1, \ldots, \nu_n)) \cong \begin{cases} \mathbb{Z}^{2-n+\sum \nu_j} & \bullet = even \\ \mathbb{Z}^{2g} & \bullet = odd \end{cases}$$

In the twisted case, one still has the equivalence $C_0(\Gamma \backslash G, \mathcal{E}) \simeq C_0(\Gamma \backslash G, \mathcal{E}_\sigma)$ when the class $\delta(\sigma) = 0$, where $\delta : H^2(\Gamma, U(1)) \to H^3(\Gamma, \mathbb{Z})$ is a surjection coming from the long exact sequence of $1 \to \mathbb{Z} \hookrightarrow \mathbb{R} \xrightarrow{\exp(2\pi i \cdot)} U(1) \to 1$.

The computation of the range of the trace (3.27) then follows from an index theorem. Let $\mathcal{E}$ be an orbifold vector bundle over $\Sigma = \Sigma(g; \nu_1, \ldots, \nu_n)$, and $[\mathcal{E}] \in K^\bullet_{orb}(\Sigma)$. Let $\tilde{\partial}^+_\mathcal{E}$ be the twisted Dirac operator on the universal cover $\mathbb{H}$. For $\nabla^2 = i\omega$ the magnetic field, the operator $\tilde{\partial}^+_\mathcal{E} \otimes \nabla$ commutes with the projective action of (Γ, σ). There is an analytic index

$$(3.28) \qquad \mathrm{ind}_{(\Gamma, \sigma)} \tilde{\partial}^+_\mathcal{E} \otimes \nabla \in K_0(C_r^*(\Gamma, \sigma)),$$

which is the image under the (twisted) Kasparov map

$$\mu_\sigma([\mathcal{E}]) = \mathrm{ind}_{(\Gamma, \sigma)} \tilde{\partial}^+_\mathcal{E} \otimes \nabla.$$

To compute the range of the trace $[\mathrm{tr}] : K_0(C_r^*(\Gamma, \sigma)) \to \mathbb{R}$ one computes then the index

$$(3.29) \qquad \mathrm{Ind}_{L^2}\left(\tilde{\partial}^+_\mathcal{E} \otimes \nabla \right) = [\mathrm{tr}]\left(\mathrm{ind}_{(\Gamma, \sigma)} \tilde{\partial}^+_\mathcal{E} \otimes \nabla \right).$$

We have

$$\mathrm{Ind}_{L^2}\left(\tilde{\partial}^+_\mathcal{E} \otimes \nabla \right) = \frac{1}{2\pi} \int_\Sigma \hat{A} \, \mathrm{tr}(e^{R^\mathcal{E}}) e^\omega.$$

Since Σ is of real dimension 2, this formula reduces to

$$\frac{\mathrm{rank}\mathcal{E}}{2\pi} \int_\Sigma \omega + \frac{1}{2\pi} \int_\Sigma \mathrm{tr}(R^\mathcal{E}).$$

The first term is computed by

$$\frac{\mathrm{rank}(\mathcal{E})}{2\pi} \int_\Sigma \omega = \frac{\mathrm{rank}(\mathcal{E})}{2\pi \#G} \int_{\Sigma_{g'}} \omega = \mathrm{rank}(\mathcal{E}) \langle [\sigma], [\Gamma] \rangle \in \theta\mathbb{Z},$$

while the second term is computed by the Kawasaki index theorem for orbifolds

$$\mathbb{Z} \ni \mathrm{ind}(\partial_{\mathcal{E}}^{+}) = \frac{1}{2\pi} \int_{\Sigma} \mathrm{tr}(R^{\mathcal{E}}) + \frac{1}{2\pi} \sum_{i=1}^{n} \frac{\beta_i}{\nu_i},$$

where (β_i, ν_i) are the Seifert invariants of $\mathcal{E}$. This implies that

$$\frac{1}{2\pi} \int_{\Sigma} \mathrm{tr}(R^{\mathcal{E}}) \in \mathbb{Z} + \sum_{i=1}^{n} \frac{1}{\nu_i} \mathbb{Z}.$$

Rieffel, and Pimsner and Voiculescu established analogous results in the case $\Gamma = \mathbb{Z}^2$. The result in the case of torsion-free Fuchsian groups was established in [5]. In more recent work, Mathai generalized this result to discrete subgroups of rank 1 groups and to all amenable groups, and more generally whenever the Baum–Connes conjecture with coefficients holds for the discrete group, [26]. By contrast, the behavior of spectral gaps when the flux is irrational is still mysterious. The problem can be formulated in terms of the following conjecture (also known as the "generalized ten Martini problem"), cf. [24] [25].

CONJECTURE 3.1. *Let Γ be a cocompact Fuchsian group and σ be a multiplier on Γ. If the flux θ is irrational, then there is a $V \in \mathbb{C}(\Gamma, \sigma)$ such that $H_{\sigma, V}$ has an infinite number of gaps in its spectrum.*

It is not yet known if *any* gaps exist at all in this case! However, using Morse–type potentials, Mathai and Shubin [27] proved that there is an arbitrarily large number of gaps in the spectrum of magnetic Schrödinger operators on covering spaces, (*i.e.* in the continuous model).

Recent work of Dodziuk, Mathai, and Yates [14] shows another interesting property of the spectrum, namely the fact that all L^2 eigenvalues of the Harper operators of surface groups Γ are *algebraic numbers*, whenever the multiplier is algebraic, that is, when $[\sigma] \in H^2(\Gamma, \overline{\mathbb{Q}}/\mathbb{Z})$. In fact the same result remains true when adding potentials V in $\overline{\mathbb{Q}}(\Gamma, \sigma)$ to the Harper operator.

4. Hall conductance

We finally come to a discussion of the quantization of the Hall conductance. This will follow again from a topological argument, and index theorem, as in the Bellissard case, but in our setting with hyperbolic geometry. We will derive, from our model, a formula for the Hall conductance in terms of values of the orbifold Euler characteristic, and we will compare the results with experimentally observed values.

A smooth subalgebra. We will consider a cyclic cocycle associated to the Connes–Kubo formula for the conductance, which will be defined in terms of certain derivations. For this reason, we need to introduce a smooth subalgebra, namely, a dense involutive subalgebra of the algebra of observables $C_r^*(\Gamma, \sigma)$. This subalgebra contains $\mathbb{C}(\Gamma, \sigma)$ and is contained in the domain of definition of the derivations. It contains the spectral projection P_E, when the Fermi level is in a gap of the energy spectrum. Moreover, it satisfies the following two key properties.

(1) The inclusion $\mathcal{R} \subset C_r^*(\Gamma, \sigma)$ induces an isomorphism in K-theory.
(2) Polynomial growth group cocycles on Γ define cyclic cocycles on $\mathbb{C}(\Gamma, \sigma)$ that extend continuously to $\mathcal{R}$.

$\mathcal{R}$ is defined as follows. Consider an operator D defined as

$$D\delta_\gamma = \ell(\gamma)\delta_\gamma \; \forall \gamma \in \Gamma,$$

where $\ell(\gamma)$ denotes the word length of γ. Let $\delta = \mathrm{ad}(D)$ denote the commutator $[D, \cdot]$. Then δ is an unbounded, but closed derivation on $C_r^*(\Gamma, \sigma)$. Define

$$\mathcal{R} := \bigcap_{k \in \mathbb{N}} \mathrm{Dom}(\delta^k).$$

It is clear that $\mathcal{R}$ contains $\delta_\gamma \; \forall \gamma \in \Gamma$ and so it contains $\mathbb{C}(\Gamma, \sigma)$. Hence it is dense in $C_r^*(\Gamma, \sigma)$. It is not hard to see that $\mathcal{R}$ is closed under the holomorphic functional calculus, and therefore by a result of Connes, property (1) above holds, and by equation (3.26), $P_E \in \mathcal{R}$.

Until now, we have not used any special property of the group Γ. But now assume that Γ is a surface group. Then it follows from a result of [20], [22] that there is a $k \in \mathbb{N}$ and a positive constant C' such that for all $f \in \mathbb{C}(\Gamma, \sigma)$, one has the *Haagerup inequality*

(4.1) $$\|f\| \le C' \, \nu_k(f),$$

where $\|f\|$ denotes the operator norm of the operator on $\ell^2(\Gamma)$ given by left convolution by f, and the bound $\nu_k(f)$ is given in terms of the L^2 norms of f and of $(1 + l^2)^{s/2}f$, for all $0 \le s \le k$. Using this, it is routine to show that property (2) holds.

Notice that the spectral projections onto gaps in the Hamiltonian H belong to the algebra of observables $\mathcal{R}$, for any choice of electric potential V.

Cyclic cocycles. Cyclic cohomology was introduced by Connes in [11]. It is a main source of invariants of noncommutative spaces, obtained by the pairing of cyclic cocycles with K-theory. Cyclic cocycles are also called multilinear traces, and the word cyclic refers to invariance under the cyclic group $\mathbb{Z}/(n+1)\mathbb{Z}$ acting on the slots of the Cartesian product. Namely, t is a cyclic n-cocycle if

$$t : \mathcal{R} \times \mathcal{R} \cdots \times \mathcal{R} \to \mathbb{C}$$

satisfies the cyclic condition

$$t(a_0, a_1, \ldots, a_n) = t(a_n, a_0, a_1, \ldots, a_{n-1}) = \cdots = t(a_1, \ldots, a_n, a_0),$$

and the cocycle condition

$$t(aa_0, a_1, \ldots, a_n) - t(a, a_0a_1, \ldots, a_n) \cdots (-1)^{n+1} t(a_n a, a_0, \ldots, a_{n-1}) = 0.$$

For instance, a *cyclic 0-cocycle* is just a trace. In fact, in this case, the condition it satisfies is $t(ab) = t(ba)$. A *cyclic 1-cocycle* satisfies $t(a, b) = t(b, a)$ and $t(ab, c) - t(a, bc) + t(ca, b) = 0$, and a *cyclic 2-cocycle* satisfies

$$t(a, b, c) = t(c, a, b) = t(b, c, a) \quad \text{and}$$

$$t(ab, c, d) - t(a, bc, d) + t(a, b, cd) - t(da, b, c) = 0.$$

Conductance cocycle. A formula for the Hall conductance is obtained from transport theory. In the case of $\Gamma = \mathbb{Z}^2$, the current density in e_1 direction corresponds to the functional derivative δ_1 of H_σ by A_1, the corresponding component of the magnetic potential. The expected value of current is the given by $\text{tr}(P\delta_1 H)$ for a state P of the system. Using $\partial_t P = i[P, H]$ and $\partial_t = \frac{\partial A_2}{\partial t} \times \delta_2$, where $e_2 \perp e_1$, one gets

$$i\text{tr}(P[\partial_t P, \delta_1 P]) = -iE_2\text{tr}(P[\delta_2 P, \delta_1 P]),$$

where the electrostatic potential has been gauged away, leaving $\mathbf{E} = -\frac{\partial \mathbf{A}}{\partial t}$. Because the charge carriers are Fermions, two different charge carriers must occupy different quantum eigenstates of the Hamiltonian H. In the zero temperature limit, charge carriers occupy all levels below the Fermi level, so that we can set $P = P_F$ in the formula above. This gives the Kubo formula for the conductance

$$\sigma_H = \text{tr}(P_F[\delta_1 P_F, \delta_2 P_F]).$$

This argument can be generalized to our setting, keeping into account the fact that, in our model, by effect of the strong multi-electron interaction, to a moving elector the directions $\{e_1, e_2\}$ appear split into $\{e_i, e_{i+g}\}_{i=1,\ldots,g}$ corresponding to a_i, b_i, for some lattice in the hyperbolic plane. The following is a general mathematical formulation of the result.

Given a 1-cocycle a on the discrete group Γ, $i.e.$

$$a(\gamma_1\gamma_2) = a(\gamma_1) + a(\gamma_2) \qquad \forall \gamma_1, \gamma_2 \in \Gamma,$$

one can define a linear functional δ_a on the twisted group algebra $\mathbb{C}(\Gamma, \sigma)$

$$\delta_a(f)(\gamma) = a(\gamma)f(\gamma).$$

Then one verifies that δ_a is a derivation:

$$\delta_a(fg)(\gamma) = a(\gamma)fg(\gamma)$$
$$= a(\gamma)\sum_{\gamma=\gamma_1\gamma_2} f(\gamma_1)g(\gamma_2)\sigma(\gamma_1, \gamma_2)$$
$$= \sum_{\gamma=\gamma_1\gamma_2} \Big(a(\gamma_1) + a(\gamma_2)\Big)f(\gamma_1)g(\gamma_2)\sigma(\gamma_1, \gamma_2)$$
$$= \sum_{\gamma=\gamma_1\gamma_2} \Big(\delta_a(f)(\gamma_1)g(\gamma_2)\sigma(\gamma_1, \gamma_2) + f(\gamma_1)\delta_a(g)(\gamma_2)\sigma(\gamma_1, \gamma_2)\Big)$$
$$= (\delta_a(f)g)(\gamma) + (f\delta_a g)(\gamma).$$

In the case of a Fuchsian group Γ, the first cohomology $H^1(\Gamma, \mathbb{Z})$ of the group Γ is a free Abelian group of rank $2g$, where g is the genus of $\Gamma\backslash\mathbb{H}$. The cohomology $H^1(\Gamma, \mathbb{R})$ is in fact a symplectic vector space, and we can assume that $\{a_j, b_j\}_{j=1,\ldots,g}$ is a symplectic basis.

We denote δ_{a_j} by δ_j and δ_{b_j} by δ_{j+g}. These derivations give rise to a cyclic 2-cocycle on the twisted group algebra $\mathbb{C}(\Gamma, \sigma)$,

$$(4.2) \qquad \text{tr}_K(f_0, f_1, f_2) = \sum_{j=1}^{g} \text{tr}(f_0(\delta_j(f_1)\delta_{j+g}(f_2) - \delta_{j+g}(f_1)\delta_j(f_2))).$$

tr_K is called the *conductance 2-cocycle*.

Let P_E denote denote the spectral projection associated to the Fermi level, $i.e.$ $P_E = \chi_{(-\infty, E]}(H)$. Then, in the zero temperature limit, the Hall conductance is given by

$$\sigma_E = \text{tr}_K(P_E, P_E, P_E).$$

Quantum adiabatic limit. We recall briefly the justification of (4.2) in terms of the quantum adiabatic limit for a slowly varying time dependent Hamiltonian, *cf.* [7].

If $H(s)$ is a smooth family of self-adjoint Hamiltonians and $P(s)$ are spectral projections on a gap in the spectrum of $H(s)$, then

$$X(s) = \frac{1}{2\pi i} \oint_C R(z,s)\partial_s P(s)R(z,s)dz,$$

with $R(z,s) = (H(s) - z)^{-1}$, satisfies the commutation relations

$$[\partial_s P(s), P(s)] = [H(s), X(s)].$$

The quantum adiabatic limit theorem (*cf.* [2]) then shows that the adiabatic evolution approximates well the physical evolution, for large values of the adiabatic parameter $\tau \to \infty$, via an estimate of the form

$$\|(U_\tau(s) - U_a(s))P(0)\| \leq$$

$$\frac{1}{\tau} \max_{s\in[0,\infty)} \{2\|X(s)P(s)\| + \|\partial_s(X(s)P(s))P(s)\|\}.$$

Here the physical evolution satisfies

$$i\partial_s U_\tau(s) = \tau H(s)U_\tau(s),$$

$U_\tau(0) = 1$, where $s = t/\tau$ is a scaled time, and the adiabatic evolution is defined by the equation

$$P(s) = U_a(s)P(0)U_a(s)^*$$

with $U_a(0) = 1$.

In our setting, the functional derivative $\delta_k H$, with respect to a component A_k of the magnetic potential, gives a current density J_k. Its expectation value in a state described by a projection P on a gap in the spectrum of the Hamiltonian is then computed by $\mathrm{tr}(P\delta_k H)$. In the quantum adiabatic limit, one can replace $\delta_k H$ with $\delta_k H_a$, where the adiabatic Hamiltonian H_a satisfies

$$i\partial_s U_a(s) = \tau H_a(s)U_a(s)$$

and the equation of motion

$$[H_a(s), P(s)] = \frac{i}{\tau}\partial_s P(s).$$

This implies that the relation

(4.3) $$\mathrm{tr}(P[\partial_t P, \delta_k P]) = i\mathrm{tr}(\delta_k(PH_a)) - i\mathrm{tr}(P\delta_k H_a).$$

We make some simplifying assumptions. If the trace is invariant under variations of A_k, then the first term in the right hand side of (4.3) vanishes. We also assume that the only time dependence of H and P is in the adiabatic variation of a component A_j distinct from A_k, and we work in the Landau gauge, so that the electrostatic potential vanishes and the electric field is given by $\mathbf{E} = -\partial \mathbf{A}/\partial t$. Then we have $\partial_t = -E_j\delta_j$, so that the expectation of the current J_k is given by

$$\mathrm{tr}(P\delta_k H) = i\mathrm{tr}(P[\partial_t P, \delta_k P])$$

$$= -iE_j\mathrm{tr}(P[\delta_j P, \delta_k P]),$$

hence the conductance for a current in the k direction induced by an electric field in the j direction is given by $-i\text{tr}(P[\delta_j P, \delta_k P])$. The analytic aspects of this formal argument can be made rigorous following the techniques used in [**33**].

Area cocycle. Our conclusion above, as in the case of the integer Hall effect, is that one can compute the Hall conductance by evaluating a certain cyclic cocycle on a projection, namely on some element in K-theory. It is often the case that, in order to compute the pairing of a cyclic cocycle with K-theory, one can simplify the problem by passing to another cocycle in the same cohomology class, *i.e.* that differs by a coboundary. This is what will happen in our case.

We introduce another cyclic cocycle, which has a more direct geometric meaning. On $G = \text{PSL}(2, \mathbb{R})$, there is an *area cocycle* (*cf.* [**12**]). This is the 2-cocycle

$$C : G \times G \to \mathbb{R}$$

$$C(\gamma_1, \gamma_2) = \text{(oriented) hyperbolic area of the}$$

$$\text{geodesic triangle with}$$

$$\text{vertices at} (z_0, \gamma_1^{-1} z_0, \gamma_2 z_0), \quad z_0 \in \mathbb{H}$$

The restriction of this cocycle to a discrete subgroup $\Gamma \subset \text{PSL}(2, \mathbb{R})$ gives the area group cocycle on Γ. This in turn defines a cyclic 2-cocycle on $\mathbb{C}(\Gamma, \sigma)$ by

$$(4.4) \qquad \text{tr}_C(f_0, f_1, f_2) = \sum_{\gamma_0 \gamma_1 \gamma_2 = 1} f_0(\gamma_0) f_1(\gamma_1) f_2(\gamma_2) C(\gamma_1, \gamma_2) \sigma(\gamma_1, \gamma_2).$$

Since C is (polynomially) bounded, tr_C can be shown to extend to the smooth subalgebra $\mathcal{R}$.

Comparison. Two cyclic 2-cocycles t_1 and t_2 differ by a coboundary (that is, they define the same cyclic cohomology class) iff

$$t_1(a_0, a_1, a_2) - t_2(a_0, a_1, a_2) = \lambda(a_0 a_1, a_2) - \lambda(a_0, a_1 a_2) + \lambda(a_2 a_0, a_1),$$

where λ is a cyclic 1-cocycle.

As in [**5**], [**24**], the difference between the conductance cocycle tr_K and the area cocycle tr_C can be evaluated in terms of the difference between the hyperbolic area of a geodesic triangle and the Euclidean area of its image under the Abel-Jacobi map. This difference can be expressed as a sum of three terms

$$(4.5) \qquad U(\gamma_1, \gamma_2) = h(\gamma_2^{-1}, 1) - h(\gamma_1^{-1}, \gamma_2) + h(1, \gamma_1),$$

where each term is a difference of line integrals, one along a geodesic segment in $\mathbb{H}$ and one along a straight line in the Jacobian variety. The cocycles correspondingly differ by

$$\text{tr}_K(f_0, f_1, f_2) - \text{tr}_C(f_0, f_1, f_2) = \sum_{\gamma_0 \gamma_1 \gamma_2 = 1} f_0(\gamma_0) f_1(\gamma_1) f_2(\gamma_2) U(\gamma_1, \gamma_2) \sigma(\gamma_1, \gamma_2).$$

This expression can be written as $\lambda(f_0 f_1, f_2) - \lambda(f_0, f_1 f_2) + \lambda(f_2 f_0, f_1)$ where

$$\lambda(f_0, f_1) = \sum_{\gamma_0 \gamma_1 = 1} f_0(\gamma_0) f_1(\gamma_1) h(1, \gamma_1) \sigma(\gamma_0, \gamma_1),$$

with h as in (4.5).

Thus, the cocycles tr_K and tr_C differ by a coboundary. Since they are cohomologous, tr_K and tr_C induce the same map on K-theory.

Values of the Hall conductance. The problem of deriving the values of the Hall conductance is now reduced to computing the pairing of the area cyclic 2-cocycle with K-theory. The computation is again done through an index theorem. This time the appropriate framework is (a twisted version of) the Connes–Moscovici higher index theorem [13]. We have the following result, [24]:

THEOREM 4.1. *The values of the Hall conductance are given by the twisted higher index formula*

$$(4.6) \qquad \mathrm{Ind}_{c,\Gamma,\sigma}(\bar{\partial}_{\mathcal{E}}^{+} \otimes \nabla) = \frac{1}{2\pi\#G} \int_{\Sigma_{g'}} \hat{A}\,\mathrm{tr}(e^{R_{\mathcal{E}}})e^{\omega}u_c,$$

where $\omega = d\eta$ is the 2-form of the magnetic field, $\nabla^2 = i\omega$, c is a cyclic cocycle c and u_c is its lift, as in [13], to a 2-form on $\Sigma_{g'}$.

Again, since Σ is 2-dimensional, the formula (4.6) reduces to just the term

$$(4.7) \qquad \mathrm{Ind}_{c,\Gamma,\sigma}(\bar{\partial}_{\mathcal{E}}^{+} \otimes \nabla) = \frac{\mathrm{rank}\mathcal{E}}{2\pi\#G} \int_{\Sigma_{g'}} u_c.$$

Notice that, while it seems at first that in (4.7) all dependence on the magnetic field has disappeared in this formula, in fact it is still present through the orbifold vector bundle $\mathcal{E}$ that corresponds (through Baum–Connes) to the class of the spectral projection $P_{\mathcal{E}}$ in $K_0(C_r^*(\Gamma,\sigma))$, of the Fermi level.

When c is the area cocycle, the corresponding 2-form u_c is just the hyperbolic volume form, hence the right hand side of (4.7) is computed by the Gauss–Bonnet formula $\int_{\Sigma_{g'}} u_c = 2\pi(2g' - 2)$, so that

$$(4.8) \qquad \frac{\mathrm{rank}(\mathcal{E})}{2\pi\#G} \int_{\Sigma_{g'}} u_c = \mathrm{rank}(\mathcal{E})\frac{(2g' - 2)}{\#G} = -\mathrm{rank}(\mathcal{E})\chi_{orb}(\Sigma) \in \mathbb{Q}$$

which yields an integer multiple of the orbifold Euler characteristic.

The conclusion is that, in our model, the Hall conductance takes rational values that are integer multiples of orbifold Euler characteristics, Rational values of the conductance

$$\sigma_H = \mathrm{tr}^K(P_F, P_F, P_F) = \mathrm{tr}^C(P_F, P_F, P_F) \in \mathbb{Z}\chi_{orb}(\Sigma).$$

Discussion of the model. A first important observation, in terms of physical predictions, is that our model of FQHE predicts the existence of an *absolute lower bound* on the fractional values of the Hall conductance. The lower bound is imposed by the orbifold geometry, and does not have an analog in other theoretical models, hence it appears to be an excellent possible experimental test of the validity of our theoretical model. The lower bound is obtained from the Hurwitz theorem, which states that the maximal order of a finite group G acting by isometries on a smooth Riemann surface $\Sigma_{g'}$ is $\#G = 84(g' - 1)$. This imposes the constraint on the possible quantum Hall fractions:

$$\phi \geq \frac{2(g' - 1)}{84(g' - 1)} = \frac{1}{42}.$$

The lower bound is realized by $1/42 = -\chi_{orb}(\Sigma(0; 2, 3, 7))$.

A key advantage of our hyperbolic model is that it treats the FQHE within the same framework developed by Bellissard et al. for the IQHE, with hyperbolic geometry replacing Euclidean geometry, to account for the effect of electron correlation, while remaining formally within a single particle model.

The fractions for the Hall conductance that we get are obtained from an equivariant index theorem and are thus *topological* in nature. Consequently, the Hall conductance is seen to be stable under small deformations of the Hamiltonian. Thus, this model can be generalized to systems with disorder as in [6], and then the hypothesis that the Fermi level is in a spectral gap of the Hamiltonian can be relaxed to the assumption that it is in a gap of extended states. This is a necessary step in order to establish the presence of plateaux.

In fact, this solves the apparent paradox that we still have a FQHE, even though the Hamiltonian $H_{\sigma,V}$ may not have any spectral gaps. The reason is that, as explained in [6], the domains of the cyclic 2-cocycles tr_C and tr_K are in fact larger than the smooth subalgebra $\mathcal{R}$. More precisely, there is a *-subalgebra $\mathcal{A}$ such that $\mathcal{R} \subset \mathcal{A} \subset \mathcal{U}(\Gamma,\sigma)$ and $\mathcal{A}$ is contained in the domains of tr_C and tr_K. $\mathcal{A}$ is closed under the Besov space functional calculus, and the spectral projections P_E of the Hamiltonian $H_{\sigma,V}$ that lie in $\mathcal{A}$ are called *gaps in extended states*. They include all the spectral projections onto gaps in the energy spectrum, but contain many more spectral projections. In particular, even though the Hamiltonian $H_{\sigma,V}$ may not have any spectral gaps, it may still have *gaps in extended states*. The results extend in a straightforward way to the case with disorder, where one allows the potential V to be random, *cf.* [6].

Let us discuss the comparison with experimental data on the quantum Hall effect. Our model recovers the observed fractions (including the elusive 1/2). Table 1 below illustrates how low genus orbifolds with a small number of cone points are sufficient to recover many observed fractions. In this first table, we consider experimentally observed fractions, which we recover in our model. Notice how fractions like 1/3, 2/5, 2/3, which experimentally appear with a wider and more clearly marked plateau, also correspond to the fractions realized by a larger number of orbifolds (we only checked the number of solutions for small values $\nu_j \leq 20$, $n = 3$, $g = 0$, and $\phi < 1$). These observations should be compared with the experimental data, *cf. e.g.* [29] [8].

Regarding the varying width of the plateaux, what appears promising in Table 1 is the fact that the fractions that are more easily observed experimentally, *i.e.* those that appear with a larger and more clearly marked plateau (*cf. e.g.* [29], [8]), also correspond to orbifold Euler characteristics that are realized by a large number of orbifolds. We can derive a corresponding qualitative graph of the widths, to be compared with the experimental ones. Table 2 shows how to obtain some experimentally observed fractions with $\phi > 1$ (without counting multiplicities).

The main limitation of our model is that it seems to predict too many fractions, which at present do not seem to correspond to experimentally observed values. To our knowledge, however, this is also a limitation in the other theoretical models available in the literature. Another serious limitation is the fact that this model does not explain why even denominator fractions are more difficult to observe than odd ones. In fact, even for small number of cone points and low genus, one obtains a large number of orbifold Euler characteristics with even denominator, which are not justified experimentally. On the occurrence of even denominators in the fractional quantum Hall effect experiments, *cf.* [32] [10] [15]. Table 3 provide a list of odd and even denominator fractions predicted by our model, using genus zero orbifolds with three cone points.

Questions and directions. We have discussed the transition from classical Bloch theory to noncommutative Bloch theory, as effect of the presence of a magnetic field. In particular, we have seen that the Brillouin zone becomes a noncommutative space. It would be interesting to investigate, using this point of view based on noncommutative geometry, what happens to the algebro-geometric theory of Fermi curves and periods. Another natural question related to the results discussed here is whether a Chern–Simons approach to the fractional quantum Hall effect may give a different justification for the presence of the orbifolds $\Sigma(g; \nu_1, \ldots, \nu_n)$. In fact, these and their symmetric products appear as spaces of vortices in Chern–Simons (or Seiberg–Witten) theory.

Tables 1 and 2: experimental fractions.

experimental	$g = 0 \ n = 3$	experimental	$g = 0 \ n = 3$
1/3	$\Sigma(0; 3, 6, 6)$	2/5	$\Sigma(0; 5, 5, 5)$
	$\Sigma(0; 4, 4, 6)$		$\Sigma(0; 4, 4, 10)$
	$\Sigma(0; 3, 4, 12)$		$\Sigma(0; 3, 6, 10)$
	$\Sigma(0; 2, 12, 12)$		$\Sigma(0; 3, 6, 10)$
	$\Sigma(0; 2, 10, 15)$		$\Sigma(0; 3, 5, 15)$
	$\Sigma(0; 2, 9, 18)$		$\Sigma(0; 2, 20, 20)$
2/3	$\Sigma(0; 9, 9, 9)$	3/5	$\Sigma(0; 5, 10, 10)$
	$\Sigma(0; 8, 8, 12)$		$\Sigma(0; 6, 6, 15)$
	$\Sigma(0; 6, 12, 12)$		$\Sigma(0; 4, 12, 15)$
	$\Sigma(0; 6, 10, 15)$		$\Sigma(0; 4, 10, 20)$
	$\Sigma(0; 6, 9, 18)$		
	$\Sigma(0; 5, 15, 15)$		
	$\Sigma(0; 5, 12, 20)$		
4/9	$\Sigma(0; 3, 9, 9)$	5/9	$\Sigma(0; 6, 6, 9)$
	$\Sigma(0; 4, 4, 18)$		$\Sigma(0; 4, 9, 12)$
	$\Sigma(0; 3, 6, 18)$		$\Sigma(0; 3, 18, 18)$
4/5	$\Sigma(0; 15, 15, 15)$	3/7	$\Sigma(0; 4, 4, 14)$
	$\Sigma(0; 12, 15, 20)$		$\Sigma(0; 3, 6, 14)$
	$\Sigma(0; 10, 20, 20)$		
4/7	$\Sigma(0; 7, 7, 7)$	5/7	$\Sigma(0; 7, 14, 14)$

experimental	$g = 0$ or $g = 1$
8/5	$\Sigma(0; 2, 4, 4, 5, 5)$
11/7	$\Sigma(0; 2, 2, 7, 7, 7)$
14/9	$\Sigma(1; 3, 9)$
4/3	$\Sigma(1; 3, 3)$
7/5	$\Sigma(0; 5, 5, 10, 10)$
10/7	$\Sigma(0; 7, 7, 7, 7)$
13/9	$\Sigma(0; 6, 6, 9, 9)$
5/2	$\Sigma(1; 6, 6, 6)$

Table 3: predicted fractions.

odd	$g = 0\ n = 3$	even	$g = 0\ n = 3$
8/15	$\Sigma(0; 5, 6, 10)\ \Sigma(0; 5, 5, 15)$ $\Sigma(0; 4, 6, 20)\ \Sigma(0; 3, 15, 15)$ $\Sigma(0; 3, 12, 20)$	1/2	$\Sigma(0; 6, 6, 6)\ \Sigma(0; 5, 5, 10)$ $\Sigma(0; 4, 8, 8)\ \Sigma(0; 4, 6, 12)$ $\Sigma(0; 4, 5, 20)\ \Sigma(0; 3, 12, 12)$ $\Sigma(0; 3, 10, 15)\ \Sigma(0; 3, 9, 18)$
7/9	$\Sigma(0; 12, 12, 18)\ \Sigma(0; 10, 15, 18)$ $\Sigma(0; 9, 18, 18)$	1/4	$\Sigma(0; 4, 4, 4)\ \Sigma(0; 3, 4, 6)$ $\Sigma(0; 3, 3, 12)\ \Sigma(0; 2, 8, 8)$ $\Sigma(0; 3, 3, 12)\ \Sigma(0; 2, 8, 8)$ $\Sigma(0; 2, 6, 12)\ \Sigma(0; 2, 5, 20)$
11/21	$\Sigma(0; 6, 6, 7)\ \Sigma(0; 4, 7, 12)$ $\Sigma(0; 3, 14, 14)$	7/12	$\Sigma(0; 6, 8, 8)\ \Sigma(0; 6, 6, 12)$ $\Sigma(0; 5, 6, 20)\ \Sigma(0; 4, 12, 12)$ $\Sigma(0; 4, 10, 15)\ \Sigma(0; 4, 9, 18)$
16/21	$\Sigma(0; 12, 12, 14)\ \Sigma(0; 10, 14, 15)$ $\Sigma(0; 9, 14, 18)$		
11/15	$\Sigma(0; 10, 10, 15)\ \Sigma(0; 10, 12, 12)$ $\Sigma(0; 9, 10, 18)\ \Sigma(0; 6, 20, 20)$		

Acknowledgment. The first author is supported by the Humboldt Foundation Sofja Kovalevskaya Award. The second author is supported by the Australian Research Council.

References

[1] J. Avron, R. Seiler, B. Simon, *Charge deficiency, charge transport and comparison of dimensions*, Comm. Math. Phys. Vol.**159** (1994), no. 2, 399–422.

[2] J. Avron, R. Seiler, I. Yaffe, *Adiabatic theorems and applications to the integer quantum Hall effect*, Commun. Math. Phys. Vol.**110** (1987) 33–49.

[3] J. Bellissard, A. van Elst, H. Schulz-Baldes, *The noncommutative geometry of the quantum Hall effect*, J.Math.Phys. **35** (1994) 5373–5451.

[4] J. Bellissard, *The noncommutative geometry of aperiodic solids*, in "Geometric and topological methods for quantum field theory (Villa de Leyva, 2001)", 86–156, World Scientific, 2003.

[5] A. Carey, K. Hannabuss, V. Mathai, P. McCann, *Quantum Hall Effect on the hyperbolic plane*, Commun. Math. Physics, Vol.**190**, no. 3 (1998) 629–673.

[6] A. Carey, K. Hannabuss, V. Mathai, *Quantum Hall Effect on the Hyperbolic Plane in the presence of disorder*, Letters in Mathematical Physics, Vol. **47** (1999) 215–236.

[7] A. Carey, K. Hannabuss, V. Mathai, *Quantum Hall effect and noncommutative geometry*, arXiv:math.OA/0008115.

[8] T. Chakraborti, P. Pietilänen, *The Quantum Hall Effects*, Second Edition, Springer 1995.

[9] T.-S. Choy, J. Naset, J. Chen, S. Hershfield, and C. Stanton. *A database of fermi surface in virtual reality modeling language (vrml)*, Bulletin of The American Physical Society, **45**(1):L36 42, 2000.

[10] R.G.Clark, R.J.Nicholas, A.Usher, C.T.Foxon, J.J.Harris, *Surf.Sci.* **170** (1986) 141.

[11] A. Connes, *Non–commutative differential geometry*, Publ.Math. IHES, Vol.**62** (1985) 257–360.

[12] A. Connes, *Noncommutative geometry.* Academic Press, Inc., San Diego, CA, 1994.

[13] A. Connes, H. Moscovici, *Cyclic cohomology, the Novikov conjecture and hyperbolic groups*, Topology, Vol. **29** (1990) no. 3, 345–388.

[14] J. Dodziuk, V. Mathai, S. Yates, *Arithmetic properties of eigenvalues of generalized Harper operators on graphs*, arXiv math.SP/0311315

[15] J.P.Eisenstein, G.S.Boebinger, L.N.Pfeiffer, K.W.West, S.He, *Phys. Rev. Lett.* **68** (1992) 1383; S.Q. Murphy, J.P.Eisenstein, G.S.Boebinger, L.N.Pfeiffer, K.W.West, *Phys. Rev. Lett.* 72 (1994) 728.

[16] D. Gieseker, H. Knörrer, E. Trubowitz, *The geometry of algebraic Fermi curves*, Perspectives in Mathematics, Vol.**14**. Academic Press, 1993. viii+236 pp.

[17] D. Gieseker, H. Knörrer, E. Trubowitz, *An overview of the geometry of algebraic Fermi curves*, in "Algebraic geometry: Sundance 1988", 19–46, Contemp. Math. Vol.**116**, Amer. Math. Soc. 1991.

[18] M. Gruber, *Noncommutative Bloch theory*, J.Math.Phys. Vol.**42** (2001), no. 6, 2438–2465.

[19] E.H. Hall, *On a new action of the magnet on electric currents*, Amer. J. of Math. Vol.**287**, (1879) N.2.

[20] R. Ji, *Smooth dense subalgebras of reduced group C^*-algebras, Schwartz cohomology of groups, and cyclic cohomology*, J. Funct. Anal. **107** (1992), no. 1, 1–33.

[21] K. von Klitzing, G. Dorda, and M. Pepper, *New method for high-accuracy determination of the fine-structure constant based on quantized hall resistance*, Phys. Rev. Lett., Vol. **45** (1980) N.6, 494–497.

[22] Y. Kordyukov, V. Mathai and M.A. Shubin, *Equivalence of spectral projections in semi-classical limit and a vanishing theorem for higher traces in K-theory*, J.Reine Angew.Math. (Crelle), Vol.**581** (2005) 44 pages (to appear).

[23] B. Laughlin, *Quantized hall conductivity in two dimensions*, Phys. Rev. B, Vol.**23** (1981) 5232.

[24] M. Marcolli and V. Mathai, *Twisted index theory on good orbifolds, II: fractional quantum numbers*, Communications in Mathematical Physics, Vol.**217**, no.1 (2001) 55–87.

[25] M. Marcolli and V. Mathai, *Twisted index theory on good orbifolds, I: noncommutative Bloch theory*, Communications in Contemporary Mathematics, Vol.**1** (1999) 553–587.

[26] V. Mathai, *On positivity of the Kadison constant and noncommutative Bloch theory*, Tohoku Mathematical Publications, Vol.**20** (2001) 107–124.

[27] V. Mathai, M. Shubin, *Semiclassical asymptotics and gaps in the spectra of magnetic Schrödinger operators*, Geometriae Dedicata, Vol. **91**, no. 1, (2002) 155–173.

[28] R.G.MUΦ, *The fractional quantum Hall effect, Chern-Simons theory, and integral lattices*, in "Proceedings of the International Congress of Mathematicians", Vol. **1, 2** (Zrich, 1994), 75–105, Birkhäuser, 1995.

[29] H.L. Störmer, *Advances in solid state physics*, ed. P.Grosse, vol.**24**, Vieweg 1984.

[30] T. Sunada, *A discrete analogue of periodic magnetic Schrödinger operators*, Contemp. Math. Vol.**173** (1994) 283–299.

[31] D.J. Thouless, M. Kohmono, M.P. Nightingale, M. den Nijs, *Quantized Hall conductance in a two-dimensional periodic potential*, Phys. Rev. Lett. **49** (1982) N.6, 405–408.

[32] R.Willett, J.P.Eisenstein, H.L.Störmer, D.C.Tsui, A.C.Gossard, J.H.English, *Phys. Rev. Lett.* **59** (1987) 1776.

[33] J. Xia, *Geometric invariants of the quantum hall effect*, Commun. Math. Phys. Vol. **119** (1988), 29–50.

M. MARCOLLI: MAX–PLANCK INSTITUT FÜR MATHEMATIK, VIVATSGASSE 7, BONN, D-53111 GERMANY
E-mail address: marcolli@mpim-bonn.mpg.de

V. MATHAI: DEPARTMENT OF PURE MATHEMATICS, UNIVERSITY OF ADELAIDE, 5005 AUSTRALIA
E-mail address: vmathai@maths.adelaide.edu.au

Homological algebra for Schwartz algebras of reductive p-adic groups

Ralf Meyer

ABSTRACT. Let G be a reductive group over a non-Archimedean local field. Then the canonical functor from the derived category of smooth tempered representations of G to the derived category of all smooth representations of G is fully faithful. Here we consider representations on bornological vector spaces. As a consequence, if G is semi-simple, V and W are tempered irreducible representations of G, and V or W is square-integrable, then $\text{Ext}_G^n(V, W) \cong 0$ for all $n \geq 1$. We use this to prove in full generality a formula for the formal dimension of square-integrable representations due to Schneider and Stuhler.

1. Introduction

Let G be a linear algebraic group over a non-Archimedean local field whose connected component of the identity element is reductive; we briefly call such groups *reductive p-adic groups*. For the purposes of exposition, we assume throughout the introduction that the connected centre of G is trivial, although we treat groups with arbitrary centre in the main body of this article.

We are going to compare homological and cohomological computations for the Hecke algebra $\mathcal{H}(G)$ and the Harish-Chandra Schwartz algebra $\mathcal{S}(G)$. Our main result asserts that the derived category of $\mathcal{S}(G)$ is a full subcategory of the derived category of $\mathcal{H}(G)$. These derived categories incorporate a certain amount of functional analysis because $\mathcal{S}(G)$ is more than just an algebra. Before we discuss this, we sketch two purely algebraic applications of our main theorem.

Let $\text{Mod}_{\text{alg}}(G)$ be the category of smooth representations of G on complex vector spaces. We compute some extension spaces in this Abelian category. If both V and W are irreducible tempered representations and one of them is square-integrable, then $\text{Ext}_G^n(V, W) = 0$ for $n \geq 1$. If the local field underlying G has characteristic 0, this is proven by very different means in [20]. We get a more transparent proof that also works in prime characteristic.

The vanishing of $\text{Ext}_G^n(V, W)$ is almost trivial if V or W is supercuspidal because then V or W is both projective and injective in $\text{Mod}_{\text{alg}}(G)$. This is related to the fact that supercuspidal representations are isolated points in the admissible dual. Square-integrable representations are isolated points in the tempered dual. Hence they are projective and injective in an appropriate category $\text{Mod}(\mathcal{S}(G))$ of tempered smooth representations of G. Both $\text{Mod}(\mathcal{S}(G))$ and $\text{Mod}_{\text{alg}}(G)$ are full subcategories in a larger category $\mathsf{Mod}(G)$. That is, we have fully faithful functors

$$\mathsf{Mod}\big(\mathcal{S}(G)\big) \to \mathsf{Mod}(G) \leftarrow \mathsf{Mod}_{\mathrm{alg}}(G).$$

We will show that the induced functors between the derived categories,

$$\mathsf{Der}\big(\mathcal{S}(G)\big) \to \mathsf{Der}(G) \leftarrow \mathsf{Der}_{\mathrm{alg}}(G),$$

are still fully faithful. This contains the vanishing result for Ext as a special case.

Another application involves Euler characteristics for square-integrable representations. Let V be an irreducible square-integrable representation of G. By a theorem of Joseph Bernstein, any finitely generated smooth representation of G, such as V, has a finite type projective resolution $P_\bullet \to V$. Its Euler characteristic is defined as

$$\chi(V) := \sum (-1)^n [P_n] \in \mathrm{K}_0\big(\mathcal{H}(G)\big).$$

Since V is square-integrable, it is a projective $\mathcal{S}(G)$-module and therefore defines a class $[V] \in \mathrm{K}_0\big(\mathcal{S}(G)\big)$. We show that the map $\mathrm{K}_0\big(\mathcal{H}(G)\big) \to \mathrm{K}_0\big(\mathcal{S}(G)\big)$ induced by the embedding $\mathcal{H}(G) \to \mathcal{S}(G)$ maps $\chi(V)$ to $[V]$. This is useful because in [20] Peter Schneider and Ulrich Stuhler construct very explicit finite type projective resolutions, so that we get a nice formula for $[V] \in \mathrm{K}_0\big(\mathcal{S}(G)\big)$. This implies an explicit formula for the formal dimension of V, which is proven in [20] if V is supercuspidal or if G has characteristic 0. One consequence of this formula is that the formal dimensions are quantised, that is, they are all multiples of some $\alpha > 0$. This allows to estimate the number of irreducible square-integrable representations that contain a given representation of a compact open subgroup of G.

Although these applications can be stated purely algebraically, their proofs require functional analysis. We may view $\mathcal{S}(G)$ just as an algebra and consider the category $\mathsf{Mod}_{\mathrm{alg}}\big(\mathcal{S}(G)\big)$ of modules over $\mathcal{S}(G)$ in the algebraic sense as in [21]. However, the functor $\mathsf{Der}_{\mathrm{alg}}\big(\mathcal{S}(G)\big) \to \mathsf{Der}_{\mathrm{alg}}(G)$ fails to be fully faithful. This problem already occurs for $G = \mathbb{Z}$. The issue is that the tensor product of $\mathcal{S}(G)$ with itself plays a crucial role. If we work in $\mathsf{Mod}_{\mathrm{alg}}\big(\mathcal{S}(G)\big)$, we have to deal with $\mathcal{S}(G) \otimes \mathcal{S}(G)$, which appears quite intractable. In $\mathsf{Mod}\big(\mathcal{S}(G)\big)$ we meet instead the much simpler completion $\mathcal{S}(G \times G)$ of this space.

Now it is time to explain briefly how we do analysis. I am an advocate of bornologies as opposed to topologies. This means working with bounded subsets and bounded maps instead of open subsets and continuous maps. General bornological vector spaces behave better than general topological vector spaces for purposes of representation theory and homological algebra (see [15, 17]). The spaces that we shall use here carry both a bornology and a topology, and both structures determine each other. Therefore, readers who are familiar with topological vector spaces may be able to follow this article without learning much about bornologies. We explain some notions of bornological analysis along the way because they may be unfamiliar to many readers.

We let $\mathsf{Mod}(G)$ be the category of smooth representations of G on bornological vector spaces as in [15]. The algebras $\mathcal{H}(G)$ and $\mathcal{S}(G)$ are bornological algebras in a natural way. A smooth representation $\pi \colon G \to \mathrm{Aut}(V)$ on a bornological vector space V is called tempered if its integrated form extends to a bounded algebra homomorphism $\mathcal{S}(G) \to \mathrm{End}(V)$. We may identify $\mathsf{Mod}(G)$ with the category of essential (or non-degenerate) bornological left modules over $\mathcal{H}(G)$ ([15]). As our notation suggests, this identifies the subcategory $\mathsf{Mod}\big(\mathcal{S}(G)\big)$ with the category of essential bornological modules over $\mathcal{S}(G)$. We turn $\mathsf{Mod}(G)$ and $\mathsf{Mod}(\mathcal{S}(G))$ into exact categories using the class of extensions with a bounded linear section. The exact category structure allows us to form the derived categories $\mathsf{Der}(G)$ and $\mathsf{Der}(\mathcal{S}(G))$ as in [11]. Actually, the passage to derived categories is rather easy in both cases because our categories have enough projective and injective objects.

Equipping a vector space with the finest possible bornology, we identify the category of vector spaces with a full subcategory of the category of bornological vector spaces. Thus $\mathsf{Mod}_{\mathrm{alg}}(G)$ becomes a full subcategory of $\mathsf{Mod}(G)$. Moreover, this embedding maps projective objects again to projective objects. Therefore, the induced functor $\mathsf{Der}_{\mathrm{alg}}(G) \to \mathsf{Der}(G)$ is still fully faithful. Our main theorem asserts that the canonical functor $\mathsf{Der}(\mathcal{S}(G)) \to \mathsf{Der}(G)$ is fully faithful as well. The basic technology for its proof is already contained in [16, 17].

In [17], I define the category $\mathsf{Mod}(A)$ of essential modules and its derived category $\mathsf{Der}(A)$ for a "quasi-unital" bornological algebra A and extend some homological machinery to this setting. A morphism $A \to B$ is called *isocohomological* if the induced functor $\mathsf{Der}(B) \to \mathsf{Der}(A)$ is fully faithful. [17] gives several equivalent characterisations of isocohomological morphisms. The criterion that is most easy to verify is the following: let $P_\bullet \to A$ be a projective A-bimodule resolution of A; then $A \to B$ is isocohomological if and only if $B \hat{\otimes}_A P_\bullet \hat{\otimes}_A B$ is a resolution of B (in both cases, resolution means that there is a bounded contracting homotopy).

The article [16] deals with the special case of the embedding $\mathbb{C}[G] \to \mathcal{S}_1(G)$ for a finitely generated discrete group G and a certain Schwartz algebra $\mathcal{S}_1(G)$, which is defined by ℓ_1-estimates. The chain complex whose contractibility decides whether this embedding is isocohomological turns out to be a *coarse geometric invariant* of G. That is, it depends only on the quasi-isometry class of a word-length function on G. If the group G admits a sufficiently nice combing, then I construct an explicit contracting homotopy of this chain complex. Thus $\mathbb{C}[G] \to \mathcal{S}_1(G)$ is isocohomological for such groups.

The argument for Schwartz algebras of reductive p-adic groups follows the same pattern. Let $H \subseteq G$ be some compact open subgroup and let $X := G/H$. This is a discrete space which inherits a canonical coarse geometric structure from G. Since G is reductive, it acts properly and cocompactly on a Euclidean building, namely, its affine Bruhat-Tits building. Such buildings are CAT(0) spaces and hence combable. Since X is coarsely equivalent to the building, it is combable as well. Thus the geometric condition of [16] is easily fulfilled for all reductive p-adic groups. However, we also have to check that the constructions in [16] are compatible with uniform smoothness of functions because G is no longer discrete. This forces us to look more carefully at the geometry of the building.

2. Bornological analysis

Algebras like the Schwartz algebra $\mathcal{S}(G)$ of a reductive p-adic group carry an additional structure that allows to do analysis in them. The homological algebra for modules over such algebras *simplifies* if we take this additional structure into account. One reason is that the complete tensor product $\mathcal{S}(G) \hat{\otimes} \mathcal{S}(G)$ can be identified with $\mathcal{S}(G^2)$ (Lemma 2).

It is customary to describe this additional structure using a locally convex topology. We prefer to use bornologies instead. This means that we work with bounded subsets and bounded operators instead of open subsets and continuous operators. A basic reference on bornologies is [9]. We use bornologies because of their advantages in connection with homological algebra (see [17]).

We mainly need bornological vector spaces that are complete and convex. Therefore, we drop these adjectives and tacitly require all bornologies to be complete and convex. When we use incomplete bornologies, we explicitly say so.

We need two classes of examples: fine bornologies and von Neumann bornologies. Let V be a vector space over $\mathbb{C}$. The *fine bornology* Fine(V) is the finest possible bornology on V. A subset $T \subseteq V$ is bounded in Fine(V) if and only if there is a finite-dimensional subspace $V_T \subseteq V$ such that T is a bounded subset of $V_T \cong \mathbb{R}^n$ in the usual sense. We also write Fine(V) for V equipped with the fine bornology.

Any linear map Fine(V) $\to W$ is bounded. This means that Fine is a fully faithful functor from the category of vector spaces to the category of bornological vector spaces that is left-adjoint to the forgetful functor in the opposite direction.

Let V be a (quasi)complete locally convex topological vector space. A subset $T \subseteq V$ is called *von Neumann bounded* if it is absorbed by all neighbourhoods of zero. These subsets form a bornology on V called the *von Neumann bornology* (following [9]). We write vN(V) for V equipped with this bornology.

This defines a functor vN from topological to bornological vector spaces. Its restriction to the full subcategory of Fréchet spaces or, more generally, of LF-spaces, is fully faithful. That is, a linear map between such spaces is bounded if and only if it is continuous. A crucial advantage of bornologies is that joint boundedness is much weaker than joint continuity for multilinear maps: if $V_1, \ldots, V_n, W$ are (quasi)complete locally convex topological vector spaces, then any separately continuous n-linear map $V_1 \times \cdots \times V_n \to W$ is (jointly) bounded. The converse also holds under mild hypotheses.

Let G be a reductive p-adic group. We carefully explain how the Schwartz algebra $\mathcal{S}(G)$ looks like as a bornological algebra. The most convenient definition for our purposes is due to Marie-France Vignéras ([25]). Let $\sigma \colon G \to \mathbb{N}$ be the usual scale on G. It can be defined using a representation of G. Let $L_2(G)$ be the Hilbert space of square-integrable functions with respect to some Haar measure on G. Let

$$L_2^\sigma(G) := \{f \colon G \to \mathbb{C} \mid f \cdot \sigma^k \in L_2(G) \text{ for all } k \in \mathbb{N}\}.$$

A subset $T \subseteq L_2^\sigma(G)$ is bounded if for all $k \in \mathbb{N}$ there exists a constant $C_k \in \mathbb{R}_+$ such that $\|f \cdot \sigma^k\|_{L_2(G)} \le C_k$ for all $f \in T$. This is the von Neumann bornology with respect to the Fréchet topology on $L_2^\sigma(G)$ defined by the sequence of semi-norms

$$\|f\|_2^k := \|f \cdot \sigma^k\|_{L_2(G)}.$$

Let $\mathsf{CO}(G)$ be the set of compact open subgroups of G, ordered by inclusion. For $U \in \mathsf{CO}(G)$, let $\mathcal{S}(G/\!/U) = L_2^\sigma(G/\!/U)$ be the subspace of U-bi-invariant functions in $L_2^\sigma(G)$. We give $\mathcal{S}(G/\!/U)$ the subspace bornology, that is, a subset is bounded if and only if it is bounded in $L_2^\sigma(G)$. Finally, we let

$$\mathcal{S}(G) := \varinjlim \mathcal{S}(G/\!/U),$$

where U runs through $\mathsf{CO}(G)$. We equip $\mathcal{S}(G)$ with the direct-limit bornology. That is, a subset of $\mathcal{S}(G)$ is bounded if and only if it is a bounded subset of $\mathcal{S}(G/\!/U)$ for some $U \in \mathsf{CO}(G)$. We may also characterise this bornology as the von Neumann bornology with respect to the direct-limit topology on $\mathcal{S}(G)$, using the well-known description of bounded subsets in LF-spaces (see [24, Proposition 14.6]).

LEMMA 1 ([25]). *The definition of the Schwartz algebra above agrees with the one of Harish-Chandra in [22, 26].*

PROOF. The first crucial point is that the space of *double* cosets $G/\!/U$—as opposed to the group G itself—has polynomial growth with respect to the scale σ. It suffices to check this for a good maximal compact subgroup K because the map $G/\!/U \to G/\!/K$ is finite-to-one. By the Iwasawa decomposition, the double cosets in $G/\!/K$ can be parametrised by points in a maximal split torus. The scale on G restricts to a standard word-length function on this torus, so that we get the desired polynomial growth. As a result, there exists $d > 0$ such that $\sum_{x \in G/\!/U} \sigma^{-d}(x)$ is bounded.

Moreover, we need the following relationship between the growth of the double cosets UxU and the Harish-Chandra spherical function Ξ: there are constants $C, r > 0$ such that

$$\mathrm{vol}(UxU) \le C\sigma(x)^r \cdot \Xi(x)^{-2}, \qquad \Xi(x)^{-2} \le C\sigma(x)^r \cdot \mathrm{vol}(UxU).$$

This follows from Equation I.1.(5) and Lemma II.1.1 in [26]. Hence

$$\int_G |f(x)|^2 \sigma(x)^s \, dx = \sum_{x \in G/\!/U} |f(UxU)|^2 \sigma(x)^s \, \mathrm{vol}(UxU)$$

$$\le \sum_{x \in G/\!/U} |f(UxU)|^2 \Xi(x)^{-2} C\sigma(x)^{r+s}$$

$$\le \max_{x \in G} |f(x)|^2 \Xi(x)^{-2} \sigma(x)^{r+s+d} \sum_{y \in G/\!/U} C\sigma^{-d}(y).$$

A similar computation shows

$$\int_G |f(x)|^2 \sigma(x)^s \, dx \ge \max_{x \in G} |f(x)|^2 \Xi(x)^{-2} C^{-1} \sigma(x)^{s-r}.$$

Therefore, the sequences of semi-norms $\|f\sigma^s\|_2$ and $\|f\Xi^{-1}\sigma^s\|_\infty$ for $s \in \mathbb{N}$ are equivalent and define the same function space $\mathcal{S}(G/\!/U)$. □

Convolution defines a continuous bilinear map $\mathcal{S}(G/\!/U) \times \mathcal{S}(G/\!/U) \to \mathcal{S}(G/\!/U)$ for any $U \in \mathrm{CO}(G)$ by [26, Lemme III.6.1]. Since $\mathcal{S}(G/\!/U)$ is a Fréchet space, boundedness and continuity of the convolution are equivalent. Since any bounded subset of $\mathcal{S}(G)$ is already contained in $\mathcal{S}(G/\!/U)$ for some U, the convolution is a bounded bilinear map on $\mathcal{S}(G)$, so that $\mathcal{S}(G)$ is a bornological algebra. In contrast, the convolution on $\mathcal{S}(G)$ is only separately continuous.

Now we return to the general theory and define the Hom functor and the tensor product. Let $\mathrm{Hom}(V, W)$ be the vector space of bounded linear maps $V \to W$. A subset T of $\mathrm{Hom}(V, W)$ is bounded if and only if it is *equibounded*, that is, $\{f(v) \mid f \in T, v \in S\}$ is bounded for any bounded subset $S \subseteq V$. This bornology is automatically complete if W is.

The *complete projective bornological tensor product* $\hat{\otimes}$ is defined in [8] by the expected universal property: it is a bornological vector space $V \hat{\otimes} W$ together with a bounded bilinear map $b \colon V \times W \to V \hat{\otimes} W$ such that $l \mapsto l \circ b$ is a bijection between bounded linear maps $V \hat{\otimes} W \to X$ and bounded bilinear maps $V \times W \to X$. This tensor product enjoys many useful properties. It is commutative, associative, and commutes with direct limits. It satisfies the adjoint associativity relation

(1) $$\mathrm{Hom}(V \hat{\otimes} W, X) \cong \mathrm{Hom}\big(V, \mathrm{Hom}(W, X)\big).$$

Therefore, a bornological module over a bornological algebra A can be defined in three equivalent ways, using a bounded linear map $A \to \text{End}(V)$, a bounded bilinear map $A \times V \to V$, or a bounded linear map $A \mathbin{\hat{\otimes}} V \to V$.

Let $\otimes$ be the usual tensor product of vector spaces. The fine bornology functor is compatible with tensor products; that is, the obvious map $V \otimes W \to V \mathbin{\hat{\otimes}} W$ is a bornological isomorphism

$$(2) \qquad\qquad \text{Fine}(V \otimes W) \cong \text{Fine}(V) \mathbin{\hat{\otimes}} \text{Fine}(W)$$

for any two vector spaces V and W. More generally, if W is any bornological vector space, then the underlying vector space of $\text{Fine}(V) \mathbin{\hat{\otimes}} W$ is equal to the purely algebraic tensor product $V \otimes W$. A subset $T \subseteq V \otimes W$ is bounded if and only if there is a finite-dimensional subspace $V_T \subseteq V$ such that T is contained in and bounded in $V_T \otimes W \cong \mathbb{R}^n \otimes W \cong W^n$. Here W^n carries the direct-sum bornology. The reason for this is that $\mathbin{\hat{\otimes}}$ commutes with direct limits.

If V_1 and V_2 are Fréchet-Montel spaces, then we have a natural isomorphism

$$(3) \qquad\qquad \text{vN}(V_1 \mathbin{\hat{\otimes}_\pi} V_2) \cong \text{vN}(V_1) \mathbin{\hat{\otimes}} \text{vN}(V_2),$$

where $\hat{\otimes}_\pi$ denotes the *complete projective topological tensor product* (see [7, 24]). This isomorphism is proven in [13, Appendix A.1.4], based on results of Alexander Grothendieck. The Montel condition means that all von Neumann bounded subsets are precompact (equivalently, relatively compact).

LEMMA 2. *Let G be a reductive p-adic group. Then $\mathcal{S}(G) \mathbin{\hat{\otimes}} \mathcal{S}(G) \cong \mathcal{S}(G \times G)$.*

PROOF. It is shown in [25] that $\mathcal{S}(G/\!/U)$ is a nuclear Fréchet space for all $U \in \text{CO}(G)$; in fact, this follows easily from the proof of Lemma 1. Equip G^2 with the scale $\sigma(a, b) := \sigma(a)\sigma(b)$ for all $a, b \in G$. By definition, $\mathcal{S}(G^2) \cong \varinjlim \mathcal{S}(G^2/\!/U^2)$. Since $\mathbin{\hat{\otimes}}$ commutes with direct limits,

$$\mathcal{S}(G) \mathbin{\hat{\otimes}} \mathcal{S}(G) \cong \varinjlim \mathcal{S}(G/\!/U) \mathbin{\hat{\otimes}} \mathcal{S}(G/\!/U)$$

as well. It remains to prove $\mathcal{S}(G/\!/U)^{\hat{\otimes}2} \cong \mathcal{S}(G^2/\!/U^2)$. Since these Fréchet spaces are nuclear, they are Montel spaces. Hence (3) allows us to replace $\mathbin{\hat{\otimes}}$ by $\hat{\otimes}_\pi$. Now we merely have to recall the definition of nuclearity (see [7, 24]).

Let V and W be Fréchet spaces. The natural map $V \otimes W \to \text{Hom}(V', W)$ defines another topology on $V \otimes W$, which may be weaker than the projective tensor product topology. A Fréchet space is nuclear if and only if this topology coincides with the projective tensor product topology. Equivalently, there is only one topology on $V \otimes W$ for which the canonical maps $V \times W \to V \otimes W$ and $V \otimes W \to \text{Hom}(V', W)$ are continuous. It is clear that the subspace topology from $\mathcal{S}(G^2/\!/U^2)$ on $\mathcal{S}(G/\!/U) \otimes \mathcal{S}(G/\!/U)$ has these two properties. Hence it agrees with the projective tensor product topology. Now the assertion follows because $\mathcal{S}(G/\!/U)^{\otimes 2}$ is dense in $\mathcal{S}(G^2/\!/U^2)$.   □

3. Basic homological algebra over the Hecke algebra

Throughout this section, G denotes a totally disconnected, locally compact group, H denotes a fixed compact open subgroup of G, and $X := G/H$.

Let V be a bornological vector space and let $\pi \colon G \to \text{Aut}(V)$ be a representation of G by bounded linear operators. The representation π is called *smooth* if for any bounded subset $T \subseteq V$ there exists an open subgroup $U \subseteq G$ such that

$\pi(g, v) = v$ for all $g \in U$, $v \in T$ (see [15]). For example, the left and right regular representations of G on $\mathcal{S}(G)$ are smooth. If V carries the fine bornology, the definition above is equivalent to the usual notion of a smooth representation on a vector space.

Let $\mathsf{Mod}(G)$ be the category of smooth representations of G on bornological vector spaces; its morphisms are the G-equivariant bounded linear maps. Let $\mathsf{Mod}_{\mathrm{alg}}(G)$ be the category of smooth representations of G on $\mathbb{C}$-vector spaces. The fine bornology functor identifies $\mathsf{Mod}_{\mathrm{alg}}(G)$ with a full subcategory of $\mathsf{Mod}(G)$.

Let $\mathcal{H}(G)$ be the Hecke algebra of G; its elements are the locally constant, compactly supported functions on G. The convolution is defined by

$$f_1 * f_2(g) = \int_G f_1(x) f_2(x^{-1}g) \, dx$$

for some left-invariant Haar measure dx; we normalise it so that $\mathrm{vol}(H) = 1$. We equip $\mathcal{H}(G)$ with the fine bornology, so that $\mathcal{H}(G) \in \mathsf{Mod}_{\mathrm{alg}}(G) \subseteq \mathsf{Mod}(G)$. More generally, given any bornological vector space V, we let $\mathcal{H}(G, V) := \mathcal{H}(G) \mathbin{\hat{\otimes}} V$. The underlying vector space of $\mathcal{H}(G, V)$ is just $\mathcal{H}(G) \otimes V$ because $\mathcal{H}(G)$ carries the fine bornology. Hence $\mathcal{H}(G, V)$ is the space of locally constant, compactly supported functions $G \to V$. The *left regular representation* λ and the *right regular representation* ρ of G on $\mathcal{H}(G, V)$ are defined by

$$\lambda_g f(x) := f(g^{-1}x), \qquad \rho_g f(x) := f(xg)$$

as usual. They are both smooth.

Any continuous representation $\pi \colon G \to \mathrm{Aut}(V)$ on a bornological vector space V can be integrated to a bounded algebra homomorphism $\mathcal{H}(G) \to \mathrm{End}(V)$, which we again denote by π. By adjoint associativity, this corresponds to a map

$$\pi_* \colon \mathcal{H}(G, V) = \mathcal{H}(G) \mathbin{\hat{\otimes}} V \to V, \qquad f \mapsto \int_G \pi(g, f(g)) \, dg.$$

The map π_* is G-equivariant if G acts on $\mathcal{H}(G, V)$ by λ. By [15, Proposition 4.7], the representation π is smooth if and only if π_* is a bornological quotient map, that is, any bounded subset of V is of the form $\pi_*(T)$ for some bounded subset $T \subseteq \mathcal{H}(G, V)$. Even more, if π is smooth, then π_* has a bounded linear section. Namely, we can use

(4) $$\sigma_H \colon V \to \mathcal{H}(G, V), \qquad \sigma_H v(g) = \pi(g^{-1}, v) 1_H(g),$$

where 1_H denotes the characteristic function of H. Thus the category $\mathsf{Mod}(G)$ becomes isomorphic to the category $\mathsf{Mod}(\mathcal{H}(G))$ of essential modules over $\mathcal{H}(G)$ (see [15, Theorem 4.8]). The term "essential" is a synonym for "non-degenerate" that is not as widely used for other purposes.

Let Ext be the class of all extensions in $\mathsf{Mod}(G)$ that have a bounded linear section. This turns $\mathsf{Mod}(G)$ into an exact category in the sense of Daniel Quillen. Hence the usual machinery of homological algebra applies to $\mathsf{Mod}(G)$: we can form a derived category $\mathsf{Der}(G)$ and derived functors (see [11, 17]). The exact category $\mathsf{Mod}(G)$ has enough projective and injective objects, so that the usual recipes for computing derived functors apply. We shall use the following standard projective resolution in $\mathsf{Mod}(G)$, which already occurs in [16].

The homogeneous space $X := G/H$ is discrete because H is open in G. Let

(5) $$X_n := \{(x_0, \dots, x_n) \in X^{n+1} \mid x_0 \neq x_1, \dots, x_{n-1} \neq x_n\}.$$

We equip $\mathbb{C}[X_n]$ with the fine bornology. We let G act diagonally on X_n and equip $\mathbb{C}[X_n]$ with the induced representation

$$g \cdot f(x_0, \ldots, x_n) := f(g^{-1}x_0, \ldots, g^{-1}x_n).$$

The stabilisers of points

$$\text{Stab}(x_0, \ldots, x_n) = \bigcap_{j=0}^{n} x_j H x_j^{-1}$$

are compact open subgroups of G for each $(x_0, \ldots, x_n) \in X_n$. Let $X_n' \subseteq X_n$ be a subset that contains exactly one representative from each orbit. We get

$$(6) \qquad X_n = \coprod_{\xi \in X_n'} G/\text{Stab}(\xi), \qquad \mathbb{C}[X_n] = \bigoplus_{\xi \in X_n'} \mathbb{C}[G/\text{Stab}(\xi)].$$

If $U \in \text{CO}(G)$, then we have a natural isomorphism

$$\text{Hom}_G(\mathbb{C}[G/U], V) \xrightarrow{\cong} \text{Fix}(U, V), \qquad f \mapsto f(1_U).$$

Since U is compact, this is an exact functor of V, so that $\mathbb{C}[G/U]$ is a projective object of $\text{Mod}(G)$. Therefore, $\mathbb{C}[X_n]$ is projective by (6).

In the following, we view X_n as a subset of $\mathbb{C}[X_n]$ in the usual way. We let $(x_0, \ldots, x_n) = 0$ if $x_j = x_{j+1}$ for some $j \in \{0, \ldots, n-1\}$. Thus $(x_0, \ldots, x_n) \in \mathbb{C}[X_n]$ is defined for all $(x_0, \ldots, x_n) \in X^{n+1}$.

We define the boundary map $\delta = \delta_n \colon \mathbb{C}[X_{n+1}] \to \mathbb{C}[X_n]$ for $n \in \mathbb{N}$ by

$$\delta\big((x_0, \ldots, x_{n+1})\big) := \sum_{j=0}^{n+1} (-1)^j \cdot (x_0, \ldots, \widehat{x_j}, \ldots, x_{n+1}),$$

where $\widehat{x_j}$ means that x_j is omitted. In terms of functions, we can write

$$(7) \qquad \delta\phi(x_0, \ldots, x_n) = \sum_{j=0}^{n+1} (-1)^j \sum_{y \in X} \phi(x_0, \ldots, x_{j-1}, y, x_j, \ldots, x_n).$$

The operators δ_n are G-equivariant for all $n \in \mathbb{N}$. We define the augmentation map $\alpha \colon \mathbb{C}[X_0] \to \mathbb{C}$ by $\alpha(x) = 1$ for all $x \in X_0 = X$. It is G-equivariant with respect to the trivial representation of G on $\mathbb{C}$. It is easy to see that $\delta^2 = 0$ and $\alpha \circ \delta_0 = 0$. Hence we get a chain complex

$$C_\bullet(X) := (\mathbb{C}[X_n], \delta_n)_{n \in \mathbb{N}}$$

over $\mathbb{C}$. We also form the reduced complex $\tilde{C}_\bullet(X)$, which has $\mathbb{C}[X_n]$ in degree $n \geq 1$ and $\ker(\alpha \colon \mathbb{C}[X] \to \mathbb{C})$ in degree 0. The complex $\tilde{C}_\bullet(X)$ is exact. Thus $C_\bullet(X) \to \mathbb{C}$ is a projective resolution of $\mathbb{C}$ in $\text{Mod}_{\text{alg}}(G)$.

Next we define *bivariant co-invariant spaces*. For $V, W \in \text{Mod}(G)$, let $V \hat{\otimes}_G W$ be the quotient of $V \hat{\otimes} W$ by the closed linear span of $v \otimes w - gv \otimes gw$ for $v \in V$, $w \in W$, $g \in G$. Thus $V \hat{\otimes}_G W$ is again a complete bornological vector space. By definition, we have

$$V \hat{\otimes}_G W \cong W \hat{\otimes}_G V \cong (V \hat{\otimes} W) \hat{\otimes}_G \mathbb{C},$$

where we equip $V \hat{\otimes} W$ with the diagonal representation and $\mathbb{C}$ with the trivial representation of G. If X is another bornological vector space, then we may identify $\text{Hom}(V \hat{\otimes}_G W, X)$ with the space of bounded bilinear maps $f \colon V \times W \to X$ that satisfy $f(gv, gw) = f(v, w)$ for all $g \in G$, $v \in V$, $w \in W$. This universal property characterises $V \hat{\otimes}_G W$ uniquely. It follows from the defining property of $\hat{\otimes}$.

There is an alternative description of $V \hat{\otimes}_G W$ in terms of $\mathcal{H}(G)$-modules. Turn V into a right and W into a left bornological $\mathcal{H}(G)$-module by

$$v * f := \int_G f(g)\, g^{-1}v\, dg, \qquad f * w := \int_G f(g)\, gw\, dg$$

for all $f \in \mathcal{H}(G)$, $v \in V$, $w \in W$. Let $V \hat{\otimes}_{\mathcal{H}(G)} W$ be the quotient of $V \hat{\otimes} W$ by the closed linear span of $v * f \otimes w - v \otimes f * w$ for $v \in V$, $f \in \mathcal{H}(G)$, $w \in W$.

LEMMA 3. $V \hat{\otimes}_G W = V \hat{\otimes}_{\mathcal{H}(G)} W$, that is, the elements $gv \otimes gw - v \otimes w$ and $v * f \otimes w - v \otimes f * w$ generate the same closed linear subspace.

PROOF. We have to show $\mathrm{Hom}(V \hat{\otimes}_G W, X) = \mathrm{Hom}(V \hat{\otimes}_{\mathcal{H}(G)} W, X)$ for all bornological vector spaces X. By definition, $\mathrm{Hom}(V \hat{\otimes}_G W, X)$ is the space of bounded bilinear maps $l : V \times W \to X$ that satisfy $l(g^{-1}v, w) = l(v, gw)$ for all $v \in V$, $w \in W$, $g \in G$. This implies $l(v * f, w) = l(v, f * w)$ for all $v \in V$, $w \in W$, $f \in \mathcal{H}(G)$. Conversely, suppose $l(v * f, w) = l(v, f * w)$. Then

$$l\big(gv, g \cdot (f * w)\big) = l(gv, (\delta_g * f) * w)$$
$$= l(gv * (\delta_g * f), w) = l\big((g^{-1} \cdot gv) * f, w\big) = l(v, f * w)$$

for all $v \in V$, $w \in W$, $g \in G$, $f \in \mathcal{H}(G)$. This implies $l(gv, gw) = l(v, w)$ for all $v \in V$, $w \in W$, $g \in G$ because any $w \in W$ is fixed by some $U \in \mathrm{CO}(G)$ and therefore of the form $\mu_U * w$, where μ_U is the normalised Haar measure of U. $\square$

Since $\hat{\otimes}_G$ is functorial in both variables, we can apply it to chain complexes. Especially, we get a chain complex of bornological vector spaces $V \hat{\otimes}_G C_\bullet(X)$. Since $C_\bullet(X)$ is a projective resolution of the trivial representation, we denote the chain homotopy type of $V \hat{\otimes}_G C_\bullet(X)$ by $V \hat{\otimes}_G^{\mathrm{L}} \mathbb{C}$. The homology vector spaces of $V \hat{\otimes}_G^{\mathrm{L}} \mathbb{C}$ may be denoted $\mathrm{Tor}_n^G(V, \mathbb{C})$ or $\mathrm{Tor}_n^{\mathcal{H}(G)}(V, \mathbb{C})$. However, this passage to homology forgets an important part of the structure, namely, the bornology. Therefore, it is better to work with $V \hat{\otimes}_G^{\mathrm{L}} \mathbb{C}$ instead.

If V and W are just vector spaces, we can identify $V \otimes_G W$ with a purely algebraic construction. Let $V \otimes_G W$ be the quotient of $V \otimes W$ by the linear span of $v \otimes w - gv \otimes gw$ for $v \in V$, $w \in W$, $g \in G$. Then

$$\mathrm{Fine}(V \otimes_G W) \cong \mathrm{Fine}(V) \hat{\otimes}_G \mathrm{Fine}(W).$$

This follows from (2) and the fact that any linear subspace of a fine bornological vector space is closed. Therefore, $\mathrm{Fine}(V) \hat{\otimes}_G C_\bullet(X) \cong \mathrm{Fine}(V \otimes_G C_\bullet(X))$, and $\mathrm{Tor}_*^G(V, \mathbb{C})$ is the homology of the chain complex $V \otimes_G C_\bullet(X)$. In this case, passage to homology is harmless because $V \hat{\otimes}_G^{\mathrm{L}} \mathbb{C}$ carries the fine bornology; this implies that it is quasi-isomorphic to its homology viewed as a complex with vanishing boundary map.

Our next goal is to describe $V \hat{\otimes}_G C_\bullet(X)$ (Proposition 6). This requires some geometric preparations.

DEFINITION 4. Given a finite subset $F \subseteq X$, we define the relation $\sim_F$ on X by

$$(8) \qquad x \sim_F y \iff (x, y) \in \bigcup_{g \in G} g \cdot (\{H\} \times F) \iff x^{-1}y \in HFH.$$

Here we view $x^{-1}y \in G /\!/ H$ and $HFH \subseteq G /\!/ H$.

A subset $S \subseteq X_n$ is *controlled by* F if $x_i \sim_F x_j$ for all $(x_0, \ldots, x_n) \in S$ and all $i, j \in \{0, \ldots, n\}$. We call $S \subseteq X_n$ *controlled* if it is controlled by some finite F. Roughly speaking, this means that all entries of S are uniformly close.

A subset $S \subseteq X_n$ is controlled if and only if S is G-finite, that is, there is a finite subset $F \subseteq X_n$ such that $S \subseteq G \cdot F$. This alternative characterisation will be used frequently. Definition 4 emphasises a crucial link between the controlled support condition and geometric group theory.

A *coarse (geometric) structure* on a locally compact space such as X is a family of relations on X satisfying some natural axioms due to John Roe (see also [6]). The subrelations of the relations $\sim_F$ above define a coarse geometric structure on X in this sense. Since it is generated by G-invariant relations, it renders the action of G on X isometric. This property already characterises the coarse structure uniquely: whenever a locally compact group acts properly and cocompactly on a locally compact space, there is a unique coarse structure for which this action is isometric (see [6, Example 6]). Moreover, with this coarse structure, the space X is coarsely equivalent to G.

By definition, the notion of a controlled subset of X_n depends only on the coarse geometric structure of X. Thus the space of functions on X_n of controlled support only depends on the large scale geometry of X.

Although our main examples, reductive groups, are unimodular, we want to treat groups with non-trivial modular function as well. Therefore, we have to decorate several formulas with modular functions. We define the modular homomorphism $\Delta_G \colon G \to \mathbb{R}_{>0}$ by $\Delta_G(g)d(g^{-1}) = dg$ and $d(gh) = \Delta_G(h)\,dg$ for all $h \in G$.

DEFINITION 5. Let $\mathcal{C}(X_n, V)^\Delta$ be the space of all maps $\phi \colon X_n \to V$ that have controlled support and satisfy the covariance condition

(9) $$\phi(g\xi) = \Delta_G(g)^{-1}\pi\big(g, \phi(\xi)\big)$$

for all $\xi \in X_n$, $g \in G$. A subset $T \subseteq \mathcal{C}(X_n, V)^\Delta$ is *bounded* if $\{\phi(\xi) \mid \phi \in T\}$ is bounded in V for all $\xi \in X_n$ and the supports of all $\phi \in T$ are controlled by the same finite subset $F \subseteq X$.

PROPOSITION 6. *For any* $V \in \mathrm{Mod}(G)$, *there is a natural bornological isomorphism*

$$V \hat{\otimes}_G \mathbb{C}[X_n] \cong \mathcal{C}(X_n, V)^\Delta.$$

The induced boundary map on $V \hat{\otimes}_G \mathbb{C}[X_n]$ *corresponds to the boundary map*

$$\delta = \delta_n \colon \mathcal{C}(X_{n+1}, V)^\Delta \to \mathcal{C}(X_n, V)^\Delta$$

defined by (7).

We denote the resulting chain complex $(\mathcal{C}(X_n, V)^\Delta, \delta_n)_{n \in \mathbb{N}}$ by $\mathcal{C}(X_\bullet, V)^\Delta$.

PROOF. The bifunctor $\hat{\otimes}_G$ commutes with direct limits and in particular with direct sums. Hence (6) yields

(10) $$V \hat{\otimes}_G \mathbb{C}[X_n] \cong \bigoplus_{\xi \in X_n'} V \hat{\otimes}_G \mathbb{C}[G\xi] \cong \bigoplus_{\xi \in X_n'} V \hat{\otimes}_G \mathbb{C}[G/\mathrm{Stab}(\xi)].$$

Fix $\xi \in X'_n$ and let $\mathrm{Map}(G \cdot \xi, V)^\Delta$ be the space of all maps from $G \cdot \xi$ to V that satisfy the covariance condition (9). We equip $\mathrm{Map}(G \cdot \xi, V)^\Delta$ with the product bornology as in Definition 5. We claim that the map

$$I \colon V \hat{\otimes} \mathbb{C}[G\xi] \to \mathrm{Map}(G\xi, V), \qquad v \otimes \phi \mapsto [\eta \mapsto \int_G \pi(h^{-1}, v) \cdot \phi(h\eta) \, dh],$$

yields a bornological isomorphism

$$V \hat{\otimes}_G \mathbb{C}[G\xi] \cong \mathrm{Map}_G(G\xi, V)^\Delta.$$

We check that I descends to $V \hat{\otimes}_G \mathbb{C}[G\xi]$ and maps into $\mathrm{Map}(G\xi, V)^\Delta$:

$$I(gv \otimes g\phi)(\eta) = \int_G \pi(h^{-1}, gv) \cdot g\phi(h\eta) \, dh = \int_G \pi(h^{-1}g, v) \cdot \phi(g^{-1}h\eta) \, dh$$

$$= \int_G \pi(h^{-1}, v) \cdot \phi(h\eta) \, dh = I(v \otimes \phi),$$

$$I(v \otimes \phi)(g\eta) = \int_G \pi(h^{-1}, v) \cdot \phi(hg\eta) \, dh$$

$$= \int_G \pi(gh^{-1}, v)\phi(h\eta) \, d(hg^{-1}) = \Delta_G(g^{-1})\pi(g, I(v \otimes \phi)(\eta)).$$

Thus we get a well-defined map $V \hat{\otimes}_G \mathbb{C}[G\xi] \to \mathrm{Map}(G\xi, V)^\Delta$. Evaluation at ξ defines a bornological isomorphism $\mathrm{Map}(G\xi, V)^\Delta \cong \mathrm{Fix}(\mathrm{Stab}(\xi), V)$. We claim that the latter is isomorphic to $V \hat{\otimes}_G \mathbb{C}[G/\mathrm{Stab}(\xi)]$. Since $\mathrm{Stab}(\xi)$ is compact and open, the Haar measure $\mu_{\mathrm{Stab}(\xi)}$ of $\mathrm{Stab}(\xi)$ is an element of $\mathcal{H}(G)$. Convolution on the right with $\mu_{\mathrm{Stab}(\xi)}$ is an idempotent left module homomorphism on $\mathcal{H}(G)$, whose range is $\mathbb{C}[G/\mathrm{Stab}(\xi)]$. Since $V \hat{\otimes}_G \mathcal{H}(G) \cong V$ for all V, additivity implies that $V \hat{\otimes}_G \mathbb{C}[G/\mathrm{Stab}(\xi)]$ is equal to the range of $\mu_{\mathrm{Stab}(\xi)}$ on V, that is, to $\mathrm{Fix}(\mathrm{Stab}(\xi), V)$. Thus we obtain an isomorphism $V \hat{\otimes}_G \mathbb{C}[G\xi] \cong \mathrm{Map}_G(G\xi, V)^\Delta$, which can easily be identified with the map I.

Recall that a subset of X_n is controlled if and only if it meets only finitely many G-orbits. Therefore, we get the counterpart $\mathcal{C}(X_n, V)^\Delta \cong \bigoplus_{\xi \in X'_n} \mathrm{Map}(G\xi, V)^\Delta$ to (10). We can piece our isomorphisms on orbits together to an isomorphism

$$I \colon V \hat{\otimes}_G \mathbb{C}[X_n] \to \mathcal{C}(X_n, V)^\Delta, \qquad v \otimes \phi \mapsto [\xi \mapsto \int_G \pi(g^{-1}, v)\phi(g\xi) \, dg].$$

A straightforward computation yields $\delta \circ I(v \otimes \phi) = I(v \otimes \delta\phi)$ for all $v \in V$, $\phi \in \mathbb{C}[X_n]$ with δ as in (7). Therefore, I intertwines $\mathrm{id} \hat{\otimes}_G \delta$ and δ. $\qquad \square$

Now let $C_\bullet(X, V) := C_\bullet(X) \hat{\otimes} V$, equipped with the diagonal representation of G. Since $C_\bullet(X)$ carries the fine bornology, the underlying vector space of $\mathbb{C}[X_n] \hat{\otimes} V$ may be identified with the space of functions $X_n \to V$ with finite support.

LEMMA 7. *The chain complex $C_\bullet(X, V)$ is a projective resolution of V in* $\mathrm{Mod}(G)$.

PROOF. The complex $C_\bullet(X, V)$ is exact because $\hat{\otimes}$ is exact on extensions with a bounded linear section. We have $\mathrm{Hom}_G(\mathbb{C}[G/U] \hat{\otimes} V, W) \cong \mathrm{Hom}_U(V, W)$ for any $U \in \mathrm{CO}(G)$ and any smooth representation W. Since this is an exact functor of W, $\mathbb{C}[G/U] \hat{\otimes} V$ is projective. Equation (6) shows that $\mathbb{C}[X_n] \hat{\otimes} V$ is a direct sum of such representations and therefore projective as well. $\qquad \square$

We view $\mathcal{H}(G)$ as a bimodule over itself in the usual way, by convolution on the left and right. Since right convolution commutes with the left regular representation, the complex $C_\bullet(X, \mathcal{H}(G))$ is a complex of $\mathcal{H}(G)$-bimodules. The same reasoning as in the proof of Lemma 7 shows that it is a projective $\mathcal{H}(G)$-bimodule resolution of $\mathcal{H}(G)$.

For $V, W \in \mathrm{Mod}(G)$, we let $\mathrm{Hom}_G(V, W)$ be the space of bounded G-equivariant linear maps $V \to W$, equipped with the equibounded bornology. It agrees with the space $\mathrm{Hom}_{\mathcal{H}(G)}(V, W)$ of bounded linear $\mathcal{H}(G)$-module homomorphisms. We also apply the bifunctor Hom_G to chain complexes. In particular, we can plug in the projective resolution $C_\bullet(X, V)$ of Lemma 7. The homotopy type of the resulting cochain complex of bornological vector spaces $\mathrm{Hom}_G(C_\bullet(X, V), W)$ is denoted by $\mathbb{R}\,\mathrm{Hom}_G(V, W)$. Its nth cohomology vector space is $\mathrm{Ext}_G^n(V, W)$. As with $V \hat{\otimes}_G^{\mathbb{L}} \mathbb{C}$, it is preferable to retain the cochain complex itself.

If V and W carry the fine bornology, then $C_\bullet(X, V) = C_\bullet(X) \otimes V$ with the fine bornology. Therefore, $\mathbb{R}\,\mathrm{Hom}_G(V, W)$ is equal to the space of all G-equivariant linear maps $C_\bullet(X) \otimes V \to W$. Hence the Ext spaces above agree with the purely algebraic Ext spaces. In more fancy language, the embedding $\mathrm{Mod}_{\mathrm{alg}}(G) \to \mathrm{Mod}(G)$ induces a fully faithful functor between the derived categories $\mathrm{Der}_{\mathrm{alg}}(G) \to \mathrm{Der}(G)$. This allows us to apply results proven using analysis in a purely algebraic context.

4. Isocohomological smooth convolution algebras

We introduce a class of convolution algebras on totally disconnected, locally compact groups G. These have the technical properties that allow us to formulate the problem. Then we examine the notion of an isocohomological embedding and formulate a necessary and sufficient condition for $\mathcal{H}(G) \to \mathcal{T}(G)$ to be isocohomological. This criterion involves the contractibility of a certain bornological chain complex, which is quite close to the one that arises in [16].

4.1. Unconditional smooth convolution algebras with rapid decay.
Let G be a totally disconnected, locally compact group. Let $\sigma \colon G \to \mathbb{R}_{\geq 1}$ be a scale with the following properties: $\sigma(ab) \leq \sigma(a)\sigma(b)$ and $\sigma(a) = \sigma(a^{-1})$; σ is U-bi-invariant for some $U \in \mathrm{CO}(G)$; the map σ is proper, that is, the subsets

$$(11) \qquad\qquad B_R(G) := \{g \in G \mid \sigma(g) \leq R\}$$

are compact for all $R \geq 1$. The usual scale on a reductive p-adic groups has these properties. If the group G is finitely generated and discrete, then $\sigma = 1 + \ell$ or 2^ℓ for a word-length function ℓ are good, inequivalent choices.

Let $U \in \mathrm{CO}(G)$. Given sets S, S' of functions $G /\!\!/ U \to \mathbb{C}$ we say that S' *dominates* S if for any $\phi \in S$ there exists $\phi' \in S'$ with $|\phi'(g)| \geq |\phi(g)|$ for all $g \in G$.

DEFINITION 8. Let $\mathcal{T}(G)$ be a bornological vector space of functions $\phi \colon G \to \mathbb{C}$. We call $\mathcal{T}(G)$ an *unconditional smooth convolution algebra of rapid decay* if it satisfies the following conditions:

8.1. $\mathcal{T}(G)$ contains $\mathcal{H}(G)$;

8.2. $\mathcal{H}(G)$ is dense in $\mathcal{T}(G)$;

8.3. the convolution extends to a bounded bilinear map $\mathcal{T}(G) \times \mathcal{T}(G) \to \mathcal{T}(G)$;

8.4. $\mathcal{T}(G) = \varinjlim \mathcal{T}(G /\!\!/ U)$ as bornological vector spaces, where U runs through $\mathrm{CO}(G)$ and $\mathcal{T}(G /\!\!/ U)$ is the space of U-bi-invariant functions in $\mathcal{T}(G)$;

8.5. if a set of functions $G/\!\!/U \to \mathbb{C}$ is dominated by a bounded subset of $T(G/\!\!/U)$, then it is itself a bounded subset of $T(G/\!\!/U)$;

8.6. M_σ is a bounded linear operator on $T(G)$.

The first four conditions define a *smooth convolution algebra*, the fifth condition means that the convolution algebra is *unconditional*, the last one means that it has *rapid decay*.

An example of such a convolution algebra is the Schwartz algebra of a reductive p-adic group.

Let $T(G)$ be an unconditional smooth convolution algebra of rapid decay. A representation $\pi\colon G \to \mathrm{Aut}(V)$ is called $T(G)$-*tempered* if its integrated form extends to a bounded algebra homomorphism $T(G) \to \mathrm{End}(V)$ or, equivalently, to a bounded bilinear map $T(G) \times V \to V$. The density of $\mathcal{H}(G)$ in $T(G)$ implies that this extension is unique once it exists. Furthermore, G-equivariant maps are $T(G)$-module homomorphisms. Since the subalgebras $T(G/\!\!/U)$ are unital, the algebra $T(G)$ is "quasi-unital" in the notation of [17], so that the category $\mathsf{Mod}(T(G))$ of essential bornological left $T(G)$-modules is defined. This category is naturally isomorphic to the category of $T(G)$-tempered smooth representations of G (see [17]). Thus $\mathsf{Mod}(T(G))$ is a full subcategory of $\mathsf{Mod}(G)$.

The following lemmas prove some technical properties of $T(G)$ that are obvious in most examples, anyway. Define $P_R\colon T(G) \to \mathcal{H}(G)$ by $P_R\phi(x) = \phi(x)$ for $x \in B_R(G)$ and $P_R\phi(x) = 0$ otherwise, with $B_R(G)$ as in (11).

LEMMA 9. $\lim_{R\to\infty} P_R(\phi) = \phi$ *uniformly for ϕ in a bounded subset of $T(G)$.*

PROOF. If $T \subseteq T(G)$ is bounded, then $T \subseteq T(G/\!\!/U)$ for some $U \in \mathsf{CO}(G)$. Shrinking U further, we achieve that the scale σ is U-bi-invariant. We may further assume that $\phi' \in T$ whenever $\phi'\colon G/\!\!/U \to \mathbb{C}$ is dominated by some $\phi \in T$ because $T(G)$ is unconditional. Since M_σ is bounded, the subset $M_\sigma(T) \subseteq T(G/\!\!/U)$ is bounded as well. For any $\phi \in T$, we have $|\phi - P_R\phi| \le R^{-1}|M_\sigma\phi|$, so that $\phi - P_R\phi \in R^{-1}M_\sigma(T)$. This implies uniform convergence $P_R(\phi) \to \phi$ for $\phi \in T$. □

In the following, we briefly write

$$T(G^2) := T(G) \mathbin{\hat{\otimes}} T(G).$$

Lemma 2 justifies this notation for Schwartz algebras of reductive groups. In general, consider the bilinear maps

$$T(G) \times T(G) \to \mathbb{C}, \qquad (\phi_1, \phi_2) \mapsto \phi_1(x)\phi_2(y)$$

for $(x,y) \in G^2$. They extend to bounded linear functionals on $T(G^2)$ and hence map $T(G^2)$ to a space of smooth functions on G^2.

LEMMA 10. *This representation of $T(G^2)$ by functions on G^2 is faithful, that is, $\phi \in T(G^2)$ vanishes once $\phi(x,y) = 0$ for all $x,y \in G$.*

PROOF. The claim follows easily from Lemma 9 (this is a well-known argument in connection with Grothendieck's Approximation Property). If $\phi(x,y) = 0$ for all $x,y \in G$, then also $(P_R \mathbin{\hat{\otimes}} P_R)\phi(x,y) = 0$ for all $R \in \mathbb{N}$, $x,y \in G$. Since $P_R \mathbin{\hat{\otimes}} P_R(\phi) \in \mathcal{H}(G^2)$, this implies $P_R \mathbin{\hat{\otimes}} P_R(\phi) = 0$ for all $R \in \mathbb{N}$. Lemma 9 implies that $P_R \mathbin{\hat{\otimes}} P_R$ converges towards the identity operator on $T(G) \mathbin{\hat{\otimes}} T(G)$. This yields $\phi = 0$ as desired. □

Hence we may view $\mathcal{T}(G^2)$ as a space of functions on G^2. It is easy to see that $\mathcal{T}(G^2)$ is again a smooth convolution algebra on G^2. Equip G^2 with the scale $\sigma_2(a, b) := \sigma(a)\sigma(b)$ for $a, b \in G$. Then the operator $M_{\sigma_2} = M_\sigma \hat{\otimes} M_\sigma$ is bounded, that is, $\mathcal{T}(G^2)$ also satisfies the rapid decay condition. However, $\mathcal{T}(G^2)$ need not be unconditional. We *assume* $\mathcal{T}(G^2)$ *to be unconditional* in the following. This is needed for the proof of our main theorem.

Let $\mathcal{T}_c(G)$ be $\mathcal{H}(G)$ equipped with the subspace bornology from $\mathcal{T}(G)$. This bornology is incomplete, of course. Similarly, we let $\mathcal{T}_c(G^2)$ be $\mathcal{H}(G^2)$ equipped with the subspace bornology from $\mathcal{T}(G^2)$.

LEMMA 11. *The completions of $\mathcal{T}_c(G)$ and $\mathcal{T}_c(G^2)$ are naturally isomorphic to $\mathcal{T}(G)$ and $\mathcal{T}(G^2)$.*

PROOF. It suffices to prove this for $\mathcal{T}_c(G)$. We verify by hand that $\mathcal{T}(G)$ satisfies the universal property that defines the completion of $\mathcal{T}_c(G)$. Alternatively, we could use general characterisations of completions in [14, Section 4]. We must show that any bounded linear map $f: \mathcal{T}_c(G) \to W$ into a complete bornological vector space W extends uniquely to a bounded linear map on $\mathcal{T}(G)$. By Lemma 9, the sequence of operators $P_R: \mathcal{T}(G) \to \mathcal{T}_c(G)$ converges uniformly on bounded subsets towards the identity map on $\mathcal{T}(G)$. Hence any bounded extension $\bar{f}$ of f satisfies $\bar{f}(\phi) = \lim_{R \to \infty} f \circ P_R(\phi)$ for all $\phi \in \mathcal{T}(G)$. Conversely, this prescription defines a bounded linear extension of f. $\qquad \square$

4.2. Isocohomological convolution algebras. Let A be a quasi-unital algebra such as $\mathcal{H}(G)$ or $\mathcal{T}(G)$. In [17] I define the exact category $\mathsf{Mod}(A)$ of essential bornological left A-modules and its derived category $\mathsf{Der}(A)$. A bounded algebra homomorphism $f: A \to B$ between two quasi-unital bornological algebras induces functors $f^*: \mathsf{Mod}(B) \to \mathsf{Mod}(A)$ and $f^*: \mathsf{Der}(B) \to \mathsf{Der}(A)$. Trivially, if f has dense range then $f^*: \mathsf{Mod}(B) \to \mathsf{Mod}(A)$ is fully faithful. We call f *isocohomological* if $f^*: \mathsf{Der}(B) \to \mathsf{Der}(A)$ is fully faithful as well ([17]). We are interested in the embedding $\mathcal{H}(G) \to \mathcal{T}(G)$. If it is isocohomological, we briefly say that $\mathcal{T}(G)$ is *isocohomological*. The following conditions are proven in [17, Theorem 35] to be equivalent to $\mathcal{T}(G)$ being isocohomological:

- $V \hat{\otimes}^{\mathbb{L}}_{\mathcal{T}(G)} W \cong V \hat{\otimes}^{\mathbb{L}}_G W$ for all $V, W \in \mathsf{Der}(\mathcal{T}(G))$ (recall that $\hat{\otimes}^{\mathbb{L}}_{\mathcal{H}(G)} \cong \hat{\otimes}^{\mathbb{L}}_G$);
- $\mathbb{R} \operatorname{Hom}_{\mathcal{T}(G)}(V, W) \cong \mathbb{R} \operatorname{Hom}_G(V, W)$ for all $V, W \in \mathsf{Der}(\mathcal{T}(G))$;
- the functor $f^*: \mathsf{Der}(\mathcal{T}(G)) \to \mathsf{Der}(G)$ is fully faithful;
- $\mathcal{T}(G) \hat{\otimes}^{\mathbb{L}}_G V \cong V$ for all $V \in \mathsf{Mod}(\mathcal{T}(G))$;
- $\mathcal{T}(G) \hat{\otimes}^{\mathbb{L}}_G \mathcal{T}(G) \cong \mathcal{T}(G)$.

The last condition tends to be the easiest one to verify in practice. We will formulate it more concretely below. The signs "$\cong$" in these statements mean isomorphism in the homotopy category of chain complexes of bornological vector spaces. This is stronger than an isomorphism of homology groups. As a consequence, we have $\operatorname{Tor}^G_n(V, W) \cong \operatorname{Tor}^{\mathcal{T}(G)}_n(V, W)$ and $\operatorname{Ext}^n_G(V, W) \cong \operatorname{Ext}^n_{\mathcal{T}(G)}(V, W)$ for all $V, W \in \mathsf{Mod}(\mathcal{T}(G))$ if $\mathcal{T}(G)$ is isocohomological.

Notions equivalent to that of an isocohomological embedding have been defined independently by several authors, as kindly pointed out to me by A. Yu. Pirkovskii (see [18] and the references given there). We warn the reader that in categories of topological algebras some of the conditions above are no longer equivalent. Namely, the cohomological conditions in terms of the derived category and $\mathbb{R}\operatorname{Hom}$ are weaker than the homological conditions involving $\hat{\otimes}^{\mathrm{L}}$.

We have seen in Section 3 that $V\hat{\otimes}_G W \cong (V\hat{\otimes}W)\hat{\otimes}_G \mathbb{C}$ for all $V, W \in \operatorname{Mod}(G)$, where $V\hat{\otimes}W$ is equipped with the diagonal representation of G. Since $V\hat{\otimes}W$ is projective if V or W is projective, this implies an isomorphism

$$V\hat{\otimes}_G^{\mathrm{L}} W \cong (V\hat{\otimes}W)\hat{\otimes}_G^{\mathrm{L}} \mathbb{C}$$

for all $V, W \in \operatorname{Der}(G)$. Thus $\mathcal{T}(G)$ is isocohomological if and only if

$$(\mathcal{T}(G)\hat{\otimes}\mathcal{T}(G))\hat{\otimes}_G^{\mathrm{L}} \mathbb{C} = \mathcal{T}(G^2)\hat{\otimes}_G^{\mathrm{L}} \mathbb{C} \cong \mathcal{T}(G).$$

Here we equip $\mathcal{T}(G^2)$ with the diagonal representation of G, which is given by

$$g\cdot f(x,y) := \Delta_G(g)f(xg, g^{-1}y)$$

for all $g \in G$, $f \in \mathcal{T}(G^2)$, $x, y \in G$ because the left and right $\mathcal{H}(G)$-module structures on $\mathcal{T}(G)$ are the integrated forms of the left regular representation λ and the *twisted* right regular representation $\rho\cdot\Delta_G$. The convolution map

$$\mathcal{T}(G^2) = \mathcal{T}(G)\hat{\otimes}\mathcal{T}(G) \overset{*}{\to} \mathcal{T}(G)$$

descends to a bounded linear map $\mathcal{T}(G^2)\hat{\otimes}_G \mathbb{C} \to \mathcal{T}(G)$. The latter map is a bornological isomorphism because $A\hat{\otimes}_A A \cong A$ for any quasi-unital bornological algebra by [17, Proposition 16]. Moreover, the convolution map $\mathcal{T}(G^2) \to \mathcal{T}(G)$ has a bounded linear section, namely, the map $\mathcal{T}(G) \to \mathcal{H}(G, \mathcal{T}(G)) \subseteq \mathcal{T}(G^2)$ defined in (4).

We may use the projective resolution $C_\bullet(X) \to \mathbb{C}$ to compute $\mathcal{T}(G^2)\hat{\otimes}_G^{\mathrm{L}} \mathbb{C}$. Proposition 6 identifies $\mathcal{T}(G^2)\hat{\otimes}_G C_\bullet(X)$ with $\mathcal{C}(X_\bullet, \mathcal{T}(G^2))^\Delta$. We augment this chain complex by the map

(12) $$\alpha\colon \mathcal{T}(G^2)\hat{\otimes}_G \mathbb{C}[X_0] \xrightarrow{\operatorname{id}\hat{\otimes}_G \alpha} \mathcal{T}(G^2)\hat{\otimes}_G \mathbb{C} \overset{*}{\underset{\cong}{\to}} \mathcal{T}(G)$$

We let $\tilde{\mathcal{C}}(X_\bullet, \mathcal{T}(G^2))^\Delta$ be the subcomplex of $\mathcal{C}(X_\bullet, \mathcal{T}(G^2))^\Delta$ that we get if we replace $\mathcal{C}(X_0, \mathcal{T}(G^2))^\Delta$ by $\tilde{\mathcal{C}}(X_0, \mathcal{T}(G^2))^\Delta := \ker\alpha$.

PROPOSITION 12. *$\mathcal{T}(G)$ is isocohomological if and only if $\tilde{\mathcal{C}}(X_\bullet, \mathcal{T}(G^2))^\Delta$ has a bounded contracting homotopy.*

PROOF. Our discussion of the convolution map implies that the augmentation map in (12) is a surjection with a bounded linear section. Hence it is a chain homotopy equivalence $\mathcal{T}(G^2)\hat{\otimes}_G^{\mathrm{L}} \mathbb{C} \to \mathcal{T}(G)$ if and only if its kernel $\tilde{\mathcal{C}}(X_0, \mathcal{T}(G^2))^\Delta$ is contractible. □

To give the reader an idea why the various characterisations of isocohomological embeddings listed above are equivalent, we explain how the contractibility of $\tilde{\mathcal{C}}(X_\bullet, \mathcal{T}(G^2))^\Delta$ yields isomorphisms $\mathbb{R}\operatorname{Hom}_{\mathcal{T}(G)}(V, W) \cong \mathbb{R}\operatorname{Hom}_G(V, W)$ for $V, W \in \operatorname{Mod}(\mathcal{T}(G))$. Almost the same argument yields $V\hat{\otimes}_{\mathcal{T}(G)}^{\mathrm{L}} W \cong V\hat{\otimes}_G^{\mathrm{L}} W$. The extension to objects of the derived categories is a mere formality.

The space $\mathcal{T}(G^2)$ carries a $\mathcal{T}(G)$-bimodule structure via $f_1 * (f_2 \otimes f_3) * f_4 = (f_1 * f_2) \otimes (f_3 * f_4)$. This structure commutes with the inner conjugation action, so that $P_\bullet := T(G^2) \hat\otimes_G C_\bullet(X)$ becomes a chain complex of bornological $T(G)$-bimodules over $T(G)$. As above, we can compute these spaces explicitly:

$$T(G^2) \hat\otimes_G \mathbb{C}[X_n] \cong \bigoplus_{\xi \in X'_n} \mathrm{Fix}(\mathrm{Stab}\,\xi, T(G^2)).$$

It is not hard to see that $T(G^2)$ is a projective object of $\mathrm{Mod}(T(G^2))$. That is, $T(G^2)$ is a projective bimodule. Since the summands of $T(G^2) \hat\otimes_G \mathbb{C}[X_n]$ are all retracts of $T(G^2)$, we conclude that $T(G^2)\hat\otimes_G\mathbb{C}[X_n]$ is a projective $T(G)$-bimodule.

Suppose now that $P_\bullet$ is a resolution of $T(G)$. Then it is a projective $T(G^2)$-bimodule resolution. Since $T(G)$ is projective as a right module, the contracting homotopy of $P_\bullet$ can be improved to consist of bounded right $T(G)$-module homomorphisms. Therefore, $P_\bullet \hat\otimes_{T(G)} V$ is again a resolution of V. Explicitly,

$$P_n \hat\otimes_{T(G)} V \cong \bigoplus_{\xi \in X'_n} \mathrm{Fix}(\mathrm{Stab}(\xi), T(G) \hat\otimes V)$$

because $T(G)\hat\otimes_{T(G)}V \cong V$. The summands are retracts of the projective left $T(G)$-module $T(G) \hat\otimes V$. Hence $P_\bullet \hat\otimes_{T(G)} V$ is a projective left $T(G)$-module resolution of V. We use it to compute

$$\mathbb{R}\mathrm{Hom}_{T(G)}(V, W) = \mathrm{Hom}_{T(G)}(P_\bullet \hat\otimes_{T(G)} V, W).$$

Let $U \in \mathrm{CO}(G)$ act on $T(G) \hat\otimes V$ by $\Delta_G|_U \cdot \rho \otimes \pi = \rho \otimes \pi$. If $f \colon V \to W$ is bounded and U-equivariant, then $\phi \otimes v \mapsto \phi * f(v)$ defines a bounded G-equivariant linear map $\mathrm{Fix}(U, T(G) \hat\otimes V) \to W$. One can show that this establishes a bornological isomorphism

$$\mathrm{Hom}_{T(G)}(\mathrm{Fix}(U, T(G) \hat\otimes V), W) \cong \mathrm{Hom}_U(V, W).$$

This yields a natural isomorphism

$$\mathrm{Hom}_{T(G)}(P_n \hat\otimes_{T(G)} V, W) \cong \bigoplus_{\xi \in X'_n} \mathrm{Hom}_{\mathrm{Stab}(\xi)}(V, W).$$

The right hand side no longer depends on $T(G)$! Thus $\mathrm{Hom}_G(C_\bullet(X, V), W)$ is isomorphic to the same complex, and $\mathbb{R}\mathrm{Hom}_{T(G)}(V, W) \cong \mathbb{R}\mathrm{Hom}_G(V, W)$ as asserted.

Next we simplify the chain complex $\mathcal{C}(X_\bullet, T(G^2))^\Delta$. To $\phi \in \mathcal{C}(X_n, T(G^2))^\Delta$ we associate a function $\phi_* \colon G \times G \times X_n \to \mathbb{C}$ by $\phi_*(g, h, \xi) := \phi(\xi)(g, h)$. This identifies $\mathcal{C}(X_n, T(G^2))^\Delta$ with a space of functions on $G^2 \times X_n$ by Lemma 10. More precisely, we get the space of functions $\phi \colon G^2 \times X_n \to \mathbb{C}$ with the following properties:

- $\mathrm{supp}\,\phi \subseteq G^2 \times S$ for some controlled subset $S \subseteq X_n$;
- the function $(a, b) \mapsto \phi(a, b, \xi)$ belongs to $T(G^2)$ for all $\xi \in X_n$;
- $\phi(ag, g^{-1}b, g^{-1}\xi) = \phi(a, b, \xi)$ for all $\xi \in X_n$, $g, a, b \in G$ (the two modular functions cancel).

The last condition means that ϕ is determined by its restriction to $\{1\} \times G \times X_n$ by $\phi(a, b, \xi) = \phi(1, ab, a\xi)$. Thus we identify $\mathcal{C}(X_n, T(G^2))^\Delta$ with the following function space on $G \times X_n$:

DEFINITION 13. Let $\mathcal{C}(G \times X_n, T)$ be the space of all functions $\phi \colon G \times X_n \to \mathbb{C}$ with the following properties:

13.1. $\mathrm{supp}\,\phi \subseteq G \times S$ for some controlled subset $S \subseteq X_n$;

13.2. the function $(a, b) \mapsto \phi(ab, a\xi)$ belongs to $T(G^2)$ for all $\xi \in X_n$.

A subset $T \subseteq \mathcal{C}(G \times X_n, T)$ is bounded if there is a controlled subset $S \subseteq G$ such that $\operatorname{supp} \phi \subseteq G \times S$ for all $\phi \in T$ and if for any $\xi \in X_n$, the set of functions $(a, b) \mapsto \phi(ab, a\xi)$ for $\phi \in T$ is bounded in $T(G^2)$.

The boundary map δ on $\mathcal{C}(X_n, T(G^2))^{\Delta}$ corresponds to the boundary map

$$\delta \colon \mathcal{C}(G \times X_{n+1}, T) \to \mathcal{C}(G \times X_n, T),$$

$$\delta\phi(g, x_0, \dots, x_n) = \sum_{j=0}^{n+1} (-1)^j \sum_{y \in X} \phi(g, x_0, \dots, x_{j-1}, y, x_j, \dots, x_n).$$

The augmentation map $\mathcal{C}(X, T(G^2))^{\Delta} \to T(G)$ corresponds to

(13) $$\alpha \colon \mathcal{C}(G \times X, T) \to T(G), \qquad \alpha\phi(g) = \sum_{x \in X} \phi(g, x).$$

The proofs are easy computations, which we omit. Let $\tilde{\mathcal{C}}(G \times X_0, T) \subseteq \mathcal{C}(G \times X_0, T)$ be the kernel of α and let $\tilde{\mathcal{C}}(G \times X_\bullet, T)$ be the bornological chain complex that we get if we replace $\mathcal{C}(G \times X_0, T)$ by $\tilde{\mathcal{C}}(G \times X_0, T)$. Thus

$$\tilde{\mathcal{C}}(G \times X_\bullet, T) \cong \tilde{\mathcal{C}}(X_\bullet, T(G^2))^{\Delta}.$$

Smooth functions of compact support automatically satisfy both conditions in Definition 13, so that $\mathcal{H}(G) \otimes \mathbb{C}[X_n] \subseteq \mathcal{C}(G \times X_n, T)$. These embeddings are compatible with the boundary and augmentation maps. Thus $\mathcal{H}(G) \otimes \tilde{C}_\bullet(X)$ becomes a subcomplex of $\tilde{\mathcal{C}}(G \times X_\bullet, T)$. We write $\tilde{\mathcal{C}}(G \times X_\bullet, T_c)$ and $\mathcal{C}(G \times X_\bullet, T_c)$ for the chain complexes $\mathcal{H}(G) \otimes \tilde{C}_\bullet(X)$ and $\mathcal{H}(G) \otimes C_\bullet(X)$ equipped with the incomplete subspace bornologies from $\mathcal{C}(G \times X_\bullet, T)$. The complex $\tilde{\mathcal{C}}(G \times X_\bullet, T_c)$ is contractible because $\tilde{C}_\bullet(X)$ is. However, the obvious contracting homotopy is unbounded.

LEMMA 14. *Suppose that there is a contracting homotopy D for $\tilde{C}_\bullet(X)$ such that $\operatorname{id}_{\mathcal{H}(G)} \otimes D$ is bounded on $\tilde{\mathcal{C}}(G \times X_\bullet, T_c)$. Then $T(G)$ is isocohomological.*

PROOF. We claim that $\tilde{\mathcal{C}}(G \times X_\bullet, T)$ is the completion of $\tilde{\mathcal{C}}(G \times X_\bullet, T_c)$. Then $\tilde{\mathcal{C}}(G \times X_\bullet, T) \cong \tilde{\mathcal{C}}(X_\bullet, T(G^2))^{\Delta}$ inherits a bounded contracting homotopy because completion is functorial. Proposition 12 yields that $T(G)$ is isocohomological. It remains to prove the claim. We do this by reducing the assertion to Lemma 11. Since $\tilde{\mathcal{C}}(G \times X_\bullet, T)$ is a direct summand in $\mathcal{C}(G \times X_\bullet, T)$, it suffices to prove that $\mathcal{C}(G \times X_\bullet, T)$ is the completion of $\mathcal{C}(G \times X_\bullet, T_c)$. Recall that $X'_\bullet$ denotes a subset of $X_\bullet$ containing one point from each G-orbit. The decomposition of $X_\bullet$ into G-orbits yields a direct-sum decomposition

(14) $$\mathcal{C}(G \times X_\bullet, T_c) \cong \bigoplus_{\xi \in X'_\bullet} \operatorname{Fix}\big(\operatorname{Stab} \xi, T_c(G^2)\big),$$

and a similar decomposition for $\mathcal{C}(G \times X_\bullet, T)$. Here direct sums are equipped with the canonical bornology: a subset is bounded if it is contained in and bounded in a finite sub-sum. The reason for (14) is that a subset of X_n is controlled if and only if it meets only finitely many G-orbits. Since completion commutes with direct sums, the assertion now follows from Lemma 11. $\qquad\square$

5. Contracting homotopies constructed from combings

In order to apply Lemma 14, we have to construct contracting homotopies of $\tilde{C}_\bullet(X)$. For this we use the geometric recipes of [16]. The only ingredient is a sequence of maps $p_k \colon X \to X$ with certain properties. We first explain how such a sequence of maps gives rise to a contracting homotopy D of $\tilde{C}_\bullet(X)$. Then we formulate conditions on (p_k) and prove that they imply boundedness of D.

5.1. A recipe for contracting homotopies.
The construction of $C_\bullet(X)$ and $\tilde{C}_\bullet(X)$ is natural: a map $f \colon X \to X$ induces a chain map $f_* \colon C_\bullet(X) \to C_\bullet(X)$ by $f_*((x_0, \ldots, x_n)) := (f(x_0), \ldots, f(x_n))$ or, equivalently,

$$(15) \qquad f_*\phi(x_0, \ldots, x_n) = \sum_{y_j \colon f(y_j) = x_j} \phi(y_0, \ldots, y_n).$$

Since $\alpha \circ f_* = \alpha$, this restricts to a chain map on $\tilde{C}_\bullet(X)$. We have $\mathrm{id}_* = \mathrm{id}$ and $(fg)_* = f_* g_*$. Let p_0 be be the constant map $x \mapsto H$ for all $x \in X$. We claim that $(p_0)_* = 0$ on $\tilde{C}_\bullet(X)$. On $\mathbb{C}[X_n]$ for $n \geq 1$ this is due to our convention that $(x_0, \ldots, x_n) = 0$ if $x_i = x_{i+1}$ for some i. For $\phi \in \mathbb{C}[X_0]$, we get $(p_0)_* \phi = \alpha(\phi) \cdot (H)$, where $(H) \in \mathbb{C}[X]$ is the characteristic function of $H \in X$. This implies the claim.

Given maps $f, f' \colon X \to X$, we define operators $D_j(f, f') \colon \mathbb{C}[X_n] \to \mathbb{C}[X_{n+1}]$ for $j \in \{0, \ldots, n\}$ by

$$(16) \qquad D_j(f, f')((x_0, \ldots, x_n)) := (f(x_0), \ldots, f(x_j), f'(x_j), \ldots, f'(x_n))$$

and let $D(f, f') := \sum_{j=0}^{n} (-1)^j D_j(f, f')$. It is checked in [16] that

$$[\delta, D(f, f')] := \delta \circ D(f, f') + D(f, f') \circ \delta = f'_* - f_*.$$

Thus the chain maps f_* on $\tilde{C}_\bullet(X)$ for $f \colon X \to X$ are all chain homotopic. In particular, $D(\mathrm{id}, p_0)$ is a contracting homotopy of $\tilde{C}_\bullet(X)$ because $(p_0)_* = 0$.

However, this trivial contracting homotopy does not work for Lemma 14. Instead, we use a sequence of maps $(p_k)_{k \in \mathbb{N}}$ with p_0 as above and $\lim_{k \to \infty} p_k = \mathrm{id}$, that is, for each $x \in X$ there is $k_0 \in \mathbb{N}$ such that $p_k(x) = x$ for all $k \geq k_0$. We let

$$D_k := D(p_k, p_{k+1}), \qquad D_{jk} := D_j(p_k, p_{k+1}).$$

Observe that D_{jk} vanishes on the basis vector $(x_0, \ldots, x_n)$ unless $p_k(x_j) \neq p_{k+1}(x_j)$. Therefore, all but finitely many summands of

$$D := \sum_{k=0}^{\infty} D_k$$

vanish on any given basis vector. Thus D is a well-defined operator on $C_\bullet(X)$.

The operator D is a contracting homotopy of $\tilde{C}_\bullet(X)$ because

$$[D, \delta] = \sum_{k=0}^{\infty} [D(p_k, p_{k+1}), \delta] = \sum_{k=0}^{\infty} (p_{k+1})_* - (p_k)_* = \lim_{k \to \infty} (p_k)_* - (p_0)_* = \mathrm{id}.$$

To verify this computation, plug in a basis vector and use that all but finitely many terms vanish. This is the operator we want to use in Lemma 14.

5.2. Sufficient conditions for boundedness. Construct D as above and let $D' := \mathrm{id}_{\mathcal{H}(G)} \otimes D$. We want this to be a bounded operator on $\mathcal{C}(G \times X_\bullet, \mathcal{T}_c)$. For this, we impose three further conditions on (p_k). First, (p_k) should be a *combing*. This notion comes from geometric group theory and is already used in [16]. It allows us to control the support of $D'\phi$ for ϕ with controlled support. Secondly, the combing (p_k) should be *smooth*. This allows us to control the smoothness of the functions $(a, b) \mapsto D'\phi(ab, a\xi)$ on G^2 for $\xi \in X_n$. Only the third condition involves the convolution algebra $\mathcal{T}(G)$. It asks for a certain sequence of operators $\mathcal{T}_c(G) \to \mathcal{T}_c(G^2)$ to be equibounded.

The smoothness condition is vacuous for discrete groups. The third condition is almost vacuous for ℓ_1-Schwartz algebras of discrete groups. Hence these two conditions are not needed in [16]. In our application to reductive groups, we construct the operators (p_k) using the retraction of the affine Bruhat-Tits building of the group along geodesic paths. This is a combing because Euclidean buildings are CAT(0) spaces. Its smoothness amounts to the existence of congruence subgroups. The third condition follows easily from Lemma 2.

We now formulate the above conditions on (p_k) in detail and state the main result. We use the relation $\sim_F$ for a finite subset $F \subseteq X = G/H$ defined in (8) by
$$x \sim_F y \iff x^{-1}y \in HFH.$$

DEFINITION 15. A sequence of maps $(p_k)_{k \in \mathbb{N}}$ as above is called a *combing* of X if it has the following additional two properties:

15.1. there is a finite subset $F \subseteq X$ such that $p_k(x) \sim_F p_{k+1}(x)$ for all $k \in \mathbb{N}$, $x \in X$;

15.2. for any finite subset $F \subseteq X$ there is a finite subset $\bar{F} \subseteq X$ such that $p_k(x) \sim_{\bar{F}} p_k(y)$ for all $k \in \mathbb{N}$ and $x, y \in X$ with $x \sim_F y$.

We say that the combing has *polynomial growth* (with respect to the scale σ) if the least k_0 such that $p_k(gH) = gH$ for all $k \geq k_0$ grows at most polynomially in $\sigma(g)$. (This definition of growth differs slightly from the one in [16].)

We may view the sequence $(p_k(x))$ as a path from H to x. The conditions on a combing mean that these paths do not jump too far in each step and that nearby elements have nearby paths $(p_k(x))$.

DEFINITION 16. A combing $(p_k)_{k \in G}$ of G/H is called *smooth* if it has the following two properties:

16.1. all maps p_k are $\tilde{H}$-equivariant for some open subgroup $\tilde{H} \subseteq H$;

16.2. for any $U \in \mathrm{CO}(G)$, there exists $V \in \mathrm{CO}(G)$ such that $aVb \subseteq UabU$ for all $a, b \in G$ with $p_k(abH) = aH$.

DEFINITION 17. Let (p_k) be a smooth combing of polynomial growth. Define
$$R_k \colon \mathcal{T}_c(G) \to \mathcal{T}_c(G^2), \qquad R_k\phi(a, b) := \begin{cases} \phi(ab) & \text{if } p_k(abH) = aH; \\ 0 & \text{otherwise.} \end{cases}$$

We say that (p_k) is *compatible with* $\mathcal{T}(G)$ if the sequence of operators (R_k) is equibounded.

THEOREM 18. *Let G be a totally disconnected, locally compact group and let $\mathcal{T}(G)$ be an unconditional smooth convolution algebra of rapid decay on G. Suppose also that the function space $\mathcal{T}(G^2)$ on G^2 is unconditional. If G/H for some* *compact open subgroup $H \subseteq G$ admits a smooth combing of polynomial growth that is compatible with $T(G)$, then $T(G)$ is isocohomological.*

5.3. Proof of Theorem 18.

LEMMA 19. *Suppose that (p_k) is a combing. Then for any controlled subset $S \subseteq X_n$ there is a controlled subset $\bar{S} \subseteq X_{n+1}$ such that $\operatorname{supp} \phi \subseteq S$ implies $\operatorname{supp} D(\phi) \subseteq \bar{S}$ for all $\phi \in \mathbb{C}[X_n]$.*

PROOF. Since S is controlled, there is a finite subset $F \subseteq X$ such that $x_i \sim_F x_j$ for all $i, j \in \{0, \ldots, n\}$, $(x_i) \in S$. Since (p_k) is a combing, we can find a finite subset $\bar{F} \subseteq X$ such that $p_k(x_i) \sim_{\bar{F}} p_{k+1}(x_i)$ and $p_k(x_i) \sim_{\bar{F}} p_k(x_j)$ for all $k \in \mathbb{N}$ and all $(x_j) \in S$. Let $F' := H\bar{F}H\bar{F}H \subseteq G/H$. If $x \sim_{\bar{F}} y \sim_{\bar{F}} z$, then $x \sim_{F'} z$. Hence $(p_k(x_0), \ldots, p_k(x_j), p_{k+1}(x_j), \ldots, p_{k+1}(x_n))$ is controlled by F' for all $(x_i) \in S$. This means that all summands in $D(x_0, \ldots, x_n)$ are controlled by F'. $\square$

Hence $D' := \operatorname{id}_{\mathcal{H}(G)} \otimes D$ preserves controlled supports in $\mathcal{C}(G \times X_\bullet, T_c)$. We have seen in (14) that

$$\mathcal{C}(G \times X_\bullet, T_c) \cong \bigoplus_{\xi \in X'_\bullet} \operatorname{Fix}\left(\operatorname{Stab} \xi, T_c(G^2)\right).$$

The isomorphism sends $\phi \colon G \times X_n \to \mathbb{C}$ to the family of functions $(\phi_\xi)_{\xi \in X'_n}$ defined by $\phi_\xi(a, b) := \phi(ab, a\xi)$. Thus we may describe any operator on $\mathcal{C}(G \times X_\bullet, T_c)$ by a block matrix. In particular, we get $D' = (D'_{\xi\eta})_{\xi, \eta \in X'_\bullet}$ with certain operators

$$D'_{\xi\eta} \colon \operatorname{Fix}\left(\operatorname{Stab} \eta, T_c(G^2)\right) \to \operatorname{Fix}\left(\operatorname{Stab} \xi, T_c(G^2)\right).$$

The fact that D' preserves controlled supports means that for fixed η we have $D'_{\xi\eta} = 0$ for all but finitely many ξ. Thus the whole operator D' is bounded if and only if all its matrix entries $D'_{\xi\eta}$ are bounded.

For $j, n \in \mathbb{N}$, $n \geq j$, define $p_{jk} \colon X_n \to X^{n+2}$ by

$$p_{jk}((x_0, \ldots, x_n)) := (p_k(x_0), \ldots, p_k(x_j), p_{k+1}(x_j), \ldots, p_{k+1}(x_n)).$$

Then the operator $D_{jk} \colon \mathbb{C}[X_n] \to \mathbb{C}[X_{n+1}]$ is given by

$$D_{jk}\phi(\xi) = \sum_{\eta \in p_{jk}^{-1}(\xi)} \phi(\eta).$$

Let $D'_{jk} = \operatorname{id}_{\mathcal{H}(G)} \otimes D_{jk}$ and let $D'_{jk,\xi\eta}$ be the matrix entries of D'_{jk} with respect to the decomposition (14). Thus $D'_{\xi\eta} = \sum_{k \in \mathbb{N}} \sum_{j=0}^n (-1)^j D'_{jk,\xi\eta}$. Writing $\phi_\xi(a, b) = \phi(ab, a\xi)$ and $\phi_\eta(ag, g^{-1}b) = \phi(ab, ag\eta)$, we get

$$(17) \quad D'_{jk,\xi\eta}\psi(a, b) = \sum_{\{g \in G/\operatorname{Stab}(\eta) \mid p_{jk}(ag\eta) = a\xi\}} \psi(ag, g^{-1}b)$$

$$= \operatorname{vol}(\operatorname{Stab} \eta)^{-1} \int_{\{g \in G \mid p_{jk}(ag\eta) = a\xi\}} \psi(ag, g^{-1}b) \, dg$$

for $\psi \in \operatorname{Fix}\left(\operatorname{Stab} \eta, T_c(G^2)\right)$. The right hand side of (17) makes sense for arbitrary $\psi \in T_c(G^2)$ and extends $D'_{jk,\xi\eta}$ to an operator on $T_c(G^2)$. Now we fix ξ, η until further notice and sometimes omit them from our notation.

Let $U \subseteq \mathrm{Stab}(\eta)$ be an open subgroup and $\phi \in \mathcal{T}_c(G/\!\!/U)$. Let $\mu_U \in \mathcal{H}(G)$ be the normalised Haar measure of U, that is, $\mathrm{supp}\,\mu_U = U$ and $\mu_U(g) = \mathrm{vol}(U)^{-1}$ for $g \in U$. Equation (17) yields

$$\mathrm{vol}(\mathrm{Stab}\,\eta)D'_{jk,\xi\eta}(\phi \otimes \mu_U)(a,b)$$

$$= \mathrm{vol}(U)^{-1} \int_{\{g \in bU \mid p_{jk}(ag\eta) = a\xi\}} \phi(ag)\,dg = \begin{cases} \phi(ab) & \text{if } p_{jk}(ab\eta) = a\xi; \\ 0 & \text{otherwise.} \end{cases}$$

Let $\chi_{j,\xi\eta}(a,b)$ be the number of $k \in \mathbb{N}$ with $p_{jk}(ab\eta) = a\xi$ and let

$$\chi(a,b) = \chi_{\xi\eta}(a,b) := \sum_{j=0}^{n}(-1)^j \chi_{j,\xi\eta}(a,b).$$

These numbers are finite for any $a, b \in G$ for the same reason that guarantees that the sum defining D is finite on each basis vector. Define

$$A = A_{\xi\eta}\colon \mathcal{T}_c(G) \to \mathcal{T}_c(G^2), \qquad A\phi(a,b) = \chi(a,b) \cdot \phi(ab).$$

Our computation shows that $A\phi = D'_{\xi\eta}(\phi \otimes \mu_U) \cdot \mathrm{vol}(\mathrm{Stab}\,\eta)$ if U is an open subgroup of $\mathrm{Stab}\,\eta$ and $\phi \in \mathcal{T}_c(G/\!\!/U)$.

LEMMA 20. *The operator $D'_{\xi\eta}$ is bounded if and only if $A_{\xi\eta}$ is bounded.*

PROOF. The boundedness of $D'_{\xi\eta}$ implies that A is bounded on $\mathcal{T}_c(G/\!\!/U)$ for sufficiently small U and hence on all of $\mathcal{T}_c(G)$. Suppose conversely that A is bounded. We turn $\mathcal{T}_c(G^2)$ into an (incomplete) bornological right $\mathcal{T}_c(G)$-module by

$$(f_1 \otimes f_2) * f_3 := f_1 \otimes (f_2 * f_3).$$

This bilinear map $\mathcal{T}_c(G^2) \times \mathcal{T}_c(G) \to \mathcal{T}_c(G^2)$ is bounded because the convolution in $\mathcal{T}(G)$ is bounded. The operators $D'_{jk,\xi\eta}$ and hence $D'_{\xi\eta}$ are $\mathcal{T}_c(G)$-module homomorphisms by (17). Let $U \subseteq \mathrm{Stab}(\eta)$ be open and $\phi_1, \phi_2 \in \mathcal{T}_c(G/\!\!/U)$. Then

$$D'_{\xi\eta}(\phi_1 \otimes \phi_2) = D'_{\xi\eta}(\phi_1 \otimes \mu_U) * \phi_2 = \mathrm{vol}(\mathrm{Stab}\,\eta)^{-1} A(\phi_1) * \phi_2.$$

This implies the boundedness of $D'_{\xi\eta}$ because U is arbitrarily small and A and the convolution $\mathcal{T}_c(G^2) \to \mathcal{T}_c(G)$ are bounded.   □

LEMMA 21. *If the combing is smooth, then for any $U \in \mathrm{CO}(G)$ there is $V \in \mathrm{CO}(G)$ such that A maps $\mathcal{T}_c(G/\!\!/U)$ into $\mathcal{T}_c(G^2/\!\!/V^2)$.*

PROOF. Let $a, b \in G$. Clearly, $\chi(a, by) = \chi(a,b)$ for $y \in \mathrm{Stab}(\eta)$. Since p_k is $\tilde{H}$-equivariant, so is p_{jk}. Hence $\chi(ha,b) = \chi(a,b)$ for $h \in \tilde{H}$. Therefore, $A\phi(ua,b) = A\phi(a,b) = A\phi(a,bu)$ for $\phi \in \mathcal{T}_c(G/\!\!/U)$ provided $U \subseteq \tilde{H} \cap \mathrm{Stab}(\eta)$. Moreover, we have $A\phi(a,xb) = A\phi(ax,b)$ if $x \in \mathrm{Stab}(\xi)$.

We may assume that the zeroth components η_0 and ξ_0 are H: any G-orbit on X_n has such a representative. Then $p_{jk}(ab\eta) = a\xi$ implies $p_k(abH) = aH$. By the definition of a smooth combing, there is $V \in \mathrm{CO}(G)$ such that $aVb \subseteq UabU$ whenever $p_{jk}(ab\eta) = a\xi$ for some $k \in \mathbb{N}$. Hence $\phi(avb) = \phi(ab)$ for $(a,b) \in \mathrm{supp}\,\chi$ and $v \in V$. We may shrink V such that $V \subseteq \mathrm{Stab}(\xi) \cap U$. Then $\chi(a, vb) = \chi(a,b) = \chi(av,b)$ as well, so that A maps $\mathcal{T}_c(G/\!\!/U)$ to $\mathcal{T}_c(G^2/\!\!/V^2)$.   □

Definition 17 requires the following sequence of operators to be equibounded:

$$R_k \colon \mathcal{T}_c(G) \to \mathcal{T}_c(G^2), \qquad R_k\phi(a,b) := \begin{cases} \phi(ab) & \text{if } p_k(abH) = aH; \\ 0 & \text{otherwise.} \end{cases}$$

Hence $R := \sum_{k\in\mathbb{N}}(k+1)^{-2}R_k$ is bounded. We have $R\phi(a,b) = \phi(ab)\chi'(a,b)$, where

$$\chi'(a,b) = \sum_{\{k \in \mathbb{N} \mid p_k(abH) = aH\}} (k+1)^{-2}.$$

Now let $S \subseteq \mathcal{T}_c(G)$ be bounded. Then $S \subseteq \mathcal{T}_c(G/\!/U)$ for some $U \in \mathsf{CO}(G)$. We have already found $V \in \mathsf{CO}(G)$ such that $A\phi \in \mathcal{T}_c(G^2/\!/V^2)$ for all $\phi \in \mathcal{T}_c(G/\!/U)$. Since pointwise multiplication by the scale σ is bounded on $\mathcal{T}_c(G/\!/V)$, it follows that the set of functions $\sigma(a)^N\sigma(b)^NR(S)$ is bounded in $\mathcal{T}_c(G^2/\!/V^2)$. We claim that $\sigma(a)^N\sigma(b)^NR(S)$ dominates $A(S)$ for sufficiently large N. Since $\mathcal{T}(G^2)$ is unconditional, this implies the boundedness of A.

Since A and R are multiplication operators, the claim follows if $\chi'\sigma^N$ dominates $\chi_{j,\xi\eta}$ for any fixed j,ξ,η. This is what we are going to prove. Let $k_0(ab)$ be the least k_0 such that $p_k(ab\eta_j) = ab\eta_j$ for all $k \geq k_0$. The polynomial growth of the combing implies that $k_0(ab)$ is dominated by $C\sigma(ab)^N \leq C\sigma(a)^N\sigma(b)^N$ for sufficiently large C, N. Since $\xi_j \neq \xi_{j+1}$, we have $p_{jk}(ab\eta) \notin G\xi$ for $k \geq k_0$. Hence, if $p_{jk}(ab\eta) = a\xi$, then $k < k_0$. We choose the set of representatives $X'_\bullet$ such that $\xi_0 = H$ for all $\xi \in X'_\bullet$. Then $p_{jk}(ab\eta) = a\xi$ implies $p_k(abH) = aH$. Therefore, for each summand 1 in $\chi_{j,\xi\eta}(a,b)$ there is a summand $1/(k+1)^2 \geq 1/k_0^2$ in $\chi'(a,b)$. This yields the desired estimate $\chi_{j,\xi\eta}(a,b) \leq k_0(a,b)^2\chi'(a,b) \leq C^2\sigma(a)^{2N}\sigma(b)^{2N}\chi'(a,b)$. Thus the operators $A_{\xi\eta}$ are bounded for all ξ,η. This implies boundedness of $D'_{\xi\eta}$ by Lemma 20. By Lemma 19, it follows that D' is bounded. Finally, Lemma 14 yields that $\mathcal{T}(G)$ is isocohomological. This finishes the proof of Theorem 18.

6. A smooth combing for reductive p-adic groups

The following theorem is the main goal of this section. In addition, we prove a variant (Theorem 28) that deals with the subcategories of χ-homogeneous representations for a character $\chi \colon C(G) \to \mathrm{U}(1)$ of the connected centre of G.

THEOREM 22. *The Schwartz algebra $\mathcal{S}(G)$ of a reductive p-adic group G is isocohomological.*

PROOF. We are going to apply Theorem 18. Let σ be the standard scale as in the definition of the Schwartz algebra and let $\mathcal{T}(G) := \mathcal{S}(G)$. The space $\mathcal{T}(G^2)$ is defined as $\mathcal{T}(G) \hat{\otimes} \mathcal{T}(G)$. This notation is permitted because of Lemma 2, which identifies $\mathcal{S}(G) \hat{\otimes} \mathcal{S}(G)$ with the Schwartz algebra of G^2 (which is again a reductive p-adic group). Clearly, the Schwartz algebras $\mathcal{S}(G)$ and $\mathcal{S}(G^2)$ are unconditional smooth convolution algebras of rapid decay.

We let $\mathcal{BT} = \mathcal{BT}(G)$ be the *affine Bruhat-Tits building* of G, as defined in [5, 23]. This is a Euclidean building on which G acts isometrically, properly, and cocompactly. Let $C(G)$ be the connected centre of G, so that the quotient $G/C(G)$ is semi-simple. In Section 7, we will also use the variant $\mathcal{BT}(G/C(G))$ of $\mathcal{BT}(G)$, which we call the *semi-simple affine Bruhat-Tits building* of G.

Let $G_\circ$ be the connected component of G as an algebraic group. Thus $G_\circ$ is a reductive group and G is a finite extension of $G_\circ$. Inspection of the definition in [23] shows that the buildings for G and $G_\circ$ are equal. We remark that it is not hard to reduce the case of general reductive p-adic groups to the special case of connected semi-simple groups, or even connected simple groups. At first I followed this route myself. Eventually, it turned out that this intermediate step is unnecessary because all arguments work directly in the generality we need.

Let $\xi_0 \in \mathcal{BT}$, $H := \mathrm{Stab}(\xi_0)$, and $X := G/H$. We have $H \in \mathrm{CO}(G)$, and X may be identified with the discrete subset $G\xi_0 \subseteq \mathcal{BT}$. We need a combing of X. As a preparation, we construct a combing of $\mathcal{BT}$, using that Euclidean buildings are CAT(0) spaces, that is, have "non-positive curvature" (see [4,5]). In particular, any two points in $\mathcal{BT}$ are joined by a unique geodesic. For $\xi \in \mathcal{BT}$, let

$$p(\xi) \colon [0, d(\xi, \xi_0)] \to \mathcal{BT}, \qquad t \mapsto p_t(\xi),$$

be the unit speed geodesic segment from ξ_0 to ξ; extend this by $p_t(\xi) := \xi$ for $t > d(\xi, \xi_0)$. Restricting to $t \in \mathbb{N}$, we get a sequence of maps $p_k \colon \mathcal{BT} \to \mathcal{BT}$.

LEMMA 23. *The maps p_k for $k \in \mathbb{N}$ form a combing of linear growth of $\mathcal{BT}$.*

This means $d\big(p_k(\xi), p_k(\eta)\big) \le R \cdot d(\xi, \eta) + R$ and $d\big(p_k(\xi), p_{k+1}(\xi)\big) \le R$ for all $\xi, \eta \in \mathcal{BT}$, $k \in \mathbb{N}$, for some $R > 0$. Linear growth means that the least k_0 such that $p_k(\xi)$ is constant for $k \ge k_0$ grows at most linearly in $d(\xi, \xi_0)$.

PROOF. By construction, $d\big(p_s(\xi), p_t(\xi)\big) \le |t - s|$ for all $s, t \in \mathbb{R}_+$, $\xi \in \mathcal{BT}$, and $p_s(\xi)$ is constant for $s \ge d(\xi, \xi_0)$. The lemma follows if we prove the following claim: $d\big(p_t(\xi), p_t(\eta)\big) \le d(\xi, \eta)$ for all $\xi, \eta \in \mathcal{BT}$, $t \in \mathbb{R}_+$.

Fix $\xi, \eta \in \mathcal{BT}$ and $t \in \mathbb{R}_+$. We may assume $d(\xi, \xi_0) \ge d(\eta, \xi_0)$ (otherwise exchange ξ and η) and $d(\xi, \xi_0) \ge t$ (otherwise $p_t(\xi) = \xi$ and $p_t(\eta) = \eta$). Let d^* be the usual flat Euclidean metric on $\mathbb{R}^2$. Let ξ^* and η^* be points in $\mathbb{R}^2$ with

$$d^*(\xi^*, 0) = d(\xi, \xi_0), \qquad d^*(\eta^*, 0) = d(\eta, \xi_0), \qquad d^*(\xi^*, \eta^*) = d(\xi, \eta).$$

The CAT(0) condition means that distances between points on the boundary of the geodesic triangle (ξ, η, ξ_0) are dominated by the distances between the corresponding points in the comparison triangle $(\xi^*, \eta^*, 0)$. Here the point $p_t(\xi)$ corresponds to the point $p_t(\xi)^*$ on $[0, \xi^*]$ of distance t from the origin. The point $p_t(\eta)$ corresponds to the point $p_t(\eta)^*$ of distance $\min\{t, d(\eta, \xi_0)\}$ from the origin. An easy

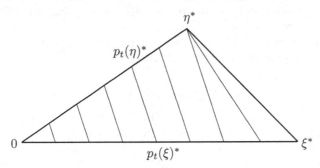

FIGURE 1. *A comparison triangle*

computation or a glance at Figure 1 shows $d^*\big(p_t(\xi)^*, p_t(\eta)^*\big) \le d^*(\xi^*, \eta^*)$. By the CAT(0) condition, this implies $d\big(p_t(\xi), p_t(\eta)\big) \le d(\xi, \eta)$. $\qquad \square$

We identify G/H with the orbit $G\xi_0 \subseteq \mathcal{BT}$. Since the group action is cocompact, there is some $R > 0$ such that for any $\xi \in \mathcal{BT}$ there exists $\xi' \in G\xi_0$ with

$d(\xi, \xi') < R$. We let $p'_k(\xi)$ for $\xi \in \mathcal{BT}$ be a point in $G\xi_0$ with $d\big(p'_k(\xi), p_k(\xi)\big) < R$. We claim that any such choice defines a combing of G/H.

If we equip $G\xi_0$ with the metric d from $\mathcal{BT}$, the maps p'_k on $G\xi_0$ still form a combing in the metric sense because they are "close" to the combing (p_k). The subspace metric from $\mathcal{BT}$ and the relations $\sim_F$ in Definition 15 generate the same coarse geometric structure on X. That is, for any $R > 0$ there is a finite subset $F \subseteq X$ such that $d(x, y) \leq R$ implies $x \sim_F y$, and for any finite subset $F \subseteq X$ there is $R > 0$ such that $x \sim_F y$ implies $d(x, y) \leq R$. This is easy to verify by hand. Alternatively, it follows from the uniqueness of coarse structures mentioned after Definition 15. Hence (p'_k) is a combing of G/H in the sense we need.

We also need the combing to be smooth. To get this, we must choose the base point ξ_0 and the approximations $p'_k(\xi)$ more carefully. This requires some geometric facts about the apartments in the building. Let K be the non-Archimedean local field over which G is defined. Let $S \subseteq G$ be a maximal K-split torus of G. We do not distinguish in our notation between the algebraic groups S and G and their locally compact groups of K-rational points. Let $X^*(S)$ and $X_*(S)$ be the groups of algebraic characters and cocharacters of S, respectively. The $\mathbb{R}$-vector space $A := X_*(S) \otimes \mathbb{R}$ is the *basic apartment* of (G, S).

Let $\Phi \subseteq X^*(S)$ be the set of roots of G relative to S. Choose a simple system of roots $\Delta \subseteq \Phi$ and let $A_+ \subseteq A$ be the corresponding closed Weyl chamber:

$$A_+ := \{x \in A \mid \alpha(x) \geq 0 \text{ for all } \alpha \in \Delta\}.$$

Let W be the *Weyl group* of the root system Φ. It is the Coxeter group generated by orthogonal reflections in the hyperplanes $\alpha(x) = 0$ for $\alpha \in \Delta$. The positive cone A_+ is a fundamental domain for this action, that is, $W(A_+) = A$ (see [10]).

Let $Z \subseteq G$ be the centraliser of S. There is a canonical homomorphism $\nu \colon Z \to A$ (see [23, (1.2)]). Its kernel is compact and its range is a lattice $\Lambda \subseteq A$; that is, Λ is a discrete and cocompact subgroup of A. Moreover, let $\Lambda_+ := \Lambda \cap A_+$. Since Λ is free Abelian, we can lift it to a subgroup of Z and view $\Lambda \subseteq Z \subseteq G$. Let Φ_{af} be the set of *affine roots* as in [23, (1.6)]. These are affine functions $\alpha \colon A \to \mathbb{R}$ of the form $\alpha(x) = \alpha_0(x) + \gamma$ with $\alpha_0 \in \Phi$ and certain $\gamma \in \mathbb{R}$. Recall that $\Phi \subseteq \Phi_{\mathrm{af}}$ and that Φ_{af} is invariant under translation by Λ. The subsets of A of the form $\{x \in A \mid \alpha(x) = 0\}$ and $\{x \in A \mid \alpha(x) \geq 0\}$ for $\alpha \in \Phi_{\mathrm{af}}$ are called *walls* and *half-apartments*, respectively.

We define the *closure* $\mathrm{cl}(\Omega) \subseteq A$ of a non-empty subset $\Omega \subseteq A$ as the intersection of all closed half-apartments containing Ω (see [5, (7.1.2)]). We claim that

$$(18) \qquad\qquad \mathrm{cl}(\{0, \xi\}) = A_+ \cap (\xi - A_+)$$

for all $\xi \in \Lambda_+$. Let $B := A_+ \cap (\xi - A_+)$. This is an intersection of half-apartments containing 0 and ξ because $\Phi \cup (\xi - \Phi) \subseteq \Phi_{\mathrm{af}}$. Hence $\mathrm{cl}(\{0, \xi\}) \subseteq B$. It remains to show that any half-apartment C containing 0 and ξ also contains B. Let C be defined by the equation $\alpha \geq 0$ for some affine root α with linear part $\alpha_0 \in \Phi$. We distinguish the cases $\alpha_0 > 0$ and $\alpha_0 < 0$. If $\alpha_0 > 0$, then $\alpha(x) = \alpha_0(x) + \alpha(0) \geq \alpha_0(x)$ is non-negative on A_+; if $\alpha_0 < 0$, then $\alpha(\xi - x) = -\alpha_0(x) + \alpha(\xi) \geq -\alpha_0(x)$ is non-negative on A_+, so that α is non-negative on $\xi - A_+$. Thus $B \subseteq C$ in either case. This finishes the proof that $\mathrm{cl}(\{0, \xi\}) = B$.

LEMMA 24. *There is $R > 0$ such that for all $\xi \in \Lambda_+$ and all $\eta \in \mathrm{cl}(\{0, \xi\})$, there is $\eta' \in \Lambda \cap \mathrm{cl}(\{0, \xi\})$ with $d(\eta, \eta') \leq R$.*

PROOF. Let $\Delta = \{\alpha_1, \ldots, \alpha_s\}$ be the system of simple roots that determines A_+. If G is semi-simple, these roots form a basis of A. In general, they are linearly independent, so that we can extend them to a basis by certain α_j for $s < j \leq r$. Define a vector space isomorphism $\gamma \colon A \to \mathbb{R}^r$ by $\gamma(\eta)_j = \alpha_j(\eta)$ for $j = 1, \ldots, r$. This identifies A_+ with the set of $(x_j) \in \mathbb{R}^r$ with $x_j \geq 0$ for $1 \leq j \leq s$. Equation (18) identifies $B := \mathrm{cl}(\{0, \xi\})$ with

$$\gamma(B) = \{(x_j) \in \mathbb{R}^r \mid 0 \leq x_j \leq \gamma(\xi)_j, \; j = 1, \ldots, s\}.$$

We may assume $\alpha_j \in X_*(S) \otimes \mathbb{Q} \subseteq A$, so that $\gamma(\Lambda) \subseteq \mathbb{Q}^r$. Replacing γ by $n^{-1}\gamma$ for some $n \in \mathbb{N}^*$, we can achieve $\mathbb{Z}^r \subseteq \gamma(\Lambda)$. Hence if $(x_j) \in \gamma(B)$, then the truncated vector $\lfloor (x_j) \rfloor := (\lfloor x_j \rfloor)$ belongs to $\gamma(B \cap \Lambda)$. It satisfies $|x_j - \lfloor x_j \rfloor| < 1$ for all j. Since the norm $\|\gamma(\eta)\|_\infty$ is equivalent to the Euclidean norm on A, we have $d(\gamma^{-1}\lfloor \gamma(\eta) \rfloor, \eta) < R$ for all $\eta \in B$ for some $R > 0$. $\qquad\square$

The building $\mathcal{BT}$ can be defined as the quotient of $G \times A$ by a certain equivalence relation. We may view gA for $g \in G$ as a subspace of $\mathcal{BT}$; these are the *apartments* of $\mathcal{BT}$. We now choose ξ_0 to be the origin in $A \subseteq \mathcal{BT}$. Recall that $H \subseteq G$ denotes the stabiliser of ξ_0. We have the *Cartan decomposition* $G = H\Lambda_+ H$ by [23, (3.3.3)], so that $G\xi_0 = H\Lambda_+\xi_0$. Let $G_\circ \subseteq G$ be the connected component of the identity (as an algebraic variety) and let $H_\circ := H \cap G_\circ$. These are open normal subgroups of finite index in G and H, respectively, and $\mathcal{BT}(G)$ is isomorphic to $\mathcal{BT}(G_\circ)$ equipped with a canonical action of G.

Choose $\Lambda'_+ \subseteq \Lambda_+$ to contain one representative for each H-orbit in $G\xi_0$. Fix $\xi \in \Lambda'_+$ and $k \in \mathbb{N}$. We further decompose $H\xi$ as a disjoint union of finitely many $H_\circ$-orbits $H_\circ h_j \xi = h_j H_\circ \xi$ for suitable $h_1, \ldots, h_N \in H$. We let $p'_k(\xi)$ be some point in $\Lambda \cap \mathrm{cl}(\{0, \xi\}) \subseteq A \subseteq \mathcal{BT}$ that has minimal distance from $p_k(\xi)$.

PROPOSITION 25. *Let* $\Omega \subseteq A \subseteq$, $\Omega \neq \emptyset$. *If* $g \in G_\circ$ *satisfies* $gx = x$ *for all* $x \in \Omega$, *then* $gx = x$ *for all* $x \in \mathrm{cl}(\Omega)$. *(This may fail if we allow* $g \in G$.)

PROOF. The proof requires some facts about stabilisers of points in $\mathcal{BT}$, which are conveniently summarised in [20, Section I.1]. The subgroups $P_\Omega \subseteq G_\circ$ defined there manifestly satisfy $P_\Omega = P_{\mathrm{cl}(\Omega)}$. This implies our claim because

$$P_\Omega = \{g \in G_\circ \mid gx = x \; \forall x \in \Omega\}. \qquad\square$$

Proposition 25 allows us to define $p'_k(h_j h_\circ \xi) := h_j h_\circ p'_k(\xi)$ for all $h_\circ \in H_\circ$. Letting ξ, j vary, we get a map $p'_k \colon G\xi_0 \to G\xi_0$. It is $H_\circ$-equivariant because $H_\circ$ is normal in H. Since $p_k(\xi) \in \mathrm{cl}(\{0, \xi\})$, Lemma 24 yields $R > 0$ such that $d\big(p_k(\xi), p'_k(\xi)\big) \leq R$ for all $k \in \mathbb{N}$, $\xi \in \Lambda'_+$. The same holds for $\xi \in G\xi_0$ because p_k is H-equivariant and G acts isometrically on $\mathcal{BT}$. Moreover, $p'_0(\xi) = \xi_0$ and $p'_k(\xi) = \xi$ for $k \geq d(\xi, \xi_0)$. Thus (p'_k) is a combing of G/H of linear growth.

LEMMA 26. *The combing* (p'_k) *is smooth.*

PROOF. There is a decreasing sequence $(U_n)_{n \in \mathbb{N}}$ in $\mathrm{CO}(G)$ such that each U_n is normal in H and can be written as $U_n^+ \cdot U_n^- = U_n^- \cdot U_n^+$ with

$$(19) \qquad \lambda U_n^+ \subseteq U_n^+ \lambda, \qquad U_n^- \lambda \subseteq \lambda U_n^-$$

for all $\lambda \in \Lambda_+$ (see [20, Section I.2]). Let $a, b \in G$ satisfy $p'_k(abH) = aH$. Write $ab = h_1 \lambda h_2$ with $h_1, h_2 \in H$, $\lambda \in \Lambda'_+$, and h_1 chosen as carefully as above if G is disconnected. Then $a = h_1 p'_k(\lambda) h_3$ with the same h_1 and some $h_3 \in H$. Hence

$b = h_3^{-1}p_k'(\lambda)^{-1}\lambda h_2$. Since $p_k'(\lambda) \in \mathrm{cl}(\{0, \lambda\})$, equation (18) yields $p_k'(\lambda) \in \Lambda_+$ and $p_k'(\lambda)^{-1}\lambda \in \Lambda_+$. Using (19) and that U_n is normal in H, we get

$$aU_n b = h_1 p_k'(\lambda)U_n^+ U_n^- p_k'(\lambda)^{-1}\lambda h_2 \subseteq h_1 U_n^+ p_k'(\lambda)p_k'(\lambda)^{-1}\lambda U_n^- h_2$$
$$\subseteq h_1 U_n \lambda U_n h_2 = U_n h_1 \lambda h_2 U_n = U_n ab U_n.$$

Thus the combing (p_k') is smooth in the sense of Definition 16. $\qquad\square$

Finally, the compatibility condition of Definition 17 is easy to check using the explicit description of $\mathcal{S}(G^2)$ in Lemma 2. In order to cover also the Schwartz algebras for discrete groups, which are defined by ℓ_1-estimates, we define spaces

$$L_p^\sigma(G) := \{f\colon G \to \mathbb{C} \mid f \cdot \sigma^k \in L_p(G)\ \forall k \in \mathbb{N}\}$$

for all $1 \le p < \infty$, and equip them with the evident bornology: a subset $T \subseteq L_p^\sigma(G)$ is bounded if and only if for any $k \in \mathbb{N}$ there is C_k such that $\|f \cdot \sigma^k\|_{L_p(G)} \le C_k$ for all $f \in T$. Let $L_p^\sigma(G/\!\!/U)$ be the subspace of U-bi-invariant functions in $L_p^\sigma(G)$.

For $p = 2$, this agrees with the previous definition, so that Lemma 2 yields

$$\mathcal{S}(G) = \varinjlim L_2^\sigma(G/\!\!/U), \qquad \mathcal{S}(G^2) = \varinjlim L_2^\sigma(G^2/\!\!/U^2).$$

LEMMA 27. *Let (p_k) be a combing of polynomial growth on G/H. Then the sequence of operators (R_k) used in Definition 17 is uniformly bounded as operators $L_p^\sigma(G) \to L_p^\sigma(G^2)$. Here we use the scale $\sigma(a,b) := \sigma(a)\sigma(b)$ on G^2.*

PROOF. The operator $W\phi(x,y) := \phi(x, x^{-1}y)$ is an isometry of $L_p(G^2)$. It is also sufficiently compatible with the scale on G^2 for W and its inverse to be bounded linear operators on $L_p^\sigma(G^2)$. We have $R_k\phi(a,b) = \phi(ab)1_{p_k(abH)}(a)$, where $1_{p_k(abH)}$ denotes the characteristic function of $p_k(abH) \subseteq G$. Hence

$$WR_k\phi(x,y) = \phi(y) \cdot 1_{p_k(yH)}(x).$$

Since the combing (p_k) has polynomial growth, $\sigma(p_k(yH))$ is controlled by a polynomial in $\sigma(y)$. The boundedness of $W \circ R_k$ is now immediate because all cosets xH have volume 1. This implies the boundedness of R_k. $\qquad\square$

Since the combing (p_k') is smooth, for any $U \in \mathrm{CO}(G)$ there exists $V \in \mathrm{CO}(G)$ such that R_k maps $\mathcal{T}_\mathrm{c}(G/\!\!/U)$ into $\mathcal{T}_\mathrm{c}(G^2/\!\!/V^2)$. Together with Lemma 27 for $p = 2$, this yields that the combing (p_k') is compatible with $\mathcal{S}(G)$ in the sense of Definition 17. We have now verified all the hypotheses of Theorem 18. Thus $\mathcal{S}(G)$ is isocohomological. $\qquad\square$

6.1. Decomposition with respect to the centre of G. As before, we let G be a reductive p-adic group. Let $C(G) \subseteq G$ be the connected centre of G and let $\chi\colon C(G) \to \mathrm{U}(1)$ be a unitary character on $C(G)$. Let $\mathrm{Mod}_\chi(G)$ be the full subcategory of $\mathrm{Mod}(G)$ whose objects are the representations $\pi\colon G \to \mathrm{Aut}(V)$ that satisfy $\pi(z) = \chi(z)\mathrm{id}_V$ for all $z \in C(G)$. Let

$$\mathrm{Mod}_\chi\big(\mathcal{S}(G)\big) := \mathrm{Mod}\big(\mathcal{S}(G)\big) \cap \mathrm{Mod}_\chi(G)$$

be the subcategory of tempered representations in $\mathrm{Mod}_\chi(G)$. The class of extensions with a bounded linear section turns $\mathrm{Mod}_\chi(G)$ and $\mathrm{Mod}_\chi(\mathcal{S}(G))$ into exact categories, so that we can form the derived categories $\mathrm{Der}_\chi(G)$ and $\mathrm{Der}_\chi\big(\mathcal{S}(G)\big)$. Let $G_\mathrm{ss} := G/C(G)$, this is again a reductive p-adic group. If $\chi = 1$, then

$\mathrm{Mod}_\chi(G) = \mathrm{Mod}(G_{\mathrm{ss}})$ and $\mathrm{Mod}_\chi(\mathcal{S}(G)) = \mathrm{Mod}(\mathcal{S}(G_{\mathrm{ss}}))$. In general, there are quasi-unital bornological algebras $\mathcal{H}_\chi(G)$ and $\mathcal{S}_\chi(G)$ such that

$$\mathrm{Mod}_\chi(G) \cong \mathrm{Mod}(\mathcal{H}_\chi(G)), \qquad \mathrm{Mod}_\chi(\mathcal{S}(G)) \cong \mathrm{Mod}(\mathcal{S}_\chi(G)).$$

We briefly recall their well-known definitions. A $C(G)$-invariant subset of G is called $C(G)$-*compact* if its image in G_{ss} is compact. Let $\mathcal{H}_\chi(G)$ be the space of locally constant functions $f\colon G \to \mathbb{C}$ with $C(G)$-compact support such that $f(z^{-1}g) = \chi(z)f(g)$ for all $g \in G$, $z \in C(G)$. If $f_1, f_2 \in \mathcal{H}_\chi(G)$, then the function $h \mapsto f_1(h)f_2(h^{-1}g)$ is $C(G)$-invariant, so that

$$f_1 * f_2(g) := \int_{G_{\mathrm{ss}}} f_1(h)f_2(h^{-1}g)\,dh$$

makes sense; here dh denotes the Haar measure on G_{ss}. This turns $\mathcal{H}_\chi(G)$ into an algebra, which we equip with the fine bornology.

Let $L_2^\sigma(G)_\chi$ be the space of functions $f\colon G \to \mathbb{C}$ that satisfy $f(z^{-1}g) = \chi(z)f(g)$ for all $g \in G$, $z \in C(G)$, and such that $zg \mapsto |f(g)|$ is an element of $L_2^\sigma(G_{\mathrm{ss}})$. Let

$$\mathcal{S}_\chi(G) := \varinjlim L_2^\sigma(G/\!/U)_\chi$$

where U runs through the set of compact open subgroups with $U \cap C(G) \subseteq \ker \chi$. The same estimates as for $\mathcal{S}(G_{\mathrm{ss}})$ show that the convolution on $\mathcal{H}_\chi(G)$ extends to a bounded multiplication on $\mathcal{S}_\chi(G)$.

Consider the map

$$\rho\colon \mathcal{H}(G) \to \mathcal{H}_\chi(G), \qquad \rho f(g) := \int_{C(G)} \chi(z)f(zg)\,dz.$$

For appropriately normalised Haar measures, this is a surjective, bounded algebra homomorphism; that is, $\mathcal{H}_\chi(G)$ is a quotient algebra of $\mathcal{H}(G)$. Using ρ, we can pull back $\mathcal{H}_\chi(G)$-modules to $\mathcal{H}(G)$-modules. This construction maps essential modules again to essential modules (ρ is a proper morphism in the notation of [17]). Thus we have got a functor $\rho^*\colon \mathrm{Mod}(\mathcal{H}_\chi(G)) \to \mathrm{Mod}(\mathcal{H}(G))$. Since ρ is surjective, ρ^* is fully faithful. Thus $\mathrm{Mod}(\mathcal{H}_\chi(G))$ becomes a full subcategory of $\mathrm{Mod}(G)$. It is easy to identify this subcategory with $\mathrm{Mod}_\chi(G)$. If $(V, \pi) \in \mathrm{Mod}_\chi(G)$, then V becomes an essential $\mathcal{H}_\chi(G)$-module by

$$\pi(f) := \int_{G_{\mathrm{ss}}} f(g)\pi(g)\,dg.$$

This is well-defined because $f(gz)\pi(gz) = f(g)\pi(g)$ for all $g \in G$, $z \in C(G)$.

We can extend ρ to a bounded algebra homomorphism $\rho^{\mathcal{S}}\colon \mathcal{S}(G) \to \mathcal{S}_\chi(G)$. The map $\rho^{\mathcal{S}}$ has a bounded linear section and its kernel is the closure of $\ker\rho \subseteq \mathcal{H}(G)$. Therefore, bounded algebra homomorphisms $\mathcal{S}_\chi(G) \to \mathrm{End}(V)$ correspond to bounded algebra homomorphisms $\mathcal{S}(G) \to \mathrm{End}(V)$ whose restriction to $\mathcal{H}(G)$ vanishes on $\ker\rho$. Equivalently, $\mathrm{Mod}(\mathcal{S}_\chi(G)) \cong \mathrm{Mod}(\mathcal{S}(G)) \cap \mathrm{Mod}_\chi(G)$. Thus $\mathcal{H}_\chi(G)$ and $\mathcal{S}_\chi(G)$ have the required properties.

THEOREM 28. *The embedding* $\mathcal{H}_\chi(G) \to \mathcal{S}_\chi(G)$ *is isocohomological.*

PROOF. Let $\mathcal{H}_\chi(G)^{\mathrm{op}}$ be the opposite algebra of $\mathcal{H}_\chi(G)$, so that $\mathrm{Mod}(\mathcal{H}_\chi(G)^{\mathrm{op}})$ is the category of *right* $\mathcal{H}_\chi(G)$-modules. Since $\mathcal{H}_\chi(G)^{\mathrm{op}} \cong \mathcal{H}_{\chi^{-1}}(G)$, we have an isomorphism of categories $\mathrm{Mod}(\mathcal{H}_\chi(G)^{\mathrm{op}}) \cong \mathrm{Mod}_{\chi^{-1}}(G)$.

Equip $X \in \mathrm{Mod}(\mathcal{H}_\chi(G)^{\mathrm{op}})$, $V \in \mathrm{Mod}(\mathcal{H}_\chi(G))$ with the associated representations of G. We equip $X \mathbin{\hat\otimes} V$ with the diagonal representation. Since χ and χ^{-1} cancel, $C(G)$ acts trivially on $X \mathbin{\hat\otimes} V$. Thus we obtain a bifunctor

$$(20) \qquad \mathrm{Mod}(\mathcal{H}_\chi(G)^{\mathrm{op}}) \times \mathrm{Mod}(\mathcal{H}_\chi(G)) \to \mathrm{Mod}(G_{\mathrm{ss}}), \qquad (X,V) \mapsto X \mathbin{\hat\otimes} V.$$

This functor is evidently exact for extensions with a bounded linear section. Moreover, we claim that $X \mathbin{\hat\otimes} V$ is projective if X or V are projective. It suffices to treat the case where X is projective. We may even assume that X is a free essential module $X_0 \mathbin{\hat\otimes} \mathcal{H}_\chi(G)$. The diagonal representation on $X_0 \mathbin{\hat\otimes} \mathcal{H}_\chi(G) \mathbin{\hat\otimes} Y$ is isomorphic to the regular representation $\rho_g \otimes 1 \otimes 1$ on $\mathcal{H}(G_{\mathrm{ss}}) \mathbin{\hat\otimes} X_0 \mathbin{\hat\otimes} Y$. The intertwining operator is given by $\Phi(x \otimes f \otimes y)(g) := f(g)x \otimes gy$ for all $g \in G$; this function only depends on the class of g in G_{ss}.

Let $X \in \mathrm{Mod}(\mathcal{H}_\chi(G)^{\mathrm{op}})$, $V \in \mathrm{Mod}(\mathcal{H}_\chi(G))$. Then $X \mathbin{\hat\otimes}_{\mathcal{H}_\chi(G)} V$ is defined as the quotient of $X \mathbin{\hat\otimes} V$ by the closed linear span of $x * f \otimes v - x \otimes f * v$ for $x \in X$, $f \in \mathcal{H}_\chi(G)$, $v \in V$. Since $\rho \colon \mathcal{H}(G) \to \mathcal{H}_\chi(G)$ is surjective, this is the same as $X \mathbin{\hat\otimes}_{\mathcal{H}(G)} V$, which we have identified with $X \mathbin{\hat\otimes}_G V$ in Section 3. Thus

$$X \mathbin{\hat\otimes}_{\mathcal{H}_\chi(G)} V \cong \mathbb{C} \mathbin{\hat\otimes}_{G_{\mathrm{ss}}} (X \mathbin{\hat\otimes} V).$$

The same assertion holds for the total derived functors because the bifunctor in (20) is exact and preserves projectives. Especially, we get

$$\mathcal{S}_\chi(G) \mathbin{\hat\otimes}^{\mathbb{L}}_{\mathcal{H}_\chi(G)} \mathcal{S}_\chi(G) \cong \mathbb{C} \mathbin{\hat\otimes}^{\mathbb{L}}_{G_{\mathrm{ss}}} (\mathcal{S}_\chi(G) \mathbin{\hat\otimes} \mathcal{S}_\chi(G)).$$

Here $\mathcal{S}_\chi(G) \mathbin{\hat\otimes} \mathcal{S}_\chi(G)$ is equipped with the inner conjugation action of G_{ss}. We identify $\mathcal{S}_\chi(G) \mathbin{\hat\otimes} \mathcal{S}_\chi(G) \cong \mathcal{S}_{\chi \times \chi}(G \times G)$ as in Lemma 2. By [17, Theorem 35.2], the embedding $\mathcal{H}_\chi(G) \to \mathcal{S}_\chi(G)$ is isocohomological if and only if $\mathcal{S}_\chi(G) \mathbin{\hat\otimes}^{\mathbb{L}}_{\mathcal{H}_\chi(G)} \mathcal{S}_\chi(G)$ is a resolution of $\mathcal{S}_\chi(G)$. Thus the assertion that we have to prove is equivalent to

$$\mathbb{C} \mathbin{\hat\otimes}^{\mathbb{L}}_{G_{\mathrm{ss}}} \mathcal{S}_{\chi \times \chi}(G \times G) \cong \mathcal{S}_\chi(G).$$

We already know $\mathbb{C} \mathbin{\hat\otimes}^{\mathbb{L}}_{G_{\mathrm{ss}}} \mathcal{S}(G_{\mathrm{ss}} \times G_{\mathrm{ss}}) \cong \mathcal{S}(G_{\mathrm{ss}})$ because $\mathcal{S}(G_{\mathrm{ss}})$ is isocohomological (Theorem 22) and this condition is equivalent to $\mathcal{S}(G_{\mathrm{ss}})$ being isocohomological.

Now we choose a continuous section $s \colon G_{\mathrm{ss}} \to G$; this is possible because G is totally disconnected. It yields bornological isomorphisms

$$\Psi' \colon \mathcal{S}_\chi(G) \to \mathcal{S}(G_{\mathrm{ss}}), \qquad\qquad \Psi' f(g) := f \circ s(g),$$

$$\Psi \colon \mathcal{S}_{\chi \times \chi}(G \times G) \to \mathcal{S}(G_{\mathrm{ss}} \times G_{\mathrm{ss}}), \qquad \Psi f(g,h) := f\big(s(g), s(g)^{-1} s(gh)\big).$$

The isomorphism Ψ intertwines the inner conjugation actions of G_{ss} on $\mathcal{S}_{\chi \times \chi}(G \times G)$ and $\mathcal{S}(G_{\mathrm{ss}} \times G_{\mathrm{ss}})$. Thus we get isomorphisms

$$(21) \qquad \mathbb{C} \mathbin{\hat\otimes}^{\mathbb{L}}_{G_{\mathrm{ss}}} \mathcal{S}_{\chi \times \chi}(G \times G) \cong \mathbb{C} \mathbin{\hat\otimes}^{\mathbb{L}}_{G_{\mathrm{ss}}} \mathcal{S}(G_{\mathrm{ss}} \times G_{\mathrm{ss}}) \cong \mathcal{S}(G_{\mathrm{ss}}) \cong \mathcal{S}_\chi(G).$$

It is easy to see that the composite isomorphism is induced by the convolution map in $\mathcal{S}_\chi(G)$. $\qquad\square$

7. Applications to representation theory

Let G be a reductive p-adic group, let $C(G)$ be its centre, and let $G_{\mathrm{ss}} := G/C(G)$. So far we have used very large projective $\mathcal{H}(G)$-module resolutions, which offer great flexibility for writing down contracting homotopies. Now we consider much smaller projective resolutions, which are useful for explicit calculations. We write $\mathrm{Mod}_{(\chi)}(G)$ if it makes no difference whether we work in $\mathrm{Mod}(G)$ or $\mathrm{Mod}_\chi(G)$

for some character $\chi\colon C(G) \to \mathrm{U}(1)$. Similarly, we write $\mathcal{H}_{(\chi)}(G)$ and $\mathcal{S}_{(\chi)}(G)$. The actual applications of our main theorem are contained in Sections 7.1, 7.7 and 7.8. The other subsections contain small variations on known results. Our presentation differs somewhat from the accounts in [20, 25] because we want to exhibit connections with K-theory and assembly maps.

7.1. Cohomological dimension. Let $\mathrm{rk}\,G = \dim \mathcal{BT}(G)$ be the *rank* of G.

THEOREM 29. *The cohomological dimensions of the exact categories* $\mathsf{Mod}(G)$ *and* $\mathsf{Mod}(\mathcal{S}(G))$ *are (at most)* $\mathrm{rk}\,G$; *that is, any object has a projective resolution of length* $\mathrm{rk}\,G$. *Similarly, the cohomological dimensions of* $\mathsf{Mod}_\chi(G)$ *and* $\mathsf{Mod}_\chi(\mathcal{S}(G))$ *for a character* $\chi\colon C(G) \to \mathrm{U}(1)$ *are at most* $\mathrm{rk}\,G_{\mathrm{ss}}$.

PROOF. The assertions are well-known for $\mathsf{Mod}(G)$ and $\mathsf{Mod}_\chi(G)$. For the proof, equip $\mathcal{BT} = \mathcal{BT}(G)$ with a CW-complex structure for which G acts by cellular maps. Then the cellular chain complex $C_\bullet(\mathcal{BT})$ is a projective $\mathcal{H}(G)$-module resolution of the trivial representation of length $\mathrm{rk}\,G$. The chain complex $C_\bullet(\mathcal{BT}) \hat{\otimes} V$ with the diagonal representation of G is a projective $\mathcal{H}(G)$-module resolution of V for arbitrary $V \in \mathsf{Mod}(G)$. If $V \in \mathsf{Mod}_\chi(G)$, we use the building $\mathcal{BT}(G_{\mathrm{ss}})$ instead; $C_\bullet(\mathcal{BT}(G_{\mathrm{ss}})) \hat{\otimes} V$ is a projective resolution of V in $\mathsf{Mod}_\chi(G)$.

What is new is that we get the same assertions for $\mathsf{Mod}(\mathcal{S}(G))$ and $\mathsf{Mod}(\mathcal{S}_\chi(G))$. Since the argument is the same in both cases, we only write it down for $\mathcal{S}(G)$. Let $V \in \mathsf{Mod}(\mathcal{S}(G))$ and let $P_\bullet \to V$ be a projective resolution in $\mathsf{Mod}(G)$ of length $\mathrm{rk}\,G$. Then $\mathcal{S}(G) \hat{\otimes}_G P_\bullet$ has the homotopy type of

$$\mathcal{S}(G) \hat{\otimes}^{\mathrm{L}}_G V \cong \mathcal{S}(G) \hat{\otimes}^{\mathrm{L}}_{\mathcal{S}(G)} V \cong V \tag{22}$$

because $\mathcal{S}(G)$ is isocohomological (Theorem 22); here we use one of the equivalent characterisations of isocohomological embeddings listed in Section 4.2. Equation (22) means that $\mathcal{S}(G) \hat{\otimes}_G P_\bullet$ is a resolution of V. This resolution is projective and has length $\mathrm{rk}\,G$. □

Conversely, there is $V \in \mathsf{Mod}(G)$ with $\mathrm{Ext}^{\mathrm{rk}\,G}_G(V,V) \neq 0$. Hence the cohomological dimension of $\mathsf{Mod}(G)$ is equal to $\mathrm{rk}\,G$. We can even take V tempered and irreducible. Hence $\mathrm{Ext}^{\mathrm{rk}\,G}_{\mathcal{S}(G)}(V,V) \neq 0$ as well because $\mathcal{S}(G)$ is isocohomological. Thus $\mathsf{Mod}(\mathcal{S}(G))$ also has cohomological dimension equal to $\mathrm{rk}\,G$. Similarly, the cohomological dimension of $\mathsf{Mod}_\chi(G)$ and $\mathsf{Mod}_\chi(\mathcal{S}(G))$ is equal to $\mathrm{rk}\,G_{\mathrm{ss}}$.

7.2. Finite projective resolutions. We use a result of Joseph Bernstein to attach an Euler characteristic $\mathrm{Eul}(V)$ in $\mathrm{K}_0(\mathcal{H}_{(\chi)}(G))$ to a finitely generated representation $V \in \mathsf{Mod}_{(\chi)}(G)$.

DEFINITION 30. A smooth representation V is called *finitely generated* if there exist finitely many elements $v_1, \ldots, v_n$ such that the map

$$\mathcal{H}(G)^n \to V, \qquad (f_1, \ldots, f_n) \mapsto \sum_{j=1}^n f_j * v_j$$

is a bornological quotient map.

An admissible representation is finitely generated if and only if it has *finite length*, that is, it has a Jordan-Hölder series of finite length.

Since $\mathcal{H}(G)^n$ carries the fine bornology, the same is true for its quotients. Hence a finitely generated representation necessarily belongs to $\mathrm{Mod}_{\mathrm{alg}}(G)$. In the situation of Definition 30, there exists $U \in \mathrm{CO}(G)$ fixing v_j for all $j \in \{1, \ldots, n\}$. Thus we get a bornological quotient map $\mathcal{H}(G/U)^n \twoheadrightarrow V$. Conversely, $\mathcal{H}(G/U)^n$ is finitely generated and projective. Thus a smooth representation is finitely generated if and only if it is a quotient of $\mathcal{H}(G/U)^n$ for some $U \in \mathrm{CO}(G)$. If $V \in \mathrm{Mod}_\chi(G)$, then we may replace $\mathcal{H}(G/U)^n$ by $\mathcal{H}_\chi(G/U)^n$ for some $U \in \mathrm{CO}(G)$ with $\chi|_{U \cap C(G)} = 1$.

DEFINITION 31. An object of $\mathrm{Mod}_{(\chi)}(G)$ has type *(FP)* if it admits a resolution of finite length by finitely generated projective objects of $\mathrm{Mod}_{(\chi)}(G)$. Such a resolution is called a *finite projective resolution*.

THEOREM 32 (Joseph Bernstein). *An object of* $\mathrm{Mod}_{(\chi)}(G)$ *has type (FP) if and only if it is finitely generated.*

PROOF. It is trivial that representations of type (FP) are finitely generated. Conversely, if V is finitely generated, then V is a quotient of a finitely generated projective representation, say, $\partial_0 \colon \mathcal{H}_{(\chi)}(G/U)^n \twoheadrightarrow V$. By [3, Remark 3.12], subrepresentations of finitely generated representations are again finitely generated. Especially, $\ker \partial_0$ is finitely generated. By induction, we get a resolution (P_n, ∂_n) of V by finitely generated projective objects. By Theorem 29, the kernel of $\partial_n \colon P_n \to P_{n-1}$ is projective for sufficiently large n. Hence

$$0 \to \ker \partial_n \to P_n \to \ldots \to P_0 \to V$$

is a finite projective resolution.   □

The *algebraic K-theory* $\mathrm{K}_0\big(\mathcal{H}_{(\chi)}(G)\big)$ is the Grothendieck group of the monoid of finitely generated projective $\mathcal{H}_{(\chi)}(G)$-modules. This is so because $\mathcal{H}_{(\chi)}(G)$ is a union of unital subalgebras.

DEFINITION 33. Let $V \in \mathrm{Mod}_{(\chi)}(G)$ be finitely generated. Then V is of type (FP) by Bernstein's Theorem 32. Choose a finite projective resolution

$$0 \to P_n \to \ldots \to P_0 \to V \to 0.$$

The *Euler characteristic* of V is defined by

$$\mathrm{Eul}(V) := \sum_{j=0}^n (-1)^j [P_j] \in \mathrm{K}_0\big(\mathcal{H}_{(\chi)}(G)\big).$$

We check that this does not depend on the resolution (see also [19, Section 1.7]). Define the Euler characteristic $\mathrm{Eul}(P_\bullet)$ for finite projective complexes in the obvious fashion. Let $P_\bullet$ and $P'_\bullet$ be two finite projective resolutions of V. The identity map on V lifts to a chain homotopy equivalence $f \colon P_\bullet \to P'_\bullet$. Hence the mapping cone C_f of f is contractible. The Euler characteristic vanishes for contractible complexes. Hence $\mathrm{Eul}(C_f) = 0$. This is equivalent to $\mathrm{Eul}(P_\bullet) = \mathrm{Eul}(P'_\bullet)$.

DEFINITION 34. Let

$$\mathrm{HH}_0\big(\mathcal{H}_{(\chi)}(G)\big) := \mathcal{H}_{(\chi)}(G)/[\mathcal{H}_{(\chi)}(G), \mathcal{H}_{(\chi)}(G)].$$

The *universal trace* is a map

$$\mathrm{tr}_{\mathrm{univ}} \colon \mathrm{K}_0\big(\mathcal{H}_{(\chi)}(G)\big) \to \mathrm{HH}_0\big(\mathcal{H}_{(\chi)}(G)\big).$$

If $(p_{ij}) \in M_n\big(\mathcal{H}_{(\chi)}(G)\big)$ is an idempotent with $\mathcal{H}_{(\chi)}(G)^n \cdot (p_{ij}) \cong V$, then we have $\mathrm{tr}_{\mathrm{univ}}[V] = \big[\sum p_{ii}\big]$.

The above definitions are inspired by constructions of Hyman Bass in [2], where $\mathrm{tr}_{\mathrm{univ}} \mathrm{Eul}(V)$ is constructed for modules of type (FP) over unital algebras.

7.3. Traces from admissible representations. Let $W \in \mathrm{Mod}_{(\chi)}(G)$ be an admissible representation. Its integrated form is an algebra homomorphism ρ from $\mathcal{H}_{(\chi)}(G)$ to the algebra $\mathrm{End}_{\mathrm{fin}}(W) := \tilde{W} \otimes W$ of smooth finite rank operators on W. Here $\tilde{W}$ denotes the contragradient representation and $\mathrm{End}_{\mathrm{fin}}(W)$ carries the fine bornology. Composing ρ with the standard trace on $\mathrm{End}_{\mathrm{fin}}(W)$, we get a trace $\mathrm{tr}_W \colon \mathrm{HH}_0(\mathcal{H}_{(\chi)}(G)) \to \mathbb{C}$ and a functional

$$\tau_W \colon \mathrm{K}_0(\mathcal{H}_{(\chi)}(G)) \to \mathbb{Z}.$$

The following computation of τ_W is a variant of [2, Proposition 4.2].

PROPOSITION 35. *Let $V, W \in \mathrm{Mod}_{(\chi)}(G)$, let V be finitely generated projective and let W be admissible. Then $\tau_W[V] = \dim \mathrm{Hom}_G(V, W)$ and $\mathrm{Hom}_G(V, W)$ is finite-dimensional.*

PROOF. The functoriality of K_0 for the homomorphism $\mathcal{H}_{(\chi)}(G) \to \mathrm{End}_{\mathrm{fin}}(W)$ maps $[V]$ to the class of the finitely generated projective module

$$V' := \mathrm{End}_{\mathrm{fin}}(W) \hat{\otimes}_{\mathcal{H}_{(\chi)}(G)} V \cong W \otimes (\tilde{W} \otimes_G V) \cong W^{\dim \tilde{W} \otimes_G V}$$

over $\mathrm{End}_{\mathrm{fin}}(W)$. Thus $\tau_W[V] = \dim \tilde{W} \otimes_G V$. By adjoint associativity,

$$\mathrm{Hom}(\tilde{W} \otimes_G V, \mathbb{C}) \cong \mathrm{Hom}_G(V, \mathrm{Hom}(\tilde{W}, \mathbb{C})) \cong \mathrm{Hom}_G(V, \tilde{\tilde{W}}).$$

We have $W \cong \tilde{\tilde{W}}$ because W is admissible. Thus $\tau_W[V] = \dim \mathrm{Hom}_G(V, W)$. $\square$

Let $V, W \in \mathrm{Mod}_{(\chi)}(G)$, let W be admissible and let V be finitely generated. By Bernstein's Theorem 32, there is a finite projective resolution $P_\bullet \to V$ in $\mathrm{Mod}_{(\chi)}(G)$. By Proposition 35, $\mathrm{Hom}_G(P_\bullet, W)$ is a chain complex of finite-dimensional vector spaces. Hence its homology $\mathrm{Ext}^n_{\mathcal{H}_{(\chi)}(G)}(V, W)$ is finite-dimensional as well and

$$(23) \quad \sum_{n=0}^{\infty} (-1)^n \dim \mathrm{Ext}^n_{\mathcal{H}_{(\chi)}(G)}(V, W) = \sum_{n=0}^{\infty} (-1)^n \dim \mathrm{Hom}_G(P_n, W)$$

$$= \sum_{n=0}^{\infty} (-1)^n \tau_W[P_n] = \tau_W(\mathrm{Eul}(V)).$$

We call this the *Euler-Poincaré characteristic* $\mathrm{EP}_{(\chi)}(V, W)$ of V and W (compare [20, page 135]).

7.4. Formal dimensions. Evaluation at $1 \in G$ is a trace on $\mathcal{H}_{(\chi)}(G)$, that is, a linear functional $\tau_1 \colon \mathrm{HH}_0(\mathcal{H}_{(\chi)}(G)) \to \mathbb{C}$. The functional

$$\dim := \tau_1 \circ \mathrm{tr}_{\mathrm{univ}} \colon \mathrm{K}_0(\mathcal{H}_{(\chi)}(G)) \to \mathbb{C}$$

computes the *formal dimension* for finitely generated projective $\mathcal{H}_{(\chi)}(G)$-modules. Recall that an irreducible representation in $\mathrm{Mod}_\chi(G)$ is projective if and only if it is supercuspidal. Unless $C(G)$ is compact, $\mathrm{Mod}(G)$ has no irreducible projective objects.

We can also define the formal dimension for representations that are square-integrable (see [22]). Let $(V, \pi) \in \mathrm{Mod}_\chi(G)$ be irreducible and square-integrable (or, more precisely, square-integrable *modulo the centre* $C(G)$). Since irreducible representations are admissible, V carries the fine bornology. Moreover, square-integrable representations are tempered. Thus the integrated form of π extends to a bounded homomorphism $\pi\colon \mathcal{S}_\chi(G) \to \mathrm{End}_{\mathrm{fin}}(V)$. Since V is irreducible, this homomorphism is surjective. The crucial property of irreducible square-integrable representation is that there is an ideal $I \subseteq \mathcal{S}_\chi(G)$ such that $I \oplus \ker\pi \cong \mathcal{S}_\chi(G)$. Thus $\pi|_I\colon I \to \mathrm{End}_{\mathrm{fin}}(V)$ is an algebra isomorphism. It is necessarily a bornological isomorphism because it is bounded and $\mathrm{End}_{\mathrm{fin}}(V)$ carries the fine bornology.

PROPOSITION 36. *Let $V \in \mathrm{Mod}_\chi(G)$ be irreducible and square-integrable. Then V is both projective and injective as an object of $\mathrm{Mod}_\chi(\mathcal{S}(G))$.*

PROOF. The direct-sum decomposition $\mathcal{S}_\chi(G) \cong \ker\pi \oplus \mathrm{End}_{\mathrm{fin}}(V)$ gives rise to an equivalence of exact categories

$$\mathrm{Mod}_\chi(\mathcal{S}(G)) \cong \mathrm{Mod}(\ker\pi) \times \mathrm{Mod}(\mathrm{End}_{\mathrm{fin}}(V)).$$

The representation V belongs to the second factor. The algebra $\mathrm{End}_{\mathrm{fin}}(V)$ is canonically Morita equivalent to $\mathbb{C}$, so that $\mathrm{Mod}(\mathrm{End}_{\mathrm{fin}}(V))$ and $\mathrm{Mod}(\mathbb{C})$ are equivalent exact categories. The easiest way to get this Morita equivalence uses a basis in V to identify $\mathrm{End}_{\mathrm{fin}}(V) \cong \bigcup_{n=1}^\infty M_n(\mathbb{C})$. Since any extension in $\mathrm{Mod}(\mathbb{C})$ splits, any object of $\mathrm{Mod}(\mathbb{C})$ is both injective and projective. $\square$

We can also define $\mathrm{K}_0(\mathcal{S}_{(\chi)}(G))$, $\mathrm{HH}_0(\mathcal{S}_{(\chi)}(G))$, and

$$\mathrm{tr}_{\mathrm{univ}}\colon \mathrm{K}_0(\mathcal{S}_{(\chi)}(G)) \to \mathrm{HH}_0(\mathcal{S}_{(\chi)}(G)).$$

It is irrelevant for the following whether we divide by the linear or closed linear span of the commutators in the definition of $\mathrm{HH}_0(\mathcal{S}_{(\chi)}(G))$. The trace τ_1 extends to a bounded trace $\tau_1^{\mathcal{S}}\colon \mathrm{HH}_0(\mathcal{S}_{(\chi)}(G)) \to \mathbb{C}$. This induces a functional $\dim^{\mathcal{S}}\colon \mathrm{K}_0(\mathcal{S}_{(\chi)}(G)) \to \mathbb{Z}$. An irreducible square-integrable representation V defines a class $[V] \in \mathrm{K}_0(\mathcal{S}_{(\chi)}(G))$ by Proposition 36; we define its *formal dimension* by $\dim^{\mathcal{S}} V := \dim^{\mathcal{S}}[V]$.

The embedding $\mathcal{H}_{(\chi)}(G) \to \mathcal{S}_{(\chi)}(G)$ induces natural maps

$$\iota\colon \mathrm{K}_0(\mathcal{H}_{(\chi)}(G)) \to \mathrm{K}_0(\mathcal{S}_{(\chi)}(G)), \qquad [V] \mapsto [\mathcal{S}_{(\chi)}(V) \,\hat{\otimes}_{\mathcal{H}_{(\chi)}(G)}\, V],$$
$$\iota\colon \mathrm{HH}_0(\mathcal{H}_{(\chi)}(G)) \to \mathrm{HH}_0(\mathcal{S}_{(\chi)}(G)). \qquad [f] \mapsto [f].$$

These maps are compatible with the universal traces and satisfy $\tau_1^{\mathcal{S}} \circ \iota = \tau_1$ and $\dim^{\mathcal{S}} \circ \iota = \dim$.

It is shown in [25] that $\mathcal{S}_{(\chi)}(G)$ is closed under holomorphic functional calculus in the C^*-algebra $C^*_{\mathrm{red},(\chi)}(G)$. Hence

$$\mathrm{K}_0(\mathcal{S}_{(\chi)}(G)) \cong \mathrm{K}_0(C^*_{\mathrm{red},(\chi)}(G)).$$

It follows also that any finitely generated projective module V over $\mathcal{S}_{(\chi)}(G)$ is the range of a *self-adjoint* idempotent element in $M_n(\mathcal{S}_{(\chi)}(G))$ for some $n \in \mathbb{N}$. Since the trace $\tau_1^{\mathcal{S}}$ is positive, we get $\dim^{\mathcal{S}} V > 0$ unless $V = 0$.

Yet another notion of formal dimension comes from the theory of von Neumann algebras. The *(χ-twisted) group von Neumann algebra* of G is the closure $N_{(\chi)}(G)$ of $\mathcal{H}_{(\chi)}(G)$ or $\mathcal{S}_{(\chi)}(G)$ in the weak operator topology on $L_2(G)_{(\chi)}$. We may extend $\tau_1^{\mathcal{S}}$ to a positive unbounded trace τ_1^N on $N_{(\chi)}(G)$. Any normal $*$-representation ρ of $N_{(\chi)}(G)$ on a separable Hilbert space is isomorphic to the left regular representation on the Hilbert space $(L_2(G)_{(\chi)} \,\bar{\otimes}\, \ell_2(\mathbb{N})) \cdot p_\rho$ for some projection $p_\rho \in N_{(\chi)}(G) \,\bar{\otimes}$

$B(\ell_2\mathbb{N})$ where $\bar{\otimes}$ denotes spatial tensor products of Hilbert spaces and von Neumann algebras, respectively; the projection p_ρ is unique up to unitary equivalence. We define the *formal dimension* $\dim^N(\rho)$ to be $\tau_1^N(p_\rho) \in [0, \infty]$; this does not depend on the choice of p_ρ.

By definition, we have $\dim^N(L_2(G)_{(\chi)}^n \cdot e) = \dim^S[e]$ if $e \in M_n(\mathcal{S}_{(\chi)}(G))$ is a *self-adjoint* idempotent. Since $\mathcal{S}_{(\chi)}(G)$ is closed under holomorphic functional calculus in the reduced group C^*-algebra $C^*_{\mathrm{red},(\chi)}(G)$, any idempotent element of $\mathcal{S}_{(\chi)}(G)$ is similar to a self-adjoint idempotent. Hence $\dim^N(L_2(G)_{(\chi)}^n \cdot e) = \dim^S[e]$ holds for any idempotent $e \in M_n(\mathcal{S}_{(\chi)}(G))$. We have

$$L_2(G)_{(\chi)} \, \hat{\otimes}_{\mathcal{S}_{(\chi)}(G)} \, \mathcal{S}_{(\chi)}(G)^n \cdot e \cong L_2(G)_{(\chi)}^n \cdot e.$$

Therefore, if V is a finitely generated projective left $\mathcal{S}_{(\chi)}(G)$-module V, then we may view $L_2(G)_{(\chi)} \, \hat{\otimes}_{\mathcal{S}_{(\chi)}(G)} \, V$ as a Hilbert space equipped with a faithful normal $*$-representation of $N_{(\chi)}(G)$; the resulting representation is uniquely determined up to unitary equivalence because any two self-adjoint idempotents realising V are unitarily equivalent in $\mathcal{S}_{(\chi)}(G)$. The formal dimensions from the Schwartz algebra and the von Neumann algebra are compatible in the following sense:

$$\dim^N(L_2(G)_{(\chi)} \, \hat{\otimes}_{\mathcal{S}_{(\chi)}(G)} \, V) \cong \dim^S(V).$$

Thus $\dim^S V$ only depends on the unitary equivalence class of the associated unitary representation $L_2(G)_{(\chi)} \, \hat{\otimes}_{\mathcal{S}_{(\chi)}(G)} \, V$.

7.5. Compactly induced representations. Equip $U \in \mathsf{CO}(G)$ with the restriction of the Haar measure from G. The map $i_U^G \colon \mathcal{H}(U) \to \mathcal{H}(G)$ that extends functions by 0 outside U is an algebra homomorphism. Hence it induces a map

$$(i_U^G)_! \colon \mathrm{Rep}(U) \cong \mathrm{K}_0\big(\mathcal{H}(U)\big) \to \mathrm{K}_0\big(\mathcal{H}(G)\big), \qquad [V] \mapsto [\mathcal{H}(G) \otimes_{\mathcal{H}(U)} V].$$

This is the standard functoriality of K-theory. We denote it by $(i_U^G)_!$ because this notation is used in [17]. We call representations of the form $(i_U^G)_!(V)$ *compactly induced* because $\mathcal{H}(G) \otimes_{\mathcal{H}(U)} V \cong \mathrm{c\text{-}Ind}_U^G(V)$ (see [15]).

Let $U, V \in \mathsf{CO}(G)$ and suppose that $gUg^{-1} \subseteq V$ for some $g \in G$. Then we have $i_U^G = \gamma_g^{-1} \circ i_V^G \circ i_{gUg^{-1}}^V \circ \gamma_g$, where γ_g denotes conjugation by g. One checks that γ_g acts trivially on $\mathrm{K}_0\big(\mathcal{H}(G)\big)$. Hence $(i_U^G)_!$ is the composite of $(i_V^G)_!$ and the map $\mathrm{Rep}(U) \overset{\cong}{\to} \mathrm{Rep}(gUg^{-1}) \to \mathrm{Rep}(V)$ that is associated to the group homomorphism $U \to V$, $x \mapsto gxg^{-1}$. Let $\mathsf{Sub}(G)$ be the category whose objects are the compact open subgroups of G and whose morphisms are these special group homomorphisms. We have exhibited that $U \mapsto \mathrm{Rep}(U)$ is a module over this category. The various maps $(i_U^G)_!$ combine to a natural map

$$\varinjlim_{\mathsf{Sub}(G)} \mathrm{Rep}(U) \to \mathrm{K}_0\big(\mathcal{H}(G)\big).$$

We call this map the *assembly map* for $\mathrm{K}_0\big(\mathcal{H}(G)\big)$ because it is a variant of the Farrell-Jones assembly map for discrete groups (see [12, Conjecture 3.3]), which is in turn closely related to the Baum-Connes assembly map.

The above definitions carry over to $\mathcal{H}_\chi(G)$ in a straightforward fashion. Let $\mathsf{CO}\big(G; C(G)\big)$ be the set of $C(G)$-compact open subgroups of G containing $C(G)$. The projection to G_{ss} identifies $\mathsf{CO}\big(G; C(G)\big)$ with $\mathsf{CO}(G_{\mathrm{ss}})$. As above, we get algebra homomorphisms $i_U^G \colon \mathcal{H}_\chi(U) \to \mathcal{H}_\chi(G)$ for $U \in \mathsf{CO}\big(G; C(G)\big)$. There is an analogue of the Peter-Weyl theorem for $\mathcal{H}_\chi(U)$; that is, $\mathcal{H}_\chi(U)$ is a direct sum of matrix algebras. Therefore, finitely generated projective modules over $\mathcal{H}_\chi(U)$ are the same as finite-dimensional representations in $\mathsf{Mod}_\chi(U)$. This justifies defining $\mathrm{Rep}_\chi(U) := \mathrm{K}_0(\mathcal{H}_\chi(U))$. As above, we can factor $(i_U^G)_!$ through $(i_V^G)_!$ if U is subconjugate to V. The relevant category organising these subconjugations is the category $\mathsf{Sub}(G; C(G))$ whose set of objects is $\mathsf{CO}(G; C(G))$ and whose morphisms are the group homomorphisms $U \to V$ of the form $x \mapsto gxg^{-1}$ for some $g \in G$. Thus we get an assembly map

$$(24) \qquad \varinjlim_{\mathsf{Sub}(G;C(G))} \mathrm{Rep}_\chi(U) \to \mathrm{K}_0(\mathcal{H}_\chi(G)).$$

Let $\Gamma(G_{\mathrm{ss}}, dg) \subseteq \mathbb{R}$ be the subgroup generated by $\mathrm{vol}(U)^{-1}$ for $U \in \mathsf{CO}(G_{\mathrm{ss}})$. Since $\mathrm{vol}(U)/\mathrm{vol}(V) \in \mathbb{N}$ for $V \subseteq U$, this group is already generated by $\mathrm{vol}(U)^{-1}$ for maximal compact subgroups $U \subseteq G_{\mathrm{ss}}$. We have $\Gamma(G_{\mathrm{ss}}, dg) = \alpha\mathbb{Z}$ for some $\alpha > 0$ because there are only finitely many maximal compact subgroups and $\mathrm{vol}(U)/\mathrm{vol}(V) \in \mathbb{Q}$ for all $U, V \in \mathsf{CO}(G_{\mathrm{ss}})$. The number α depends on the choice of the Haar measure, of course. We let $\mathrm{size}(U) := \alpha\,\mathrm{vol}(U/C(G))$, so that $\mathrm{size}(U)^{-1} \in \mathbb{N}$ for all $U \in \mathsf{CO}(G; C(G))$.

LEMMA 37. *Let $U \in \mathsf{CO}(G; C(G))$, and let $W \in \mathsf{Mod}_\chi(U)$ be finite-dimensional. Let $c_W \colon U \to \mathbb{C}$ be the character of W. Then $\mathrm{tr}_{\mathrm{univ}}(i_U^G)_![W] \in \mathrm{HH}_0(\mathcal{H}_\chi(G))$ is represented by the function*

$$(25) \qquad c_{U,W}^G(g) := \begin{cases} \alpha\,\mathrm{size}(U)^{-1}\overline{c_W(g)} & \text{for } g \in U, \\ 0 & \text{for } g \notin U. \end{cases}$$

Moreover, $\dim(i_U^G)_![W] = \alpha\,\mathrm{size}(U)^{-1}\dim(W)$. Thus $\dim x \in \alpha\mathbb{Z}$ for all x in the range of the assembly map (24).

A similar result holds for compactly induced projective objects of $\mathsf{Mod}(G)$.

PROOF. Since the universal trace is compatible with the functoriality of K_0 and HH_0, the first assertion follows if $\mathrm{tr}_{\mathrm{univ}}[W] = \mathrm{vol}(U/C(G))^{-1}\overline{c_W(g)}$ in $\mathrm{HH}_0(\mathcal{H}_\chi(U))$. We briefly recall how this well-known identity is proved. We may assume that W is irreducible. Hence there is an idempotent $p_W \in \mathcal{H}_\chi(U)$ with $W \cong \mathcal{H}_\chi(U)p_W$. Thus $\mathrm{tr}_{\mathrm{univ}}[W] = [p_W]$. We can compute $c_W(g)$ for $g \in G$ as the trace of the finite rank operator $f \mapsto \lambda(g)f * p_W$ on $\mathcal{H}_\chi(U)$. This operator has the integral kernel $(x, y) \mapsto p_W(y^{-1}g^{-1}x)$, so that

$$\overline{c_W(g)} = c_W(g^{-1}) = \int_{U/C(G)} p_W(x^{-1}gx)\,dx.$$

This implies $[\overline{c_W}] = \int_{U/C(G)}[W]\,dx = \mathrm{vol}(U/C(G))[W]$ because conjugation does not change the class in $\mathrm{HH}_0(\mathcal{H}_\chi(U))$. We get the formula for formal dimensions because $\dim x = \mathrm{tr}_{\mathrm{univ}}(x)(1)$ and $c_W(1) = \dim W$. This lies in $\alpha\mathbb{Z}$ by construction of α. By additivity, we get $\dim x \in \alpha\mathbb{Z}$ for all x in the range of the assembly map (24). $\qquad\square$

7.6. Explicit finite projective resolutions. Let $V \in \mathsf{Mod}_\chi(G)$ be of finite length. Peter Schneider and Ulrich Stuhler construct an explicit finite projective resolution for such V in [20]. We only sketch the construction very briefly. Let $\mathcal{BT}(G_{\mathrm{ss}})$ be the affine Bruhat-Tits building of G_{ss}. One defines a coefficient system $\gamma_e(V)$ on

$\mathcal{BT}(G_{\mathrm{ss}})$, which depends on an auxiliary parameter $e \in \mathbb{N}$ (see [20, Section II.2]); its value on a facet F of $\mathcal{BT}(G_{\mathrm{ss}})$ is the—finite-dimensional—space $\mathrm{Fix}(U_F^e, V)$ for certain $U_F^e \in \mathrm{CO}(G)$. The cellular chain complex $C_\bullet(\mathcal{BT}(G_{\mathrm{ss}}), \gamma_e(V))$ with values in $\gamma_e(V)$ is a resolution of V for sufficiently large e ([20, Theorem II.3.1]). It is a finite projective resolution of V in $\mathrm{Mod}_\chi(G)$ because the stabilisers of facets belong to $\mathrm{CO}(G; C(G))$ and the set of facets is G_{ss}-finite.

PROPOSITION 38. *If* $V \in \mathrm{Mod}_\chi(G)$ *has finite length, then* $\mathrm{Eul}(V)$ *belongs to the range of the assembly map* (24). *Hence* $\dim(\mathrm{Eul}(V)) \in \alpha\mathbb{Z}$.

Define the Euler-Poincaré function $f_{\mathrm{EP}}^V \in \mathcal{H}_\chi(G)$ *of* V *as in [20, page 135]. Then* $[f_{\mathrm{EP}}^V] = \mathrm{tr}_{\mathrm{univ}} \mathrm{Eul}(V) \in \mathrm{HH}_0(\mathcal{H}_\chi(G))$. *Thus* $\dim(\mathrm{Eul}(V)) = f_{\mathrm{EP}}(1)$ *and* $\mathrm{EP}_\chi(V, W) = \mathrm{tr}_W(f_{\mathrm{EP}}^V)$ *for all admissible* $W \in \mathrm{Mod}_\chi(G)$.

See also [20, Proposition III.4.22] and [20, Proposition III.4.1].

PROOF. The finite projective resolution $C_\bullet(\mathcal{BT}(G_{\mathrm{ss}}), \gamma_e(V))$ is explicitly built out of compactly induced representations. Hence $\mathrm{Eul}(V)$ belongs to the range of the assembly map. Lemma 37 yields $\dim(\mathrm{Eul}(V)) \in \alpha\mathbb{Z}$ and allows us to compute $\mathrm{tr}_{\mathrm{univ}} \mathrm{Eul}(V)$. Inspection shows that this is exactly $[f_{\mathrm{EP}}^V]$. $\square$

Proposition 38 yields $\dim V = \dim \mathrm{Eul}(V) \in \alpha\mathbb{Z}$ if V is irreducible supercuspidal. This rationality result is due to Marie-France Vignéras ([25]).

7.7. Euler characteristics and formal dimensions for square-integrable representations.

THEOREM 39. *Let* $V \in \mathrm{Mod}_\chi(G)$ *be irreducible and square-integrable. Let*
$$\iota \colon \mathrm{K}_0(\mathcal{H}_\chi(G)) \to \mathrm{K}_0(\mathcal{S}_\chi(G))$$
be induced by the embedding $\mathcal{H}_\chi(G) \to \mathcal{S}_\chi(G)$. *Then* $\iota(\mathrm{Eul}(V)) = [V]$. *Hence* $[V]$ *lies in the range of the assembly map*
$$\varinjlim_{\mathrm{Sub}(G; C(G))} \mathrm{Rep}_\chi(U) \to \mathrm{K}_0(\mathcal{H}_\chi(G)) \to \mathrm{K}_0(\mathcal{S}_\chi(G))$$
and $f_{\mathrm{EP}}^V(1) = \dim(\mathrm{Eul}(V)) = \dim^{\mathcal{S}}(V)$. *This number belongs to* $\alpha \cdot \mathbb{N}_{\geq 1}$ *with* α *as in Lemma 37.*

PROOF. Choose $e \in \mathbb{N}$ large enough such that $C_\bullet(\mathcal{BT}(G_{\mathrm{ss}}), \gamma_e(V))$ is a projective $\mathcal{H}_\chi(G)$-module resolution of V. Then
$$\iota \mathrm{Eul}(V) = \sum_{n=0}^{\infty} (-1)^n [\mathcal{S}_\chi(G) \hat{\otimes}_{\mathcal{H}_\chi(G)} C_n(\mathcal{BT}(G_{\mathrm{ss}}), \gamma_e(V))].$$

Since $\mathcal{S}_\chi(G)$ is isocohomological (Theorem 28), $\mathcal{S}_\chi(G) \hat{\otimes}_{\mathcal{H}_\chi(G)}^{\mathrm{L}} V \cong V$. Therefore, $\mathcal{S}_\chi(G) \hat{\otimes}_{\mathcal{H}_\chi(G)} C_\bullet(\mathcal{BT}(G_{\mathrm{ss}}), \gamma_e(V))$ is still a projective $\mathcal{S}_\chi(G)$-module resolution of V. Since V is projective as well (Proposition 36), this resolution splits by bounded $\mathcal{S}_\chi(G)$-module homomorphisms. This implies $[V] = \iota \mathrm{Eul}(V)$. The remaining assertions now follow from Proposition 38 and $\dim^{\mathcal{S}}(V) > 0$. $\square$

THEOREM 40. *Let* $U \in \mathrm{CO}(G; C(G))$ *and let* $W \in \mathrm{Mod}_\chi(U)$ *be finite-dimensional. Then there are at most* $\dim(W) \cdot \mathrm{size}(U)^{-1}$ *different irreducible square-integrable representations whose restriction to* U *contains the representation* W.

PROOF. Let $V_1, \ldots, V_N$ be pairwise non-isomorphic irreducible square-integrable representations whose restriction to U contains W. Let $X := \mathcal{S}_\chi(G) \hat{\otimes}_{\mathcal{H}_\chi(U)} W$. There are natural adjoint associativity isomorphisms

$$\mathrm{Hom}_{\mathcal{H}_\chi(U)}(W, V_j) \cong \mathrm{Hom}_{\mathcal{S}_\chi(G)}(X, V_j)$$

for all j (see [17]). Thus we get non-zero maps $X \to V_j$. They are surjective and admit bounded linear sections because the representations V_j are irreducible and carry the fine bornology; since the representations V_j are projective (Proposition 36), they even admit G-equivariant bounded linear sections. Thus $V_1, \ldots, V_N$ are direct summands of X. Since they are not isomorphic, $\bigoplus V_j$ is a direct summand of X as well. Therefore, $\sum_{j=1}^N \dim^{\mathcal{S}}(V_j) \leq \dim^{\mathcal{S}} X$. Lemma 37 and Theorem 39 yield $\dim^{\mathcal{S}} X = \alpha \dim(W) \operatorname{size}(U)^{-1}$ and $\dim^{\mathcal{S}}(V_j) \geq \alpha$ for all j. Hence $N \leq \dim(W) \operatorname{size}(U)^{-1}$. $\qquad\square$

An irreducible square-integrable representation that is not supercuspidal is a subquotient of a representation that we get by Jacquet induction from a proper Levi subgroup. It is desirable in this situation to compute the formal dimension (and other invariants) of V from its cuspidal data. This gives rise to some rather intricate computations; these are carried out in [1] for representations of $\mathrm{Gl}_m(D)$ for a division algebra D.

7.8. Some vanishing results.

THEOREM 41. Let $V, W \in \mathrm{Mod}_\chi(G)$ be irreducible and tempered. If V or W is square-integrable, then $\mathrm{Ext}_G^n(V, W) = 0$ for all $n \geq 1$ and

$$\mathrm{EP}_\chi(V, W) = \begin{cases} 1 & \text{if } V \cong W; \\ 0 & \text{otherwise.} \end{cases}$$

PROOF. Since V and W are tempered and $\mathcal{S}_\chi(G)$ is isocohomological, we have

$$\mathrm{Ext}_{\mathcal{H}_\chi(G)}^n(V, W) \cong \mathrm{Ext}_{\mathcal{S}_\chi(G)}^n(V, W).$$

The latter vanishes for $n \geq 1$ by Proposition 36. For $n = 0$ we are dealing with $\mathrm{Hom}_G(V, W)$, which is computed by Schur's Lemma. $\qquad\square$

THEOREM 42. If V is irreducible and tempered but not square-integrable, then $f_{\mathrm{EP}}^V(1) = \dim \mathrm{Eul}(V) = 0$.

PROOF. This follows from the abstract Plancherel Theorem and Theorem 41 as in the proof of [20, Corollary III.4.7]. We merely outline the proof. The abstract Plancherel theorem applied to the type I C^*-algebra $C_{\mathrm{red},\chi}^*(G)$ yields that $f_{\mathrm{EP}}^V(1)$ is the integral of its Fourier transform $W \mapsto \mathrm{tr}_W(f_{\mathrm{EP}}^V)$ with respect to some measure μ, which is called the Plancherel measure. Here W runs through the tempered irreducible representations in $\mathrm{Mod}_\chi(G)$. Proposition 38 asserts that $\mathrm{tr}_W(f_{\mathrm{EP}}^V) = \mathrm{EP}_\chi(V, W)$.

We have $\mathrm{Ext}_{\mathcal{H}_\chi(G)}^n(V, W) = 0$ for all $n \in \mathbb{N}$ and hence $\mathrm{EP}_\chi(V, W) = 0$ unless V and W have the same infinitesimal character. Since the infinitesimal character is finite-to-one, the support of the function $W \mapsto \mathrm{EP}_\chi(V, W)$ is finite. Hence only atoms of the Plancherel measure μ contribute to the integral

$$f_{\mathrm{EP}}^V(1) = \int \mathrm{EP}_\chi(V, W) \, d\mu(W).$$

These atoms are exactly the square-integrable representations. Now Theorem 41 yields $f_{\mathrm{EP}}^V(1) = 0$ unless V is square-integrable. In addition, this computation shows that $f_{\mathrm{EP}}^V(1) = \dim^{\mathcal{S}}(V)$ if V is square-integrable (compare Theorem 39). $\qquad \Box$

Acknowledgment. This research was supported by the EU-Network *Quantum Spaces and Noncommutative Geometry* (Contract HPRN-CT-2002-00280) and the *Deutsche Forschungsgemeinschaft* (SFB 478).

References

[1] Anne-Marie Aubert and Roger Plymen, *Plancherel measure for* GL(n, F) *and* GL(m, D): *explicit formulas and Bernstein decomposition*, J. Number Theory **112** (2005), 26–66. MR2131140

[2] Hyman Bass, *Euler characteristics and characters of discrete groups*, Invent. Math. **35** (1976), 155–196. MR0432781 (55 #5764)

[3] J. N. Bernstein, *Le "centre" de Bernstein*, 1–32, Edited by P. Deligne. MR771671 (86e:22028)

[4] Martin R. Bridson and André Haefliger, *Metric spaces of non-positive curvature*, Grundlehren der Mathematischen Wissenschaften, vol. 319, Springer-Verlag, Berlin, 1999, ISBN 3-540-64324-9. MR1744486 (2000k:53038)

[5] F. Bruhat and J. Tits, *Groupes réductifs sur un corps local*, Inst. Hautes Études Sci. Publ. Math. (1972), 5–251. MR0327923 (48 #6265) (French)

[6] Heath Emerson and Ralf Meyer, *Dualizing the coarse assembly map*, J. Inst. Math. Jussieu (2004), http://arxiv.org/math.OA/0401227 (to appear).

[7] Alexander Grothendieck, *Produits tensoriels topologiques et espaces nucléaires*, Mem. Amer. Math. Soc., vol. 16, 1955. MR0075539 (17,763c) (French)

[8] Henri Hogbe-Nlend, *Complétion, tenseurs et nucléarité en bornologie*, J. Math. Pures Appl. (9) **49** (1970), 193–288. MR0279557 (43 #5279) (French)

[9] _____, *Bornologies and functional analysis*, North-Holland Mathematics Studies, vol. 26, North-Holland Publishing Co., Amsterdam, 1977, ISBN 0-7204-0712-5. MR0500064 (58 #17774)

[10] James E. Humphreys, *Reflection groups and Coxeter groups*, Cambridge Studies in Advanced Mathematics, vol. 29, Cambridge University Press, Cambridge, 1990, ISBN 0-521-37510-X. MR1066460 (92h:20002)

[11] Bernhard Keller, *Derived categories and their uses*, 1996, pp. 671–701. MR1421815 (98h:18013)

[12] Wolfgang Lück and Holger Reich, *The Baum-Connes and the Farrell-Jones Conjectures in K- and L-theory*, Preprintreihe SFB 478 **324** (2004), Universität Münster.

[13] Ralf Meyer, *Analytic cyclic cohomology*, Ph.D. Thesis, Westfälische Wilhelms-Universität Münster, 1999.

[14] _____, *Bornological versus topological analysis in metrizable spaces*, 249–278. MR2097966

[15] _____, *Smooth group representations on bornological vector spaces*, Bull. Sci. Math. **128** (2004), 127–166. MR2039113 (2005c:22013) (English, with English and French summaries)

[16] _____, *Combable groups have group cohomology of polynomial growth* (2004), http://arxiv.org/math.KT/0410597 (to appear in Q. J. Math.)

[17] _____, *Embeddings of derived categories of bornological modules* (2004), http://arxiv.org/math.FA/0410596 (eprint).

[18] A. Yu. Pirkovskii, *Stably flat completions of universal enveloping algebras* (2003), http://arxiv.org/math.FA/0311492 (eprint).

[19] Jonathan Rosenberg, *Algebraic K-theory and its applications*, Graduate Texts in Mathematics, vol. 147, Springer-Verlag, New York, 1994, ISBN 0-387-94248-3. MR1282290 (95e:19001)

[20] Peter Schneider and Ulrich Stuhler, *Representation theory and sheaves on the Bruhat-Tits building*, Inst. Hautes Études Sci. Publ. Math. (1997), 97–191. MR1471867 (98m:22023)

[21] P. Schneider and E.-W. Zink, *K-types for the tempered components of a p-adic general linear group*, J. Reine Angew. Math. **517** (1999), 161–208, With an appendix by P. Schneider and U. Stuhler. MR1728541 (2001f:22029)

[22] Allan J. Silberger, *Introduction to harmonic analysis on reductive p-adic groups*, Mathematical Notes, vol. 23, Princeton University Press, Princeton, N.J., 1979, ISBN 0-691-08246-4, Based on lectures by Harish-Chandra at the Institute for Advanced Study, 1971–1973. MR544991 (81m:22025)

[23] J. Tits, *Reductive groups over local fields*, 29–69. MR546588 (80h:20064)

[24] François Trèves, *Topological vector spaces, distributions and kernels*, Academic Press, New York, 1967. MR0225131 (37 #726)

[25] Marie-France Vignéras, *On formal dimensions for reductive p-adic groups*, 225–266. MR1159104 (93c:22034)

[26] J.-L. Waldspurger, *La formule de Plancherel pour les groupes p-adiques (d'après Harish-Chandra)*, J. Inst. Math. Jussieu **2** (2003), 235–333. MR1989693 (2004d:22009) (French)

E-mail address: `rameyer@math.uni-muenster.de`

MATHEMATISCHES INSTITUT, WESTFÄLISCHE WILHELMS-UNIVERSITÄT MÜNSTER, EINSTEINSTR. 62, 48149 MÜNSTER, GERMANY

A non-commutative geometry approach to the representation theory of reductive p-adic groups: Homology of Hecke algebras, a survey and some new results

Victor Nistor

ABSTRACT. We survey some of the known results on the relation between the homology of the *full* Hecke algebra of a reductive p-adic group G, and the representation theory of G. Let us denote by $\mathcal{C}_c^\infty(G)$ the full Hecke algebra of G and by $\mathrm{HP}_*(\mathcal{C}_c^\infty(G))$ its periodic cyclic homology groups. Let $\hat{G}$ denote the admissible dual of G. One of the main points of this paper is that the groups $\mathrm{HP}_*(\mathcal{C}_c^\infty(G))$ are, on the one hand, directly related to the topology of $\hat{G}$ and, on the other hand, the groups $\mathrm{HP}_*(\mathcal{C}_c^\infty(G))$ are explicitly computable in terms of G (essentially, in terms of the conjugacy classes of G and the cohomology of their stabilizers). The relation between $\mathrm{HP}_*(\mathcal{C}_c^\infty(G))$ and the topology of $\hat{G}$ is established as part of a more general principle relating $\mathrm{HP}_*(A)$ to the topology of $\mathrm{Prim}(A)$, the primitive ideal spectrum of A, for any finite typee algebra A. We provide several new examples illustrating in detail this principle. We also prove in this paper a few new results, mostly in order to better explain and tie together the results that are presented here. For example, we compute the Hochschild homology of $\mathcal{O}(X) \rtimes \Gamma$, the crossed product of the ring of regular functions on a smooth, complex algebraic variety X by a finite group Γ. We also outline a very tentative program to use these results to construct and classify the cuspidal representations of G. At the end of the paper, we also recall the definitions of Hochschild and cyclic homology.

Introduction

To motivate the results surveyed in this paper, let us look at the following simple example. Precise definitions will be given below. Let G be a finite group and $A := \mathbb{C}[G]$ be its complex group algebra. Then A is a finite dimensional, semi-simple complex algebra, and hence $A \simeq \oplus_{j=1}^d M_{n_j}(\mathbb{C})$. (This is an elementary result that can be found in [31]; see also [50]). The Hochschild homology of A is then, on the one hand,

$$\mathrm{HH}^0(A) \simeq \oplus_{j=1}^d \mathrm{HH}^0(M_{n_j}(\mathbb{C})) \simeq \mathbb{C}^d. \tag{1}$$

On the other hand, $\mathrm{HH}^0(A)$ is the space of traces on A, and hence it identifies with the space of class functions on G. Let $\langle G \rangle$ denote the set of conjugacy classes of G and $\#S$ denote the number of elements in a set S. Then $\mathrm{HH}^0(A)$ has dimension $\#\langle G \rangle$. In other words,

PROPOSITION 0.1 (Classical). *Let G be a finite group. Then G has as many (equivalence classes of) irreducible, complex representations as conjugacy classes.*

One of our goals was to investigate to what extent Proposition 0.1 extends to other groups. It is clear that the formulation of any possible extension of Proposition 0.1 will depend on the class of groups considered and will not be as simple as in the finite group case. Moreover, this question will not be answered in a few papers and is more of a program (going back to Gelfand, Langlands, Manin, and other people) than an explicit question. Nevertheless, something from Proposition 0.1 does remain true in certain cases. An example is the theory of characters for compact Lie groups.

In this paper, we will investigate a possible analog of Proposition 0.1 for the case of a reductive p–adic group G. Recall that a p–adic group $G = \mathbb{G}(\mathbb{F})$ is the set of $\mathbb{F}$–rational points of a linear algebraic group $\mathbb{G}$ defined over a non-archemedean, non-discrete, locally compact field $\mathbb{F}$ of characteristic zero (so $\mathbb{F}$ is a finite algebraic extension of the field $\mathbb{Q}_l$ of l-adic numbers, for some prime l). A vague formulation of our main result is as follows.

Let $\mathrm{HP}_k(A)$ denote the periodic cyclic homology groups of the algebra A. Also, let $\mathcal{C}_c^\infty(G)$ denote the space of compactly supported, locally constant functions on G and let $\hat{G}$ denote the admissible dual of G with the Jacobson topology. Then the results of [2, 3, 29, 47] give the following result that will be made more precise below.

THEOREM 0.2. *The groups* $\mathrm{HP}_j(\mathcal{C}_c^\infty(G))$ *are explicitly determined by the geometry of the conjugacy classes of G and the cohomology of their stabilizers and they are (essentially) isomorphic to the singular cohomology groups of $\hat{G}$.*

One of the main purposes of this paper is to explain the above theorem. This theorem is useful especially because it is much easier to determine the groups $\mathrm{HP}_j(\mathcal{C}_c^\infty(G))$ (and hence, to a large extent, the algebraic cohomology of $\hat{G}$) than it is to determine $\hat{G}$ itself. Moreover, we will briefly sketch a plan to say more about the actual structure of $\hat{G}$ using the knowledge of the topology of $\hat{G}$ acquired from the determination of $\mathrm{HP}_j(\mathcal{C}_c^\infty(G))$ in terms of the geometry of the conjugacy classes of G and the cohomology of their stabilizers. See also [1] for a survey of the applications of non-commutative geometry to the representation theory of reductive p-adic groups.

The paper is divided into two parts. The first part, consisting of Sections 1–4 is more advanced, whereas the last three sections review some basic material. In Section 1 we review the basic result relating the cohomology of the maximal spectrum of a commutative algebra A to its periodic cyclic homology groups $\mathrm{HP}_*(A)$. The relation between forms on $\mathrm{Max}(A)$ and the Hochschild homology groups $\mathrm{HH}_*(A)$ are also discussed here. These are basic results due to Connes [16], Feigin and Tsygan [20], and Loday and Quillen [33]. We also discuss the Excisision principle in periodic cyclic homology [19] and it's relation with K-theory. In Section 2 we discuss generalizations of these results to finite type algebras, a class of algebras directly relevant to the representation theory of p-adic groups that was introduced in [29]. We also use these results to compute the periodic cyclic homology and the Hochschild homology of several typical examples of finite type algebras. In the following section, Section 3, we introduce spectrum preserving morphisms, which were shown in [3] to induce isomorphisms on periodic cyclic homology. This then led to a determination of the periodic cyclic homology of Iwahori-Hecke algebras in that paper. In Section 4, we recall the explicit calculation of the Hochschild and periodic cyclic homology groups of the full Hecke algebra $\mathcal{C}_c^\infty(G)$. The last three sections briefly review for the benefit of the reader the definitions of Hochschild homology, cyclic and periodic cyclic homology, and, respectively, the Chern character.

This first part of the paper follows fairly closely the structure of my talk given at the conference "Non-commutative geometry and number theory" organized by Yuri Manin and Matilde Marcolli, whom I thank for their great work and for the opportunity to present my results. I have included, however, some new results, mostly to better explain and illustrate the results surveyed.

1. Periodic cyclic homology versus singular cohomology

Let us discuss first to what extent the periodic cyclic homology groups $\mathrm{HP}(\mathcal{C}_c^\infty(G))$ are related to the topology – more precisely to the singular cohomology – of $\hat{G}$. In the next section, we will discuss this again in the more general framework of "finite type algebras" (Definition 2.1). The definitions of the homology groups considered in this paper and of the Connes-Karoubi character are recalled in the last three sections of this paper.

One of the main goals of non-commutative geometry is to generalize the correspondence (more precisely, contravariant equivalence of categories)

(2)   "Space" $X \leftrightarrow \mathcal{F}(X) :=$ "the algebra of functions on X"

to allow for non-commutative algebras on the right–hand side of this correspondence. This philosophy was developed in many papers, including [**5, 13, 4, 15, 16, 17, 25, 30, 32, 33, 40, 41, 42, 43, 44**], to mention only some of the more recent ones. The study of the K-theory of C^*-algebras, a field on its own, certainly fits into this philosophy. The extension of the correspondence in Equation 2 would lead to methods to study (possibly) non-commutative algebras using our geometric intuition. In Algebraic geometry, this philosophy is illustrated by the correspondence (*i.e.* contravariant equivalence of categories) between affine algebraic varieties over a field $\mathfrak{k}$ and commutative, reduced, finitely generated algebras over $\mathfrak{k}$. In Functional analysis, this principle is illustrated by the Gelfand–Naimark equivalence between the category of compact topological spaces and the category of commutative, unital C^*-algebras. In all these cases, the study of the "space" then proceeds through the study of the "algebra of functions on that space." For this approach to be useful, one should be able to define many invariants of X in terms of $\mathcal{F}(X)$ alone, preferably without using the commutativity of $\mathcal{F}(X)$.

It is a remarkable fact that one can give completely algebraic definitions for $\Omega^q(X)$, the space of differential forms on X (for suitable X) just in terms of $\mathcal{F}(X)$. Even more remarkable is that the singular cohomology of X (again for suitable X) can be defined in purely algebraic terms using only $\mathcal{F}(X)$. In these definitions, we can then replace $\mathcal{F}(X)$ with a non-commutative algebra A. Let us now recall these results.

We denote by $\mathrm{HH}_j(A)$ the Hochschild homology groups of an algebra A (see Section 5 for the definition). As we will see below, for applications to representation theory we are mostly interested in the algebraic case, so we state those first and then we state the results on smooth manifolds. We begin with a result of Loday-Quillen [**33**], in this result $\mathcal{F}(X) = \mathcal{O}(X)$, the ring of regular (*i.e.* polynomial) functions on the algebraic variety X.

THEOREM 1.1 (Loday-Quillen). *Let X be a smooth, complex, affine algebraic variety. Then*

$$\mathrm{HH}_j(\mathcal{O}(X)) \simeq \Omega^j(X),$$

the space of algebraic forms on X.

A similar results holds when X is a smooth compact manifold and $\mathcal{F}(X) = \mathcal{C}^\infty(X)$ is the algebra of smooth functions on X [16]. See also the Hochschild–Kostant–Rosenberg paper [23].

THEOREM 1.2 (Connes). *Let X be a compact, smooth manifold. Then*

$$\mathrm{HH}_j(\mathcal{C}^\infty(X)) \simeq \Omega^j(X),$$

the space of smooth forms on X.

These results extend to recover the singular cohomology of (suitable) spaces, as seen in the following two results due to Feigin-Tsygan [20] and Connes [16]. For any functor F_j [respectively, F^j], we shall denote by $F_{[j]} = \oplus_k F_{j+2k}$ [respectively, $F^{[j]} = \oplus_k F^{j+2k}$]. This will mostly be used for $F^j(X) = \mathrm{H}^j(X)$, the singular cohomology of X.

THEOREM 1.3 (Feigin-Tsygan). *Let X be a complex, affine algebraic variety and $\mathcal{O}(X)$ be the ring of regular (i.e. polynomial) functions on X. Then*

$$\mathrm{HP}_j(\mathcal{O}(X)) \simeq \mathrm{H}^{[j]}(X).$$

For smooth algebraic varieties, this result follows from the Loday-Quillen result on Hochschild homology mentioned above. See [29] for a proof of this theorem that proceed by reducing it to the case of smooth varieties. For smooth manifolds, the result again follows from the corresponding result on Hoschschild homology.

THEOREM 1.4 (Connes). *Let X be a compact, smooth manifold and $\mathcal{C}^\infty(X)$ be the algebra of smooth functions on X. Then*

$$\mathrm{HP}_j(\mathcal{C}^\infty(X)) \simeq \mathrm{H}^{[j]}(X).$$

These results are already enough justification for declaring periodic cyclic homology to be the "right" extension of singular cohomology for the category (suitable) spaces to suitable categories of algebras. However, the most remarkable result justifying this is the "Excision property" in periodic cyclic homology, a breakthrough result of Cuntz and Quillen [19].

THEOREM 1.5 (Cuntz-Quillen). *Any two-sided ideal J of an algebra A over a characteristic 0 field gives rise to a periodic six-term exact sequence*

(3)
$$
\begin{array}{ccc}
\mathrm{HP}_0(J) \longrightarrow \mathrm{HP}_0(A) \longrightarrow \mathrm{HP}_0(A/J) \\
\partial \uparrow \qquad\qquad\qquad\qquad\qquad \downarrow \partial \\
\mathrm{HP}_1(A/J) \longleftarrow \mathrm{HP}_1(A) \longleftarrow \mathrm{HP}_1(J).
\end{array}
$$

A similar results holds for Hochschild and cyclic homology, provided that the ideal J is an H-unital algebra in the sense of Wodzicki [53], see Section 5. We shall refer to the following result of Wodzicki as the "Excision principle in Hochschild homology."

THEOREM 1.6 (Wodzicki). *Let $J \subset A$ be a H-unital ideal of a complex algebra A. Then there exists a long exact sequence*

$$0 \leftarrow HH_0(A/J) \leftarrow HH_0(A) \leftarrow HH_0(J) \overset{\partial}{\leftarrow} HH_1(A/J)$$

$$\leftarrow HH_k(A/J) \leftarrow HH_k(A) \leftarrow HH_k(J) \overset{\partial}{\leftarrow} HH_{k+1}(A/J) \leftarrow \dots$$

The same result remains valid if we replace Hochschild homology with cyclic homology.

Also, there exist excision results for topological algebras [12, 18]. An important part of the proof of the Excision property is to provide a different definition of cyclic homology in terms of X-complexes. Then the proof is ingeniously reduced to Wodzicki's result on the excision in Hochschild homology, using also an important theorem of Goodwillie [22] that we now recall.

THEOREM 1.7 (Goodwillie). *If $I \subset A$ is a nilpotent two-sided ideal, then the quotient morphism $A \to A/I$ induces an isomorphism $\mathrm{HP}_*(A) \to \mathrm{HP}_*(A/I)$. In particular, $\mathrm{HP}_*(I) = 0$ whenever I is nilpotent.*

One of the main original motivations for the study of cyclic homology was the need for a generalization of the classical Chern character $Ch : K^j(X) \to \mathrm{H}^{[j]}(X)$ [16, 17, 25, 32]. Indeed, an extension is obtained in the form of the Connes-Karoubi character

(4) $$Ch : K_j(A) \to \mathrm{HP}_j(A).$$

It is interesting then to notice that excision in periodic cyclic homology is compatible with excision in K-theory, which is seen from the following result [46], originally motivated by questions in the analysis of elliptic operators (more precisely, Index theory).

THEOREM 1.8 (Nistor). *Let $I \subset A$ the a two-sided ideal of a complex algebra A. Then the following diagram commutes*

$$\begin{array}{ccccccccccc}
K_1(I) & \to & K_1(A) & \to & K_1(A/I) & \overset{\partial}{\to} & K_0(I) & \to & K_0(A) & \to & K_0(A/I) \\
\downarrow & & \downarrow & & \downarrow & & \downarrow & & \downarrow & & \downarrow \\
\mathrm{HP}_1(I) & \to & \mathrm{HP}_1(A) & \to & \mathrm{HP}_1(A/I) & \overset{\partial}{\to} & \mathrm{HP}_0(I) & \to & \mathrm{HP}_0(A) & \to & \mathrm{HP}_0(A/I).
\end{array}$$

Let X be a complex, affine algebraic variety, $\mathcal{O}(X)$ the ring of polynomial functions on X, X^{an} the underlying locally compact topological space, and $Y \subset X$ be a subvariety. Let $I \subset \mathcal{O}(X)$ be the ideal of functions vanishing on Y. Then the above theorem shows, in particular, that the periodic six term exact sequence of periodic cyclic homology groups associated to the exact sequence

(5) $$0 \longrightarrow I \longrightarrow \mathcal{O}(X) \longrightarrow \mathcal{O}(Y) \longrightarrow 0,$$

of algebras by the Excision principle is obtained from the long exact sequence in singular cohomology of the pair $(X^{\mathrm{an}}, Y^{\mathrm{an}})$ by making the groups periodic of period two [29, 46]. The same result holds true for the exact sequence of algebras associated to a closed submanifold Y of a smooth manifold X.

2. Periodic cyclic homology and $\hat{G}$

Let A be an arbitrary complex algebra. The kernel of an irreducible representation of A is called a *primitive* ideal of A. We shall denote by $\mathrm{Prim}(A)$ the *primitive ideal spectrum* of A, consisting of all primitive ideals of A. We endow $\mathrm{Prim}(A)$ with the Jacobson topology. Thus, a set $V \subset \mathrm{Prim}(A)$ is open if, and only if, V is the set of primitive ideals not containing some fixed ideal I of A. We have

$$\mathrm{Prim}(\mathcal{O}(X)) =: \mathrm{Max}(\mathcal{O}(X)) = X,$$

the set of maximal ideals of $A = \mathcal{O}(X)$ with the Zariski topology. If $A = \mathcal{C}^\infty(X)$, where X is a smooth compact manifold, then again $\mathrm{Prim}(A) = \mathrm{Max}(A) = X$ with the usual (*i.e.* locally compact, Hausdorff) topology on X. We are interested in the primitive ideal spectra of algebras because

$$(6) \qquad\qquad \mathrm{Prim}(\mathcal{C}_c^\infty(G)) = \hat{G},$$

a deep result due to Bernstein [7]. For the purpose of this paper, we could as well take $\mathrm{Prim}(\mathcal{C}_c^\infty(G))$ to be the actual definition of $\hat{G}$.

In view of the results presented in the previous section, it is reasonable to assume that the determination of the groups $\mathrm{HP}_j(\mathcal{C}_c^\infty(G))$ would give us some insight into the topology of $\hat{G}$. I do not know any general result relating the singular cohomology of $\hat{G}$ to $\mathrm{HP}_j(\mathcal{C}_c^\infty(G))$, although it is likely that they coincide. Anyway, due to the fact that the topology on $\hat{G}$ is highly non-Hausdorff topology, it is not clear that the knowledge of the groups $\mathrm{H}^j(\hat{G})$ would be more helpful than the knowledge of the groups $\mathrm{HP}_j(\mathcal{C}_c^\infty(G))$.

For reasons that we will explain below, it will be convenient in what follows to work in the framework of "finite typee algebras" [29]. All our rings have a unit (*i.e.* they are unital), but the algebras are not required to have a unit.

DEFINITION 2.1. Let $\mathfrak{k}$ be a finitely generated commutative, complex ring. A finite typee $\mathfrak{k}$-algebra is a $\mathfrak{k}$-algebra that is a finitely generated $\mathfrak{k}$-module.

The study of $\hat{G}$ as well as that of $A = \mathcal{C}_c^\infty(G)$ reduces to the study of finite typee algebras by considering the connected components of $\hat{G}$ and their commuting algebras, in view of some results of Bernstein [7] that we now recall. Let $D \subset \hat{G}$ be a connected component of $\hat{G}$. Then D corresponds to a cuspidal representation σ of a Levi subgroup $M \subset G$. Let M_0 be the subgroup of M generated by the compact subgroups of M and H_D be the representation of G induced from the restriction of σ to M_0. The space H_D can be thought of as the holomorphic family of induced representations of $\mathrm{ind}_M^G(\sigma\chi)$, where χ ranges through the caracters of M/M_0. Let A_D be the algebra of G-endomorphisms of H_D. This is Bernstein's celebrated "commuting algebra." The annihilator of H_D turns out to be a direct summand of $\mathcal{C}_c^\infty(G)$ with complement the two-sided ideal $\mathcal{C}_c^\infty(G)_D \subset \mathcal{C}_c^\infty(G)$. Then the category of modules over $\mathcal{C}_c^\infty(G)_D$ is equivalent to the category of modules over A_D. Our main reason for introducing finite typee algebras is that the algebra A_D is a unital finite typee algebra and

$$(7) \qquad\qquad D = \mathrm{Prim}(A_D).$$

Moreover,

$$(8) \qquad\qquad \mathcal{C}_c^\infty(G)) = \oplus_D \mathcal{C}_c^\infty(G)_D.$$

To get consequences for the periodic cyclic homology, we shall need the following result.

PROPOSITION 2.2. *Let B be a complex algebra such that $B \simeq \oplus_{n \in \mathbb{N}} B_n$ for some two-sided ideals $B_n \subset B$. Assume that $B = \cup e_k B e_k$, for a sequence of idempotents $e_k \in B$ and that $\mathrm{HH}_q(B) = 0$ for $q > N$, for some given N. Then*

$$\mathrm{HP}_q(B) \simeq \oplus_{n \in \mathbb{N}} \mathrm{HP}_q(B_n).$$

PROOF. The algebra B and each of the two-sided ideals B_n are H-unital. Then

$$\mathrm{HH}_q(B) \simeq \oplus_{n \in \mathbb{N}} \mathrm{HH}_q(B_n),$$

by Wodzicki's excision theorem and the continuity of Hochschild homology (*i.e.* the compatibility of Hochschild homology with inductive limits). Then $\mathrm{HH}_q(B_n) = 0$ for $q > N$ and any n. Therefore $\mathrm{HP}_k(B_n) = \mathrm{HC}_k(B_n)$ for $k \geq N$, by the SBI-long exact sequence (this exact sequence is recalled in Section 6). Unlike periodic cyclic homology, cyclic homology is continuous (*i.e.* it is compatible with inductive limits). Using again Wodzicki's excision theorem, we obtain

$$\mathrm{HP}_k = \mathrm{HC}_k(B) \simeq \oplus_{n \in \mathbb{N}} \mathrm{HC}_k(B_n) = \oplus_{n \in \mathbb{N}} \mathrm{HP}_k(B_n),$$

for $k \geq N$. This completes the proof. $\square$

The above discussion gives the following result mentioned in [**29**] without proof. For the proof, we shal also use Theorem 4.1, which implies, in particular, that $\mathrm{HH}_q(\mathcal{C}_c^\infty(G))$ vanishes for q greater than the split rank of G. This is, in fact, a quite non-trivial property of $\mathcal{C}_c^\infty(G)$ and of the finite type algebras A_D, as we shall see below in Example 2.14.

THEOREM 2.3. *Let D be the set of connected components of $\hat{G}$, then*

$$\mathrm{HP}_q(\mathcal{C}_c^\infty(G)) = \oplus_D \mathrm{HP}_q(\mathcal{C}_c^\infty(G)_D) \simeq \oplus_D \mathrm{HP}_q(A_D).$$

PROOF. The first part follows directly from the results above, namely from Equation (8), Proposition 2.2, and Theorem 4.1 (which implies that $\mathrm{HH}_q(\mathcal{C}_c^\infty(G)) = 0$ for $q > N$, for N large).

To complete the proof, we need to check that

(9) $$\mathrm{HP}_q(\mathcal{C}_c^\infty(G)_D) \simeq \mathrm{HP}_q(A_D).$$

Let e_k be a sequence of idempotents of $\mathcal{C}_c^\infty(G)$ corresponding to a basis of neighborhood of the identity of G consisting of compact open subgroups. Then, for k large, the unital algebra $e_k \mathcal{C}_c^\infty(G) e_k$ is Morita equivalent to A_D, an imprimitivity module being given by $e_k H_D$. (Recall from above that H_D is the induced representation from a cuspidal representation of M_0, where M is a Levi subfactor defining the connected component D and M_0 is the subgroup of M generated by its compact subgroups.)

In particular, $\mathrm{HH}_q(e_k \mathcal{C}_c^\infty(G)_D e_k)$ vanish for q large. The same argument as above (using Theorem 4.1 below) shows that $\mathrm{HP}_q(\mathcal{C}_c^\infty(G)_D) \simeq \mathrm{HP}_q(e_k \mathcal{C}_c^\infty(G)_D e_k)$, for k large. The isomorphism $\mathrm{HP}_q(\mathcal{C}_c^\infty(G)_D) \simeq \mathrm{HP}_q(A_D)$ then follows from the invariance of Hochschild homology with respect to Morita equivalence. $\square$

For suitable G and D,

(10) $$A_D = H_q,$$

that is, the commuting algebra of D is the Iwahori-Hecke algebra associated to G (or to its extended affine Weyl group), [9]. The periodic cyclic homology groups of H_q were determined in [2, 3], and will be recalled in Section 3.

In view of Equations (7) and (8) and of the Theorem 2.3, we see that in order to relate the groups $\mathrm{HP}_*(\mathcal{C}_c^\infty(G))$ to the topology of $\hat{G}$, it is enough to relate $\mathrm{HP}_*(A)$ to the topology of $\mathrm{Prim}(A)$ for an arbitrary finite typee algebra. For the rest of this and the following section, we shall therefore concentrate on finite typee algebras and their periodic cyclic homology.

If $\mathcal{I} \subset A$ is a primitive ideal of the finite typee $\mathfrak{k}$-algebra A, then the intersection $\mathcal{I} \cap Z(A)$ is a maximal ideal of $Z(A)$. The resulting map

$$\Theta : \mathrm{Prim}(A) \to \mathrm{Max}(Z(A))$$

is called the *central character* map. We similarly obtain a map $\Theta : \mathrm{Prim}(A) \to \mathrm{Max}(\mathfrak{k})$, also called the central character map.

The topology on $\mathrm{Prim}(A)$ and the groups $\mathrm{HP}_j(A)$ are related through a spectral sequence whose E^2 are given by the singular cohomology of various strata of $\mathrm{Prim}(A)$ that are better behaved than $\mathrm{Prim}(A)$ itself, Theorem 2.5 below. To state our next result, due to Kazhdan, Nistor, and Schneider [29], we need to introduce some notation and definitions.

We shall use the customary notation to denote by 0 the ideal $\{0\}$. Recall that an algebra B is called *semiprimitive* if the intersection of all its primitive ideals is 0. Also, we shall denote by $Z(B)$ the center of an algebra B. We shall need the following definition from [29]

DEFINITION 2.4. A finite decreasing sequence

$$A = \mathcal{I}_0 \supset \mathcal{I}_1 \supset \ldots \supset \mathcal{I}_{n-1} \supset \mathcal{I}_n$$

of two-sided ideals of a unital finite type algebra A is an *abelian filtration* if and only if the following three conditions are satisfied for each k:

(i) The quotient $A/\mathcal{I}_k$ is semiprimitive.

(ii) For each maximal ideal $\mathfrak{p} \subset Z_k := Z(A/\mathcal{I}_k)$ not containing $I_{k-1} := Z_k \cap (\mathcal{I}_{k-1}/\mathcal{I}_k)$, the localization $(A/\mathcal{I}_k)_\mathfrak{p} = (Z_k \setminus \mathfrak{p})^{-1}(A/\mathcal{I}_k)$ is an Azumaya algebra over $(Z_k)_\mathfrak{p}$,

(iii) The quotient $(\mathcal{I}_{k-1}/\mathcal{I}_k)/I_{k-1}(A/\mathcal{I}_k)$ is nilpotent and $\mathcal{I}_n$ is the intersection of all primitive ideals of A.

Consider an abelian filtration $(\mathcal{I}_k)$, $k = 0, \ldots, n$ of a finite type $\mathfrak{k}$-algebra A. Then, for each k, the center Z_k of $A/\mathcal{I}_k$ is a finitely generated complex algebra, and hence it is isomorphic to the ring of regular functions on an affine, complex algebraic variety X_k. Let $Y_k \subset X_k$ be the subvariety defined by $I_{k-1} := Z_k \cap (\mathcal{I}_{k-1}/\mathcal{I}_k)$. For any complex algebraic variety X, we shall denote by X^{an} the topological space obtained by endowing the set X with the locally compact topology induced by some embedding $X \subset \mathbb{C}^N$, N large. We shall call X^{an} the analytic space underlying X. For instance, we shall refer to X_k^{an} and Y_k^{an} as the *analytic spaces* associated to the filtration $(\mathcal{I}_k)$ of A. Then we have the following result from [29]

THEOREM 2.5 (Kazhdan-Nistor-Schneider). *If $Y_p^{\mathrm{an}} \subset X_p^{\mathrm{an}}$ are the analytic spaces associated to an abelian filtration of a finite type algebra A, then there exists a natural spectral sequence with*

$$E^1_{-p,q} = \mathrm{H}^{[q-p]}(X_p^{\mathrm{an}}, Y_p^{\mathrm{an}})$$

convergent to $HP_{q-p}(A)$.

If the algebra A in the above theorem is commutative, then any decreasing filtration of A (by radical ideals) is abelian, which explains the terminology "abelian filtration." Moroever, in this case our spectral sequence reduces to the spectral sequence in singular cohomology associated to the filtration of a space by closed subsets.

Every finite type algebra has an abelian filtration. Indeed, we can take $\mathfrak{I}_k$ to be the intersection of the kernels of all irreducible representations of dimension at most k. This filtration is called the *standard filtration*.

EXAMPLE 2.6. Let $a_1, a_2, \ldots, a_l \in \mathbb{C}$ be distinct points and $v_1, v_2, \ldots, v_l \in \mathbb{C}^2$ be non-zero (column) vectors. Let $\mathbb{C}[x] = \mathcal{O}(\mathbb{C})$ denote the algebra of polynomials in one variable and define

(11) $$A_1 := \{P \in M_2(\mathbb{C}[x]), \; F(a_j)v_j \in \mathbb{C}v_j\}.$$

Then A_1 is a finite type algebra with center $Z = Z(A_1)$, the subalgebra of matrices of the form PE_2, where $P \in \mathbb{C}[x]$ and E_2 is the identity matrix in $M_2(\mathbb{C})$.

We filter A_1 by the ideals

$$\mathfrak{I}_1 = \{P \in A_1, \; F(a_j) = 0\}$$

and $\mathfrak{I}_2 = 0$. We have

$$A_1/\mathfrak{I}_1 \simeq \mathbb{C}^{2k}$$

a semi-primitive (*i.e.* reduced) commutative algebra with center

$$Z_1 := Z(A_1/\mathfrak{I}_1) = A_1/\mathfrak{I}_1 = \mathbb{C}^{2k}.$$

Then $I_0 := Z(A_1/\mathfrak{I}_1) \cap \mathfrak{I}_0 = A_1$ and hence no maximal ideal $\mathfrak{p}$ contains I_0. On the other hand, the algebra $A_1/\mathfrak{I}_1$ is an Azumaya algebra, so Condition (ii) of Definition 2.4 is automatically satisfied for $k = 1$. Similarly, the quotient $(\mathfrak{I}_0/\mathfrak{I}_1)/I_0(A_1/\mathfrak{I}_1) = 0$, and hence it is nilpotent. Hence Condition (iii) of Definition 2.4 is also satisfied for $k = 1$. The space X_1 consists of $2l$ points (two copies of each a_j) and Y_1 is empty.

Next, $Z_2 = Z \simeq \mathbb{C}[x]$ and $A_1 = A_1/\mathfrak{I}_2$ is semi-primitive. The ideal $I_1 := Z_2 \cap \mathfrak{I}_1$ consists of the polynomials vanishing at the given values a_j. In this case, $X_2 = \mathbb{C}$ and $Y_2 = \{a_1, \ldots, a_l\}$. The Conditions (i–iii) of Definition 2.4 are easily checked for $k = 2$. This also follows from the fact that $\mathcal{I}_j$ is the standard filtration of A_1.

The spectral sequence of Theorem 2.5 then becomes $E^1_{-p,q} = 0$, unless $p = 1$ or $p = 2$, in which case we get

$$E^1_{-1,q} = \mathrm{H}^{[q-1]}(X_1, Y_1) = \begin{cases} \mathbb{C}^{2l} & \text{if } q \text{ is odd} \\ 0 & \text{otherwise} \end{cases}$$

and

$$E^1_{-2,q} = \mathrm{H}^{[q-2]}(X_2, Y_2) = \begin{cases} \mathbb{C}^{l-1} & \text{if } q \text{ is odd} \\ 0 & \text{otherwise.} \end{cases}$$

The differential $d^1 : E^1_{-p,q} \to E^1_{-p-1,q}$ turns out to be surjective for $p = 1$ and q odd. For some obvious geometric reasons, the spectral sequence $E^r_{p,q}$ at $r = 2$ then collapses and gives

(12) $$HP_0(A_1) \simeq E^2_{-1,1} \simeq \mathbb{C}^{l+1}$$

and $\mathrm{HP}_1(A_1) = 0$. We shall look again at the algebra A_2, from a different point of view, in Example 3.5.

Let us consider now the following related example.

EXAMPLE 2.7.   Let $A_2 = M_2(\mathbb{C}[x]) \oplus \mathbb{C}^l$. Then the standard filtration of A_2 is $\mathfrak{I}_0 = A_2$, $\mathfrak{I}_1 = M_2(\mathbb{C}[x])$, and $\mathfrak{I}_2 = 0$. The center of $A_2/\mathfrak{I}_1$ is $Z_1 \simeq \mathbb{C}^l$. We have that X_1 consists of l points and $Y_1 = 0$. The center of $A_2/\mathfrak{I}_2$ is $Z_2 \simeq \mathbb{C}[x] \oplus \mathbb{C}^l$. Then $X_2 = \mathbb{C} \cup X_1$ and $Y_2 = X_1$, where $X_1 \cap \mathbb{C} = \emptyset$.

The spectral sequence of Theorem 2.5 then becomes $E^1_{-p,q} = 0$, unless $p = 1$ or $p = 2$, in which case we get

$$E^1_{-1,q} = \mathrm{H}^{[q-1]}(X_1, Y_1) = \begin{cases} \mathbb{C}^l & \text{if } q \text{ is odd} \\ 0 & \text{otherwise} \end{cases}$$

and

$$E^1_{-2,q} = \mathrm{H}^{[q-2]}(X_2, Y_2) = \begin{cases} 0 & \text{if } q \text{ is odd} \\ \mathbb{C} & \text{otherwise.} \end{cases}$$

The spectral sequence collapses at the E^1 term for geometric reasons and hence

$$(13) \qquad\qquad \mathrm{HP}_0(A_2) \simeq E^1_{-1,1} \oplus E^1_{-2,2} \simeq \mathbb{C}^{l+1}$$

and $\mathrm{HP}_1(A_1) = 0$.

The algebras A_1 and A_2 in the above examples turned out to have the same periodic cyclic homology groups. These algebras are simple, but representative, of the finite type algebras arising in the representation theory of reductive p-adic groups. Clearly, the periodic cyclic homology groups of these algebras provide important information on the structure of these algebras, but fails to distinguish them. At a heuristical level, distinguishing between A_1 and A_2 is the same problem as distinguishing between square integrable representations and supercuspidal representations. This issue arises because both these types of representations provide similar homology classes in HP_0 (through the Chern characters of the idempotents defining them). It is then an important question to distinguish between these homology classes.

Let us complete our discussion with some related results on the Hochschild homology of finite type algebras. We begin with the Hochschild homology of certain cross product algebras.

Let Γ be a finite group acting on a smooth complex algebraic variety X. For any $\gamma \in \Gamma$, let us denote by $X^\gamma \subset X$ the points of X fixed by γ. Let

$$\Gamma_\gamma := \{g \in \Gamma, g\gamma = \gamma g\}$$

denote the centralizer of γ in Γ. Let C_γ be the (finite, cyclic) subgroup generated by γ. There exists a natural Γ-invariant map

$$(14) \qquad\qquad \hat{X} := \bigcup_{\gamma \in \Gamma} \{\gamma\} \times X^\gamma \times (\Gamma_\gamma/C_\gamma) \to X,$$

given simply by the projection onto the second component. This gives then rise to a Γ-equivariant morphism $\mathcal{O}(X) \to \mathcal{O}(\hat{X})$. Choose a representative $x \in \Gamma$ from each conjugacy class $\langle x \rangle$ of Γ and denote by m_x the number of elements in the conjugacy class of m_x. Denote by C^*_γ the dual of C_γ, that is the set of multiplicative maps

$\pi : C_\gamma \to \mathbb{C}^*$. Recall that $\langle \Gamma \rangle$ denotes the set of conjugacy classes of Γ. We are finally ready to define the morphisms

$$\psi_{\gamma,\pi} : \mathcal{O}(X) \rtimes \Gamma \to M_{m_x}(\mathcal{O}(X^\gamma)),$$

where γ runs through a system of representatives of the conjugacy classes of Γ and $\pi \in C_\gamma^*$

$$(15) \quad \psi = \bigoplus_{\langle \gamma \rangle \in \langle \Gamma \rangle, \pi \in C_\gamma^*} \psi_{\gamma,\pi} : \mathcal{O}(X) \rtimes \Gamma \to \mathcal{O}(\hat{X}) \rtimes \Gamma$$

$$\simeq \bigoplus_{\langle \gamma \rangle \in \langle \Gamma \rangle} M_{m_x}(\mathcal{O}(X^\gamma) \otimes \mathbb{C}[C_\gamma]) \simeq \bigoplus_{\langle \gamma \rangle \in \langle \Gamma \rangle, \pi \in C_\gamma^*} M_{m_x}(\mathcal{O}(X^\gamma)).$$

The equations above give rise to a map

$$(16) \quad \phi = \bigoplus_{\langle \gamma \rangle \in \langle \Gamma \rangle} \phi_\gamma : \mathrm{HH}_q(\mathcal{O}(X) \rtimes \Gamma) \to \bigoplus_{\langle \gamma \rangle \in \langle \Gamma \rangle} \mathrm{H}^q(X^\gamma),$$

defined using $\mathrm{HH}_q(M_m(\mathcal{O}(X^\gamma))) \simeq \mathrm{HH}_q(\mathcal{O}(X^\gamma)) \simeq \mathrm{H}^q(X^\gamma)$ and

$$(17) \quad \phi_\gamma = \sum_{\pi \in C_\gamma^*} \frac{\overline{\pi(\gamma)}}{\#C_\gamma} \mathrm{HH}_q(\psi_{\gamma,\pi}) : \mathrm{HH}_q(\mathcal{O}(X) \rtimes \Gamma) \to \Omega^q(X^\gamma).$$

(This map was defined in a joint work in progress with J. Brodzki [11].)

Then we have the following lemma, which is a particular case of the Theorem 2.11 below.

LEMMA 2.8. *Assume that $X = \mathbb{C}^n$ and that Γ acts linearly on $\mathbb{C}^n$. Then the map ϕ of Equation 16 defines an isomorphism*

$$\phi : \mathrm{HH}_q(\mathcal{O}(X) \rtimes \Gamma) \to \bigoplus_{\langle \gamma \rangle \in \langle \Gamma \rangle} \Omega^q(X^\gamma)^{\Gamma_\gamma}.$$

PROOF. Let A be a complex algebra acted upon by automorphisms by a group Γ. Let us recall [21, 44] that the groups $\mathrm{HH}_q(A \rtimes \Gamma)$ decompose naturally as a direct sum

$$(18) \quad \mathrm{HH}_q(A \rtimes \Gamma) \simeq \oplus_{\langle \gamma \rangle \in \langle \Gamma \rangle} \mathrm{HH}_q(A \rtimes \Gamma)_\gamma.$$

The components $\mathrm{HH}_q(A \rtimes \Gamma)_\gamma$ are then identified as follows. Let A_γ be the A–A bimodule with action $a \cdot b \cdot c = ab\gamma(c)$. Let $\mathrm{HH}_q(A, M)$ denote the Hochschild homology groups of an A–bimodule M [38, 39, 52]. Then

$$(19) \quad \mathrm{HH}_q(A \rtimes \Gamma)_\gamma \simeq \mathrm{HH}_q(A, A_\gamma)^{\Gamma_\gamma}.$$

This follows from example from [44][Lemma 3.3] (take $G = \Gamma$ in that lemma). This will also be discussed in [11, 49].

Let now $A = \mathcal{O}(X)$, with $X = \mathbb{C}^n$ and Γ acting linearly on X. The same method as the one used in the proof of [15][Lemma 5.2] shows that

$$(20) \quad \mathrm{HH}_q(A, A_\gamma) \simeq \mathrm{HH}_q(\mathcal{O}(X^\gamma), \mathcal{O}(X^\gamma)) = \mathrm{HH}_q(\mathcal{O}(X^\gamma)) = \Omega^q(X^\gamma),$$

the isomorphism being given by the restriction morphism $\mathcal{O}(X) \to \mathcal{O}(X^\gamma)$.

A direct calculation based on the formula for the map J in [44][Lemma 3.3] shows that the composition of the morphisms of Equations (19) and (20) is the map ϕ_γ of Equation (17). This completes the proof. □

We now extend the above result to X an arbitrary smooth, complex algebraic variety. To do this, we shall need two results on Hochschild homology. We begin with the following result of Brylinski from [**14**].

PROPOSITION 2.9 (Brylinski). *Let S be a multiplicative subset of the center Z of the algebra A. Then $\mathrm{HH}_*(S^{-1}A) \simeq S^{-1}\mathrm{HH}_*(A)$.*

A related result, from [**29**], studies the completion of Hochschild homology. In the following result, we shall use *topological Hochschild homology*, whose definition is similar to that of the usual Hochschild homology, except that one completes with respect to the powers of an ideal (below, we shall use this for the ideals I and IA).

THEOREM 2.10. *Suppose that A is a unital finite type $\mathfrak{k}$-algebra and $I \subset \mathfrak{k}$ is an ideal. Then the natural map $\mathrm{HH}_*(A) \to \mathrm{HH}_*^{\mathrm{top}}(\hat{A})$ and the $\mathfrak{k}$-module structure on $\mathrm{HH}_*(A)$ define an isomorphism*

$$\mathrm{HH}_*(A) \otimes_{\mathfrak{k}} \hat{\mathfrak{k}} \simeq \mathrm{HH}_*^{\mathrm{top}}(\hat{A})$$

of $\hat{\mathfrak{k}}$-modules.

We are ready now to prove the following theorem.

THEOREM 2.11. *Assume that X is a smooth, complex algebraic variety and that Γ acts on X by algebraic automorphisms. Then the map ϕ of Equation 16 defines an isomorphism*

$$\phi : \mathrm{HH}_q(\mathcal{O}(X) \rtimes \Gamma) \to \bigoplus_{\langle \gamma \rangle \in \langle \Gamma \rangle} \Omega^q(X^\gamma)^{\Gamma_\gamma}.$$

PROOF. Let

$$Z := \mathcal{O}(X)^\Gamma = \mathcal{O}(X/\Gamma).$$

Then the morphism ψ of Equation (15) is Z-linear. It follows that the map ϕ of Equation (16) is also Z-linear. It is enough hence to prove that the localization of this map to any maximal ideal of Z is an isomorphism. It is also enough to prove that the completion of any of these localizations with respect to that maximal ideal is an isomorphism. Since X is smooth (and hence the completion of the local ring of X at any point is a power series ring) this reduces to the case of $X = \mathbb{C}^n$ acted upon linearly by Γ. The result hence follows from Lemma 2.8.   □

Let us include the following corollary of the above proof.

COROLLARY 2.12. *The map $\phi_\gamma : \mathrm{HH}_q(\mathcal{O}(X) \rtimes \Gamma) \to \Omega^q(X^\gamma)$ of (17) is such that $\phi_\gamma = 0$ on $\mathrm{HH}_q(\mathcal{O}(X) \rtimes \Gamma)_g$ if g and γ are not in the same conjugacy class and induces an isomorphism*

$$\phi_\gamma : \mathrm{HH}_q(\mathcal{O}(X) \rtimes \Gamma)_\gamma \to \Omega^q(X^\gamma)^{\Gamma_\gamma}.$$

Let us apply these results to the algebra A_1 of Example 2.6.

EXAMPLE 2.13. Let $\Gamma = \mathbb{Z}/2\mathbb{Z}$ act by $z \to -z$ on $\mathbb{C}$. Chose $l = 1$ and $a_1 = 0$ in Example 2.6. Then $\mathcal{O}(\mathbb{C}) \rtimes \Gamma \simeq A_1$, and hence

$$\mathrm{HH}_q(A_1) \simeq \begin{cases} \mathcal{O}(\mathbb{C})^\Gamma \oplus \mathbb{C} & \text{if } q = 0 \\ \mathcal{O}(\mathbb{C})^\Gamma & \text{if } q = 1 \\ 0 & \text{otherwise.} \end{cases}$$

In particular, $HH_q(A_1)$ vanishes for q large. Also, note that

$$\mathcal{O}(\mathbb{C})^\Gamma = \mathbb{C}[x]^\Gamma = \mathbb{C}[x^2] \simeq \mathbb{C}[x].$$

Let us see now an example of a finite type algebra for which Hochschild homology does not vanish in all degrees.

EXAMPLE 2.14. Let $A_3 \subset M_2(\mathcal{O}(\mathbb{C})) = M_2(\mathbb{C}[x])$ be the subalgebra of those matrices

$$P = \begin{bmatrix} P_{11} & P_{12} \\ P_{21} & P_{22} \end{bmatrix} =: P_{11}e_{11} + P_{12}e_{12}P_{21}e_{21} + P_{22},$$

with the property that $P_{21}(0) = P'_{21}(0) = 0$. For a suitable choice of v_1, this is a subalgebra of the algebra A_1 considered in Example 2.13. Let $V_1 := A_3e_{11}$ and $V_2 := A_3e_{22}$. Then $M := V_2/V_1 \simeq \mathbb{C}[x]/(x^2)$. The modules V_1 and $V_2^\tau := e_{22}A_3$ can be used to produced a projective resolution of A_3 with free A_3–A_3 bimodules that gives

(21) $$HH_q(A_3) \simeq \begin{cases} \mathbb{C}[x] \oplus M & \text{if } q = 0 \\ \mathbb{C}[x] \oplus \mathbb{C} & \text{if } q = 1 \\ \mathbb{C} & \text{otherwise.} \end{cases}$$

3. Spectrum preserving morphisms

We shall now give more evidence for the close relationship between the topology of $\mathrm{Prim}(A)$ and $HP_*(A)$ by studying a class of morphisms implicitly appearing in Lusztig's work on the representation of Iwahori-Hecke algebras, see [34, 35, 36, 37].

Let L and J be two finite type $\mathfrak{k}$-algebras. If $\phi : L \to J$ is a $\mathfrak{k}$-linear morphism, we define

(22) $$\mathcal{R}_\phi := \{(\mathfrak{P}', \mathfrak{P}) \subset \mathrm{Prim}(J) \times \mathrm{Prim}(L), \ \phi^{-1}(\mathfrak{P}') \subset \mathfrak{P}\}.$$

We now introduce the class of morphisms we are interested in.

DEFINITION 3.1. Let $\phi : L \to J$ be a $\mathfrak{k}$-linear morphism of unital, finite type $\mathfrak{k}$-algebras. We say that ϕ is a *spectrum preserving morphism* if, and only if, the set $\mathcal{R}_\phi$ defined in Equation (22) is the graph of a bijective function

$$\phi^* : \mathrm{Prim}(J) \to \mathrm{Prim}(L).$$

More concretely, we see that $\phi : L \to J$ is spectrum preserving if, and only if, the following two conditions are satisfied:

(1) *For any primitive ideal $\mathfrak{P}$ of J, the ideal $\phi^{-1}(\mathfrak{P})$ is contained in a unique primitive ideal of L, namely $\phi^*(\mathfrak{P})$, and*
(2) *The resulting map $\phi^* : \mathrm{Prim}(J) \to \mathrm{Prim}(L)$ is a bijection.*

We have the following result combining two theorems from [3].

THEOREM 3.2 (Baum-Nistor). *Let L and J be finite type $\mathfrak{k}$–algebras and $\phi : L \to J$ be a $\mathfrak{k}$–linear spectrum preserving morphism. Then the induced map $\phi^* : \mathrm{Prim}(J) \to \mathrm{Prim}(L)$ between primitive ideal spectra is a homeomorphism and the induced map $\phi_* : HP_*(L) \to HP_*(J)$ between periodic cyclic homology groups is an isomorphism.*

We also obtain an isomorphism on periodic cyclic homology for a slightly more general class of algebra morphisms.

DEFINITION 3.3. A morphism $\phi : L \to J$ of finite type algebras is called *weakly spectrum preserving* if, and only if, there exist increasing filtrations

$$(0) = L_0 \subset L_1 \subset \ldots \subset L_n = L \quad \text{and} \quad (0) = J_0 \subset J_1 \subset \ldots \subset J_n = J$$

of two-sided ideals such that $\phi(L_k) \subset J_k$ and the induced morphisms $L_k/L_{k-1} \to J_k/J_{k-1}$ are spectrum preserving.

Combining the above theorem with the excision property, we obtain the following result from [3].

THEOREM 3.4 (Baum-Nistor). *Let L and J be finite type $\mathfrak{k}$-algebras and $\phi : L \to J$ be a $\mathfrak{k}$-linear weakly spectrum preserving morphism. Then the induced map $\phi_* : \mathrm{HP}_*(L) \to \mathrm{HP}_*(J)$ between periodic cyclic homology groups is an isomorphism.*

The main application of this theorem is the determination of the periodic cyclic homology of Iwahori-Hecke algebras in [2, 3]. Let H_q be an Iwahori-Hecke algebra and J the corresponding asymptotic Hecke algebra associated to an extended affine Weyl group $\widehat{W}$ [36] (their definition is recalled in [3], where more details and more complete references are given). Then there exists a morphism $\phi : H_q \to J$ of $\mathfrak{k}$-finite type algebras, $\mathfrak{k} = Z(H_q)$ that is weakly spectrum-preserving provided that q is not a root of unity or $q = 1$. Therefore

$$(23) \qquad\qquad \phi_* : \mathrm{HP}_*(H_q) \to \mathrm{HP}_*(J)$$

is an isomorphism. The algebra H_1 is a group algebra, and hence its periodic cyclic homology can be calculated directly.

The above theorem also helps us clarify the Examples 2.6 and 2.7.

EXAMPLE 3.5. Let us assume that $v_j = v \in \mathbb{C}^2$ in Example 2.6. Let $e \in M_2(\mathbb{C})$ be the projection onto the vector e. The morphism $\phi : \mathbb{C}[x] \ni P \to Pe \in A_1 \subset M_2(\mathbb{C}[x])$ is not weakly spectrum preserving. However, $\phi : \mathbb{C}[x] \to \mathfrak{I}_1$ is a spectrum preserving morphism of $\mathbb{C}[x]$-algebras. Combining with the inclusion $\mathfrak{I}_1 \subset M_2(\mathbb{C}[x])$, we see that $\phi_* : \mathrm{HP}_*(\mathbb{C}[x]) \to \mathrm{HP}_*(\mathfrak{I}_1)$ is an isomorphism and $\mathrm{HP}_*(\mathfrak{I}_1)$ is a direct summand of $\mathrm{HP}_*(A_1)$. The excision theorem then gives

$$\mathrm{HP}_q(A_1) \simeq \mathrm{HP}_q(\mathfrak{I}_1) \oplus \mathrm{HP}_q(A/\mathfrak{I}_1) \simeq \mathrm{HP}_q(\mathbb{C}[x]) \oplus \mathrm{HP}_q(\mathbb{C}^l) = \begin{cases} \mathbb{C}^{l+1} & \text{if } q \text{ is even} \\ 0 & \text{otherwise.} \end{cases}$$

The case of the algebra A_2 of example is even simpler.

EXAMPLE 3.6. The inclusion $Z(A_2) \to A_2$ is a spectrum preserving morphism of $Z(A_2)$-algebras. Consequently,

$$\mathrm{HP}_q(A_2) \simeq \mathrm{HP}_q(\mathbb{C}[x]) \oplus \mathrm{HP}_q(\mathbb{C}^l) = \begin{cases} \mathbb{C}^{l+1} & \text{if } q \text{ is even} \\ 0 & \text{otherwise.} \end{cases}$$

Let us notices that by considering the action of the natural morphisms

$$\mathbb{C}[x] \to Z(A_1) \subset A_1, \ A_1 \to M_2(\mathbb{C}[x]), \ \mathbb{C}[x] \to Z(A_2) \subset A_2, \ or A_2 \to M_2(\mathbb{C}[x])$$

on periodic cyclic homology, we will still not be able to distinguish between A_1 and A_2. However, the natural products $\mathrm{HP}_i(\mathbb{C}[x]) \otimes \mathrm{HP}_j(A_k) \to \mathrm{HP}_*(A_k)$ (see [26, 27]) *will* distinguish between these algebras.

4. The periodic cyclic homology of $\mathcal{C}_c^\infty(G)$

Having discussed the relation between $HP_*(\mathcal{C}_c^\infty(G))$ and the admissible spectrum $\hat{G} = \mathrm{Prim}(\mathcal{C}_c^\infty(G))$, let us recall the explicit calculation of $HP_*(\mathcal{C}_c^\infty(G))$ from [47]. The calculation of $HP_*(\mathcal{C}_c^\infty(G))$ in [47] follows right away from the calculation of the Hochschild homology groups of $\mathcal{C}_c^\infty(G)$. The calculations of presented in this section complement the results on the cyclic homology of p–adic groups in [45, 48].

To state the main result of [47] on the Hochschild homology of the algebra $\mathcal{C}_c^\infty(G)$, we need to introduce first the concepts of a "standard subgroup" and of a "relatively regular element" of a standard subgroup. For any group G and any subset $A \subset G$, we shall denote

$$C_G(A) := \{g \in G, ga = ag, \ \forall a \in A\}, \quad N_G(A) := \{g \in G, gA = Ag\},$$

$W_G(A) := N_G(A)/C_G(A)$, and $Z(A) := A \cap C_G(A)$. This latter notation will be used only when A is a subgroup of G. The subscript G will be dropped from the notation whenever the group G is understood. A commutative subgroup S of G is called *standard* if S is the group of semisimple elements of the center of $C(s)$ for some semi-simple element $s \in G$. An element $s \in S$ with this property will be called *regular relative to S*, or *S-regular*. The set of S-regular elements will be denoted by S^{reg}.

We fix from now on a p-adic group G. (Recall that a p–adic group $G = \mathbb{G}(\mathbb{F})$ is the set of $\mathbb{F}$–rational points of a linear algebraic group $\mathbb{G}$ defined over a non-archemedean, non-discrete, locally compact field $\mathbb{F}$ of characteristic zero.) Our results will be stated in terms of standard subgroups of G. We shall denote by H_u the set of unipotent elements of a subgroup H. Sometimes, the set $C(S)_u$ is also denoted by $\mathcal{U}_S$, in order to avoid having to many paranthesis in our formulae. Let $\Delta_{C(S)}$ denote the modular function of the group $C(S)$ and let

$$\mathcal{C}_c^\infty(\mathcal{U}_S)_\delta := \mathcal{C}_c^\infty(C(S)_u) \otimes \Delta_{C(S)},$$

be $\mathcal{C}_c^\infty(\mathcal{U}_S)$ as a vector space, but with the product $C(S)$-module structure, that is

$$\gamma(f)(u) = \Delta_{C(S)}(\gamma)f(\gamma^{-1}u\gamma),$$

for all $\gamma \in C(S)$, $f \in \mathcal{C}_c^\infty(\mathcal{U}_S)_\delta$ and $u \in \mathcal{U}_S$.

The groups $HH_*(\mathcal{C}_c^\infty(G))$ are determined in terms of the following data:

(1) the set Σ of conjugacy classes of standard subgroups S of G;
(2) the subsets $S^{\mathrm{reg}} \subset S$ of S-regular elements;
(3) the actions of the Weyl groups $W(S)$ on $\mathcal{C}_c^\infty(S)$; and
(4) the continuous cohomology of the $C(S)$–modules $\mathcal{C}_c^\infty(\mathcal{U}_S)_\delta$.

Combining this proposition with Corollary 2.3, we obtain the main result of this section. Also, recall that $\mathcal{U}_S$ is the set of unipotent elements commuting with the standard subgroup S, and that the action of $C(S)$ on $\mathcal{C}_c^\infty(\mathcal{U}_S)$ is twisted by the modular function of $C(S)$, yielding the module $\mathcal{C}_c^\infty(\mathcal{U}_S)_\delta = \mathcal{C}_c^\infty(\mathcal{U}_S) \otimes \Delta_{C(S)}$.

THEOREM 4.1. *Let G be a p–adic group. Let Σ be a set of representative of conjugacy classes of standard subgroups of $S \subset G$ and $W(S) = N(S)/C(S)$, then we have an isomorphism*

$$HH_q(\mathcal{C}_c^\infty(G)) \simeq \bigoplus_{S \in \Sigma} \mathcal{C}_c^\infty(S^{\mathrm{reg}})^{W(S)} \otimes H_q(C(S), \mathcal{C}_c^\infty(\mathcal{U}_S)_\delta).$$

The isomorphism of this theorem was obtained by identifying the E^∞-term of a spectral sequence convergent to $\mathrm{HH}_q(\mathcal{C}_c^\infty(G))$, and hence it is not natural. This isomorphism can be made natural by using a generalization of the Shalika germs [47].

The periodic cyclic homology groups of $\mathcal{C}_c^\infty(G)$ are then determined as follows. Recall that our convention is that $\mathrm{HH}_{[q]} := \oplus_{k \in \mathbb{Z}} \mathrm{HH}_{q+2k}$. An element $\gamma \in G$ is called *compact* if the closure of the set $\{\gamma^n\}$ is compact. The set G_{comp} of compact elements of G is open and closed in G, if we endow G with the locally compact, Hausdorff topology obtained from an embedding $G \subset \mathbb{F}^N$. We clearly have $\gamma G_{comp} \gamma^{-1} = G_{comp}$, that is, G_{comp} is G-invariant for the action of G on itself by conjugation. Also, we shall denote by $\mathrm{HH}_{[q]}(\mathcal{C}_c^\infty(G))_{comp}$ the localization of the homology group $\mathrm{HH}_{[q]}(\mathcal{C}_c^\infty(G))$ to the set of compact elements of G (see [8] or [44]). This localization is defined as follows. Let $R^\infty(G)$ be the ring of locally constant Ad_G-invariant functions on G with the pointwise product. If $\omega = f_0 \otimes f_1 \otimes \ldots f_n \in \mathcal{C}_c^\infty(G)^{\otimes(n+1)} = \mathcal{C}_c^\infty(G^{n+1})$ and $z \in R^\infty(G)$, then we define

$$[z\omega](g_0, g_1, \ldots, g_n) = z(g_0 g_1 \ldots g_n)\omega(g_0, g_1, \ldots, g_n) \in \mathcal{C}_c^\infty(G^{n+1}).$$

Let χ be the characteristic function of G_{comp} (so $\chi = 1$ on G_{comp} and $\chi = 0$ otherwise). Then $\mathrm{HH}_{[q]}(\mathcal{C}_c^\infty(G))_{comp} = \chi \mathrm{HH}_{[q]}(\mathcal{C}_c^\infty(G))$.

THEOREM 4.2. *We have*

(24) $$\mathrm{HP}_q(\mathcal{C}_c^\infty(G)) \simeq \mathrm{HH}_{[q]}(\mathcal{C}_c^\infty(G))_{comp}.$$

Let S_{comp} be the set of compact elements of a standard subgroup S, then

(25) $$\mathrm{HP}_q(\mathcal{C}_c^\infty(G)) \simeq \bigoplus_{S \in \Sigma} \mathcal{C}_c^\infty(S_{comp}^{reg})^{W(S)} \otimes \mathrm{H}_{[q]}(C(S), \mathcal{C}_c^\infty(\mathcal{U}_S)_\delta).$$

The Equation (24) follows also from the results in [45].

It is conceivable that a next step would be to study the "discrete parts" of the groups $\mathrm{HH}_q(\mathcal{C}_c^\infty(G))$, following the philosophy of [6, 28]. This can be defined as follows. In [47], we have defined "induction morphisms"

(26) $$\phi_M^G : \mathrm{HH}_q(\mathcal{C}_c^\infty(G)) \to \mathrm{HH}_q(\mathcal{C}_c^\infty(M))$$

for every Levi subgroup $M \subset G$. We define $\mathrm{HH}_q(\mathcal{C}_c^\infty(G))_0$ to be the intersection of all kernels of ϕ_M^G, for M a proper Levi subgroup of G and call this the discrete part of $\mathrm{HH}_q(\mathcal{C}_c^\infty(G))_0$.

Assuming that one has established, by induction, a procedure to construct the cuspidal representations of all p–adic groups of lower split rank, then one can study the action of the center of $\mathcal{C}_c^\infty(G)$ corresponding to the cuspidal associated to proper Levi subgroups on the discrete part of $\mathrm{HH}_q(\mathcal{C}_c^\infty(G))$, which would hopefully allow us to distinguish between the cuspidal part of $\mathrm{HH}_q(\mathcal{C}_c^\infty(G))_0$ and its part coming from square integrable representations that are not super-cuspidal.

5. Hochschild homology

We include now three short sections that recall some of the definitions used above. Nothing in this and the next section is new, and the reader interested in more details as well as precise references should consult the following standard references [10, 5, 16, 17, 25, 33, 32, 51].

We begin by recalling the definitions of Hochschild homology groups of a complex algebra A, not necessarily with unit. We define b and b' define two linear maps

(27)
$$b, b' : A^{\otimes n+1} \to A^{\otimes n}.$$

where $A^{\otimes n} := A \otimes A \otimes \ldots A$ (n times) by the formulas

(28)
$$b'(a_0 \otimes a_1 \otimes \ldots \otimes a_n) = \sum_{i=0}^{n-1}(-1)^i a_0 \otimes \ldots \otimes a_i a_{i+1} \otimes \ldots \otimes a_n,$$

$$b(a_0 \otimes a_1 \otimes \ldots \otimes a_n) = b'(a_0 \otimes a_1 \otimes \ldots \otimes a_n) + (-1)^n a_n a_0 \otimes \ldots \otimes a_{n-1},$$

where $a_0, a_1, \ldots, a_n \in A$. Let

(29)
$$\mathcal{H}_n(A) = \mathcal{H}'_n(A) := A \otimes A \otimes \ldots A \; (n+1 \text{ times }).$$

Also, let $\mathcal{H}(A) := (\mathcal{H}_n(A), b)$ and $\mathcal{H}'(A) := (\mathcal{H}/_n(A), b')$. The homology groups of $\mathcal{H}(A)$ are, by definition, the *Hochschild homology* groups of A and are non-zero. The nth Hochschild homology group of A is denoted $\mathrm{HH}_n(A)$. By dualizing, we obtain the Hochschild cohomology groups $\mathrm{HH}^n(A)$.

If A has a unit, then the complex $\mathcal{H}'(A)$ is acyclic (*i.e.* it has vanishing homology groups) because $b's + sb' = 1$, where

(30)
$$s(a_0 \otimes a_1 \otimes \ldots \otimes a_n) = 1 \otimes a_0 \otimes a_1 \otimes \ldots \otimes a_n.$$

Therefore, if A has a unit, the complex $\mathcal{H}'(A)$ is a resolution of A by free A-bimodules.

Recall that *a trace* on A is a linear map $\tau : A \to \mathbb{C}$ such that $\tau(a_0 a_1 - a_1 a_0) = 0$ for all $a_0, a_1 \in A$. The space of all traces on A is then isomorphic to $\mathrm{HH}^0(A)$.

An algebra A such that $\mathcal{H}'(A)$ is acyclic is called *H-unital*, following Wodzicki [53].

Clearly the groups $\mathrm{HH}_n(A)$ are *covariant* functors in A, in the sense that any algebra morphism $\phi : A \to B$ induces a morphism

$$\phi_* = \mathrm{HH}_n(\phi) : \mathrm{HH}_n(A) \to \mathrm{HH}_n(B)$$

for any integer $n \geq 0$. Similarly, we also obtain a morphism $\phi^* = \mathrm{HH}^n(\phi) :$ $\mathrm{HH}^n(B) \to \mathrm{HH}^n(A)$. In other words, Hochschild cohomology is a *contravariant* functor. It is interesting to note that if Z is the center of A, then $\mathrm{HH}_n(A)$ is also a Z-module, where, at the level of complexes the action is given by

(31)
$$z(a_0 \otimes a_1 \otimes \ldots \otimes a_n) = z a_0 \otimes a_1 \otimes \ldots \otimes a_n.$$

for all $z \in Z$. As z is in the center of A, this action will commute with the Hochschild differential b.

6. Cyclic homology

Let A be a unital algebra. We shall denote by t the (signed) generator of cyclic permutations:

(32)
$$t(a_0 \otimes a_1 \otimes \ldots \otimes a_n) = (-1)^n a_n \otimes a_0 \otimes \ldots \otimes a_{n-1}$$

Using the operator t and the contracting homotopy s of the complex $\mathcal{H}'(A)$, Equation (30), we construct a new differential $B := (1-t)B_0$, of degree $+1$, where

(33)
$$B_0(a_0 \otimes a_1 \otimes \ldots \otimes a_n) = s \sum_{k=0}^{n} t^k(a_0 \otimes a_1 \otimes \ldots \otimes a_n).$$

It is easy to check that $B^2 = 0$ and that $[b, B]_+ := bB + Bb = 0$.

The differentials b and B give rise to the following complex

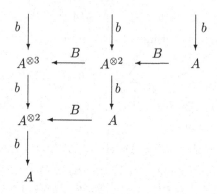

FIGURE 1. *The cyclic bicomplex of the algebra A.*

We notice that columns the above complex are copies of the Hochschild complex $\mathcal{H}(A)$. The cyclic complex $\mathcal{C}(A)$ is by definition the total complex of this double complex. Thus the space of cyclic n-chains is defined by

$$(34) \qquad \mathcal{C}(A)_n = \bigoplus_{k \geq 0} \mathcal{H}_{n-2k}(A),$$

we see that $(\mathcal{C}(A), b + B)$, is a complex, called *the cyclic complex* of A, whose homology is by definition the *cyclic homology* of A, denoted $\mathrm{HC}_q(A)$, $q \geq 0$.

There is a canonical operator $S : \mathcal{C}_n(A) \to \mathcal{C}_{n-2}(A)$, called the *Connes periodicity operator*, which shifts the cyclic complex left and down, explicitly defined by

$$(35) \qquad S(\omega_n, \omega_{n-2}, \omega_{n-4}, \dots) \mapsto (\omega_{n-2}, \omega_{n-4}, \dots),$$

where $\omega_k \in \mathcal{H}_k(A)$, for all k. This operator induces the short exact sequence of complexes

$$0 \to \mathcal{H}(A) \xrightarrow{I} \mathcal{C}(A) \xrightarrow{S} \mathcal{C}(A)[2] \to 0,$$

where the map I is the inclusion of the Hochschild complex as the first column of the cyclic complex. The snake Lemma in homology [**38, 39, 52**] gives the following long exact sequence, called the *SBI–exact sequence*

$$(36) \qquad \dots \to \mathrm{HH}_n(A) \xrightarrow{I} \mathrm{HC}_n(A) \xrightarrow{S} \mathrm{HC}_{n-2}(A) \xrightarrow{B} \mathrm{HH}_{n-1}(A) \xrightarrow{I} \dots ,$$

where B is the differential defined above, see [**16, 33**] for more details. The *periodic cyclic complex* of an algebra A is the complex

$$\mathcal{C}^{\mathrm{per}}(A) := \lim_{\leftarrow} \mathcal{C}(A),$$

the inverse limit being taken with respect to the periodicity morphism S. It is a $\mathbb{Z}/2\mathbb{Z}$-graded complex, whose chains are (possibly infinite) sequences of Hochschild chains with degrees of the same parity. The homology groups of the periodic cyclic complex $\mathcal{C}^{\mathrm{per}}(A)$ are, by definition, the *periodic cyclic homology groups* of A. A simple consequence of the SBI–exact sequence is that if $\phi : A \to B$ is a morphism of algebras that induces an isomorphism on Hochschild homology, then ϕ induces an isomorphism on cyclic and periodic cyclic homology groups as well. Here is an application of this simple principle.

Here is an application of this lemma. Consider

$$(37) \qquad Tr_* : \mathcal{H}_q(M_N(A)) \to \mathcal{H}_q(A), \quad q \in \mathbb{Z}_+,$$

the map defined by $Tr_*(b_0 \otimes \ldots \otimes b_q) = Tr(m_0 m_1 \ldots m_q) a_0 \otimes \ldots a_q$, if $b_k = m_k \otimes a_k \in M_N(\mathbb{C}) \otimes A = M_N(A)$. Also consider the (unital) inclusion $\iota : A \to M_N(A)$ and ι_* be the morphism induced on the Hochschild complexes.

PROPOSITION 6.1.   *The map Tr_* commutes with b and B. Both ι_* and Tr_* induce isomorphisms on Hochschild, cyclic, and periodic cyclic homologies and cohomologies such that $(\iota_*)^{-1} = N^{-1} Tr_*$.*

The cyclic and periodic cyclic homology groups of a non-unital algebra A are defined as the cokernels of the maps $\mathrm{HC}_q(\mathbb{C}) \to \mathrm{HC}_q(A^+)$ and $\mathrm{HP}_q(\mathbb{C}) \to \mathrm{HP}_q(A^+)$, where $A^+ = A \oplus \mathbb{C}$ is the algebra with adjoined unit.

More generally, cyclic homology groups can be defined for "mixed complexes," [5, 24, 27, 29].

7. The Chern character

We shall use these calculations to construct Chern characters. By taking X to be a point in Theorem 1.3, we obtain that $\mathrm{HC}_{2q}(\mathbb{C}) \simeq \mathbb{C}$ and $\mathrm{HC}_{2q+1}(\mathbb{C}) \simeq 0$. We can take these isomorphisms to be compatible with the periodicity operator S and such that for $q = 0$ it reduces to

$$\mathrm{HC}_0(\mathbb{C}) = \mathrm{HH}_0(\mathbb{C}) = \mathbb{C}/[\mathbb{C}, \mathbb{C}] = \mathbb{C}.$$

We shall denote by $\eta_q \in \mathrm{HC}_{2q}(\mathbb{C})$ the unique element such that

$$(38) \qquad S^q \eta_q = 1 \in \mathbb{C} = \mathrm{HC}_0(\mathbb{C}).$$

For any projection $e \in M_N(A)$ is a projection we obtain a (non-unital) morphism $\psi : \mathbb{C} \to M_N(A)$ by $\lambda \mapsto \lambda e$. Then *Connes-Karoubi Chern character of e in cyclic homology* [16, 17, 25, 32] is defined by

$$(39) \qquad Ch_q([e]) = Tr_*(\psi_*(\eta_q)) \in \mathrm{HC}_{2q}(A).$$

This map can be shown to depend only on the class of e in K-theory and to define a morphism

$$(40) \qquad Ch_q : K_0(A) \to \mathrm{HC}_{2q}(A).$$

One can define similarly the Chern character in periodic cyclic homology and the Chern character on K_1 (algebraic K-theory). For the Connes-Karoubi Chern character on K_1, we use instead $Y = \mathbb{C}^*$, whose algebra of regular functions is $\mathcal{O}[Y] \simeq \mathbb{C}[z, z^{-1}]$, the algebra of Laurent polynomials in z and z^{-1} (this algebra, in turn, is isomorphic to the group algebra of $\mathbb{Z}$). Then $\mathrm{HC}_q(\mathcal{O}[Y]) \simeq \mathbb{C}$, for any $q \geq 1$. We are interested in the odd groups, which will be generated by elements $v_k \in \mathrm{HC}_{2k+1}(\mathcal{O}[Y])$, which can be chosen to satisfy $v_1 = z^{-1} \otimes z$ and $S^k v_{2k+1} = v_1$.

Then, if $u \in M_N(A)$ is an invertible element, it defines a morphism $\psi : \mathbb{C}[\mathbb{C}^*] \to M_N(A)$. The *Connes-Karoubi Chern character of u in cyclic homology* is thus defined by

$$(41) \qquad Ch_q([u]) = Tr_*(\psi_*(v_q)) \in \mathrm{HC}_{2q+1}(A).$$

Again, this map can be shown to depend only on the class of u in K-theory and to define a morphism

$$(42) \qquad Ch_q : K_1(A) \to HC_{2q+1}(A).$$

Both the Chern character on K_0 and on K_1 are functorial, by construction.

Acknowledgment. Nistor was partially supported by NSF Grant DMS-0200808.

References

[1] P. Baum, N. Higson, and R. Plymen, Representation theory of p-adic groups: a view from operator algebras, The mathematical legacy of Harish-Chandra (Baltimore, MD, 1998), 111-149, Proc. Symp. Pure Math. 69, A. M. S. Providence RI, 2000.

[2] P. Baum and V. Nistor, Periodic cyclic homology of Iwahori-Hecke algebras, C. R. Acad. Sci. Paris **332** (2001), 783–788.

[3] P. Baum and V. Nistor, Periodic cyclic homology of Iwahori-Hecke algebras, K-theory **27** (2003), 329–358.

[4] M-T. Benameur and V. Nistor, Homology of complete symbols and noncommutative geometry, In Quantization of Singular Symplectic Quotients, N.P. Landsman, M. Pflaum, and M. Schlichenmaier, ed., Progress in Mathematics **198**, pages 21–46, Birkhäuser, Basel - Boston - Berlin, 2001.

[5] M. Benameur, J. Brodzki, and V. Nistor, Cyclic homology and pseudodifferential operators: a survey.

[6] J. Bernstein, P. Deligne, and D. Kazhdan, Trace Paley-Wiener theorem for reductive p–adic groups, J. Analyse Math. **47** (1986), 180–192.

[7] J. N. Bernstein, Le "centre" de Bernstein, Representations of reductive groups over a local field (Paris) (P. Deligne, ed.), Hermann, Paris, 1984, pp. 1–32.

[8] P. Blanc and J. L. Brylinski, Cyclic Homology and the Selberg Principle, J. Funct. Anal. **109** (1992), 289–330.

[9] A. Borel, Admissible representations of a semi-simple group over a local field with vectors fixed under an Iwahori subgroup, Invent. Math. **35** (1976), 233–259.

[10] J. Brodzki, An introduction to K-theory and cyclic cohomology, Advanced Topics in Mathematics. PWN—Polish Scientific Publishers, Warsaw, 1998.

[11] J. Brodzki and V. Nistor, Cyclic homology of crossed products, work in progress.

[12] J. Brodzki and Z. Lykova, Excision in cyclic type homology of Fréchet algebras, Bull. London Math. Soc. 33 (2001), no. 3, 283–291.

[13] J. Brodzki, R. Plymen, Chern character for the Schwartz algebra of p-adic GL(n), Bulletin of the LMS, 34, (2002) 219-228.

[14] J.-L. Brylinski, Central localization in Hochschild homology, J. Pure Appl. Algebra **57** (1989), 1–4.

[15] J.-L. Brylinski and V. Nistor, Cyclic cohomology of etale groupoids, K-Theory **8** (1994), 341–365.

[16] A. Connes, Noncommutative differential geometry, Publ. Math. IHES **62** (1985), 41–144.

[17] A. Connes, Noncommutative Geometry, Academic Press, New York - London, 1994.

[18] J. Cuntz, Excision in periodic cyclic theory for topological algebras, In: Cyclic cohomology and noncommutative geometry (Waterloo, ON, 1995), 43–53, Amer. Math. Soc., Providence, RI, 1997.

[19] J. Cuntz and D. Quillen, Excision in bivariant periodic cyclic cohomology, Invent. Math. **127** (1997), 67–98.

[20] B. Feigin and B. L. Tsygan, Additive K–Theory and cristaline cohomology, Funct. Anal. Appl. **19** (1985), 52–62.

[21] B. Feigin and B. L. Tsygan, Cyclic homology of algebras with quadratic relations, universal enveloping algebras and group algebras, in K-theory, arithmetic and geometry (Moscow, 1984-1986), 210–239 Lect. Notes Math. **1289**, Springer, Berlin, 1987.

[22] T. G. Goodwillie, Cyclic homology, derivations, and the free loopspace, Topology **24** (1985), 187–215.

[23] G. Hochschild, B. Kostant, and A. Rosenberg, *Differential forms on regular affine algebras*, Trans. AMS **102** (1962), 383–408.

[24] John D. Jones and C. Kassel, *Bivariant cyclic theory*, K-Theory **3** (1989), 339–365.

[25] M. Karoubi, *Homologie cyclique et K-théorie* Asterisque **149** (1987), 1–147.

[26] C. Kassel, *A Knneth formula for the cyclic cohomology of Z/2-graded algebra*, Math. Ann. **275** (1986), 683–699.

[27] C. Kassel, *Cyclic homology, comodules, and mixed complexes*, J. Algebra **107** (1987), 195–216.

[28] D. Kazhdan, *Cuspidal geometry of p-adic groups*, J. Analyse Math. **47** (1986), 1–36.

[29] D. Kazhdan, V. Nistor, and P. Schneider, *Hochschild and cyclic homology of finite type algebras*, Selecta Math. (N.S.) **4** (1998), 321–359.

[30] M. Kontsevich and A. Rosenberg, *Noncommutative smooth spaces*, The Gelfand Mathematical Seminar, 1996–1999, 85–108, Birkhäuser Boston, Boston, Ma, 2000.

[31] S. Lang, *Algebra*, Revised third edition, Springer, New York, 2002

[32] J.-L. Loday, Cyclic Homology, Springer-Verlag, Berlin-Heidelberg-New York, 1992.

[33] J.-L. Loday and D. Quillen, *Cyclic homology and the Lie homology of matrices*, Comment. Math. Helv. **59** (1984), 565–591.

[34] G. Lusztig, *Cells in affine Weyl groups*, In "Algebraic Groups and Related Topics," pp. 255–287, Advanced studies in Pure Math., vol. **6**, Kinokunia and North Holland, 1985.

[35] G. Lusztig, *Cells in affine Weyl groups II*, J. Algebra **109** (1987), 536–548.

[36] G. Lusztig, *Cells in affine Weyl groups III*, J. Fac. Sci. Univ. Tokyo Sect. IA Math. **34** (1987), 223–243.

[37] G. Lusztig, *Cells in affine Weyl groups IV*, J. Fac. Sci. Univ. Tokyo Sect. IA Math. **36** (1989), 297–328.

[38] S. Mac Lane, *Homology*, Springer-Verlag, Berlin-Heidelberg-New York, 1995.

[39] S. Mac Lane and I. Moerdijk, *Sheaves in geometry and logic. A first introduction to topos theory*, Universitext, Springer-Verlag, Berlin-Heidelberg-New York, 1994.

[40] Yu. Manin, Topics in noncommutative geometry, M. B. Porter Lectures, Princeton University Press, Princeton, NJ, 1991, viii+164pp.

[41] Yu. Manin, *Real multiplication and non-commutative geometry*, Lectures at M.P.I, May 2001.

[42] Yu. Manin and M. Marcolli, *Continued fractions, modular symbols, and noncommutative geometry*, Selecta Math. New Ser., **8** (2002), 475–521.

[43] V. Nistor, *Group cohomology and the cyclic cohomology of crossed products*, Invent. Math. **99** (1990), 411–424.

[44] V. Nistor, *Cyclic cohomology of crossed products by algebraic groups*, Invent. Math. **112** (1993), 615–638.

[45] N. Higson and V. Nistor, *Cyclic homology of totally disconnected groups acting on buildings*, J. Funct. Anal. **141** (1996), 466–495.

[46] V. Nistor, *Higher index theorems and the boundary map in cyclic homology*, Documenta **2** (1997), 263–295.

[47] V. Nistor, *Higher orbital integrals, Shalika germs, and the Hochschild homology of Hecke algebras of p-adic groups*, Int. J. Math. Math. Sci. 26 (2001), 129–160.

[48] P. Schneider, *The cyclic homology of p-adic reductive groups*, J. Reine Angew. Math. **475** (1996), 39–54.

[49] S. Dave, *Equivariant yclic homology of pseudodifferential operators*, Thesis, Pennsylvania State University, in preparation.

[50] J.-P. Serre, *Linear representations of finite groups*, Translated from the second French edition by Leonard L. Scott, Springer, New York, 1977.

[51] B. L. Tsygan, *Homology of matrix Lie algebras over rings and Hochschild homology*, Uspekhi Math. Nauk., **38** (1983), 217–218.

[52] C. Weibel, An introduction to homological algebra, Cambridge Studies in Advanced Mathematics **38**, *Cambridge University Press, Cambridge*, 1994.

[53] M. Wodzicki, *Excision in cyclic homology and in rational algebraic K-theory*, Annals of Mathematics **129** (1989), 591–640.

PENNSYLVANIA STATE UNIVERSITY, UNIVERSITY PARK, PA 16802
E-mail address: nistor@math.psu.edu

Three examples of non-commutative boundaries of Shimura varieties

Frédéric Paugam

ABSTRACT. Our modest aims in writing this paper were twofold: we first wanted to understand the linear algebra and algebraic group theoretic background of Manin's real multiplication program proposed in [**Man**]. Secondly, we wanted to find nice higher dimensional analogs of the non-commutative modular curve studied by Manin and Marcolli in [**MM02**]. These higher dimensional objects, that we call *irrational* or *non-commutative boundaries of Shimura varieties*, are double cosets spaces of the form $\Gamma \backslash G(\mathbb{R})/P(K)$, where G is a (connected) reductive $\mathbb{Q}$-algebraic group, $P(K) = M(K)AN \subset G(\mathbb{R})$ is a real parabolic subgroup corresponding to a rational parabolic subgroup $P \subset G$, and $\Gamma \subset G(\mathbb{Q})$ is an arithmetic subgroup. Along the way, it also seemed clear that the spaces

$$\Gamma \backslash G(\mathbb{R})/M(K)A$$

are of great interest, and sometimes more convenient to study. We study in this document three examples of these general spaces. These spaces describe degenerations of complex structures on tori in (multi)foliations.

Introduction

The non-commutative modular curve is the chaotic space

$$\mathrm{GL}_2(\mathbb{Z}) \backslash \mathbb{P}^1(\mathbb{R}).$$

Its first appearance in the non-commutative geometric world arose in the work [**Con80**] of Connes on differential geometry of non-commutative tori. Since the action of $\mathrm{GL}_2(\mathbb{Z})$ by homographies on $\mathbb{P}^1(\mathbb{R})$ is very chaotic, the classical quotient space, whose algebra is the one of continuous complex functions on $\mathbb{P}^1(\mathbb{R})$ that are invariant by $\mathrm{GL}_2(\mathbb{Z})$, is topologically identified with a point. The philosophy of Connes' non-commutative geometry and the related philosophy of topoi (as explained by Cartier in [**Car98**]) tell us that this quotient space is much more than that. Connes showed that the crossed product C^*-algebra $C^*(\mathbb{P}^1(\mathbb{R})) \rtimes \mathrm{GL}_2(\mathbb{Z})$ is a good analog of the algebra of continuous functions for such a chaotic space because it is possible to calculate from it nice cohomological invariants, as K-theory and cyclic cohomology (from its C^∞ version), that have a real geometric meaning. He showed, indeed, that cyclic cohomology is a good analog of De Rham cohomology of manifolds for such chaotic spaces. This cohomology theory is a very profound tool that permitted Connes and Moscovici, to cite one example among many others, to prove a local index formula for foliations, which is an analog of the Riemann-Roch theorem for the spaces of leaves [**CM95**]. This result was not accessible without non-commutative geometric intuition. Bost-Connes and Connes have also shown [**BC95**], [**Con99**], that non-commutative geometry can be very useful for arithmetic questions. We refer the reader to Marcolli [**Mar04**] for a nice survey on non-commutative arithmetic geometry.

The non-commutative modular curve also appeared some years ago in Connes, Douglas and Schwarz paper [**CDS97**] as moduli space for physical backgrounds for compactifications of string theory . It was already shown in [**SW99**] by Seiberg and Witten that non-commutative spaces can be good backgrounds for open string theory. The deformation quantization story of Kontsevich and Soibelman also gives related results. The author, being more informed of arithmetic geometry, will not discuss these physical motivations that are very important for future developments of non-commutative moduli spaces.

The arithmetic viewpoint of the non-commutative modular curve appeared first in the work of Manin and Marcolli [**MM02**], [**MM01**] and in Manin's real multiplication program [**Man**]. These works were for us the main inspirations for writing this paper.

The basic idea in our work is the following: the classical theory of Shimura varieties was completely rewritten in terms of Hodge structures by Deligne in [**Del79**] and it proved to be very helpful for arithmetics, for example in the theory of absolute Hodge motives made in [**DMOS82**] and for the construction of canonical models for all Shimura varieties, made by Milne in [**Mil83**]. One of the interests of this construction is to translate fine information about moduli spaces in terms of algebraic groups morphisms. The author wanted to know if it was possible to make such a translation for non-commutative boundaries of Shimura varieties.

As higher dimensional analogs of the boundary $\mathbb{P}^1(\mathbb{R})$ of the double half plane $\mathbb{H}^\pm := \mathbb{P}^1(\mathbb{C}) - \mathbb{P}^1(\mathbb{R})$, we chose to use the components of Satake's compactifications of symmetric spaces. These components can be written as quotients $G(\mathbb{R})/P(K)$, where G is a (connected) reductive $\mathbb{Q}$-algebraic group and $P(K) = M(K)AN \subset G(\mathbb{R})$ is a real parabolic subgroup[1] corresponding to a rational parabolic subgroup $P \subset G$. The higher dimensional analogs of the non-commutative modular curve $\mathrm{GL}_2(\mathbb{Z})\backslash\mathbb{P}^1(\mathbb{R})$, which we call *irrational* or *non-commutative boundaries of Shimura varieties*, are given by double coset spaces

$$\Gamma\backslash G(\mathbb{R})/P(K),$$

where $\Gamma \subset G(\mathbb{Q})$ is an arithmetic subgroup. Along the way, it seemed also clear that the spaces

$$\Gamma\backslash G(\mathbb{R})/M(K)A,$$

which we call *irrational* or *non-commutative shores of Shimura varieties*, are of great interest, and sometimes more convenient to study from the algebraic group theoretical viewpoint.

The plan of this document is the following. In the first part, we study special geodesics on the modular curve that are good analogs of elliptic curves with complex multiplication. We explain how to relate the counting of these geodesics to number theoretical considerations. In the second part, we construct the moduli space and universal family of non-commutative tori in a way analogous to the construction of the moduli space of elliptic curves and its universal family. In the third part,

---

[1]See the book [**BL01**] for the definition and decomposition of these real parabolic subgroups.

we give two higher dimensional examples of non-commutative shores of Shimura varieties that parametrize some degenerations of complex structures on tori in multi-foliations.

Notations

DEFINITION 0.1. A pair (P, X) consisting of a $\mathbb{Q}$-algebraic group P and a left $P(\mathbb{R})$-space X is called a *pre-Shimura datum*[2]. A *morphism of pre-Shimura data* $(P_1, X_1) \to (P_2, X_2)$ is a pair (ϕ, ψ) consisting of a morphism $\phi : P_1 \to P_2$ of groups and a $P_1(\mathbb{R})$-equivariant map $\psi : X_1 \to X_2$. If (P, X) is a pre-Shimura datum and $K \subset P(\mathbb{A}_f)$ is a compact open subgroup, the set

$$\mathrm{Sh}_K(P, X) := P(\mathbb{Q}) \backslash (X \times P(\mathbb{A}_f)/K)$$

is called the *pre-Shimura set of level K* for (P, X), and the set

$$\mathrm{Sh}(P, X) := \lim_{\leftarrow K} \mathrm{Sh}_K(P, X),$$

where K runs over all compact open subgroups in $P(\mathbb{A}_f)$, is called the *pre-Shimura tower* associated to (P, X).

We warn the reader that even if X has a nontrivial C^∞-structure, the corresponding Shimura space, viewed as a quotient topological space, can be very degenerate (even trivial). In such cases, it may be more interesting to study the crossed product algebra

$$C^\infty(X \times P(\mathbb{A}_f)/K) \rtimes P(\mathbb{Q}).$$

These are essentially the kind of non-commutative spaces we will consider in this paper.

1. Special geodesics on the modular curve

The modular curve is one of the basic examples of a Shimura variety. Its interpretation as a moduli space of Hodge structures (or simpler of complex structures) allows one to nicely define *special points* of this curve as those whose Mumford-Tate group is a torus.

We propose an analogous construction for the space $\mathrm{GL}_2(\mathbb{Z}) \backslash \mathrm{GL}_2(\mathbb{R})/D(\mathbb{R})$ of leaves of the geodesic foliation (where $D(\mathbb{R})$ is the subgroup of diagonal matrices in $\mathrm{GL}_2(\mathbb{R})$). This naive translation gives another point of view of the strong analogy between the modular curve and the space of leaves of the geodesic flow, which was already known to Gauss. Being closer to the modern (i.e., adelic) point of view of Shimura varieties increases its chances to be generalized to boundaries of higher dimensional Shimura varieties.

---

[2]Most of the pre-Shimura data that will appear in this document will be constructed using conjugacy classes of morphisms. However, the target group will not always be P and some useful morphisms between them will not be morphisms of conjugacy classes but just equivariant morphisms. This is why we use such a weak definition.

1.1. Mumford-Tate groups of elliptic curves. Let E be a complex elliptic curve. Let $M = H^1(E, \mathbb{Q})$ be its first singular homology group. We have the Hodge decomposition

$$M_\mathbb{C} = H^1(E, \mathbb{C}) \cong H^0(E, \Omega^1_E) \oplus H^1(E, \mathcal{O}_E).$$

Let $\mathbb{G}_{m,\mathbb{C}}$ be the multiplicative group of $\mathbb{C}$ as an algebraic group. We let $(x, y) \in (\mathbb{C}^*)^2$ act on $M_\mathbb{C}$ by multiplication by x on $H^0(E, \Omega^1_E)$ and y on $H^1(E, \mathcal{O}_E)$. This defines a natural morphism $h : \mathbb{G}^2_{m,\mathbb{C}} \to \mathrm{GL}(M_\mathbb{C})$. The *Mumford-Tate group* of E is the smallest $\mathbb{Q}$-algebraic subgroup of $\mathrm{GL}(M)$ that contains the image of h over $\mathbb{C}$. The following proposition serves us as a guide to define Mumford-Tate groups of geodesics.

PROPOSITION 1.1.1. *Let E be an elliptic curve over $\mathbb{C}$. The Mumford-Tate group of E is either GL_2, or $\mathrm{Res}_{K/\mathbb{Q}}\mathbb{G}_m$ (i.e., the group $K^\times$ as a $\mathbb{Q}$-algebraic group) with $K/\mathbb{Q}$ an imaginary quadratic field. In this case, we say that the curve is* special *or* with complex multiplication.

1.2. Geodesics and analogs of Hodge structures. We recall that the space of geodesics on the modular curve can be written as a double coset space $Y := \mathrm{GL}_2(\mathbb{Z})\backslash\mathrm{GL}_2(\mathbb{R})/D(\mathbb{R})$ with $D(\mathbb{R})$ the subgroup of diagonal matrices in $\mathrm{GL}_2(\mathbb{R})$. We denote by $\mathbb{G}_{m,\mathbb{R}}$ the multiplicative group $\mathbb{R}^*$ viewed as an algebraic group. Let $h_0 : \mathbb{G}^2_{m,\mathbb{R}} \to \mathrm{GL}_{2,\mathbb{R}}$ be the morphism of algebraic groups that sends a pair $(x, y) \in (\mathbb{R}^*)^2$ to the diagonal matrix $\mathrm{diag}(x, y) := \left(\begin{smallmatrix} x & 0 \\ 0 & y \end{smallmatrix}\right)$. We remark that $\mathrm{GL}_2(\mathbb{R})$ acts on the space $\mathrm{Hom}(\mathbb{G}^2_{m,\mathbb{R}}, \mathrm{GL}_{2,\mathbb{R}})$ by conjugation. Let

$$X := \mathrm{GL}_2(\mathbb{R}) \cdot h_0 := \{gh_0g^{-1}, g \in \mathrm{GL}_2(\mathbb{R})\}$$

be the conjugacy class of the morphism h_0.

An easy computation shows that the centralizer of h_0 in $\mathrm{GL}_2(\mathbb{R})$ is the subgroup $D(\mathbb{R})$. We thus obtain an identification $X \cong \mathrm{GL}_2(\mathbb{R})/D(\mathbb{R})$. There is also a left $\mathrm{GL}_2(\mathbb{Z})$-action on X, and this gives us an interpretation of the space of geodesics as the quotient $Y \cong \mathrm{GL}_2(\mathbb{Z})\backslash X$.

The reader will probably ask now: what did we win in this translation? The author's answer is: a strong analogy with the modular curve.

This allows us to view Y as the moduli space of triples $(M, F, \widetilde{F})$, where M is a free $\mathbb{Z}$-module of rank 2, and $F, \widetilde{F} \subset M_\mathbb{R}$ are two lines in the corresponding real vector space $M_\mathbb{R} := M \otimes_\mathbb{Z} \mathbb{R}$. To each $h \in X$, we associate a triple $(\mathbb{Z}^2, F_x, F_y)$ with F_x the line of weight x for h (i.e., $h(x, y) \cdot v = x \cdot v$ for all $v \in F_x$), and F_y the line of weight y. It is helpful to think of the direct sum decomposition

$$M_\mathbb{R} = F \oplus \widetilde{F}$$

as an analog of the Hodge decomposition of the first complex singular cohomology group of an elliptic curve $E/\mathbb{C}$:

$$H^1(E, \mathbb{C}) = H^1(E, \mathcal{O}_E) \oplus H^0(E, \Omega^1_E).$$

1.3. Examples of bad Mumford-Tate groups. In view of the analogy in the previous paragraph, we can ask: for geodesics, what is the analog of the Mumford-Tate group[3] of elliptic curves? Recall that an element $h \in X$ is a morphism of algebraic groups $h : \mathbb{G}^2_{m,\mathbb{R}} \to \mathrm{GL}_{2,\mathbb{R}}$.

---

[3]A kind of analytic Galois group.

We suggest two possible analogs. The first is obtained by copying the usual definition.

DEFINITION 1.3.1. Let $h \in X$. We define the *bad Mumford-Tate group* of h to be the smallest $\mathbb{Q}$-algebraic subgroup $\mathrm{BMT}(h) \subset \mathrm{GL}_{2,\mathbb{Q}}$ such that $h((\mathbb{R}^*)^2) \subset \mathrm{BMT}(h)(\mathbb{R})$.

Let us test this definition on some examples. It is clear that $\mathrm{BMT}(h_0) = D \cong \mathbb{G}_{m,\mathbb{Q}}^2$ is the group of rational diagonal matrices, the maximal torus of $\mathrm{GL}_{2,\mathbb{Q}}$.

Let $u \in \mathbb{R}$ and $g_u := \left(\begin{smallmatrix} 1 & u \\ 0 & 1 \end{smallmatrix}\right)$ be the corresponding unipotent matrix. Denote $h_u := g_u h_0 g_u^{-1} \in X$. The morphism h_u is given by the matrix $\left(\begin{smallmatrix} x & u(y-x) \\ 0 & y \end{smallmatrix}\right)$. If we suppose that u is not rational, then $\mathrm{BMT}(h_u)$ is the subgroup $B \subset \mathrm{GL}_{2,\mathbb{Q}}$ of upper triangular matrices. Otherwise, $\mathrm{BMT}(h_u) = g_u D g_u^{-1} \cong \mathbb{G}_{m,\mathbb{Q}}^2$ is a maximal torus.

Let h' be the conjugate of h_0 by the matrix $g' := \left(\begin{smallmatrix} 1 & 1 \\ u & -u \end{smallmatrix}\right)$, with $u = e \in \mathbb{R}$. Then we get $h' \in X$ such that $\mathrm{BMT}(h') = \mathrm{GL}_{2,\mathbb{Q}}$.

Now let us consider a square free positive integer $d > 1$. Let $g := \left(\begin{smallmatrix} 1 & 1 \\ \sqrt{d} & -\sqrt{d} \end{smallmatrix}\right)$. If we conjugate h_0 by g, we obtain the matrix $h' = \left(\begin{smallmatrix} a & b \\ db & a \end{smallmatrix}\right)$ with $a = \frac{x+y}{2}$ and $b = \frac{x-y}{2\sqrt{d}}$. The group of matrices of the form $\left(\begin{smallmatrix} a & b \\ db & a \end{smallmatrix}\right)$ with a and b rational is an algebraic torus over $\mathbb{Q}$, conjugated over $\mathbb{R}$ to the maximal torus $D(\mathbb{R})$. It is clearly the Mumford-Tate group of h'.

1.4. Good Mumford-Tate groups. We now want to modify the definition of bad Mumford-Tate groups to obtain "good" ones. The examples of last paragraph give us a quite precise idea of the different kinds of bad Mumford-Tate groups that can appear. However, we would like to have *reductive* Mumford-Tate groups in order to have a closer analogy with the case of elliptic curves.

DEFINITION 1.4.1. Let $h \in X$. A *Mumford-Tate group* for h is a minimal reductive subgroup of $\mathrm{GL}_{2,\mathbb{Q}}$ that contains $\mathrm{BMT}(h)$.

With this new definition, a Mumford-Tate group is always reductive whereas BMT may be, for example, the group B of upper triangular matrices.

LEMMA 1.4.2. *Let $h \in X$. Then there exists a unique Mumford-Tate group for h, which we denote by* $\mathrm{MT}(h)$.

PROOF. The group $\mathrm{BMT}_\mathbb{R}$ contains a maximal torus of $\mathrm{GL}_{2,\mathbb{R}}$ (the image of h). We thus have three possibilities, depending on the dimension (2,3 or 4) of BMT: BMT is a maximal torus, a Borel subgroup or the whole of $\mathrm{GL}_{2,\mathbb{Q}}$. If BMT is a maximal torus or $\mathrm{GL}_{2,\mathbb{Q}}$ then MT is clearly well defined and equal to BMT. If BMT is a Borel subgroup, then the smallest reductive group that contains it is the group $\mathrm{GL}_{2,\mathbb{Q}}$. □

We now arrive at the structure theorem for Mumford-Tate groups of geodesics.

THEOREM 1.4.3. *Let $h \in X$. The Mumford-Tate group of h is of one of the following types:*

(1) $\mathrm{MT}(h) = \mathrm{GL}_2$, *the corresponding geodesic is called MT-generic,*

(2) $\mathrm{MT}(h) = \mathrm{Res}_{E/\mathbb{Q}}\mathbb{G}_m$ *with $E/\mathbb{Q}$ a real quadratic field, we say that the corresponding geodesic is* special *or has* real multiplication *by E,*

(3) $\mathrm{MT}(h) = \mathbb{G}_{m,\mathbb{Q}}^2$, *the geodesic is* rational.

This theorem follows from the the fact that all maximal tori in $GL_{2,\mathbb{Q}}$ are of the form $Res_{E/\mathbb{Q}}\mathbb{G}_m$ with $E/\mathbb{Q}$ étale of dimension 2. Such an algebra E is either a real quadratic field, isomorphic to $\mathbb{Q}^2$, or an imaginary quadratic field. The imaginary quadratic case can not appear because the two lines F_x and F_y associated to h (see Subsection 1.2) are real.

The properties of Mumford-Tate groups have a simple interpretation in terms of dynamical properties of geodesics on the modular curve, as explained to the author by Etienne Ghys. The first case corresponds to non-closed geodesics, the second to closed geodesics not homotopic to the cusps, and the last corresponds to closed geodesics homotopic to the cusps.

1.5. Special geodesics and class field theory.

Let us look more closely at the second case in Theorem 1.4.3, following Gauss' ideas. So let $h \in X$ correspond to a special geodesic (i.e., one with real multiplication by a real quadratic field E) and let T be its Mumford-Tate group. By definition, h is a morphism into $T_{\mathbb{R}}$. The $T(\mathbb{R})$-conjugacy class of h is reduced to $\{h\}$. The natural map $(T, \{h\}) \to (GL_{2,\mathbb{Q}}, X)$ induces a map between the double coset spaces

$$T(\mathbb{Q})\backslash(\{h\} \times T(\mathbb{A}_f)/T(\widehat{\mathbb{Z}})) \longrightarrow GL_2(\mathbb{Q})\backslash(X \times GL_2(\mathbb{A}_f)/GL_2(\widehat{\mathbb{Z}}))$$

$$\|\|$$

$$Pic(\mathcal{O}_E) \longrightarrow GL_2(\mathbb{Z})\backslash X = Y$$

whose image gives the space of geodesics with real multiplication by E. The left term of this morphism is the ideal class group $Pic(\mathcal{O}_E)$ of this real multiplication field. The bottom arrow interprets $Pic(\mathcal{O}_E)$ in terms of geodesics, a well known result of Gauss. The top arrow is an adelic formulation of this result which may admit a generalization to higher rank spaces.

We want to arrive at a similar geodesic interpretation for the connected component of the idèle class group $\pi_0(T(\mathbb{Q})\backslash T(\mathbb{A}))$. Recall that if (G, H) is a pair consisting of a reductive group over $\mathbb{Q}$ and a $G(\mathbb{R})$-left space H, we denote by

$$Sh(G, H) = \varprojlim_K G(\mathbb{Q})\backslash(H \times G(\mathbb{A}_f)/K)$$

where $K \subset G(\mathbb{A}_f)$ runs over all compact open subgroups. In [**Del79**], 2.2.3, it is shown that

$$\pi_0(T(\mathbb{Q})\backslash T(\mathbb{A})) = Sh(T, \pi_0(T(\mathbb{R}))).$$

At this point, the morphism $(T, \pi_0(T(\mathbb{R}))) \to (GL_2, X)$ induces a map

$$(1) \qquad\qquad Sh(T, \pi_0(T(\mathbb{R}))) \to Sh(GL_2, X).$$

Unfortunately, the archimedian component $\pi_0(T(\mathbb{R})) := T(\mathbb{R})/T(\mathbb{R})^+$ is killed under this mapping because $X \cong GL_2(\mathbb{R})/T(\mathbb{R})$. Therefore, one replaces X by $X^{\pm} \cong GL_2(\mathbb{R})/T(\mathbb{R})^+$ in morphism (1) where $X^{\pm}$ is the space of morphisms $h \in X$ *oriented* by the choice of orientations s_x, s_y on the corresponding real lines F_x, F_y. This space projects naturally onto X. The resulting map

$$\pi_0(T(\mathbb{Q})\backslash T(\mathbb{A})) = Sh(T, \pi_0(T(\mathbb{R}))) \to Sh(GL_2, X^{\pm})$$

yields the geodesic interpretation of the connected component of the idèle class group.

We are however far from a theory of real multiplication because a "natural" rational structure on the space of geodesics, analogous to the coordinates on the modular curve given by the j and $\mathcal{P}$ functions, is missing.

2. Algebraic groups and moduli spaces of non-commutative tori

2.1. The moduli space of elliptic curves. This section recalls the classical construction of the universal family of elliptic curves in terms of mixed Shimura varieties [**Pin90**], 10.7. For classical Shimura varieties, we will refer to [**Del79**]. The basic definition of mixed Shimura varieties can be found in [**Pin90**], 2.1, 3.1, or [**Mil90**][VI]. We only need in this paper the notion of pre-Shimura datum defined in the Notations Section.

Let $\mathbb{S} := \mathrm{Res}_{\mathbb{C}/\mathbb{R}}\mathbb{G}_m$ be the Deligne torus. This is an $\mathbb{R}$-algebraic group such that $\mathbb{S}(\mathbb{R}) = \mathbb{C}^*$. Let $w : \mathbb{G}_{m,\mathbb{R}} \to \mathbb{S}$ be the weight morphism given by the natural inclusion $\mathbb{R}^* \subset \mathbb{C}^*$. Let $\mu : \mathbb{G}_{m,\mathbb{C}} \to \mathbb{S}_{\mathbb{C}} \cong \mathbb{G}_{m,\mathbb{C}} \times \mathbb{G}_{m,\mathbb{C}}$ be the *Hodge morphism* that sends z to the pair $(z, 1)$ and let $\widetilde{\mu}$ the morphism that sends z to the pair $(1, z)$. We call $\widetilde{\mu}$ the *anti-Hodge morphism*.

Let V be an $\mathbb{R}$-vector space. Recall that a representation $h : \mathbb{S}_{\mathbb{C}} \to \mathrm{GL}(V_{\mathbb{C}})$ in the $\mathbb{C}$-vector space $V_{\mathbb{C}} := V \otimes_{\mathbb{R}} \mathbb{C}$ gives an ascending filtration, the so called *weight filtration*, $W_{\bullet}V_{\mathbb{C}}$ given by the cocharacter $h \circ w_{\mathbb{C}}$ and a descending filtration, the so called *Hodge filtration*, $F^{\bullet}V_{\mathbb{C}}$ given by the cocharacter $h \circ \mu$. We will also be interested by another descending, that we call the *anti-Hodge filtration*, $\widetilde{F}^{\bullet}V_{\mathbb{C}}$ given by the cocharacter $h \circ \widetilde{\mu}$. Note that if h is not defined over $\mathbb{R}$, then the anti-Hodge filtration is not the complex conjugate filtration of $F^{\bullet}V_{\mathbb{C}}$.

Let $\mathbb{H}^{\pm} := \{\tau \in \mathbb{C}, \tau \notin \mathbb{R}\}$ be the Poincaré double half plane. This space can be identified with the $\mathrm{GL}_2(\mathbb{R})$-conjugacy class of the morphism of real algebraic groups $h_i : \mathbb{S} \to \mathrm{GL}_{2,\mathbb{R}}$ that maps $z = a + ib \in \mathbb{C}^* = \mathbb{S}(\mathbb{R})$ to the matrix $\left(\begin{smallmatrix} a & b \\ -b & a \end{smallmatrix}\right)$. The map between this conjugacy class and $\mathbb{H}^{\pm} \subset \mathbb{P}^1(\mathbb{C})$ is given by associating to $h : \mathbb{S} \to \mathrm{GL}_{2,\mathbb{R}}$ the line $F(h) := \mathrm{Ker}(h_{\mathbb{C}} \circ \mu(z) - z \cdot \mathrm{id}) \subset \mathbb{C}^2$.

DEFINITION 2.1.1. The pair $(\mathrm{GL}_2, \mathbb{H}^{\pm})$ is called the *classical modular Shimura datum*.

Let K be the compact open subgroup $\mathrm{GL}_2(\hat{\mathbb{Z}})$ of $\mathrm{GL}_2(\mathbb{A}_f)$. The associated Shimura variety of level K is by definition

$$\mathcal{M} = \mathrm{Sh}_K(\mathrm{GL}_{2,\mathbb{Q}}, \mathbb{H}^{\pm}) \begin{aligned} &= \mathrm{GL}_2(\mathbb{Q})\backslash(\mathbb{H}^{\pm} \times \mathrm{GL}_2(\mathbb{A}_f)/K) \\ &= \mathrm{GL}_2(\mathbb{Z})\backslash\mathbb{H}^{\pm} \\ &= \mathrm{PGL}_2(\mathbb{Z})\backslash\mathbb{H}, \end{aligned}$$

that is to say the classical modular curve.

Now we recall the construction of the universal family $\mathcal{E} \to \mathcal{M}$ of elliptic curves.

Let P be the group scheme $V \rtimes \mathrm{GL}_2$ with $V := \mathbb{G}_a^2$ the standard representation[4] of GL_2. Fix a rational splitting $s : \mathrm{GL}_2 \to P$ of the natural projection map $\pi : P \to \mathrm{GL}_2$, for example the map $g \mapsto (0, g) \in V \rtimes \mathrm{GL}_2$. All such splittings are conjugates under $V(\mathbb{Q}) \subset P(\mathbb{Q})$. Define $h' : \mathbb{S}_{\mathbb{C}} \to P_{\mathbb{C}}$ by $h' := s_{\mathbb{C}} \circ h_{i,\mathbb{C}}$. We will denote by $\mathbb{H}'$ the $P(\mathbb{R})$-orbit of h' in $\mathrm{Hom}(\mathbb{S}_{\mathbb{C}}, P_{\mathbb{C}})$ for the conjugation action.

DEFINITION 2.1.2. The pair $(P, \mathbb{H}')$ is called the *pre-Shimura datum of the universal elliptic curve*.

---

[4]i.e $V(\mathbb{Z}) = \mathbb{Z}^2$

LEMMA 2.1.3. *The space $\mathbb{H}'$ does not depend on the choice of the splitting s.*

PROOF. This follows from the fact that all such splittings are conjugates under $V(\mathbb{Q})$ and that $\mathbb{H}'$ is a $P(\mathbb{R})$-orbit. $\qquad\square$

Let $K^P := P(\hat{\mathbb{Z}}) \subset P(\mathbb{A}_f)$. The mixed Shimura variety associated to the data $(P, \mathbb{H}')$ and K^P is, by definition,

$$\mathcal{E} = \mathrm{Sh}_{K^P}(P, \mathbb{H}') = P(\mathbb{Q}) \backslash (\mathbb{H}' \times P(\mathbb{A}_f)/K^P) = P(\mathbb{Z}) \backslash \mathbb{H}'.$$

The natural projection map $\pi : P \to \mathrm{GL}_2$ induces a projection morphism of Shimura data $(P, \mathbb{H}') \to (\mathrm{GL}_2, \mathbb{H}^\pm)$ and a projection map

$$\mathcal{E} \to \mathcal{M}$$

between the corresponding mixed Shimura varieties.

Theorem 10.10 of [**Pin90**] shows that this map (up to the choice of finer level structures $K \subset \mathrm{GL}_2(\mathbb{A}_f)$ and $K^P \subset P(\mathbb{A}_f)$) gives the universal family of elliptic curves.

REMARK 2.1.4. We want to stress here that the fiber of $\mathbb{H}' \to \mathbb{H}^\pm$ over some $h : \mathbb{S} \to \mathrm{GL}_{2,\mathbb{R}}$ is given by the one dimensional $\mathbb{C}$-vector space $\mathbb{C}^2/F(h) \cong V(\mathbb{R})$. This will be useful in the next section.

2.2. The universal family of geodesics. We will now use the same ideas to construct, for the space of geodesics, an analog of the universal family of elliptic curves.

Let $h_0^Q : \mathbb{G}_{m,\mathbb{R}}^2 \to \mathrm{GL}_{2,\mathbb{R}}^2$ be the morphism that sends the pair $(x,y) \in (\mathbb{R}^*)^2$ to the pair of matrices $\left(\left(\begin{smallmatrix} x & 0 \\ 0 & 1 \end{smallmatrix} \right), \left(\begin{smallmatrix} 1 & 0 \\ 0 & y \end{smallmatrix} \right) \right)$. Let $Z_Q := \mathrm{GL}_2(\mathbb{R}) \cdot h_0^Q$ be the $\mathrm{GL}_2(\mathbb{R})$-conjugacy class of h_0^Q in $\mathrm{Hom}(\mathbb{G}_{m,\mathbb{R}}^2, \mathrm{GL}_{2,\mathbb{R}}^2)$. The multiplication map $m : \mathrm{GL}_2^2 \to \mathrm{GL}_2$ (which is not a group homomorphism) induces a natural $\mathrm{GL}_2(\mathbb{R})$-equivariant bijection

$$\begin{array}{ccc} Z_Q & \overset{m}{\to} & X \\ \left(g \left(\begin{smallmatrix} x & 0 \\ 0 & 1 \end{smallmatrix} \right) g^{-1}, g \left(\begin{smallmatrix} 1 & 0 \\ 0 & y \end{smallmatrix} \right) g^{-1} \right) & \mapsto & g \left(\begin{smallmatrix} x & 0 \\ 0 & y \end{smallmatrix} \right) g^{-1}. \end{array}$$

In other words, we have constructed an isomorphism $m : (\mathrm{GL}_2, Z_Q) \to (\mathrm{GL}_2, X)$ of pre-Shimura data.

Let $V := \mathbb{G}_a^2$ be the standard representation[5] of GL_2. Let Q' be the group scheme $V^2 \rtimes \mathrm{GL}_2^2$ and Q be the group scheme $V^2 \rtimes \mathrm{GL}_2$. We will also denote by $h_0^Q : \mathbb{G}_{m,\mathbb{R}}^2 \to Q'_\mathbb{R}$ the morphism obtained by composing h_0^Q with a rational section of the natural projection $Q' \to \mathrm{GL}_2^2$. Such a section is unique up to an element of $V^2(\mathbb{Q})$. Let $Y_Q := Q(\mathbb{R}) \cdot h_0^Q \subset \mathrm{Hom}(\mathbb{G}_{m,\mathbb{R}}^2, Q'_\mathbb{R})$ be the $Q(\mathbb{R})$-conjugacy class of h_0^Q. It does not depend on the chosen section because $V^2 \subset Q$.

DEFINITION 2.2.1. We call the pair (Q, Y_Q) the *pre-Shimura datum of the universal family of geodesics.*

The next lemma will explain this definition.

Let $\pi' : Q' \to \mathrm{GL}_2^2$ be the natural projection. This projection induces a natural map $Y_Q \to Z_Q$ that is compatible with the projection $\pi : Q \to \mathrm{GL}_2$. This yields

---

[5] i.e., $V(\mathbb{Z}) = \mathbb{Z}^2$.

a morphism of pre-Shimura data $\pi : (Q, Y_Q) \to (\mathrm{GL}_2, Z_Q)$. If we compose this morphism with m, we get a natural morphism of pre-Shimura data

$$(Q, Y_Q) \to (\mathrm{GL}_2, X)$$

which is in fact the quotient map by the additive group V^2.

The Shimura fibered space $\mathrm{Sh}(Q, Y_Q) \to \mathrm{Sh}(\mathrm{GL}_2, X)$ can be considered as a universal family of geodesics because of the following lemma.

LEMMA 2.2.2. *The fiber of the projection* $\mathrm{Sh}_{Q(\widehat{\mathbb{Z}})}(Q, Y_Q) \to \mathrm{Sh}_{\mathrm{GL}_2(\widehat{\mathbb{Z}})}(\mathrm{GL}_2, X)$ *over a point* $[(V, F_x, F_y)]$ *of the space of geodesics is the space* $\mathbb{Z}^2 \backslash \mathbb{R}^2 / F_x \times \mathbb{Z}^2 \backslash \mathbb{R}^2 / F_y$, *product of the two leaves spaces of the corresponding linear foliations on the torus* $\mathbb{Z}^2 \backslash \mathbb{R}^2 = V(\mathbb{Z}) \backslash V(\mathbb{R})$.

PROOF. We can embed Q' in GL_3^2 by considering pairs of matrices of the form

$$\left(\left(\begin{smallmatrix} A_1 & v_1 \\ 0 & 1 \end{smallmatrix} \right), \left(\begin{smallmatrix} A_2 & v_2 \\ 0 & 1 \end{smallmatrix} \right) \right)$$

with $A_1, A_2 \in \mathrm{GL}_2$ and $v_1, v_2 \in V$. The morphism h_0^Q is now given by the matrix $\left(\left(\begin{smallmatrix} Z_x & 0 \\ 0 & 1 \end{smallmatrix} \right), \left(\begin{smallmatrix} Z_y & 0 \\ 0 & 1 \end{smallmatrix} \right) \right)$ with $Z_x = \left(\begin{smallmatrix} x & 0 \\ 0 & 1 \end{smallmatrix} \right)$, $Z_y = \left(\begin{smallmatrix} 1 & 0 \\ 0 & y \end{smallmatrix} \right)$. The $Q(\mathbb{R})$-conjugacy class Y_Q of h_0^Q is given by pairs of matrices of the form

$$(M_x, M_y) := \left(\left(\begin{smallmatrix} A Z_x A^{-1} & (I - A Z_x A^{-1}) v_1 \\ 0 & 1 \end{smallmatrix} \right), \left(\begin{smallmatrix} A Z_y A^{-1} & (I - A Z_y A^{-1}) v_2 \\ 0 & 1 \end{smallmatrix} \right) \right)$$

with $A \in \mathrm{GL}_2(\mathbb{R})$. The projection $Y_Q \to Z_Q$ sends such a matrix to the pair $(A Z_x A^{-1}, A Z_y A^{-1})$. The action of $(v_1', v_2') \in V^2(\mathbb{R})$ by conjugation on a pair (M_x, M_y) as above gives

$$\left(\left(\begin{smallmatrix} A Z_x A^{-1} & (I - A Z_x A^{-1})(v_1 + v_1') \\ 0 & 1 \end{smallmatrix} \right), \left(\begin{smallmatrix} A Z_y A^{-1} & (I - A Z_y A^{-1})(v_2 + v_2') \\ 0 & 1 \end{smallmatrix} \right) \right).$$

In this way, we see directly that the fibre $Y_{Q,h}$ of this projection over some $h \in Z_Q \cong X$ is a $V^2(\mathbb{R})$-homogeneous space. The stabilizer of any point of $Y_{Q,h}$ for this $V^2(\mathbb{R})$-action is the sum $F_y \oplus F_x \subset V(\mathbb{R}) \oplus V(\mathbb{R})$. This shows that $Y_{Q,h} \cong V(\mathbb{R})/F_y \times V(\mathbb{R})/F_x$. Since the projection is equivariant with respect to the projection $Q \to \mathrm{GL}_2$, the fibre of the projection $\mathrm{Sh}_{Q(\widehat{\mathbb{Z}})}(Q, Y_Q) \to \mathrm{Sh}_{\mathrm{GL}_2(\widehat{\mathbb{Z}})}(\mathrm{GL}_2, X)$ at the point $[\mathrm{GL}_2(\mathbb{Z}) \cdot h]$ is given by the space $\mathbb{Z}^2 \backslash \mathbb{R}^2 / F_x \times \mathbb{Z}^2 \backslash \mathbb{R}^2 / F_y$. $\square$

2.3. The moduli problem for non-commutative tori. In this section, we want to understand Manin's point of view of the moduli space of non-commutative tori (in terms of pseudo-lattices) in the spirit of Hodge structures. To this end, we introduce the notion of pre-lilac.

DEFINITION 2.3.1. A (rank 2) *pre-lilac*[6] is a pair (M, F) of a free $\mathbb{Z}$-module of rank two and a real line $F \subset M_{\mathbb{R}}$. A *morphism of pre-lilacs* $(M_1, F_1) \to (M_2, F_2)$ is a morphism of Abelian groups $f : M_1 \to M_2$ such that $f_{\mathbb{R}}(F_1) \subset F_2$.

This notion is equivalent to Manin's notion of pseudo-lattice with weak morphisms (see [**Man**]) but it is easier for us to formulate our results in terms of lilacs because of their analogy with complex structures.

From a pre-lilac (M, F), one can construct a non-commutative algebra $\mathcal{A}(M, F)$, called the *Kroneker foliation algebra* $C^{\infty}(M \backslash M_{\mathbb{R}}) \rtimes F$. We will call such an algebra a *non-commutative torus* because the choice of an element e of a basis for M allows one to construct a Morita equivalent algebra $\mathcal{T}(M, F, e) = C^{\infty}(\mathbb{Z}.e \backslash M_{\mathbb{R}} / F) \rtimes$

---

[6]Abbreviation for **LI**ne in a **LA**tti**Ce**.

$[\mathbb{Z}.e\backslash M]$ which is an irrational rotation algebra, i.e., a non-commutative torus in the usual sense. The Morita equivalence $\mathcal{A}(M,F) \sim \mathcal{T}(M,F,e)$ follows from [**GBVF01**], corollary 12.20[7]. We are interested in the set of isomorphism classes of Kronecker foliation algebras. Let (**Nct**) be the category of such algebras with $*$-isomorphisms as morphisms. Let (**pre-Lilacs**) be the category of pre-lilacs with isomorphisms as morphisms. The assignment $(M,F) \mapsto \mathcal{A}(M,F)$ gives a functor

$$T: \quad \begin{array}{ccc} (\textbf{pre-Lilacs}) & \to & (\textbf{Nct}) \\ (M,F) & \mapsto & \mathcal{A}(M,F). \end{array}$$

For an object A of (**Nct**), denote by $S : \mathrm{HC}_2(A) \to \mathrm{HC}_0(A)$ the periodicity map in cyclic homology, by $\mathrm{ch} : K_0(A) \to \mathrm{HC}_2(A)$ the Chern character defined in [**Lod98**], and by $\mathrm{ch}_{\mathbb{R}} : K_0(A) \otimes_{\mathbb{Z}} \mathbb{R} \to \mathrm{HC}_2(A)$ the corresponding map of $\mathbb{R}$-vector spaces.

There is also a functor in the other direction

$$L: \quad \begin{array}{ccc} (\textbf{Nct}) & \to & (\textbf{pre-Lilacs}) \\ A & \mapsto & (K_0(A), \mathrm{ch}_{\mathbb{R}}^{-1}(\mathrm{Ker}(S)) \subset K_0(A) \otimes_{\mathbb{Z}} \mathbb{R}) \end{array}$$

The facts that ch induces an isomorphism $K_0(A) \otimes_{\mathbb{Z}} \mathbb{C} \to \mathrm{HC}_2(A)$, and that the filtration by $\mathrm{Ker}(S)$ on cyclic homology is real with respect to the real structure given by K-theory can be deduced from the explicit calculations of Lemma 54 of [**Con85**]. Notice that this functor was already present in the paper [**CDS97**].

LEMMA 2.3.2. *Two Kronecker foliation algebras $\mathcal{A}_1$ and $\mathcal{A}_2$ are Morita equivalent if and only if they are isomorphic as $*$-algebras.*

PROOF. This comes from the fact that for every objects $\mathcal{A}(M,F) \in (\textbf{Nct})$, one has $L(\mathcal{A}(M,F)) \cong (M,F)$. To prove this, we can look at the Morita equivalent algebra $\mathcal{T}(M,F,e)$, for e an element of a basis for M and use the explicit calculation of its Chern character in [**Con85**]. We thus obtain that for all object $\mathcal{A} \in (\textbf{Nct})$, $T \circ L(\mathcal{A}) \cong \mathcal{A}$. To finish the proof, if the two Kronecker algebras are Morita equivalent, then Morita invariance of cyclic cohomology, K-theory and Chern character implies $L(\mathcal{A}_1) \cong L(\mathcal{A}_2)$ and this implies $\mathcal{A}_1 \cong \mathcal{A}_2$. $\qquad\square$

The relation with Manin's definition of pseudo–lattices associated to non–commutative tori is the following. The long exact sequence of cyclic homology gives

$$HH_2(A) \xrightarrow{I} HC_2(A) \xrightarrow{S} HC_0(A) \xrightarrow{B} HH_1(A)$$

and we know that S is surjective in this case (by explicit calculation given in [**Con85**]), so it gives an isomorphism $HC_2(A)/\mathrm{Ker}(S) \xrightarrow{S} HC_0(A)$. Since the Chern character is compatible with S, the natural map $S \circ \mathrm{ch} : K_0(A) \to HC_0(A)$ induced by $\mathrm{ch} : K_0(A) \to HC_2(A)$ is equal to the Chern character $\mathrm{ch} : K_0(A) \to HC_0(A)$. The pseudo-lattices that Manin considers are given by this Chern character. So the functor that associates to a pre-lilac (M,F) the pseudo-lattice $(M, M_{\mathbb{C}}/F_{\mathbb{C}})$ has a natural interpretation in cyclic homology as the association $(K_0(A), \mathrm{ch}_{\mathbb{R}}^{-1}(\mathrm{Ker}(S))) \rightsquigarrow (K_0(A), HC_0(A) = HC_2(A)/\mathrm{Ker}(S))$.

The two functors T and L naturally identify the *set* of isomorphism classes of pre-lilacs and the *set* of isomorphism classes of non-commutative tori.

---

[7]See also [**GBVF01**], theorem 12.17 for a choice-of-basis free proof of this fact.

Fix a free $\mathbb{Z}$-module M of rank two. The projective space $\mathbb{P}(M_{\mathbb{R}})$ over $M_{\mathbb{R}}$ gives a parameter space for lines $F' \subset M_{\mathbb{R}}$. Two such lines correspond to isomorphic pre-lilacs if and only if they are exchanged by some $g \in \mathrm{GL}(M)$. So the set of isomorphism classes of pre-lilacs is $\mathrm{GL}(M)\backslash P(M_{\mathbb{R}}) \cong \mathrm{GL}_2(\mathbb{Z})\backslash\mathbb{P}^1(\mathbb{R})$. As said before, this is also the moduli set for non-commutative tori.

We also remark that there is a natural projection

$$\begin{aligned}
\mathrm{GL}_2(\mathbb{Z})\backslash X &\rightarrow \mathrm{GL}_2(\mathbb{Z})\backslash\mathbb{P}^1(\mathbb{R}) \\
(M, F_x, F_y) &\mapsto (M, F_x)
\end{aligned}$$

from the space of geodesics to the moduli space of pre-lilacs. In non-commutative geometry, the left hand side of this projection can be interpreted as the moduli space for triples

$$(\mathcal{A}_1, \mathcal{A}_2, \psi : K_0(\mathcal{A}_1) \xrightarrow{\sim} K_0(\mathcal{A}_2))$$

consisting of two Kronecker foliation algebras and an isomorphism ψ between their K_0 groups, such that if F_1 and F_2 are the real lines in K-theory constructed in this section, $\psi_{\mathbb{R}}(F_1) \oplus F_2 = K_0(\mathcal{A}_2)_{\mathbb{R}}$.

REMARK 2.3.3. As we saw in Section 1.5, the study of nontrivial level structures on geodesics (and thus on non-commutative tori) is also interesting because of class field theory. We will denote by $\mathrm{Irrat}(X)^{\pm} \subset X^{\pm}$ the space of pairs of irrational oriented lines in $\mathbb{R}^2$. Let $N > 1$ be an integer, and let K_N be the group defined by the exact sequence $1 \rightarrow K_N \rightarrow \mathrm{GL}_2(\hat{\mathbb{Z}}) \rightarrow \mathrm{GL}_2(\mathbb{Z}/N\mathbb{Z}) \rightarrow 1$. Then the Shimura space $\mathrm{Sh}_{K_N}(\mathrm{GL}_2, \mathrm{Irrat}(X)^{\pm})$ is the moduli space of tuples

$$(M, F_x, F_y, s_x, s_y, \phi : M \otimes_{\mathbb{Z}} \mathbb{Z}/N\mathbb{Z} \xrightarrow{\sim} (\mathbb{Z}/N\mathbb{Z})^2)$$

consisting of a free $\mathbb{Z}$-module M of rank 2, two irrational lines F_x, F_y in the underlying real vector space, equiped with two orientations s_x, s_y, and a level structure ϕ. Perhaps this space has a moduli interpretation in terms of tuples

$$(\mathcal{A}_1, \mathcal{A}_2, \psi : K_0(\mathcal{A}_1) \xrightarrow{\sim} K_0(\mathcal{A}_2), \phi : K_0(\mathcal{A}_1) \otimes_{\mathbb{Z}} \mathbb{Z}/N\mathbb{Z} \xrightarrow{\sim} (\mathbb{Z}/N\mathbb{Z})^2).$$

The orientation on the real lines could be given by the image of the positive cone in K-theory. The author's knowledge is, however, not sharp enough to be sure of this non-commutative moduli interpretation.

REMARK 2.3.4. Polishchuk's remarkable work [Pol03] on the relation of analytic non-commutative tori with nonstandard t-structures on derived categories of coherent sheaves on usual elliptic curves seems to be promising, because it allows one to give some rationality properties to the objects we have on hand. For example, if E is an elliptic curve over $\mathbb{Q}$ and we fix some t-structure on its associated analytic curve given by a line with real quadratic slope on $K_0(E) \otimes_{\mathbb{Z}} \mathbb{R}$, we can ask questions about the rationality of this t-structure. The Algebraic Proj construction in [Pol02] for real multiplication non-commutative tori also allows one to ask rationality questions.

2.4. The universal non-commutative torus. For at least four reasons (esthetic symmetry, class field theory, Mumford-Tate groups and the complications that appear in this paragraph), it seems to be more natural to study the moduli space for *pairs* of non-commutative tori as above (which is also the space of geodesics) rather than the moduli space of solitary non-commutative tori. However, we still want to construct a universal non-commutative torus because it permits to understand the relation with Tate mixed Hodge structures.

Let h_0' be the morphism of algebraic groups given by

$$h_0' : \quad \begin{matrix} \mathbb{G}_{m,\mathbb{R}}^2 & \to & \mathrm{GL}_{2,\mathbb{R}} \\ (x,y) & \mapsto & \left(\begin{smallmatrix} xy & 0 \\ 0 & 1 \end{smallmatrix} \right) \end{matrix} .$$

ASIDE 2.4.1 (rational boundary component). If we denote by U the group of unipotent upper triangular matrices and by P_1 the group of matrices of the form $\left(\begin{smallmatrix} * & * \\ 0 & 1 \end{smallmatrix} \right)$ in $\mathrm{GL}_{2,\mathbb{Q}}$, then the $U(\mathbb{C})$-conjugacy class Y_1' of $h_{0,\mathbb{C}}'$ is identified with $U(\mathbb{C}) = \mathbb{C}$. Let Y_1 be $Y_1' \times \{\pm 1\}$. Pink calls the pair (P_1, Y_1) a *rational boundary component* of $(\mathrm{GL}_2, \mathbb{H}^\pm)$. These kind of (mixed) Shimura data appear in the toroidal compactification of Shimura varieties. They are parameter spaces for mixed Hodge structures. In our particular case, the mixed Hodge structures are extensions

$$0 \to \mathbb{Z}(1) \to M \to \mathbb{Z} \to 0$$

that can be described algebraically by 1-motives of the form $[\mathbb{Z} \to \mathbb{G}_{m,\mathbb{C}}]$ (see [**Del74**] section 10 for a definition of 1-motives). Level structures on these one-motives are strongly related to roots of unity in $\mathbb{C}$, i.e., to generators of $\mathbb{Q}^{ab}$. In some sense (see [**Pin90**], 10.15 to give a precise meaning to this affirmation), we can say that $\mathrm{Sh}(P_1, Y_1)$ is a universal family over the moduli space $\mathrm{Sh}(\mathbb{G}_{m,\mathbb{Q}}, \{\pm 1\})$ of primitive roots of unity.

We will denote by X' the $\mathrm{GL}_2(\mathbb{R})$-conjugacy class of h_0'. Recall that X' is the set $\{(F_{xy}, F_0)\}$ of pairs of distinct lines in $\mathbb{R}^2$ of respective weights 0 and xy. There is a natural projection map $X' \to \mathbb{P}^1(\mathbb{R})$ given by $(F_{xy}, F_0) \mapsto F_0$.

Recall from Section 2.1 that P is the group scheme $V \rtimes \mathrm{GL}_2$, with $V := \mathbb{G}_a^2$ the standard representation of GL_2. As usual, we also denote by $h_0' : \mathbb{G}_{m,\mathbb{R}}^2 \to P_\mathbb{R}$ the morphism obtained by composing h_0' with a rational section of the natural projection $P \to \mathrm{GL}_2$. Such a section is unique up to an element of $V(\mathbb{Q})$. Let Y_P be the $P(\mathbb{R})$-conjugacy class of h_0'. It is independent of the choice of the section of $P \to \mathrm{GL}_2$ because $V \subset P$.

Let $V_P = \mathbb{Q}^3$ be the standard representation of P given by an embedding $P \hookrightarrow \mathrm{GL}_3$. For $h \in Y_P$, we let $F^0(h) := \{v \in V_{P,\mathbb{R}} | h(x,y) \cdot v = v\}$. On Y_P, we have the equivalence relation:

$$h \sim h' \Leftrightarrow F^0(h)V' = F^0(h')V'.$$

Let $\overline{Y_P} = Y_P/\sim$ be the corresponding quotient.

DEFINITION 2.4.2. We call the pair $(P, \overline{Y_P})$ the *pre-Shimura datum of the universal family of non-commutative tori*.

The following lemma shows us that the Shimura fibered space $\mathrm{Sh}(P, \overline{Y_P}) \to \mathrm{Sh}(\mathrm{GL}_2, \mathbb{P}^1(\mathbb{R}))$ can be considered as a *universal family of non-commutative tori*.

LEMMA 2.4.3. *The fiber of the projection*

$$\mathrm{Sh}_{P(\widehat{\mathbb{Z}})}(P, Y_P) \to \mathrm{Sh}_{\mathrm{GL}_2(\widehat{\mathbb{Z}})}(\mathrm{GL}_2, \mathbb{P}^1(\mathbb{R})) = \mathrm{PGL}_2(\mathbb{Z}) \backslash \mathbb{P}^1(\mathbb{R})$$

over a point (M, F_0) of the space of pre-lilacs is the space $M \backslash M_\mathbb{R} / F_0$ of leaves of the corresponding foliation on the two-torus $M \backslash M_\mathbb{R}$.

PROOF. The proof is essentially the same as the one of Lemma 2.2.2. The additional fact to check is that forgetting the F_{xy} part of the filtration is compatible with the quotient, which follows from the definition. $\qquad \square$

3. Two higher dimensional examples

3.1. Hilbert modular varieties. Let E be a totally real number field with ring of integer $\mathcal{O}_E$, let $I := \operatorname{Hom}(E, \mathbb{R})$ and let $n := \operatorname{card}(I)$. Denote by G the group scheme $\operatorname{Res}_{\mathcal{O}_E/\mathbb{Z}}\operatorname{GL}_2$. We then have $G_{\mathbb{R}} = \prod_{\iota: E \to \mathbb{R}} G_{\mathbb{R},\iota} \cong \prod_{\iota: E \to \mathbb{R}} \operatorname{GL}_{2,\mathbb{R}}$.

Let $h := \prod_{\iota: E \to \mathbb{R}} h_i : \mathbb{S} \to G_{\mathbb{R}}$ with $h_i : \mathbb{S} \to \operatorname{GL}_{2,\mathbb{R}}$ the map that sends $z = a + ib \in \mathbb{C}^* = \mathbb{S}(\mathbb{R})$ to the matrix $\left(\begin{smallmatrix} a & b \\ -b & a \end{smallmatrix}\right)$. Let X be the $G(\mathbb{R})$-conjugacy class of this morphism. We have an isomorphism $X \cong \prod_{\iota: E \to \mathbb{R}} \mathbb{H}^{\pm}$.

DEFINITION 3.1.1. The Shimura datum (G, X) is called the *Hilbert modular Shimura datum*.

Let $h(E)$ be the Hilbert class number of E. Let K be the compact open subgroup $G(\hat{\mathbb{Z}})$ of $G(\mathbb{A}_f)$. The associated Shimura variety is by definition

$$
\begin{aligned}
\mathcal{M} = \operatorname{Sh}_K(G, X) &= G(\mathbb{Q})\backslash(X \times G(\mathbb{A}_f)/K) \\
&= \operatorname{GL}_2(\mathcal{O}_E)\backslash X \text{ if } h(E) = 1, \\
&\cong \operatorname{GL}_2(\mathcal{O}_E)\backslash(\textstyle\prod_{\iota: E \to \mathbb{R}} \mathbb{H}^{\pm}) \text{ if } h(E) = 1,
\end{aligned}
$$

i.e., the Hilbert modular variety. Now let B' be the subgroup of upper triangular matrices in $\operatorname{GL}_{2,\mathcal{O}_E}$ and denote by B the group scheme $\operatorname{Res}_{\mathcal{O}_E/\mathbb{Z}}B'$.

ASIDE 3.1.2 (rational boundary component). The group scheme B is a maximal parabolic subgroup of G and corresponds to a rational boundary component (P_1, X_1) of (G, X) as in [**Pin90**], 4.11. The canonical model of the associated mixed Shimura variety is a moduli space defined over $\mathbb{Q}$ for 1-motives with additional structures.

Let $h_0^H : \mathbb{G}_{m,\mathbb{R}}^2 \to G_{\mathbb{R}}$ be the morphism given on each simple component of $G_{\mathbb{R}}$ by $h_0 : (x, y) \in (\mathbb{R}^*)^2 \mapsto \operatorname{diag}(x, y) := \left(\begin{smallmatrix} x & 0 \\ 0 & y \end{smallmatrix}\right)$. Let X_H be the $G(\mathbb{R})$-conjugacy class of h_0^H.

DEFINITION 3.1.3. We will call the pair (G, X_H) the *pre-Shimura datum of the moduli space of Hilbert lilacs*.

The corresponding Shimura space $\operatorname{Sh}_{G(\hat{\mathbb{Z}})}(G, X_H)$ is a moduli space for tuples (M, F_x, F_y, i) consisting of a rank $2n$ free $\mathbb{Z}$-module M, equipped with a decomposition

$$
M_{\mathbb{R}} = F_x \oplus F_y
$$

of its underlying real vector space in two n-dimensional subspaces, and with a morphism $i : E \hookrightarrow \operatorname{End}(M_{\mathbb{Q}}, F_x, F_y)$ (i.e., a morphism compatible with the decomposition). These objects are called *Hilbert lilacs*.

EXAMPLE 3.1.4. Let $E := \mathbb{Q}(\sqrt{2})$ and $F := \mathbb{Q}(\sqrt{2}, \sqrt{3})$. For each embedding ι in $\operatorname{Hom}(E, \mathbb{R})$, choose one embedding c_ι over it in $\operatorname{Hom}(F, \mathbb{R})$. Such a choice is called an *RM type* for F/E. Equip $M := \mathcal{O}_F$ with the "Hodge decomposition"

$$
M_{\mathbb{R}} = F_x \oplus F_y
$$

where $F_x := \oplus_\iota F_{c_\iota} \cong \mathbb{R}^2$, and F_y is the other component in the natural decomposition of $M_{\mathbb{R}}$. Then the lilac (M, F_x, F_y) gives a point in the space of Hilbert lilacs corresponding to E, and this point has Mumford-Tate group contained in $\operatorname{Res}_{F/\mathbb{Q}}\mathbb{G}_m \subset G = \operatorname{Res}_{E/\mathbb{Q}}\operatorname{GL}_2$. It is called a *special point* of this space.

Let h_0^{HQ} be the product morphism $\prod_{\iota:E\to\mathbb{R}} h_0^Q : \mathbb{G}_{m,\mathbb{R}}^2 \to G_\mathbb{R}^2$, where $h_0^Q :$ $\mathbb{G}_{m,\mathbb{R}}^2 \to \mathrm{GL}_{2,\mathbb{R}}^2$ is the morphism that sends the pair $(x,y) \in (\mathbb{R}^*)^2$ to the pair of matrices $\left(\left(\begin{smallmatrix} x & 0 \\ 0 & 1 \end{smallmatrix}\right), \left(\begin{smallmatrix} 1 & 0 \\ 0 & y \end{smallmatrix}\right)\right)$.

Let $Z_{HQ} := G(\mathbb{R}) \cdot h_0^{HQ}$ be the $G(\mathbb{R})$-conjugacy class of h_0^{HQ} in $\mathrm{Hom}(\mathbb{G}_{m,\mathbb{R}}^2, G_\mathbb{R}^2)$. The multiplication map $m : G^2 \to G$ (which is not a group homomorphism) induces, as in Subsection 2.2, a natural isomorphism $m : (G, Z_{HQ}) \to (G, X_H)$ of pre-Shimura data.

Let $V := \mathrm{Res}_{\mathcal{O}_E/\mathbb{Z}} \mathbb{G}_{a,E}^2$ be the standard representation[8] of G. Let Q' be the group scheme $V^2 \rtimes G^2$ and let Q be the group scheme $V^2 \rtimes G$. We will also denote by $h_0^{HQ} : \mathbb{G}_{m,\mathbb{R}}^2 \to Q_\mathbb{R}'$ the morphism obtained by composition with a rational section of the natural projection $Q' \to G^2$. Such a section is unique up to an element of $V^2(\mathbb{Q})$.

Let $Y_{HQ} := Q(\mathbb{R}) \cdot h_0^{HQ} \subset \mathrm{Hom}(\mathbb{G}_{m,\mathbb{R}}^2, Q_\mathbb{R}')$ be the $Q(\mathbb{R})$-conjugacy class of h_0^{HQ}. It does not depend on the chosen section because $V^2 \subset Q$.

DEFINITION 3.1.5. We call the pair (Q, Y_{HQ}) the *pre-Shimura datum of the universal family of Hilbert lilacs*.

The next lemma will explain this definition. Let $\pi' : Q' \to G^2$ be the natural projection. This projection induces a natural map $Y_{HQ} \to Z_{HQ}$ that is compatible with the natural projection $\pi : Q \to G$. This yields a morphism of pre-Shimura data $\pi : (Q, Y_{HQ}) \to (G, Z_{HQ})$. If we compose this morphism with m, we get a natural morphism of pre-Shimura data

$$(Q, Y_{HQ}) \to (G, X_H)$$

which is in fact the quotient map by the additive group V^2.

The Shimura fibered space $\mathrm{Sh}(Q, Y_{HQ}) \to \mathrm{Sh}(G, X_H)$ can be considered as a universal family of Hilbert lilacs because of the following lemma.

LEMMA 3.1.6. *The fiber of the projection* $\mathrm{Sh}_{Q(\widehat{\mathbb{Z}})}(Q, Y_Q) \to \mathrm{Sh}_{G(\widehat{\mathbb{Z}})}(G, X)$ *over a point* $[(V, F_x, F_y, i)]$ *of the space of Hilbert lilacs is the space* $V \backslash V_\mathbb{R}/F_x \times V \backslash V_\mathbb{R}/F_y$, *product of the two leaves spaces of the corresponding linear foliations on the torus* $V \backslash V_\mathbb{R}$.

PROOF. The proof is essentially the same as in Lemma 2.2.2.   □

REMARK 3.1.7. The centralizer $C_{G(\mathbb{R})}(h_0^H)$ of our basis morphism is isomorphic to the maximal torus $T(\mathbb{R})$ of $G(\mathbb{R})$. We define the Mumford-Tate group of some $h \in X_H$ as in Definition 1.4.1 as a reductive envelope, defined by André, Kahn and O'Sullivan, [**AKO02**]. Such a reductive envelope is well defined up to the centralizer of the enveloped group. We will now change a little bit our description of X_H in order to have a well defined Mumford-Tate group for all points in this (and other) spaces. These groups could also be of some interest in the study of dynamical properties of the corresponding foliations, as suggested to the author by Yves André and Etienne Ghys.

Let D be the basic maximal torus of G, i.e., $D := \mathrm{Res}_{E/\mathbb{Q}} \mathbb{G}_m^2$. Let $\mathbb{D} := D_\mathbb{R}$ be the corresponding real algebraic group, and let $h_{0,\mathcal{R}}^H : \mathbb{D} \to G_\mathbb{R}$ be the natural inclusion. Let $\mathcal{R}_H$ be the $G(\mathbb{R})$-conjugacy class of $h_{0,\mathcal{R}}^H$. The inclusion[9] $\mathbb{G}_{m,\mathbb{R}}^2 \subset$

---

[8]i.e., $V(\mathbb{Z}) = \mathcal{O}_E^2$.

[9]Induced by the rational inclusion $\mathbb{G}_{m,\mathbb{Q}}^2 \subset D$.

$\mathbb{D}$ induces a $G(\mathbb{R})$-equivariant bijection $\mathcal{R}_H \to X_H$, i.e., an isomorphism of pre-Shimura data

$$(G, \mathcal{R}_H) \to (G, X_H).$$

DEFINITION 3.1.8. Let $h \in \mathcal{R}_H$. We define the *bad Mumford-Tate group* of h to be the smallest $\mathbb{Q}$-algebraic subgroup $\mathrm{BMT}(h) \subset G$ such that $\mathrm{im}(h) \subset \mathrm{BMT}(h)(\mathbb{R})$. A *Mumford-Tate group* for h is a minimal reductive subgroup of G that contains $\mathrm{BMT}(h)$.

LEMMA 3.1.9. *Let $h \in \mathcal{R}_H$. There exists a unique Mumford-Tate group for h. It will be denoted by $\mathrm{MT}(h)$.*

PROOF. The bad Mumford-Tate group of h contains a maximal torus T of G (because the image of h is a maximal torus over $\mathbb{R}$). We know from [**SGA64**], XIX, 2.8 (or other classical references on algebraic groups) that the centralizer $C_G(T)$ of T in G is T itself, because G is a reductive group. We have an inclusion of centralizers $C_G(\mathrm{BMT}(h)) \subset C_G(T) = T$, and we already know that $T \subset \mathrm{BMT}(h)$. So we have $C_G(\mathrm{BMT}(h)) \subset \mathrm{BMT}(h)$, and the fact that a reductive envelope of a group is well defined up to the centralizer of the group implies that, in this case, the Mumford-Tate group is well-defined. $\square$

REMARK 3.1.10. This result gives a motivation to study the pair $(G, \mathcal{R}_H)$ itself. This pair is called the *shore datum of the Hilbert modular datum* (G, X). Another good reason to use a bigger basis group $\mathbb{D}$ for morphisms is given by the need, in Manin's real multiplication program, to take into account archimedian places in class field theory of totally real fields. This has been investigated in Section 1.5 in the case $E = \mathbb{Q}$ and in [**Pau04**] for other quite general examples.

3.2. The moduli space of Abelian surfaces. We first recall the construction of the moduli space of Abelian surfaces and then construct one of its irrational boundaries. As we will see, this gives a non-totally degenerate example, that is in some sense more general than the case of Hilbert moduli spaces.

Let $V = \mathbb{Z}^4$ and $\psi : V \times V \to \mathbb{Z}$ be the standard symplectic form given by the matrix $J := \begin{pmatrix} 0 & I_2 \\ -I_2 & 0 \end{pmatrix}$ with $I_2 \in \mathrm{M}_2(\mathbb{Z})$ the identity matrix. Let G be the corresponding group scheme of symplectic similitudes, whose points in a $\mathbb{Z}$-algebra A are given by $G(A) = \mathrm{GSp}_{4,\mathbb{Z}}(A) := \{M \in \mathrm{M}_4(A) | \exists \alpha(M) \in A^* \text{ with } MJM^{-1} = \alpha(M).J\}$.

Let $h_i^S : \mathbb{S} \to G_{\mathbb{R}}$ be the morphism that maps $z = a + ib \in \mathbb{C}^* = \mathbb{S}(\mathbb{R})$ to the matrix

$$\begin{pmatrix} aI_2 & bI_2 \\ -bI_2 & aI_2 \end{pmatrix},$$

and denote by $\mathcal{S}^{\pm}$ the $G(\mathbb{R})$-conjugacy class of h_i, that is usually called the two dimensional Siegel space.

DEFINITION 3.2.1. The datum $(\mathrm{GSp}_4, \mathcal{S}^{\pm})$ is called the *Siegel Shimura datum*.

Let K be the compact open subgroup $G(\hat{\mathbb{Z}})$ of $G(\mathbb{A}_f)$. The associated Shimura variety is by definition

$$\begin{aligned}
\mathrm{Sh}_K(\mathrm{GSp}_4, \mathcal{S}^{\pm}) &= G(\mathbb{Q}) \backslash (\mathcal{S}^{\pm} \times G(\mathbb{A}_f)/K) \\
&= \mathrm{GSp}_4(\mathbb{Z}) \backslash \mathcal{S}^{\pm},
\end{aligned}$$

i.e., the Siegel modular variety, which is the moduli space of principally polarized Abelian surfaces.

Let $h_0^S : \mathbb{G}_{m,\mathbb{C}}^2 \to G_\mathbb{C}$ be the morphism that associates to $(x,y) \in (\mathbb{C}^*)^2$ the matrix

$$\begin{pmatrix} a & 0 & b & 0 \\ 0 & x & 0 & 0 \\ -b & 0 & a & 0 \\ 0 & 0 & 0 & y \end{pmatrix},$$

with $a = \frac{x+y}{2}$ and $b = \frac{x-y}{2i}$. Let $\mathcal{R}_S := G(\mathbb{R}) \cdot h_0^S$ be the $G(\mathbb{R})$-conjugacy class of h_0^S. The centralizer of h_0^S is a (non-split) maximal torus of $G(\mathbb{R})$ (the subgroup $T(\mathbb{R})$ of $\mathbb{C}^* \times (\mathbb{R}^*)^2$ given by $z\bar{z} = xy$), which implies that $\mathcal{R}_S \cong \mathrm{GSp}_4(\mathbb{R})/T(\mathbb{R})$.

The Shimura space $\mathrm{Sh}_{G(\hat{\mathbb{Z}})}(G, \mathcal{R}_S)$ is a moduli space for tuples

$$(V, F_x, F_y, \Pi, F, \psi)$$

where V is a free $\mathbb{Z}$-module of rank 4, F_x, F_y are two distinct real lines and Π is a real plane in $V_\mathbb{R}$ such that $V_\mathbb{R} = F_x \oplus F_y \oplus \Pi$. Moreover, F is a complex line in $\Pi_\mathbb{C}$ such that $\Pi_\mathbb{C} = F \oplus \overline{F}$, and $\psi : V \times V \to \mathbb{Z}$ is a symplectic form that respects the decomposition

$$V_\mathbb{C} = (F_x \oplus F) \bigoplus (F_y \oplus \overline{F}).$$

This new kind of linear algebra object is quite strange and does not seem to have an easy non-commutative geometric interpretation because it mixes usual complex structures with foliations on tori. Such objects appear, however, quite often in number theory.

EXAMPLE 3.2.2. Let $K := \mathbb{Q}[x]/(x^4 - 2)$. We have a decomposition $K \otimes_\mathbb{Q} \mathbb{R} \cong \mathbb{R}_x \times \mathbb{R}_y \times \mathbb{C}_z$, and the $\mathbb{R}$-algebra $\mathbb{C}_z$ decomposes over $\mathbb{C}$ as a product of two copies of $\mathbb{C}$. Fixing a nice alternating form on K gives us exactly a point in $\mathrm{Sh}_{G(\hat{\mathbb{Z}})}(G, \mathcal{R}_S)$. This point is called a *special point* or a *quadratic multiplication point*. We showed in [**Pau04**] that the counting of points of this type in the Shimura space involves interesting number theoretical information, as in the case of geodesics studied in Section 1.5.

4. Some open problems

Here are some open problems[10] in our work.

- Find a higher dimensional and/or algebraic analog of non-commutative 2-tori adapted to number theoretical purposes.
- Understand, in the case of geodesics, the relation of our work with Manin's quantum theta functions, and Stark numbers.
- More generally, find an adelic formulation of Stark's conjectures for quadratic fields over totally real fields.
- Define a well-behaved and not ad hoc notion of level structure on non-commutative tori (resp., on Polishchuk's t-structures).
- Find the good higher dimensional analogs of Polishchuk's t-structures on categories of coherent sheaves of Abelian varieties.
- Study moduli spaces for stability conditions on these categories.
- Clarify, if it exists, the relationship with Darmon's work on Stark's conjecture for real quadratic fields.

---

[10]Some of them are collected from the literature.

Acknowledgments

I thank Matilde Marcolli for introducing me to non-commutative geometry and sharing with me some of her deep insights into non-commutative arithmetic geometry.

I thank the following people for interesting discussions and/or valuable help in the preparation of this paper: Y. André, D. Blottiere, J.-B. Bost, B. Calmes, S. Cantat, A. Connes, F. Dal'bo, E. Ha, E. Ghys, Y. Guivarch', Y. Manin, R. Noot, N. Ramachandran, A. Thuillier, D. Zagier. I thank the referee for useful comments and nice suggestions.

I thank the following institutions and workshops for financial and technical support: Bonn's Max Planck Institut für Mathematik, Bonn's worshops "Number theory and non-commutative geometry I and II", the Les Houches school "Frontières entre théorie des nombres, physique et géométrie", Regensburg's "Oberseminar Nichtkommutativ Geometrie", Regensburg's mathematics laboratory, Rennes's mathematics laboratory IRMAR, Rennes's workshop "Jeunes chercheurs en géométrie", the European networks "K-theory and Algebraic Groups" and "Arithmetic Algebraic Geometry".

References

[AKO02] Yves André, Bruno Kahn, and Peter O'Sullivan. Nilpotence, radicaux et structures monoïdales. *Rend. Sem. Mat. Univ. Padova*, 108:107–291, 2002.

[BC95] J.-B. Bost and A. Connes. Hecke algebras, type III factors and phase transitions with spontaneous symmetry breaking in number theory. *Selecta Math. (N.S.)*, 1(3):411–457, 1995.

[BL01] Armand Borel and Ji Lizhen. *Compactifications of locally symmetric spaces.* IAS/Michigan, 2001. Prepublication, June 2001.

[Car98] Pierre Cartier. La folle journée, de Grothendieck à Connes et Kontsevich: évolution des notions d'espace et de symétrie. In *Les relations entre les mathématiques et la physique théorique*, pages 23–42. Inst. Hautes Études Sci., Bures, 1998.

[CDS97] Alain Connes, Michael.R. Douglas, and Albert Schwarz. Noncommutative geometry and matrix theory: compactification on tori. *arXiv*, (http://fr.arXiv.org/abs/hep-th/9711162), 1997.

[CM95] A. Connes and H. Moscovici. The local index formula in noncommutative geometry. *Geom. Funct. Anal.*, 5(2):174–243, 1995.

[Con80] Alain Connes. C^* algèbres et géométrie différentielle. *C. R. Acad. Sci. Paris Sér. A-B*, 290(13):A599–A604, 1980.

[Con85] Alain Connes. Noncommutative differential geometry. *Inst. Hautes Études Sci. Publ. Math.*, (62):257–360, 1985.

[Con99] Alain Connes. Trace formula in noncommutative geometry and the zeros of the Riemann zeta function. *Selecta Math. (N.S.)*, 5(1):29–106, 1999.

[Del74] Pierre Deligne. Théorie de Hodge. III. *Inst. Hautes Études Sci. Publ. Math.*, (44):5–77, 1974.

[Del79] Pierre Deligne. Variétés de Shimura: intérpretation modulaire, et techniques de construction de modèles canoniques. In *Automorphic forms, representations and L-functions (Proc. Sympos. Pure Math., Oregon State Univ., Corvallis, Ore., 1977), Part 2*, pages 247–289. Amer. Math. Soc., Providence, R.I., 1979.

[DMOS82] Pierre Deligne, James S. Milne, Arthur Ogus, and Kuang-yen Shih. *Hodge cycles, motives, and Shimura varieties.* Springer-Verlag, Berlin, 1982. Philosophical Studies Series in Philosophy, 20.

[GBVF01] José M. Gracia-Bondía, Joseph C. Várilly, and Héctor Figueroa. *Elements of noncommutative geometry.* Birkhäuser Advanced Texts: Basler Lehrbücher. [Birkhäuser Advanced Texts: Basel Textbooks]. Birkhäuser Boston Inc., Boston, MA, 2001.

[Lod98]   Jean-Louis Loday. *Cyclic homology*, volume 301 of *Grundlehren der Mathematischen Wissenschaften [Fundamental Principles of Mathematical Sciences]*. Springer-Verlag, Berlin, second edition, 1998. Appendix E by María O. Ronco, Chapter 13 by the author in collaboration with Teimuraz Pirashvili.

[Man]   Yuri I. Manin. Real Multiplication and noncommutative geometry.

[Mar04]   Matilde Marcolli. Lectures on Arithmetic Noncommutative Geometry. *arXiv*, (math.QA/0409520), 2004.

[Mil83]   J. S. Milne. The action of an automorphism of **C** on a Shimura variety and its special points. In *Arithmetic and geometry, Vol. I*, volume 35 of *Progr. Math.*, pages 239–265. Birkhäuser Boston, Boston, MA, 1983.

[Mil90]   J. S. Milne. Canonical models of (mixed) Shimura varieties and automorphic vector bundles. In *Automorphic forms, Shimura varieties, and L-functions, Vol. I (Ann Arbor, MI, 1988)*, pages 283–414. Academic Press, Boston, MA, 1990.

[MM01]   Yuri I. Manin and Matilde Marcolli. Holography principle and arithmetic of algebraic curves. *Adv. Theor. Math. Phys.*, 5(3):617–650, 2001.

[MM02]   Yuri I. Manin and Matilde Marcolli. Continued fractions, modular symbols, and noncommutative geometry. *Selecta Math. (N.S.)*, 8(3):475–521, 2002.

[Pau04]   F Paugam. Quelques bords irrationnels de variétés de shimura. *Universitaetsverlag Goettingen, Mathematisches Institut, Seminars*, pages 1–12, 2004.

[Pin90]   Richard Pink. *Arithmetical compactification of mixed Shimura varieties*, volume 209 of *Bonner Mathematische Schriften [Bonn Mathematical Publications]*. Universität Bonn Mathematisches Institut, Bonn, 1990. Dissertation, Rheinische Friedrich-Wilhelms-Universität Bonn, Bonn, 1989.

[Pol02]   Alexander Polishchuk. Noncommutative 2-tori with real multiplication as noncommutative projective varieties. *arXiv*, (http://fr.arXiv.org/abs/math.AG/0212306), 2002.

[Pol03]   Alexander Polishchuk. Classification of holomorphic vector bundles on noncommutative two-tori. *arXiv*, (http://fr.arXiv.org/abs/math.QA/0308136), 2003.

[SGA64]   *Schémas en groupes. III: Structure des schémas en groupes réductifs*. Springer-Verlag, Berlin, 1962/1964.

[SW99]   Nathan Seiberg and Edward Witten. String theory and noncommutative geometry. *J. High Energy Phys.*, (9):Paper 32, 93 pp. (electronic), 1999.

E-mail address: fpaugam@math.univ-rennes1.fr

Holomorphic bundles on 2-dimensional noncommutative toric orbifolds

A. Polishchuk

ABSTRACT. We define the notion of a holomorphic bundle on the noncommutative toric orbifold T_θ/G associated with an action of a finite cyclic group G on an irrational rotation algebra. We prove that the category of such holomorphic bundles is abelian and its derived category is equivalent to the derived category of modules over a finite-dimensional algebra Λ. As an application we finish the computation of K_0-groups of the crossed product algebras describing the above orbifolds initiated in [18], [29], [30], [12] and [13]. Also, we describe a torsion pair in the category of Λ-modules, such that the tilting with respect to this torsion pair gives the category of holomorphic bundles on T_θ/G.

Introduction

Let A_θ be the algebra of smooth functions on the noncommutative 2-torus T_θ associated with an irrational real number θ. Recall that its elements are expressions of the form $\sum_{m,n} a_{m,n} U_1^m U_2^n$, where the coefficients $(a_{m,n})_{(m,n)\in\mathbb{Z}^2}$ rapidly decrease at infinity, and U_1 and U_2 satisfy the relation

$$U_1 U_2 = \exp(2\pi i\theta)U_2 U_1.$$

It is convenient to denote $U_v = \exp(-\pi imn\theta)U_1^m U_2^n$ for $v = (m,n) \in \mathbb{Z}^2$, so that

$$U_v U_w = \exp(\pi i\theta \det(v,w))U_{v+w}.$$

There is a natural action of $\mathrm{SL}_2(\mathbb{Z})$ on A_θ such that the matrix g acts by the automorphism $U_v \mapsto U_{gv}$. Hence, for a finite subgroup $G \subset \mathrm{SL}_2(\mathbb{Z})$ we can consider the crossed product algebra $B_\theta = A_\theta * G$.

The simplest case is when $G = \mathbb{Z}/2\mathbb{Z}$ generated by $-\operatorname{id} \subset \mathrm{SL}_2(\mathbb{Z})$ acting on A_θ by the so called *flip automorphism* . In this case the algebra B_θ was studied in the papers [8], [10], [18] and [28]. In particular, it is known that it is simple, has a unique tracial state, and is an AF-algebra. Also its K-theory has been computed: one has $K_0(B_\theta) = \mathbb{Z}^6$ and $K_1(B_\theta) = 0$. However, there are three more examples of finite subgroups $G \subset \mathrm{SL}_2(\mathbb{Z})$ for which the situation is not so well understood. Namely, we can take $G = \mathbb{Z}/3\mathbb{Z}$ generated by $\begin{pmatrix} -1 & 1 \\ -1 & 0 \end{pmatrix}$; or $G = \mathbb{Z}/4\mathbb{Z}$ generated by the "Fourier" matrix $\begin{pmatrix} 0 & 1 \\ -1 & 0 \end{pmatrix}$; or $G = \mathbb{Z}/6\mathbb{Z}$ generated by $\begin{pmatrix} -1 & 1 \\ 1 & 0 \end{pmatrix}$. In this paper we compute $K_0(B_\theta)$ in all these cases using holomorphic vector bundles on the corresponding noncommutative orbifolds.

By a *vector bundle* on the noncommutative toric orbifold T_θ/G we mean a finitely generated projective right B_θ-module. We want to define what is a holomorphic structure on such a vector bundle. As in [23], [22], let us consider a complex structure on T_θ associated with a complex number $\tau \in \mathbb{C} \setminus \mathbb{R}$. It is given by a derivation

$$\delta : A_\theta \to A_\theta : \sum_{m,n} a_{m,n} U_1^m U_2^n \mapsto 2\pi i \sum_{m,n} (m\tau + n) a_{m,n} U_1^m U_2^n$$

of A_θ that we view as an analogue of the $\bar{\partial}$-operator. To descend this structure to the orbifold T_θ/G we have to impose some compatibility between the action of G and δ. More precisely, we assume that there exists a character $\varepsilon : G \to \mathbb{C}^*$ such that the following relation holds:

$$(0.1) \qquad\qquad g\delta = \varepsilon(g)\delta g$$

for all $g \in G$. For example, for $G = \mathbb{Z}/2\mathbb{Z}$ acting by the flip automorphism ε is the unique nontrivial character of $\mathbb{Z}/2\mathbb{Z}$. In three other cases such a relation exists for a special choice of τ. Namely, let us identify $G = \mathbb{Z}/m\mathbb{Z}$ with the subgroup of m-th roots of unity in $\mathbb{C}^*$ and let G act on $\mathbb{C}$ by multiplication. Then we can choose τ in such a way that the lattice $\mathbb{Z}\tau + \mathbb{Z}$ is G-invariant: for $m = 4$ we take $\tau = i$, while for $m = 3$ and $m = 6$ we take $\tau = (1 + i\sqrt{3})/2$. Note that the embedding of G into $\mathrm{SL}_2(\mathbb{Z})$ is induced by its action on the basis $(\tau, 1)$ of $\mathbb{Z}\tau + \mathbb{Z}$. Then (0.1) will hold with $\varepsilon(g) = g^{-1} \in \mathbb{C}^*$.

Recall that in [23], [22] we studied the category $\mathrm{Hol}(T_{\theta,\tau})$ of holomorphic bundles on T_θ. By definition, these are pairs $(P, \overline{\nabla})$ consisting of a finitely generated projective right A_θ-module P and an operator $\overline{\nabla} : P \to P$ satisfying the Leibnitz identity

$$\overline{\nabla}(f \cdot a) = f \cdot \delta(a) + \overline{\nabla}(f) \cdot a,$$

where $f \in P$, $a \in A_\theta$. Now we extend δ to a *twisted derivation* $\widetilde{\delta}$ of B_θ by setting

$$\widetilde{\delta}\left(\sum_{g \in G} a_g g\right) = \sum_{g \in G} \epsilon(g)\delta(a_g)g,$$

where $a_g \in A_\theta$ for $g \in G$. This extended map satisfies the twisted Leibnitz identity

$$\widetilde{\delta}(b_1 b_2) = b_1 \widetilde{\delta}(b_2) + \widetilde{\delta}(b_1)\kappa(b_2),$$

where κ is the automorphism of B_θ given by $\kappa(\sum_{g \in G} a_g g) = \sum_{g \in G} \epsilon(g)a_g g$. We define a *holomorphic structure* on a vector bundle P on T_θ/G as an operator $\overline{\nabla} : P \to P$ satisfying the similar twisted Leibnitz identity

$$\overline{\nabla}(f \cdot b) = f \cdot \widetilde{\delta}(b) + \overline{\nabla}(f) \cdot \kappa(b),$$

where $f \in P$, $b \in B_\theta$. By definition, a *holomorphic bundle* is a pair $(P, \overline{\nabla})$ consisting of a vector bundle P equipped with a holomorphic structure $\overline{\nabla}$. One can define morphisms between holomorphic bundles in a natural way, so we obtain the category $\mathrm{Hol}(T_{\theta,\tau}/G)$ of holomorphic bundles.

Recall that the combined results of [23] and [22] imply that the category $\mathrm{Hol}(T_{\theta,\tau})$ is abelian and one has an equivalence of bounded derived categories

$$D^b(\mathrm{Hol}(T_{\theta,\tau})) \simeq D^b(\mathrm{Coh}(E)),$$

where $\mathrm{Coh}(E)$ is the category of coherent sheaves on the elliptic curve $E = \mathbb{C}/(\mathbb{Z} + \mathbb{Z}\tau)$. Furthermore, the image of the abelian category $\mathrm{Hol}(T_{\theta,\tau})$ in the derived category $D^b(\mathrm{Coh}(E))$ can be described as the heart of the tilted t-structure associated with a certain torsion pair in $\mathrm{Coh}(E)$ (depending on θ). Our main result is a similar explicit description of the category of holomorphic bundles on $T_{\theta,\tau}/G$, where $G = \mathbb{Z}/m\mathbb{Z} \subset \mathrm{SL}_2(\mathbb{Z})$ with $m \in \{2,3,4,6\}$.

THEOREM 0.1. *The category* $\mathrm{Hol}(T_{\theta,\tau}/G)$ *is abelian and one has an equivalence of bounded derived categories*

$$D^b(\mathrm{Hol}(T_{\theta,\tau}/G)) \simeq D^b(\mathrm{mod} - \Lambda),$$

where $\mathrm{mod} - \Lambda$ *is the category of finite-dimensional right modules over the algebra* $\Lambda = \mathbb{C}Q/(I)$ *of paths in a quiver* Q *without cycles with quadratic relations* I. *The number of vertices of* Q *is equal to* $6, 8, 9$ *or* 10 *for* $m = 2, 3, 4$ *or* 6, *respectively.*

The precise description of the algebra Λ will be given in section 1.2. It is derived equivalent to one of canonical tubular algebras considered by Ringel in [24]. Furthermore, the image of $\mathrm{Hol}(T_{\theta,\tau}/G)$ in $D^b(\mathrm{mod} - \Lambda)$ corresponds to the tilted t-structure for a certain explicit torsion pair in $\mathrm{mod} - \Lambda$ depending on θ that we will describe in Theorem 2.8.

We prove Theorem 0.1 in two steps: first, we relate holomorphic bundles on $T_{\theta,\tau}/G$ to the derived category of G-equivariant sheaves on the elliptic curve $E = \mathbb{C}/(\mathbb{Z}\tau + \mathbb{Z})$, and then we construct a derived equivalence with right modules over the algebra Λ. The second step is actually well known and works for arbitrary weighted projective curves considered in [15]. We present an alternative derivation working directly with equivariant sheaves. It is based on the semiorthogonal decomposition of the category of G-equivariant sheaves associated with a ramified G-covering of smooth curves (see Theorem 1.2).

Combining Theorem 0.1 with the results of [29] and [12] we derive the following result.

THEOREM 0.2. *One has* $K_0(B_\theta) \simeq \mathbb{Z}^r$, *where* $r = 6, 8, 9$ *or* 10 *for* $G = \mathbb{Z}/m\mathbb{Z}$ *with* $m = 2, 3, 4$ *or* 6, *respectively.*

Note that for $G = \mathbb{Z}/2\mathbb{Z}$ this was known (see [18]). For $G = \mathbb{Z}/4\mathbb{Z}$ and $G = \mathbb{Z}/6\mathbb{Z}$ this was proved for θ in a dense G_δ-set (see [30] and [13]). The case of $G = \mathbb{Z}/4\mathbb{Z}$ and arbitrary irrational θ was done by Lueck, Phillips and Walters in 2003 (unpublished). Our proof shows in addition that the natural forgetful map

$$K_0(\mathrm{Hol}(T_{\theta,\tau}/G)) \to K_0(B_\theta)$$

is, in fact, an isomorphism and identifies the positive cones in these groups.

1. Derived categories of G-sheaves

1.1. Generalities on G-sheaves. Let G be a finite group acting on an algebraic variety X over a field k of characteristic zero. Then we can consider the category $\mathrm{Coh}_G(X)$ of G-equivariant coherent sheaves. We denote its bounded derived category by $D^b_G(X)$. It is equivalent to the full subcategory in the bounded derived category of G-equivariant quasicoherent sheaves consisting of complexes with coherent cohomology (see Corollary 1 in [1]). We refer to [11], Section 4, for a more detailed discussion of this category and restrict ourself to several observations. Below we will use the term G-*sheaf* to denote a G-equivariant coherent sheaf.

There is a natural forgetting functor from $D_G^b(X)$ to $D^b(X)$, the usual derived category of coherent sheaves on X. For equivariant complexes of sheaves F_1 and F_2 we denote by $\operatorname{Hom}_G(F_1, F_2)$ (resp., $\operatorname{Hom}(F_1, F_2)$) morphisms between these objects in the former (resp., latter) category. There is a natural action of G on $\operatorname{Hom}(F_1, F_2)$ and one has

$$\operatorname{Hom}_G(F_1, F_2) \simeq \operatorname{Hom}(F_1, F_2)^G.$$

In particular, the cohomological dimension of $\operatorname{Coh}_G(X)$ is at most that of $\operatorname{Coh}(X)$. Let us also set $\operatorname{Hom}_G^i(F_1, F_2) = \operatorname{Hom}_G(F_1, F_2[i])$. If X is a smooth projective variety over k then we can define the bilinear form $\chi_G(\cdot, \cdot)$ on $K_0(D_G^b(X))$ by setting

$$\chi_G(F_1, F_2) = \sum_{i \in \mathbb{Z}} (-1)^i \dim \operatorname{Hom}_G^i(F_1, F_2).$$

Many natural constructions with sheaves carry easily to G-equivariant setting. For example, the tensor product of G-sheaves is defined. Also if ρ is a representation of G then there is a natural tensor product operation $F \mapsto F \otimes \rho$ on G-sheaves. If Y is another variety equipped with the action of G and if $f : X \to Y$ is a G-equivariant morphism then there are natural functors of push-forward and pull-back:

$$f_* : \operatorname{Coh}_G(X) \to \operatorname{Coh}_G(Y), \quad f^* : \operatorname{Coh}_G(Y) \to \operatorname{Coh}_G(X),$$

We can also consider the derived functor $Rf_* : D_G^b(X) \to D_G^b(Y)$ and if Y is smooth or f is flat, the derived functor $Lf^* : D_G^b(Y) \to D_G^b(X)$ (when f is flat we denote it simply by f^*). The pair (Lf^*, Rf_*) satisfies the usual adjunction property.

If X is smooth and projective then the category $D_G^b(X)$ is also equipped with the Serre duality of the form

$$\operatorname{Hom}_G(F_1, F_2)^* \simeq \operatorname{Hom}_G(F_2, F_1 \otimes \omega_X[\dim X]),$$

where the canonical bundle ω_X is equipped with the natural G-action.

The following observation will be useful to us.

LEMMA 1.1. *Let X be a smooth curve equipped with an action of a finite group G. Then the category $D_G^b(X)$ is equivalent to the category of G-equivariant objects in $D^b(X)$, i.e., the category of data (F, ϕ_g), where $F \in D^b(X)$ and $\phi_g : g^*F \widetilde{\to} F$, $g \in G$, is a collection of isomorphisms satisfying the natural compatibility condition.*

PROOF. . Note that there is a natural functor from $D_G^b(X)$ to the category of G-objects in $D^b(X)$. It is easy to see that it is fully faithful, so the only issue is to check that it is essentially surjective. The proof is based on the fact that every object $F \in D^b(X)$ is isomorphic to the direct sum of its cohomology sheaves: $F \simeq \oplus_n H^n F[-n]$. A G-structure on F is given by a compatible collection of isomorphisms $\phi = (\phi_g)$, where $\phi_g : g^*F \to F$. Note that the only nontrivial components of ϕ_g with respect to the above direct sum decompositions are maps $g^*H^n F[-n] \to H^n F[-n]$ and $g^*H^n F[-n] \to H^{n-1} F[-n+1]$. Let $\phi^0 = (\phi_g^0)$ be the G-structure on F given by the components of ϕ of the first kind (i.e., by the diagonal components). Since the decomposition of F into the direct sum of cohomology sheaves is compatible with ϕ^0, it suffices to find an isomorphism of G-objects

(1.1) $(F, \phi^0) \simeq (F, \phi).$

Let us write $\phi_g = a_g \circ \phi_g^0$, where $a_g \in \operatorname{Aut}(F)$. Note that a_g belongs to the abelian subgroup

$$A := \oplus_n \operatorname{Hom}(H^n F[-n], H^{n-1} F[-n+1]) \subset \operatorname{Aut}(F)$$

of "upper-triangular" automorphisms with identities as diagonal entries. It is easy to check that the compatibility condition on the data ϕ amounts to the 1-cocycle equation for a_g, where G acts on A in a natural way. On the other hand, existence of an isomorphism (1.1) is equivalent to $g \mapsto a_g$ being a coboundary. It remains to note that $H^1(G, A) = 0$ since A is a vector space over a field of characteristic zero. $\qquad \square$

1.2. Semiorthogonal decomposition associated with a Galois covering.

Recall (see [5], [7]) that a *semiorthogonal decomposition* of a triangulated category $\mathcal{A}$ is an ordered collection $(\mathcal{B}_1, \ldots, \mathcal{B}_r)$ of full triangulated subcategories in $\mathcal{A}$ such that $\operatorname{Hom}(B_i, B_j) = 0$ whenever $B_i \in \mathcal{B}_i$ and $B_j \in \mathcal{B}_j$, where $i > j$, and the subcategories $\mathcal{B}_1, \ldots, \mathcal{B}_r$ generate $\mathcal{A}$. In this case we write

$$\mathcal{A} = \langle \mathcal{B}_1, \ldots, \mathcal{B}_r \rangle.$$

Semiorthogonal decompositions are related to admissible triangulated subcategories. For a subcategory $\mathcal{B} \subset \mathcal{A}$ let us denote by $\mathcal{B}^\perp$ the *right orthogonal* of $\mathcal{B}$, i.e., the full subcategory of $\mathcal{A}$ consisting of all C such that $\operatorname{Hom}(B, C) = 0$ for all $B \in \mathcal{B}$. A triangulated subcategory $\mathcal{B} \subset \mathcal{A}$ is called *right admissible* if for every $X \in \mathcal{A}$ there exists a distinguished triangle $B \to X \to C \to \cdots$ with $B \in \mathcal{B}$ and $C \in \mathcal{B}^\perp$. Thus, a right admissible subcategory $\mathcal{B} \subset \mathcal{A}$ gives rise to a semiorthogonal decomposition

$$\mathcal{A} = \langle \mathcal{B}^\perp, \mathcal{B} \rangle.$$

Similarly, one can define the left orthogonal and left admissibility of a subcategory.

We are going to use also some results from the theory of exceptional collections (see [3], [24]). Let k be a field. Recall that an object E in a k-linear triangulated subcategory $\mathcal{A}$ is *exceptional* if $\operatorname{Hom}^i(E, E) = 0$ for $i \neq 0$ and $\operatorname{Hom}(E, E) = k$. An *exceptional collection* in $\mathcal{A}$ is a collection of exceptional objects $(E_1, \ldots, E_n)$ such that $\operatorname{Hom}^*(E_i, E_j) = 0$ for $i > j$. A triangulated subcategory $\langle E_1, \ldots, E_n \rangle$ generated by an exceptional collection is known to be left and right admissible (see [3], Theorem 3.2). In the case when $\langle E_1, \ldots, E_n \rangle = \mathcal{A}$ we will say that $(E_1, \ldots, E_n)$ is a *full exceptional collection*. An exceptional collection $(E_1, \ldots, E_n)$ is *strong* if $\operatorname{Hom}^a(E_i, E_j) = 0$ for $a \neq 0$ and all i, j. If an exceptional collection is full and strong then $E = \oplus_{i=1}^n E_i$ is a tilting object in $\mathcal{A}$, i.e., the functor $X \mapsto R\operatorname{Hom}(E, X)$ gives an equivalence between $\mathcal{A}$ and $D^b(\operatorname{mod} - A)$, where $A = \operatorname{End}(E)$ and $\operatorname{mod} - A$ is the category of finite-dimensional right A-modules (provided $\mathcal{A}$ satisfies some natural finiteness assumptions and is framed, see [6]).

Let $\pi : X \to Y$ be a ramified Galois covering with Galois group G, where X and Y are smooth projective curves over an algebraically closed field k of characteristic zero. In other words, a finite group G acts effectively on X and $Y = X/G$. We are going to construct a semiorthogonal decomposition of the derived category of G-sheaves $D_G^b(X)$ with $D^b(Y)$ as one of the pieces. Let $D_1, \ldots, D_r$ be all special fibers of π equipped with the reduced scheme structure and let $m_1, \ldots, m_r$ be the corresponding multiplicities. Let us also fix a point $p_i \in D_i$ for each $i = 1, \ldots, r$, and let $G_i \subset G$ be the stabilizer subgroup of p_i. Then we have G-equivariant isomorphisms $D_i \simeq G/G_i$. Hence, the category of G-sheaves on D_i is equivalent to finite dimensional representations of G_i. For every character ζ of G_i we denote by

ζ_{D_i} the corresponding G-sheaf on D_i. Note that G_i is a cyclic group of order m_i. Moreover, the representation of G_i on $\omega_X|_{p_i}$ allows to identify G_i with the group of m_i-th roots of unity in such a way that it acts on $\omega_X|_{p_i}$ via the standard character. Thus, we have an isomorphism of G-sheaves

$$\omega_X|_{D_i} \simeq \zeta(i)_{D_i},$$

where $\zeta(i)$ is a generator of the character group $\hat{G}_i$.

THEOREM 1.2. *(i) The natural functor $\pi^* : D^b(Y) \to D^b_G(X)$ is fully faithful. (ii) For every $i = 1, \ldots, r$, the collection of G-sheaves on X*

(1.2) $(\mathcal{O}_{(m_i-1)D_i}, \ldots, \mathcal{O}_{2D_i}, \mathcal{O}_{D_i})$

is exceptional. Let $\mathcal{B}_i$ be the triangulated subcategory in $D^b_G(X)$ generated by this exceptional collection. Then $\mathcal{B}_i$ and $\mathcal{B}_j$ are mutually orthogonal for $i \neq j$, i.e., $\mathrm{Hom}(B_i, B_j) = \mathrm{Hom}(B_j, B_i) = 0$ for all $B_i \in \mathcal{B}_i$ and $B_j \in \mathcal{B}_j$. (iii) One has a semiorthogonal decomposition

$$D^b_G(X) = \langle \pi^* D^b(Y), \mathcal{B}_1, \ldots, \mathcal{B}_r \rangle.$$

PROOF. . (i) For $F_1, F_2 \in D^b(X)$ we have

$$\mathrm{Hom}_G(\pi^* F_1, \pi^* F_2) \simeq \mathrm{Hom}(F_1, \pi_* \pi^* F_2)^G$$

$$\simeq \mathrm{Hom}(F_1, F_2 \otimes (\pi_* \mathcal{O}_X)^G) \simeq \mathrm{Hom}(F_1, F_2),$$

since $(\pi_* \mathcal{O}_X)^G \simeq \mathcal{O}_Y$.
(ii) Let us first prove that the collection of G-sheaves on X

(1.3) $(\mathcal{O}_{D_i}, \zeta(i)_{D_i}[1], \ldots, \zeta(i)^{m_i-2}_{D_i}[m_i - 2])$

is exceptional. Indeed, it is clear that there are no G-morphisms between $\zeta(i)^a_{D_i}$ and $\zeta(i)^b_{D_i}$ for $a \not\equiv b \bmod(m_i)$. Also, by Serre duality, for $a, b \in \mathbb{Z}/m_i\mathbb{Z}$ we have

$$\mathrm{Ext}^1_G(\zeta(i)^a_{D_i}, \zeta(i)^b_{D_i})^* \simeq \mathrm{Hom}_G(\zeta(i)^b_{D_i}, \zeta(i)^{a+1}_{D_i}) \simeq \mathrm{Hom}_{G_i}(\zeta(i)^b, \zeta(i)^{a+1}).$$

The latter space is nonzero only for $b = a+1$. This proves that (1.3) is exceptional. Now using the exact sequences

$$0 \to \zeta(i)^a_{D_i} \to \mathcal{O}_{(a+1)D_i} \to \mathcal{O}_{aD_i} \to 0$$

for $a = 1, \ldots, m_i - 2$, one can easily show that making a sequence of mutations in (1.3) one gets the sequence (1.2). The fact that $\mathcal{B}_i$ and $\mathcal{B}_j$ are mutually orthogonal follows from disjointness of D_i and D_j.
(iii) Since the subcategory $\langle \mathcal{B}_1, \ldots, \mathcal{B}_r \rangle$ is admissible it is enough to prove that $\pi^* D^b(Y)$ coincides with its right orthogonal. Since $\mathcal{B}_i$ is also generated by the exceptional collection (1.3), the condition $\mathrm{Hom}_G(\mathcal{B}_i, F) = 0$ for $F \in D^b_G(X)$ is equivalent

$$\mathrm{Hom}^*_G(\zeta(i)^a_{D_i}, F) = 0 \text{ for } a = 0, \ldots, m_i - 2.$$

Using Serre duality we can rewrite this as

$$\mathrm{Hom}_G(F, \zeta(i)^a_{D_i}) = 0 \text{ for } a = 1, \ldots, m_i - 1.$$

Equivalently, $F|_{p_i}$ should have a trivial G_i-action. By the main theorem of [27] this implies that $F \in \pi^* D^b(Y)$.   □

In the case when $Y = \mathbb{P}^1$ the semiorthogonal decomposition of Theorem 1.2 gives rise to a full exceptional collection in $D^b_G(X)$.

COROLLARY 1.3. *Assume $X/G \simeq \mathbb{P}^1$. Then for every $n \in \mathbb{Z}$ we have the following full exceptional collection in $D^b_G(X)$:*

$$(\pi^* \mathcal{O}_{\mathbb{P}^1}(n), \pi^* \mathcal{O}_{\mathbb{P}^1}(n+1), \mathcal{O}_{(m_1-1)D_1}, \ldots, \mathcal{O}_{D_1}, \ldots, \mathcal{O}_{(m_r-1)D_r}, \ldots, \mathcal{O}_{D_r}).$$

In particular, $K^0(D^b_G(X)) \simeq \mathbb{Z}^{2+\sum_{i=1}^r (m_i-1)}$.

Definition. For a collection of r distinct points $\bar{\lambda} = (\lambda_1, \ldots, \lambda_r)$ on $\mathbb{P}^1(k)$ and a sequence of weights $\bar{m} = (m_1, \ldots, m_r)$ let us define the algebra $\Lambda(\bar{\lambda}, \bar{m})$ as the path algebra of a quiver $Q_{\bar{m}}$ modulo relations $I(\bar{\lambda})$, where
(i) $Q_{\bar{m}}$ has $2 + \sum_{i=1}^r (m_i - 1)$ vertices named u, v, and

$$w_1^1, \ldots, w_1^{m_1-1}, \ldots, w_r^1, \ldots, w_r^{m_r-1};$$

(ii) $Q_{\bar{m}}$ has $2 + \sum_{i=1}^r (m_i - 1)$ arrows: 2 arrows $u \overset{x_0, x_1}{\to} v$, and chains of arrows

$$v \overset{e_i}{\to} w_i^{m_i-1} \to w_i^{m_i-2} \to \cdots \to w_i^1$$

for every $i = 1, \ldots, r$;
(iii) $I(\bar{\lambda})$ is generated by r quadratic relations: $L_i \cdot e_i = 0$, $i = 1, \ldots, m$, where $L_i \subset kx_0 \oplus kx_1$ is the line corresponding to $\lambda_i \in \mathbb{P}^1 = \mathbb{P}(kx_0 \oplus kx_1)$.

It is easy to see that the endomorphism algebra of the exceptional collection constructed in Corollary 1.3 is isomorphic to $\Lambda(\bar{\lambda}, \bar{m})$, where $D_i = \pi^{-1}(\lambda_i)$. Hence, we obtain the following description of the derived category of G-sheaves on a G-covering of $\mathbb{P}^1$. (where for a finite-dimensional algebra A we denote by $\mathrm{mod} - A$ the category of finite-dimensional right A-modules).

COROLLARY 1.4. *Let $\pi : X \to \mathbb{P}^1$ be a ramified Galois covering, where X is a smooth curve, with Galois group G. Let $\bar{\lambda} = (\lambda_1, \ldots, \lambda_r) \subset \mathbb{P}^1$ be the set of ramification points of π and let $\bar{m} = (m_1, \ldots, m_r)$ be multiplicities of the corresponding fibers. Then for every $n \in \mathbb{Z}$ one has an exact equivalence of triangulated categories*

$$\Phi_n : D^b_G(X) \widetilde{\to} D^b(\mathrm{mod} - \Lambda(\bar{\lambda}, \bar{m})) : F \to R\mathrm{Hom}_G(V_n, F),$$

where

$$V_n = \pi^* \mathcal{O}_{\mathbb{P}^1}(n) \oplus \pi^* \mathcal{O}_{\mathbb{P}^1}(n+1) \oplus \bigoplus_{1 \le i \le r, 1 \le j < m_i} \mathcal{O}_{jD_i}$$

Remark. Another natural full exceptional collection in $D^b_G(X)$ is
(1.4)
$$(\mathcal{O}_X, \mathcal{O}_X(D_1), \ldots, \mathcal{O}_X((m_1-1)D_1), \ldots, \mathcal{O}_X(D_r), \ldots, \mathcal{O}_X((m_r-1)D_r), \pi^* \mathcal{O}_{\mathbb{P}^1}(1)).$$

It is obtained from the collection of Corollary 1.3 for $n = 1$ by making the left mutation through $\pi^* \mathcal{O}_{\mathbb{P}^1}(1)$ of the part of the collection following this object.

A right module M over $\Lambda(\bar{\lambda}, \bar{m})$ can be viewed as a representation of the quiver $Q_{\bar{m}}^{op}$ in which the relations $I(\bar{\lambda})^{op}$ are satisfied. Thus, M is given by a collection of vector spaces (U, V, W_i^j), $i = 1, \ldots, r$, $j = 1, \ldots, m_i - 1$, equipped with linear maps

$$W_i^1 \to \cdots \to W_i^{m_i-1} \to V, \ i = 1, \ldots, r,$$

and $x_0, x_1 : V \to U$ satisfying the relations $I(\overline{\lambda})^{op}$. Let us define the following additive functions of M:

$$\deg_n(M) = |G| \cdot \left(n \dim U - (n-1) \dim V - \sum_{i=1}^{r} \sum_{j=1}^{m_i-1} \frac{\dim W_i^j}{m_i} \right),$$

$$\mathrm{rk}(M) = \dim U - \dim V.$$

We extend these functions to additive functions on $D^b(\Lambda(\overline{\lambda}, \overline{m}))$.

LEMMA 1.5. *In the situation of Corollary 1.4 one has* $\deg_n(\Phi_n(F)) = \deg(F)$ *and* $\mathrm{rk}(\Phi_n(F)) = \mathrm{rk}(F)$ *for every* $F \in D_G^b(X)$.

PROOF. . Let $M = \Phi_n(F) = R\mathrm{Hom}_G(V_n, F)$ for $F \in D_G^b(X)$. Then we have

$$\dim U = \chi_G(\pi^* \mathcal{O}_{\mathbb{P}^1}(n), F), \quad \dim V = \chi_G(\pi^* \mathcal{O}_{\mathbb{P}^1}(n+1), F), \quad \dim W_i^j = \chi_G(\mathcal{O}_{jD_i}, F),$$

and our task is to express $\mathrm{rk}(F)$ and $\deg(F)$ in terms of these numbers. To compute the rank we can use the equality

$$\mathrm{rk}(F) = -\chi_G(\pi^* \mathcal{O}_q, F),$$

where q is a generic point of $\mathbb{P}^1$. Since $[\mathcal{O}_q] = [\mathcal{O}_{\mathbb{P}^1}(n+1)] - [\mathcal{O}_{\mathbb{P}^1}(n)]$, this immediately implies the required formula

$$\mathrm{rk}(F) = \chi_G(\pi^* \mathcal{O}_{\mathbb{P}^1}(n), F) - \chi_G(\pi^* \mathcal{O}_{\mathbb{P}^1}(n+1), F).$$

The formula for $\deg(F)$ should have form

$$\deg(F) = a\chi_G(\pi^* \mathcal{O}_{\mathbb{P}^1}(n), F) + b\chi_G(\pi^* \mathcal{O}_{\mathbb{P}^1}(n+1), F) + \sum_{i=1}^{r} \sum_{j=1}^{m_i-1} c_i^j \chi_G(\mathcal{O}_{jD_i}, F)$$

for some constants a, b and c_i^j. The constants are determined by substituting in this formula the elements of the dual basis of $K_0(D_G^b(X))$:

$$([\pi^* \mathcal{O}_{\mathbb{P}^1}(n)], -[\pi^* \mathcal{O}_{\mathbb{P}^1}(n-1)], -[\zeta(1)_{D_1}], \ldots, -[\zeta(1)_{D_1}^{m_1-1}], \ldots, -[\zeta(r)_{D_r}], \ldots, -[\zeta(r)_{D_r}^{m_r-1}]).$$

$\square$

1.3. Elliptic Galois coverings of $\mathbb{P}^1$.

Now let us specialize to the case of a Galois covering $\pi : E \to \mathbb{P}^1$, where $E = \mathbb{C}/(\mathbb{Z}\tau + \mathbb{Z})$ is an elliptic curve (so $k = \mathbb{C}$). More precisely, we are interested in the following four cases in which G is a cyclic subgroup in $\mathbb{C}^*$ acting on E in the natural way.

(i) E is arbitrary and $G = \mathbb{Z}/2\mathbb{Z}$. The corresponding double covering $\pi : E \to \mathbb{P}^1$ is given by the Weierstrass $\wp$-function and is ramified exactly at 4 points of order 2 on E.

(ii) $E = \mathbb{C}/L_{tr}$, where $L_{tr} = \mathbb{Z}\frac{1+i\sqrt{3}}{2} + \mathbb{Z}$, and $G = \mathbb{Z}/3\mathbb{Z}$. In this case $\pi : E \to \mathbb{P}^1$ is given by $\wp'(z)$. Note that E^G consists of 3 points: $0 \bmod L_{tr}$ and $\pm\frac{3+i\sqrt{3}}{6} \bmod L_{tr}$. Hence, π is totally ramified over 3 points.

(iii) $E = \mathbb{C}/L_{sq}$, where $L_{sq} = \mathbb{Z}i + \mathbb{Z}$, and $G = \mathbb{Z}/4\mathbb{Z}$. In this case $\pi : E \to \mathbb{P}^1$ is given by $\wp(z)^2$. We have two points whose stabilizer subgroup is $\mathbb{Z}/2\mathbb{Z}$, namely, $\frac{1}{2} \bmod L_{sq}$ and $\frac{i}{2} \bmod L_{sq}$ (they get exchanged by the generator of $\mathbb{Z}/4\mathbb{Z}$). The two points in E^G are $0 \bmod L_{sq}$ and $\frac{i+1}{2} \bmod L_{sq}$. Hence, π is ramified over 3 points and the corresponding multiplicities are $(2, 4, 4)$.

(iv) $E = \mathbb{C}/L_{tr}$ (same curve as in (ii)) and $G = \mathbb{Z}/6\mathbb{Z}$. In this case $\pi : E \to \mathbb{P}^1$ is given by $\wp'(z)^2$. There are 3 points whose stabilizer subgroup is $\mathbb{Z}/2\mathbb{Z}$, namely, all nontrivial points of order 2 on E (they form one G-orbit). There is also a G-orbit consisting of two points $\pm \frac{3+i\sqrt{3}}{6} \bmod L_{tr}$ with stabilizer subgroup $\mathbb{Z}/3\mathbb{Z}$. Finally, $0 \bmod L_{tr}$ is the only point in E^G. Therefore, π is ramified over 3 points with multiplicities $(2,3,6)$.

From the above description of the ramification data and from Corollary 1.3 we get

COROLLARY 1.6. *One has* $K_0(D^b_G(E)) \simeq \mathbb{Z}^r$, *where* $r = 6, 8, 9$ *or* 10 *in the cases (i)-(iv), respectively.*

1.4. Galois coverings of $\mathbb{P}^1$ and weighted projective curves.
The results of this section are not used in the rest of the paper. Its purpose is to explain the relation between G-sheaves on ramified Galois coverings of $\mathbb{P}^1$ and coherent sheaves on weighted projective curves introduced in [15]. This relation is known to experts, however, our proof seems to be new.

Let us recall the definition of weighted curves [1] $C(\overline{m}, \overline{\lambda})$ of [15] associated with a sequence of positive integers $\overline{m} = (m_1, \ldots, m_r)$ and a sequence of points $\overline{\lambda} = (\lambda_1, \ldots, \lambda_r)$ in $\mathbb{P}^1(k)$. Let $Z(\overline{m})$ be the rank one abelian group with generators $e_1, \ldots, e_r$ and relations $m_1 e_1 = \ldots = m_r e_r$. Let us also choose for every $i = 1, \ldots, r$ a nonzero section $s_i \in H^0(\mathbb{P}^1, \mathcal{O}_{\mathbb{P}^1}(1))$ such that $s_i(\lambda_i) = 0$. Consider the algebra

$$S(\overline{m}, \overline{\lambda}) = k[x_1, \ldots, x_r]/I(\overline{m}, \overline{\lambda}),$$

where the ideal $I(\overline{m}, \overline{\lambda})$ is generated by all polynomials of the form $a_1 x_1^{m_1} + \ldots a_r x_r^{m_r}$ such that $\sum_{i=1}^r a_i s_i = 0$. Let $Z(\overline{m})_+ \subset Z(\overline{m})$ be the positive submonoid generated by $e_1, \ldots, e_r$. Note that the algebra $S(\overline{m}, \overline{\lambda})$ has a natural $Z(\overline{m})_+$-grading, where $\deg(x_i) = e_i$. The category $\mathrm{Coh}(C(\overline{m}, \overline{\lambda}))$ of coherent sheaves on $C(\overline{m}, \overline{\lambda})$ can be defined as the quotient-category of the category of finitely generated $Z(\overline{m})_+$-graded $S(\overline{m}, \overline{\lambda})$-modules by the subcategory of finite length modules.

Now assume we are given a ramified Galois covering $\pi : X \to \mathbb{P}^1$ with Galois group G, where X is a smooth connected curve. Define the associated data (D_i), $\overline{m}$ and $\overline{\lambda}$ as in the previous section. Let $\mathrm{Pic}_G(X)$ be the group of G-equivariant line bundles up to G-isomorphism. Let us consider the algebra

$$S(X, G) := \oplus_{[L] \in \mathrm{Pic}_G(X)} H^0(X, L)^G.$$

THEOREM 1.7. *One has an isomorphism of algebras*

$$S(\overline{m}, \overline{\lambda}) \simeq S(X, G),$$

compatible with gradings via an isomorphism $Z(\overline{m}) \simeq \mathrm{Pic}_G(X)$.

PROOF. . We claim that there is a natural homomorphism $S(\overline{m}, \overline{\lambda}) \to S(X, G)$ that sends x_i to a nonzero section f_i of $H^0(X, \mathcal{O}_X(D_i))^G$. Indeed, note that we have a natural isomorphism $\mathcal{O}_X(m_i D_i) \simeq \pi^* \mathcal{O}_{\mathbb{P}^1}(1)$ compatible with the action of G, and hence the induced isomorphism

$$H^0(\mathcal{O}_X(m_i D_i))^G \simeq H^0(\mathbb{P}^1, \mathcal{O}_{\mathbb{P}^1}(1)).$$

Let us rescale f_i in such a way that $f_i^{m_i}$ corresponds to $s_i \in H^0(\mathbb{P}^1, \mathcal{O}_{\mathbb{P}^1})$ under this isomorphism. Then $f_i^{m_i}$ will satisfy the same linear relations as s_i, hence we get a homomorphism $\alpha : S(\overline{m}, \overline{\lambda}) \to S(X, G)$. Note that α is compatible with

---

[1]These curves are also called weighted projective lines, however, they should not be confused with one-dimensional weighted projective spaces.

gradings via the homomorphism $Z(\overline{m}) \to \mathrm{Pic}_G(X)$ sending x_i to the class of D_i. Let us check that α is surjective. Assume we are given $L \in \mathrm{Pic}_G(X)$ and a nonzero G-invariant section f of L. If the divisor of zeroes of f contains D_i for some i then f is divisible by f_i in the algebra $S(X, G)$, so we can assume that the divisor of f is disjoint from all special fibers. Therefore, $L \simeq \pi^* \mathcal{O}_{\mathbb{P}^1}(n)$ and f corresponds to a section of $\mathcal{O}_{\mathbb{P}^1}(n)$ on $\mathbb{P}^1$. Note that $r \geq 2$ since X is connected. Therefore, every section of $\mathcal{O}_{\mathbb{P}^1}(n)$ can be expressed as a polynomial of $s_1, \ldots, s_r$. Hence, f belongs to the image of α. Injectivity of α follows easily from Proposition 1.3 of [**15**]. $\qquad\square$

Using Theorem 1.7 we can derive the following equivalence between the categories of sheaves.

THEOREM 1.8. *In the above situation one has an equivalence of categories*

$$\mathrm{Coh}_G(X) \simeq \mathrm{Coh}(C(\overline{m}, \overline{\lambda})).$$

PROOF. . This follows from Theorem 1.7 by a version of Serre's theorem. The only nontrivial fact one has to use is that for every G-sheaf F on X there exists a surjection of G-sheaves $\oplus_{i=1}^n L_i \to F$, where L_i are equivariant G-bundles. Since every G-sheaf F is covered by a G-bundle of the form $H^0(X, F \otimes L) \otimes L^{-1}$ for sufficiently ample G-equivariant line bundle L, it suffices to consider the case when F is locally free. Assume first that the action of G_i on the fiber $F|_{p_i}$ is trivial for all $i = 1, \ldots, r$. Then F is G-isomorphic to the pull-back of a vector bundle on $\mathbb{P}^1$. In this case the assertion is clear since all vector bundles on $\mathbb{P}^1$ are direct sums of line bundles. We are going to reduce to this case using elementary transformations along D_i's. Namely, let us decompose a representation of G_i on the fiber $F|_{p_i}$ into the direct sum of characters of G_i:

$$F|_{p_i} \simeq \oplus_{j=0}^{m_i - 1} V_j \otimes \zeta(i)^j$$

with some multiplicity spaces V_j. Note that if we define the G-bundle F' by the short exact triple

$$0 \to F' \to F \to V_j \otimes \zeta(i)_{D_i}^j \to 0$$

then we have an exact sequence of G_i-modules

$$0 \to V_j \otimes \zeta(i)^j \otimes \mathrm{Tor}_1(\mathcal{O}_{D_i}, \mathcal{O}_{p_i}) \to F'|_{p_i} \to F_{p_i} \to V_j \otimes \zeta(i)^j \to 0.$$

But $\mathrm{Tor}_1(\mathcal{O}_{D_i}, \mathcal{O}_{p_i}) \simeq \omega_X|_{p_i} \simeq \zeta(i)$, hence,

$$F'|_{p_i} \simeq \left(\oplus_{j' \neq j} V_{j'} \otimes \zeta(i)^{j'} \right) \oplus V_j \otimes \zeta(i)^{j+1}.$$

It is clear that using a sequence of transformations of this form we can pass from F to a vector bundle for which all fibers $F|_{p_i}$ have trivial G_i-action. It remains to check that if our claim holds for F' (i.e., there exists a G-surjection from a direct sum of G-equivariant line bundles to F') then the same is true for F. To this end we observe that for every $n \in \mathbb{Z}$ there exists a surjection of G-sheaves

$$\omega_X^j \otimes \pi^* \mathcal{O}_{\mathbb{P}^1}(n) \to \zeta(i)_{D_i}^j.$$

If n is sufficiently negative then this map lifts to a morphism $\omega_X^j \otimes \pi^* \mathcal{O}_{\mathbb{P}^1}(n) \to F$. Thus, from a surjection $\oplus_{i=1}^n L_i \to F'$ we obtain a surjection of the form

$$\oplus_{i=1}^n L_i \oplus \omega_X^j \otimes \pi^* \mathcal{O}_{\mathbb{P}^1}(n) \to F.$$

$\qquad\square$

In the case when $G = \mathbb{Z}/2\mathbb{Z}$ and X is an elliptic curve the above equivalence is considered in Example 5.8 of [15].

Note that the tilting bundle on $C(\overline{m}, \overline{\lambda})$ constructed in [15] corresponds to the exceptional collection (1.4).

2. Holomorphic bundles on toric orbifolds and derived categories of G-sheaves

2.1. Remarks on B_θ-modules and holomorphic bundles.

It is clear that a B_θ-module is finitely generated iff it is finitely generated as an A_θ-module. We claim that projectivity also can be checked over A_θ.

LEMMA 2.1. *Let M be a right B_θ-module. Then M is projective as a B_θ-module iff it is projective as an A_θ-module.*

PROOF. . The "only if" part is clear. Let M be a right B_θ-module, projective over A_θ. Then we have a natural surjection of B_θ-modules $p : M \otimes_{A_\theta} B_\theta \to M$ given by the action of B_θ. On the other hand, it is easy to check that the map

$$s : M \to M \otimes_{A_\theta} B_\theta : m \mapsto \frac{1}{|G|} \sum_{g \in G} mg \otimes g^{-1}$$

commutes with the right action of B_θ. Since $p \circ s = \mathrm{id}_M$, we derive that M is a direct summand in the projective B_θ-module $M \otimes_{A_\theta} B_\theta$. Hence, M itself is a projective B_θ-module. □

Thus, we can identify holomorphic bundles on $T_{\theta,\tau}/G$ with G-equivariant holomorphic bundles on $T_{\theta,\tau}$. Here is a more precise statement. Let us define an automorphism g^* of the category $\mathrm{Hol}(T_{\theta,\tau})$ by setting $g^*(P, \overline{\nabla}) = (P^g, \varepsilon(g)\overline{\nabla})$, where $P^g = P$ as a vector space but the A_θ-module structure is changed by the automorphism g of A_θ. The fact that we again obtain a holomorphic bundle on $T_{\theta,\tau}$ follows from (0.1).

LEMMA 2.2. *The category $\mathrm{Hol}(T_{\theta,\tau}/G)$ is equivalent to the category of G-equivariant objects of $\mathrm{Hol}(T_{\theta,\tau})$.*

PROOF. . By Lemma 2.1 a holomorphic bundle on $T_{\theta,\tau}/G$ is given by a finitely generated projective right A_θ-modules P equipped with a holomorphic structure $\overline{\nabla}$ and an action of G such that

$$g(f \cdot a) = g(f) \cdot g(a), \quad g \circ \overline{\nabla} = \varepsilon(g)\overline{\nabla} \circ g,$$

where $g \in G$, $f \in P$, $a \in A_\theta$. This immediately implies the assertion. □

PROPOSITION 2.3. *Every finitely generated projective B_θ-module admits a holomorphic structure.*

PROOF. . Let P be such a module. Considering P as an A_θ-module we can equip it with a holomorphic structure $\overline{\nabla}$ making it into a holomorphic bundle on T_θ (because P is a direct sum of basic modules and every basic module admits a standard holomorphic structure, see [23]). Now replace $\overline{\nabla}$ with

$$\frac{1}{|G|} \sum_{g \in G} \epsilon(g)^{-1} g \overline{\nabla} g^{-1}.$$

This new structure is compatible with the action of G, so that P becomes a holomorphic bundle on $T_{\theta,\tau}/G$. □

2.2. Generalities on torsion theory. Recall (see [16]) that a *torsion pair* in an exact category $\mathcal{C}$ is a pair of full subcategories $(\mathcal{T}, \mathcal{F})$ in $\mathcal{C}$ such that $\mathrm{Hom}(T, F) = 0$ for every $T \in \mathcal{T}$, $F \in \mathcal{F}$, and every object $C \in \mathcal{C}$ fits into a short exact triple

$$0 \to T \to C \to F \to 0$$

with $T \in \mathcal{T}$ and $F \in \mathcal{F}$. Note that if $(\mathcal{T}, \mathcal{F})$ is a torsion pair then $\mathcal{F}$ (resp., $\mathcal{T}$) coincides with the right (resp., left) orthogonal of $\mathcal{T}$, i.e. with the full subcategory of objects X such that $\mathrm{Hom}(T, X) = 0$ for all $T \in \mathcal{T}$ (resp., $\mathrm{Hom}(X, F) = 0$ for all $F \in \mathcal{F}$). In particular $\mathcal{T}$ and $\mathcal{F}$ are stable under extensions and passing to a direct summand.

It will be convenient for us to introduce a slight generalization of the notion of a torsion pair. Given a collection of full subcategories $(\mathcal{C}_1, \ldots, \mathcal{C}_n)$ in an exact category $\mathcal{C}$, let us denote by $[\mathcal{C}_1, \ldots, \mathcal{C}_n]$ the full subcategory in $\mathcal{C}$ consisting of objects C admitting an admissible filtration $0 = F_0 C \subset F_1 C \subset \ldots \subset F_n C = C$ such that $F_i C / F_{i-1} C \in \mathcal{C}_i$ for $i = 1, \ldots, n$.

Definition. A *torsion n-tuple* in an exact category $\mathcal{C}$ is a collection of full subcategories $(\mathcal{C}_1, \ldots, \mathcal{C}_n)$ such that $\mathrm{Hom}(C_i, C_j) = 0$ whenever $C_i \in \mathcal{C}_i$, $C_j \in \mathcal{C}_j$, $i < j$, and $[\mathcal{C}_1, \ldots, \mathcal{C}_n] = \mathcal{C}$.

Sometimes we will write the condition of absence of nontrivial morphisms in the above definition as $\mathrm{Hom}(\mathcal{C}_i, \mathcal{C}_j) = 0$ for $i < j$. For $n = 2$ we recover the notion of a torsion pair. Moreover, it is clear that if $(\mathcal{C}_1, \ldots, \mathcal{C}_n)$ is a torsion n-tuple then for every i the pair

$$([\mathcal{C}_1, \ldots, \mathcal{C}_i], [\mathcal{C}_{i+1}, \ldots, \mathcal{C}_n])$$

is a torsion pair. Note that the subcategories $\mathcal{C}_i$ in this definition are automatically stable under extensions. The main reason for introducing torsion n-tuples is because it is possible to substitute one such torsion tuple into another. Namely, if $(\mathcal{C}_1, \ldots, \mathcal{C}_n)$ is a torsion n-tuple in $\mathcal{C}$, and $(\mathcal{C}_{i,1}, \ldots, \mathcal{C}_{i,m})$ is a torsion m-tuple in $\mathcal{C}_i$ then

$$(\mathcal{C}_1, \ldots, \mathcal{C}_{i-1}, \mathcal{C}_{i,1}, \ldots, \mathcal{C}_{i,m}, \mathcal{C}_{i+1}, \ldots, \mathcal{C}_n)$$

is a torsion $(n + m - 1)$-tuple in $\mathcal{C}$.

If $\mathcal{C}$ is abelian then a torsion pair $(\mathcal{T}, \mathcal{F})$ defines a nondegenerate t-structure on the derived category $D^b(\mathcal{C})$ with the heart

$$\mathcal{C}^p := \{K \in D^b(\mathcal{C}) : H^i(K) = 0 \text{ for } i \neq 0, -1, H^0(K) \in \mathcal{T}, H^{-1}(K) \in \mathcal{F}\}$$

(see [16]). In other words,

$$\mathcal{C}^p = [\mathcal{F}[1], \mathcal{T}],$$

where for a pair of full subcategories $\mathcal{C}_1, \mathcal{C}_2$ in a triangulated category $\mathcal{D}$ we denote by $[\mathcal{C}_1, \mathcal{C}_2]$ the full subcategory in $\mathcal{D}$ consisting of objects K that fit into an exact triangle

$$C_1 \to K \to C_2 \to C_1[1]$$

with $C_1 \in \mathcal{C}_1$, $C_2 \in \mathcal{C}_2$. The process of passing from $\mathcal{C}$ to $\mathcal{C}^p$ is called *tilting* (also, we will call $\mathcal{C}^p$ a *tilt* of $\mathcal{C}$). Note that $(\mathcal{F}[1], \mathcal{T})$ is a torsion pair in $\mathcal{C}^p$ and applying tilting to this pair we pass back to $\mathcal{C}$. If $(\mathcal{C}_1, \ldots, \mathcal{C}_n)$ is a torsion n-tuple in $\mathcal{C}$ then we set

$$[\mathcal{C}_{i+1}[1], \ldots, \mathcal{C}_n[1], \mathcal{C}_1, \ldots, \mathcal{C}_i] := [[\mathcal{C}_{i+1}, \ldots, \mathcal{C}_n][1], [\mathcal{C}_1, \ldots, \mathcal{C}_i]] \subset \mathcal{D}^b(\mathcal{C}),$$

where $([\mathcal{C}_1, \ldots, \mathcal{C}_i], [\mathcal{C}_{i+1}, \ldots, \mathcal{C}_n])$ is the corresponding torsion pair in $\mathcal{C}$.

The main example relevant for us is the torsion pair $(\mathrm{Coh}_{>\theta}(X), \mathrm{Coh}_{<\theta}(X))$ in the category $\mathrm{Coh}(X)$ of coherent sheaves on a smooth projective curve X, associated with an irrational number θ. Recall that the *slope* of a vector bundle V on X is the ratio $\mu(V) := \deg(V)/\mathrm{rk}(V)$. Every vector bundle V on X is equipped with the *Harder-Narasimhan filtration* (introduced in [17]) $0 = V_0 \subset V_1 \subset \ldots \subset V_n = V$ such that all successive quotients V_i/V_{i-1} are semistable bundles and their slopes are strictly decreasing: $\mu(V_i/V_{i-1}) > \mu(V_{i+1}/V_i)$. This construction can be extended to coherent sheaves if we agree that torsion sheaves are semistable of slope $+\infty$. By definition, $\mathrm{Coh}^{<\theta}(X) \subset \mathrm{Coh}(X)$ (resp., $\mathrm{Coh}^{>\theta}(X) \subset \mathrm{Coh}(X)$) consists of all coherent sheaves F on X such that all successive quotients in the Harder-Narasimhan filtration of F have slope $< \theta$ (resp., $> \theta$). Note that these torsion pairs arise in connection with stability structures on $D^b(X)$ (see [2]).

2.3. Fourier-Mukai transform for noncommutative two-tori. Let

$$\mathcal{C}^\theta(E) = [\mathrm{Coh}^{<\theta}(E)[1], \mathrm{Coh}^{>\theta}(E)] \subset D^b(E)$$

be the tilt of the category of coherent sheaves on the elliptic curve $E = \mathbb{C}/(\mathbb{Z}\tau + \mathbb{Z})$ associated with θ. We know from [23],[22] that $\mathrm{Hol}(T_{\theta,\tau})$ is equivalent to $\mathcal{C}^\theta(E)$. In this section we will show that the construction of this equivalence can be adjusted to be compatible with the action of a finite group G.

Recall that the equivalence is given by a version of the Fourier-Mukai transform (see [23], Section 3.3) . With a holomorphic vector bundle $(P, \overline{\nabla})$ on $T_{\theta,\tau}$ this transform associates the complex $\mathcal{S}(P, \overline{\nabla})$ of $\mathcal{O}$-modules on E of the form $d : P_E \to P_E$ concentrated in degrees $[-1, 0]$, where P_E is obtained by descending the sheaf of holomorphic E-valued functions over $\mathbb{C}$ using an action of $\mathbb{Z}^2$ of the form

$$\rho_v(f)(z) = \exp(\pi i \theta c_v(z)) f(z + v) U_v, \ v \in \mathbb{Z}^2,$$

and the differential d is induced by the operator

$$f(z) \mapsto \overline{\nabla}(f(z)) + 2\pi i z f(z).$$

Here $(c_v(z))$ is a collection of holomorphic functions on $\mathbb{C}$ numbered by $\mathbb{Z}^2$ satisfying the condition

$$(2.1) \qquad c_{v_1}(z) + c_{v_2}(z + v_1) - c_{v_1+v_2}(z) = \det(v_1, v_2).$$

Note that in [23] we made one possible choice of $(c_v(z))$, however, it is not the only choice. In fact, one can easily see that the equivalences corresponding to different choices of $(c_v(z))$ differ by tensoring with a holomorphic line bundle on E. One of possible solutions of (2.1) is

$$c^0_{m\tau+n}(z) = -2mz - m(m\tau + n).$$

It follows from Proposition 3.7 of [23] that $\mathcal{S}$ is an equivalence of $\mathrm{Hol}(T_{\theta,\tau})$ with $\mathcal{C}^\theta(E)$ (note that the definition above differs from that of [23] by the shift of degree).

Now let us assume that a finite group G acts on the elliptic curve E by automorphisms (preserving 0). This means that G is a subgroup in $\mathbb{C}^*$ and multiplication by elements of G preserves the lattice $\mathbb{Z}\tau + \mathbb{Z} \subset \mathbb{C}$. Identifying $\mathbb{Z}^2$ with $\mathbb{Z}\tau + \mathbb{Z}$ by $(m, n) \mapsto m\tau + n$ we can view G also as a subgroup in $\mathrm{SL}_2(\mathbb{Z})$. One can immediately check that the corresponding action of G on A_θ satisfies (0.1) with $\varepsilon(g) = g^{-1} \in \mathbb{C}^*$.

Hence, for every $g \in G$ we have the corresponding automorphism g^* of the category $\mathrm{Hol}(T_{\theta,\tau})$ (see section 2.1). Let us make a G-invariant choice of $(c_v(z))$ by setting

$$c_v(z) = \frac{1}{|G|} \sum_{g \in G} c^0_{gv} gz,$$

so that $c_{gv}(gz) = c_v(z)$ for all $g \in G$. Then the resulting Fourier-Mukai transform S is compatible with the action of G in the standard way.

PROPOSITION 2.4. *With the above choice of $(c_v(z))$ one has natural isomorphisms of functors*

$$S \circ g^* \simeq (g^{-1})^* \circ S$$

from $\mathrm{Hol}(T_{\theta,\tau})$ to $\mathcal{C}^\theta(E)$, where $g \in G$.

PROOF. . By definition $g^*(P, \overline{\nabla}) = (P^g, \varepsilon(g)\overline{\nabla})$. Hence, $Sg^*(P, \overline{\nabla}) = [d_1 : P^g_E \to P^g_E)]$ where P^g_E is obtained from the action of $\mathbb{Z}^2$ on $P_{\mathbb{C}}$ given by

$$f(z) \mapsto \exp(\pi i \theta c_v(z)) f(z + v) U_{gv},$$

and the differential d_1 is induced by the operator

$$f(z) \mapsto \varepsilon(g)\overline{\nabla}(f(z)) + 2\pi i z f(z).$$

On the other hand, $(g^{-1})^* S(P, \overline{\nabla})$ is given by the complex $[d_2 : (g^{-1})^* P_E \to (g^{-1})^* P_E]$, where $(g^{-1})^* P_E$ is obtained from the action of $\mathbb{Z}^2$ on $P_{\mathbb{C}}$ given by

$$f(z) \mapsto \exp(\pi i \theta c_v(gz)) f(z + g^{-1}v) U_v,$$

and d_2 is induced by the operator

$$f(z) \mapsto \overline{\nabla}(f(z)) + 2\pi i g z f(z),$$

where we view g as an element of $\mathbb{C}^*$. Making a change of variables $v \mapsto gv$ we can identify two $\mathbb{Z}^2$-actions above, and hence we can identify with P^g_E with $(g^{-1})^* P_E$. Since $\varepsilon(g) = g^{-1}$, under this identification $d_1 = \varepsilon(g) d_2$, so we get the required isomorphism. $\qquad\square$

2.4. Proof of Theorem 0.1. From Lemma 2.2 and Proposition 2.4 we obtain that the category $\mathrm{Hol}(T_{\theta,\tau}/G)$ is equivalent to the category of G-equivariant objects of $\mathcal{C}^\theta(E)$.

Note that the Harder-Narasimhan filtration of a G-sheaf is stable under the action of G and hence, can be considered as a filtration in $\mathrm{Coh}_G(E)$. Therefore, we can define a torsion theory $((\mathrm{Coh}^{\geq \theta}_G(E), \mathrm{Coh}^{\leq \theta}_G(E))$ in $\mathrm{Coh}_G(E)$, where $\mathrm{Coh}^*_G(E)$ consists of G-sheaves F such that after forgetting the G-structure we have $F \in \mathrm{Coh}^*(E)$. Let

$$\mathcal{C}^\theta_G(E) = [\mathrm{Coh}^{\leq \theta}_G(E)[1], \mathrm{Coh}^{\geq \theta}_G(E)] \subset D^b_G(E)$$

be the corresponding tilted abelian subcategory. By Lemma 1.1 the category $\mathcal{C}^\theta_G(E)$ is equivalent to the category of G-equivariant objects on $\mathcal{C}^\theta(E)$, and hence, to $\mathrm{Hol}(T_{\theta,\tau}/G)$.

Let us show that $D^b_G(E)$ is equivalent to $D^b(\mathcal{C}^\theta_G(E))$. By Proposition 5.4.3 of [4] it is enough to check that that our torsion pair in $\mathrm{Coh}_G(E)$ is cotilting, i.e., for every G-sheaf F on E there exists a G-equivariant vector bundle $V \in \mathrm{Coh}^{<\theta}(E)$ and a G-equivariant surjection $V \to F$. However, this is clear since for every G-sheaf F there is a surjection

$$H^0(X, F \otimes \pi^* \mathcal{O}_{\mathbb{P}^1}(n)) \otimes \pi^* \mathcal{O}_{\mathbb{P}^1}(-n) \to F,$$

where n is large enough (and the space of global sections is equipped with the natural G-action).

Thus, we showed that $\mathrm{Hol}(T_{\theta,\tau})/G$ is abelian and its derived category is equivalent to $D^b_G(E)$. It remains to apply Corollary 1.4 to the covering $\pi : E \to E/G \simeq \mathbb{P}^1$. The statement about the number of vertices follows from the explicit description of these coverings in section 1.3.

2.5. Proof of Theorem 0.2. Using Theorem 0.1 and Corollary 1.6 we obtain an isomorphism

$$K_0(\mathrm{Hol}(T_{\theta,\tau}/G)) \simeq K_0(D^b_G(E)) \simeq \mathbb{Z}^r,$$

where $r = 6, 8, 9, 10$ for $G = \mathbb{Z}/m\mathbb{Z}$ with $m = 2, 3, 4, 6$, respectively. Now we observe that by Proposition 2.3 the natural homomorphism

$$K_0(\mathrm{Hol}(T_{\theta,\tau}/G)) \to K_0(B_\theta)$$

is surjective. To prove that this map is an isomorphism, it is enough to check that the rank of $K_0(B_\theta)$ is at least r. This was done in [8], [29] and [12] for $m = 2$, $m = 4$, and $m = 3, 6$, respectively (by explicitly constructing r elements in $K_0(B_\theta)$ and using unbounded traces to check their linear independence).

This result was known for $G = \mathbb{Z}/2\mathbb{Z}$ (see [18]), however, with a different proof. For $G = \mathbb{Z}/4\mathbb{Z}$ and $G = \mathbb{Z}/6\mathbb{Z}$ it was known for θ in a dense G_δ-set (see [30] and [13]). The case of $G = \mathbb{Z}/4\mathbb{Z}$ and general θ was done by Lueck, Walters and Phillips (unpublished).

Note that from the above proof we also get the following

COROLLARY 2.5. *The natural homomorphism* $K_0(\mathrm{Hol}(T_{\theta,\tau}/G)) \to K_0(B_\theta)$ *is an isomorphism. Moreover, the positive cones are the same.*

Remark. In [28] it was shown that for $G = \mathbb{Z}/2\mathbb{Z}$ the positive cone in $K_0(B_\theta)$ coincides with the preimage of the positive cone in $K_0(A_\theta)$ under the natural homomorphism $K_0(B_\theta) \to K_0(A_\theta)$ (in other words, it consists of all elements $x \in K_0(B_\theta)$ such that $\mathrm{tr}_*(x) > 0$, where $\mathrm{tr}_* : K_0(B_\theta) \to \mathbb{R}$ is the homomorphism induced by the trace). As was pointed to us by Chris Phillips, similar statement is also known to hold for other groups G. Namely, it follows from the fact that the corresponding crossed products are simple AH algebras with slow dimension growth and real rank zero (see Theorems 8.11 and 9.10 of [21]).

2.6. Tiltings associated with θ. Let $\pi : X \to \mathbb{P}^1$ be a ramified Galois covering with the Galois group G, and let $D^b_G(X)$ be the derived category of G-sheaves on X. As in the above proof of Theorem 0.1 we can define the torsion pair $(\mathrm{Coh}_G^{>\theta}(X), \mathrm{Coh}_G^{\leq\theta}(X))$ in $\mathrm{Coh}_G(X)$ associated with an irrational number θ. Our goal in this section is to describe the image of the corresponding tilted abelian subcategory

$$\mathcal{C}^\theta_G(X) := [\mathrm{Coh}_G^{\leq\theta}(X)[1], \mathrm{Coh}_G^{>\theta}(X)] \subset D^b_G(X)$$

under the equivalence Φ_n of Corollary 1.4 (for suitable n).

Let us start by describing the torsion pair in $\mathrm{Coh}_G(X)$ giving rise to the t-structure on $D^b_G(X)$ associated with Φ_n. By definition, the heart $\mathcal{M}_n$ of this t-structure consists of objects F such that $\mathrm{Hom}^i_G(V_n, F) = 0$ for $i \neq 0$.

PROPOSITION 2.6. *Let $\mathcal{T}_0 \subset \mathrm{Coh}_G(X)$ denote the full subcategory consisting of all G-sheaves isomorphic to a direct sum of G-sheaves from the collection*

$$(\mathcal{O}_{(m_1-1)D_1}, \ldots, \mathcal{O}_{D_1}, \ldots, \mathcal{O}_{(m_r-1)D_r}, \ldots, \mathcal{O}_{D_r}).$$

Let also $\mathcal{T}_1 \subset \mathrm{Coh}_G(X)$ be the full subcategory of torsion G-sheaves obtained by successive extensions from simple G-sheaves of the form $\zeta(i)_{D_i}^a$, where $i = 1, \ldots, r$, $a = 1, \ldots, m_i - 1$ (so $\mathcal{O}_{D_i}$ are not included). Then for every $n \in \mathbb{Z}$ we have a torsion quadruple

$$(\mathcal{T}_0, \pi^* \mathrm{Coh}^{\geq n}(\mathbb{P}^1), \pi^* \mathrm{Coh}^{\leq n-1}(\mathbb{P}^1), \mathcal{T}_1)$$

in $\mathrm{Coh}_G(X)$. Furthermore, we have the equality of abelian subcategories in $D_G^b(X)$

$$(2.2) \qquad \mathcal{M}_n = [\pi^* \mathrm{Coh}^{\leq n-1}(\mathbb{P}^1)[1], \mathcal{T}_1[1], \mathcal{T}_0, \pi^* \mathrm{Coh}^{\geq n}(\mathbb{P}^1)].$$

PROOF. . First, let us check that

$$(\mathcal{T}_0, \pi^* \mathrm{Coh}(\mathbb{P}^1), \mathcal{T}_1)$$

is a torsion triple in $\mathrm{Coh}_G(X)$. The conditions

$$\mathrm{Hom}_G(\mathcal{T}_0, \mathcal{T}_1) = 0, \ \mathrm{Hom}_G(\mathcal{T}_0, \pi^* \mathrm{Coh}(\mathbb{P}^1)) = 0, \ \text{and} \ \mathrm{Hom}(\pi^* \mathrm{Coh}(\mathbb{P}^1), \mathcal{T}_1) = 0$$

easily follow from the vanishings

$$\mathrm{Hom}_G(\mathcal{O}_{bD_i}, \zeta(i)_{D_i}^a) = 0, \qquad \mathrm{Hom}_G(\mathcal{O}_{bD_i}, \pi^* \mathcal{O}_{\pi(p_i)}) = 0,$$

$$\text{and} \qquad \mathrm{Hom}_G(\pi^* \mathrm{Coh}(\mathbb{P}^1), \zeta(i)_{D_i}^a) = 0,$$

where $a = 1, \ldots, m_i - 1$, $b = 1, \ldots, m_i - 1$. Using elementary transformations along D_i's as in the proof of Theorem 1.8 one can easily see that every G-bundle on X belongs to $[\pi^* \mathrm{Coh}(\mathbb{P}^1), \mathcal{T}_1]$. Now let F be an indecomposable torsion G-sheaf on X supported on D_i. Then there exists a filtration

$$0 = F_0 \subset F_1 \subset \ldots \subset F_n = F$$

by G-subsheaves such that $F_a/F_{a-1} \simeq \zeta(i)_{D_i}^{c-a}$ for $a = 1, \ldots, n$ (where $c \in \mathbb{Z}/m_i\mathbb{Z}$). If $c - a \not\equiv 0 \bmod(m_i)$ for all $a = 1, \ldots, n$ then $F \in \mathcal{T}_1$. Otherwise, let a_1 (resp., a_2) be the minimal (resp., maximal) a such that $c - a \equiv 0 \bmod(m_i)$. Then it is easy to see that

$$F_{a_1} \simeq \mathcal{O}_{a_1 D_i} \in \mathcal{T}_0, \ F_{a_2}/F_{a_1} \in \pi^* \mathrm{Coh}(\mathbb{P}^1), \ F/F_{a_2} \in \mathcal{T}_1,$$

and hence, $F \in [\mathcal{T}_0, \pi^* \mathrm{Coh}(\mathbb{P}^1), \mathcal{T}_1]$.

Substituting the torsion pair

$$(\pi^* \mathrm{Coh}^{\geq n}(\mathbb{P}^1), \pi^* \mathrm{Coh}^{\leq n-1}(\mathbb{P}^1))$$

into $\pi^* \mathrm{Coh}(\mathbb{P}^1)$ we obtain the required torsion quadruple. One can immediately check that each of the subcategories $\pi^* \mathrm{Coh}^{\leq n-1}(\mathbb{P}^1)[1], \mathcal{T}_1[1], \mathcal{T}_0$ and $\pi^* \mathrm{Coh}^{\geq n}(\mathbb{P}^1)$ belongs to $\mathcal{M}_n$. Hence, the RHS of (2.2) is contained in $\mathcal{M}_n$. Since both subcategories are hearts of nondegenerate t-structures, this implies the required equality. $\square$

Note that $\pi^* \mathrm{Coh}^{\geq n}(\mathbb{P}^1) \subset \mathrm{Coh}_{\widetilde{G}}^{\geq 2n}(X)$ and $\pi^* \mathrm{Coh}_{\leq n-1}(\mathbb{P}^1) \subset \mathrm{Coh}_{\widetilde{G}}^{\leq 2n-2}(X)$. Hence, the subcategories $\mathcal{M}_n$ and $\mathcal{C}_G^\theta(X)$ have a large intersection provided $2n-2 < \theta < 2n$, i.e. $n = [\theta/2] + 1$. Let us show that in this case these categories are related by tilting.

PROPOSITION 2.7. *Set* $n = [\theta/2] + 1$. *Then we have the following torsion quadruple in* $\mathcal{C}_G^\theta(X)$:

$$(2.3) \quad (\operatorname{Coh}_G^{\leq\theta}(X)[1], \mathcal{T}_0, \pi^* \operatorname{Coh}^{\geq n}(\mathbb{P}^1), [\pi^* \operatorname{Coh}^{\leq n-1}(\mathbb{P}^1), \mathcal{T}_1] \cap \operatorname{Coh}_G^{\geq\theta}(X)).$$

Furthermore, we have
$$(2.4)$$
$$\mathcal{M}_n = [[\pi^* \operatorname{Coh}^{\leq n-1}(\mathbb{P}^1)[1], \mathcal{T}_1[1]] \cap \operatorname{Coh}_G^{\geq\theta}(X)[1], \operatorname{Coh}_G^{\leq\theta}(X)[1], \mathcal{T}_0, \pi^* \operatorname{Coh}^{\geq n}(\mathbb{P}^1)].$$

PROOF. . First, we observe that

$$(\mathcal{T}_0, \pi^* \operatorname{Coh}^{\geq n}(\mathbb{P}^1), [\pi^* \operatorname{Coh}^{\leq n-1}(\mathbb{P}^1), \mathcal{T}_1] \cap \operatorname{Coh}_G^{\geq\theta}(X))$$

is a torsion triple in $\operatorname{Coh}_G^{\geq\theta}(X)$. Indeed, this follows immediately from Proposition 2.6 and from the fact that the subcategory $\operatorname{Coh}_G^{\geq\theta}(X) \subset \operatorname{Coh}_G(X)$ is stable under passing to quotients. Substituting this triple into the standard torsion pair $(\operatorname{Coh}_G^{\leq\theta}(X)[1], \operatorname{Coh}_G^{\geq\theta}(X))$ we obtain the torsion quadruple (2.3). It remains to check that all the constituents in the RHS of (2.4) belong to $\mathcal{M}_n$. For most of them this follows from (2.2). The remaining inclusion $\operatorname{Coh}_G^{\leq\theta}(X)[1] \subset \mathcal{M}_n$ is implied by the fact that $\pi^*\mathcal{O}_{\mathbb{P}^1}(n)$ and $\pi^*\mathcal{O}_{\mathbb{P}^1}(n+1)$ have slope $\geq 2n > \theta$. $\square$

In conclusion we are going to interpret the torsion pair in $\mathcal{M}_n$ arising in the above proposition in terms of right modules over the algebra $\Lambda(\bar\lambda, \bar m)$ (see Corollary 1.4) assuming that X is an elliptic curve.

THEOREM 2.8. *Assume that* $\pi : E \to \mathbb{P}^1$ *is a ramified Galois covering with the Galois group* G, *where* E *is an elliptic curve, and let* $\bar m, \bar\lambda$ *be the associated ramification data. Fix an irrational number* θ *and set* $n = [\theta/2] + 1$. *Let us define full subcategories* $\mathcal{T}^\theta, \mathcal{F}^\theta \subset \operatorname{mod} -\Lambda(\bar\lambda, \bar m)$ *as follows:* $\mathcal{T}^\theta$ *(resp.,* $\mathcal{F}^\theta$*) consists of all modules* $M \simeq \oplus_{i=1}^k M_i$, *where* M_i *are indecomposable and* $\deg_n(M) - \theta \operatorname{rk}(M) < 0$ *(resp.,* $\deg_n(M) - \theta \operatorname{rk}(M) > 0$*). Then* $(\mathcal{T}^\theta, \mathcal{F}^\theta)$ *is a torsion pair in* $\operatorname{mod} -\Lambda(\bar\lambda, \bar m)$ *and one has*

$$\Phi_n(\mathcal{C}_G^\theta(E)) = [\mathcal{F}^\theta, \mathcal{T}^\theta[-1]] \subset D^b(\Lambda(\bar\lambda, \bar m)).$$

PROOF. . From Proposition 2.7 we know that $\mathcal{C}_G^\theta(E) = [\mathcal{F}, \mathcal{T}[-1]]$ for the torsion pair $(\mathcal{T}, \mathcal{F})$ in $\mathcal{M}_n = \Phi_n^{-1}(\operatorname{mod} -\Lambda(\bar\lambda, \bar m))$ given by

$$\mathcal{T} = [\pi^* \operatorname{Coh}^{\leq n-1}(\mathbb{P}^1)[1], \mathcal{T}_1[1]] \cap \operatorname{Coh}_G^{\geq\theta}(E)[1], \quad \mathcal{F} = [\operatorname{Coh}_G^{\leq\theta}(E)[1], \mathcal{T}_0, \pi^* \operatorname{Coh}^{\geq n}(\mathbb{P}^1)].$$

We claim that one has $\operatorname{Ext}_{\mathcal{M}_n}^1(F, T) = 0$ for every $F \in \mathcal{F}$ and $T \in \mathcal{T}$. It suffices to check that $\operatorname{Hom}_G(F, T[1]) = 0$ for $T \in \mathcal{T}$ in the following three cases: (i) $F \in \operatorname{Coh}_G^{\leq\theta}(E)[1]$; (ii) $F \in \mathcal{T}_0$; (iii) $F \in \pi^* \operatorname{Coh}^{\geq n}(\mathbb{P}^1)$. Note that in cases (ii) and (iii) this is clear since cohomological dimension of $\operatorname{Coh}_G(E)$ is equal to 1. In case (i) we obtain by Serre duality (using triviality of ω_E)

$$\operatorname{Hom}(F, T[1])^* \simeq \operatorname{Hom}(T, F) = 0,$$

since $T \in \operatorname{Coh}^{>\theta}(E)[1]$ and $F \in \operatorname{Coh}^{<\theta}(E)[1]$.

It follows that every indecomposable object of $\mathcal{C}_G^\theta(E)$ is contained either in $\mathcal{T}$ or in $\mathcal{F}$. Therefore, $\mathcal{T}$ (resp., $\mathcal{F}$) coincides with the full subcategory of objects F such that $F \simeq \oplus_{i=1}^n F_i$, where F_i are indecomposable objects and $F_i \in \mathcal{T}$ (resp., $F_i \in \mathcal{F}$). Since $\deg(C) - \theta \operatorname{rk}(C) > 0$ for $C \in \mathcal{C}^\theta(E)$, it follows that $\deg(F) - \theta \operatorname{rk}(F) > 0$ for $F \in \mathcal{F}$ and $\deg(T) - \theta \operatorname{rk}(T) < 0$ for $T \in \mathcal{T}$. Taking into account Lemma 1.5 we derive that $\mathcal{T}^\theta = \Phi_n(\mathcal{T})$ and $\mathcal{F}^\theta = \Phi_n(\mathcal{F})$. $\square$

Acknowledgment. I am grateful to Julian Buck, Igor Burban, Pavel Etingof, Helmut Lenzing, Tony Pantev, Chris Phillips, Olivier Schiffman and Samuel Walters for useful discussions. During the first stages of work on this paper the author enjoyed hospitality of the Mathematisches Forschungsinstitut Oberwolfach. I'd like to thank this institution for providing excellent working conditions and a stimulating atmosphere. The author is supported in part by NSF grant.

References

[1] R. Bezrukavnikov, *Perverse coherent sheaves (after Deligne)*, preprint math.AG/0005152.

[2] T. Bridgeland, *Stability conditions on triangulated categories*, preprint math.AG/0212237.

[3] A. Bondal, *Representations of associative algebras and coherent sheaves*, Math. USSR Izvestiya 34 (1990), 23–42).

[4] A. Bondal, M. Van den Bergh, *Generators and representability of functors in commutative and noncommutative geometry*, Moscow Math. J. 3 (2003), 1–36.

[5] A. Bondal, M. Kapranov, *Representable functors, Serre functors, and mutations*, Math. USSR Izvestiya 35 (1990), 519–541.

[6] A. Bondal, M. Kapranov, *Framed triangulated categories*, Math. USSR-Sbornik 70 (1991), 93–107.

[7] A. Bondal, D. Orlov, *Semiorthogonal decompositions for algebraic varieties*, preprint alg-geom/9506012.

[8] O. Bratteli, G. Elliott, D. Evans, A. Kishimoto, *Non-commutative spheres. I*, Internat. J. Math. 2 (1991), 139–166.

[9] O. Bratteli, G. Elliott, D. Evans, A. Kishimoto, *Non-commutative spheres. II. Rational rotations*, J. Operator Theory 27 (1992), 53–85.

[10] O. Bratteli, A. Kishimoto, *Non-commutative spheres. III. Irrational rotations*, Comm. Math. Phys. 147 (1992), 605–624.

[11] T. Bridgeland, A. King, M. Reid, *The McKay correspondence as an equivalence of derived categories*, J. AMS 14 (2001), 535–554.

[12] J. Buck, S. Walters, *Connes-Chern characters of hexic and cubic modules*, preprint 2004, 43 pages.

[13] J. Buck, S. Walters, *Non commutative spheres associated with the hexic transform and their K-theory*, preprint 2004, 28 pages.

[14] P. Gabriel, *Indecomposable representations. II* Symposia Mathematica, XI (Convegno di Algebra Commutativa, INDAM, Rome, 1971), 81–104. Academic Press, London, 1973.

[15] W. Geigle, H. Lenzing, *A class of weighted projective curves arising in the representation theory of finite-dimensional algebras*, in *Singularities, representations of algebras and vector bundles*, Springer LNM 1273 (1987), 265–297.

[16] D. Happel, I. Reiten, S. O. Smalo, *Tilting in abelian categories and quasitilted algebras*, Memoirs AMS 575, 1996.

[17] G. Harder, M. S. Narasimhan, *On the cohomology groups of moduli spaces of vector bundles on curves*, Math. Ann. 212 (1974/75), 215–248.

[18] A. Kumjian, *On the K-theory of the symmetrized non-commutative torus*, C. R. Math. Rep. Acad. Sci. Canada 12(1990), 87–89.

[19] H. Lenzing, H. Meltzer, *Sheaves on a weighted projective line of genus one, and representations of a tubular algebra*, in *Representations of algebras (Ottawa, ON, 1992)*, 313–337, CMS Conf. Proc. 14, Providence, RI, 1993.

[20] H. Lenzing, H. Meltzer, *Tilting sheaves and concealed-canonical algebras*, in *Representation theory of algebras (Cocoyoc, 1994)*, 455–473, CMS Conf. Proc. 18, Providence, RI, 1996.

[21] C. Phillips, *Crossed products by finite cyclic group actions with the tracial Rokhlin property*, preprint math.QA/0306410.

[22] A. Polishchuk, *Classification of holomorphic bundles on noncommutative two-tori*, Documenta Math. 9 (2004), 163–181.

[23] A. Polishchuk, A. Schwarz, *Categories of holomorphic bundles on noncommutative two-tori*, Comm. Math. Phys. 236 (2003), 135–159

[24] C. M. Ringel, *Tame algebras and integral quadratic forms*, LNM 1099, 1984.

[25] A. N. Rudakov et al., *Helices and vector bundles*, Cambridge Univ. Press, 1990.

[26] O. Schiffmann, *Noncommutative projectivve curves and quantum loop algebras*, preprint math.QA/0205267.

[27] S. Térouanne, *Sur la catégorie $D^{b,G}(X)$ pour G réductif fini*, C.R.Math.Acad.Sci.Paris 336 (2003), 483–486.

[28] S. G. Walters, *Projective modules over the non-commutative sphere*, J. London Math. Soc. 51 (1995), 589–602.

[29] S. G. Walters, *Chern Characters of Fourier Modules*, Canad. J. Math. 52 (2000), 633–672.

[30] S. G. Walters, *K-theory of Non-Commutative Spheres Arising from the Fourier Automorphism*, Canad. J. Math. 53 (2001), 631–672.

A New short proof
of the local index formula of Atiyah-Singer

Raphaël Ponge

ABSTRACT. In this talk we present a new short proof of the local index formula of Atiyah-Singer for Dirac operators ([**AS1**], [**AS2**]) which, as a byproduct and unlike Getzler's short proof, allows us to compute the CM cyclic cocycle for Dirac spectral triples.

Introduction

The aim of the talk is to present the results of [**Po**] where a new short proof of the local index formula of Atiyah-Singer for Dirac operators ([**AS1**], [**AS2**]) was given. The point is that this proof is as simple as Getzler's short proof [**Ge1**] but, as a byproduct and unlike Getzler's proof, it allows us to compute the CM cocycle for Dirac spectral triples. This is interesting because:

- The probabilistic arguments in (part of) Getzler's short proof don't go through to compute the CM cocycle;

- The previous computations of the CM cocycle (Connes-Moscovici [**CM**], Chern-Hu [**CH**], Lescure [**Le**]) made use of the asymptotic ΨDO calculus used by Getzler in its first proof [**Ge2**], which we can bypass here.

The main ingredients of the proof are:

- Getzler's rescaling as in [**Ge2**];

- Greiner's approach of the heat kernel asymptotics [**Gr**].

The talk is divided into four parts. In the first one we recall Greiner's approach of the heat kernel asymptotics and in the second one we prove the Atiyah-Singer index formula. The third part reviews the framework for the local index formula in noncommutative geometry following [**CM**]. Then in the last part we compute the CM cocycle for Dirac spectral triples.

1. Greiner's approach of the heat kernel asymptotics

Here we let M^n be a compact Riemannian manifold, let $\mathcal{E}$ be a Hermitian bundle over M, and consider a second order selfadjoint elliptic differential operator $\Delta : C^\infty(M, \mathcal{E}) \to C^\infty(M, \mathcal{E})$ such that the principal symbol of Δ is positive definite. This has the effect that Δ is bounded from below and generates a bounded heat semi-group $e^{-t\Delta}$, $t \geq 0$, on $L^2(M, \mathcal{E})$. In fact, for $t > 0$ the operator $e^{-t\Delta}$ is smoothing, i.e. has a smooth distribution kernel.

On the other hand, the heat semigroup $e^{-t\Delta}$, $t \geq 0$, allows us to invert the heat equation, since if we consider the operator $Q_0 : C_c^\infty(M \times \mathbb{R}) \to \mathcal{D}'(M \times \mathbb{R})$ given by

$$(1.1) \qquad Q_0 u(x,t) = \int_0^\infty e^{-s\Delta} u(x, t-s) ds, \qquad u \in C_c^\infty(M \times \mathbb{R}, \mathcal{E}),$$

then, for $u \in C_c^\infty(M \times \mathbb{R}, \mathcal{E})$ we have

$$(1.2) \qquad (\Delta + \partial_t) Q_0 u = Q_0 (\Delta + \partial_t) u = u \quad \text{in } \mathcal{D}'(M \times \mathbb{R}, \mathcal{E}).$$

Observe also that at the level of distribution kernels (1.1) implies that:

1. Q_0 has the *Volterra property*, i.e. it has a distribution kernel of the form $K_{Q_0}(x, y, t-s)$ with $K_{Q_0}(x, y, t) = 0$ for $t < 0$.

2. For $t > 0$ we have

$$(1.3) \qquad K_{Q_0}(x, y, t) = k_t(x, y).$$

The equality (1.3) leads us to use ΨDO techniques for looking at the asymptotics of $k_t(x, x)$ as $t \to 0^+$. For achieving this aim the relevant ΨDO calculus is the Volterra calculus developed independently by Greiner [**Gr**] and Piriou [**Pi**].

In the sequel we let U be an open subset of $\mathbb{R}^n$ and let $\mathbb{C}_-$ be the complex halfplane $\{\Im \tau > 0\}$ with closure $\bar{\mathbb{C}}_-$. Then the Volterra symbols can be defined as follows.

DEFINITION 1.1. $S_{v,m}(U \times \mathbb{R}^{n+1})$, $m \in \mathbb{Z}$, *consists of smooth functions* $q_m(x, \xi, \tau)$ *on* $U_x \times (\mathbb{R}^{n+1}_{(\xi,\tau)} \setminus 0)$ *so that* $q_m(x, \xi, \tau)$ *can be extended to a smooth function on* $U_x \times [(\mathbb{R}^n_\xi \times \bar{\mathbb{C}}_-) \setminus 0]$ *in such way to be analytic with respect to* $\tau \in \mathbb{C}_-$ *and to be homogeneous of degree* m, *i.e.* $q_m(x, \lambda\xi, \lambda^2\tau) = \lambda^m q_m(x, \xi, \tau)$ *for any* $\lambda \in \mathbb{R} \setminus 0$.

In fact, Definition 1.1 is intimately related to the Volterra property, for we have:

LEMMA 1.2 ([**BGS**, Prop. 1.9]). *Any homogeneous Volterra symbol* $q(x, \xi, \tau) \in S_{v,m}(U \times \mathbb{R}^{n+1})$ *can be extended into a unique distribution*

$$g(x, \xi, \tau) \in C^\infty(U) \hat{\otimes} \mathcal{S}'(\mathbb{R}^{n+1})$$

in such way to be homogeneous with respect to the covariables (ξ, τ) *and so that* $\check{q}(x, y, t) := \mathcal{F}^{-1}_{(\xi,\tau) \to (y,t)}[g](x, y, t) = 0$ *for* $t < 0$.

DEFINITION 1.3. $S_v^m(U \times \mathbb{R}^{n+1})$, $m \in \mathbb{Z}$, *consists of the smooth functions* $q(x, \xi, \tau)$ *on* $U_x \times \mathbb{R}^{n+1}_{(\xi,\tau)}$ *that have an asymptotic expansion* $q \sim \sum_{j \geq 0} q_{m-j}$, $q_{m-j} \in S_{v,m-j}(U \times \mathbb{R}^{n+1})$, *where* $\sim$ *means that, for any integer* N *and for any compact* $K \subset U$, *there exists a constant* $C_{NK\alpha\beta k} > 0$ *such that*

$$(1.4) \qquad |\partial_x^\alpha \partial_\xi^\beta \partial_\tau^k (q - \sum_{j < N} q_{m-j})(x, \xi, \tau)| \leq C_{NK\alpha\beta k}(|\xi| + |\tau|^{1/2})^{m-|\beta|-wk-N},$$

for $x \in K$ *and* $|\xi| + |\tau|^{1/2} \geq 1$.

DEFINITION 1.4. *A Volterra* ΨDO *of order* m *is a continuous operator* Q *from* $C_c^\infty(M \times \mathbb{R}, \mathcal{E})$ *to* $C^\infty(M \times \mathbb{R}, \mathcal{E})$ *such that:*

- Q *has the Volterra property,* *i.e. it has a distribution kernel of the form* $K_Q(x, y, t-s)$ *with* $K_Q(x, y, t) = 0$ *for* $t < 0$;

- The distribution kernel of Q is smooth off the diagonal;

- Locally we can write $Q = q(x, D_x, D_t) + R$ for some Volterra symbol q of order m and some smoothing operator R.

EXAMPLE 1.5. The heat operator $\Delta + \partial_t$ is a Volterra ΨDO of order 2 and its principal symbol is $p_2(x, \xi) + i\tau$ (where p_2 denotes the principal symbol of Δ).

EXAMPLE 1.6. Let q_l be a homogeneous Volterra symbol on $U \times \mathbb{R}^n \times \mathbb{R}$ and define $q_l(x, D_x, D_t)$ to be the operator with kernel $\check{q}_l(x, x-y, t-s)$. Then $q_l(x, D_x, D_t)$ is a Volterra ΨDO of order l.

Using the Volterra ΨDO calculus we can prove:

PROPOSITION 1.7 ([**Gr**], [**Pi**]). *The heat operator $\Delta + \partial_t$ is invertible on smooth sections and its inverse is a Volterra ΨDO of order -2.*

On the other hand, we have the following.

LEMMA 1.8 ([**Gr**]). *Let $Q \in$ be a Volterra ΨDO of order m on $U \times \mathbb{R}$. Then, with asymptotics in $C^\infty(U)$, we have*

$$(1.5) \qquad K_Q(x, x, t) \sim_{t \to 0^+} t^{-(\frac{n}{2} + [\frac{m}{2}] + 1)} \sum t^l \check{q}_{2[\frac{m}{2}] - 2l}(x, 0, 1),$$

where $q \sim \sum_{j \geq 0} q_{m-j}$ denotes the symbol of Q.

Applying this to $Q = (\Delta + \partial_t)^{-1}$ and using (1.3) we get:

PROPOSITION 1.9 ([**Gr**]). *In $C^\infty(M, |\Lambda|(M) \otimes \operatorname{End}\mathcal{E})$ we have:*

$$(1.6) \qquad k_t(x, x) \sim_{t \to 0^+} t^{\frac{-n}{2}} \sum t^j a_l(\Delta)(x), \qquad a_l(\Delta)(x) = \check{q}_{-2-2l}(x, 0, 1),$$

where $q \sim \sum j \geq 0 q_{-2-j}$ denotes the symbol of any Volterra ΨDO parametrix for $\Delta + \partial_t$.

The interest of this approach is twofold. First, to determine the asymptotics of $k_t(x, x)$ at a point x_0 we only need a Volterra parametrix Q for $\Delta + \partial_t$ near x_0, for we have:

$$(1.7) \qquad k_t(x_0, x_0) = K_Q(x_0, x_0, t) + O(t^\infty).$$

Second, we can differentiate the heat kernel asymptotics as follows. Let P be a differential operator of order m and let $h_t(x, y)$ denote the kernel of $Pe^{-t\Delta}$, $t > 0$. Then:

$$(1.8) \qquad h_t = P_x k_t = P_x K_{(\Delta + \partial_t)^{-1}} = K_{P(\Delta + \partial_t)^{-1}}.$$

Therefore by applying Lemma 1.8 to the operator $P(\Delta + \partial_t)^{-1}$ we get a differentiable version of Proposition 1.9 as follows.

PROPOSITION 1.10. *With asymptotics in $C^\infty(M, |\Lambda|(M))$ we have*
$$(1.9)$$
$$h_t(x, x) \sim_{t \to 0^+} t^{-(\frac{n}{2} + [\frac{m}{2}])} \sum_{l \geq 0} t^l b_l(P, \Delta)(x), \qquad b_l(P, \Delta)(x) = \check{q}_{2[\frac{m}{2}] - 2 - 2l}(x, 0, 1),$$

where $q \sim \sum_{j \geq 0} q_{m-2-j}$ denotes the symbol of $P(\Delta + \partial_t)^{-1}$.

2. Proof of the Atiyah-Singer index formula

Let M^n be a compact Riemannian spin manifold with spin bundle $\mathcal{S} = \mathcal{S}^+ \oplus \mathcal{S}^-$, let $\mathcal{E}$ be a Hermitian bundle equipped with a unitary connection and consider the Dirac operator with coefficients in $\mathcal{E}$,

$$(2.1) \qquad \not{D}_{\mathcal{E}} : C^\infty(M, \mathcal{S} \otimes \mathcal{E}) \to C^\infty(M, \mathcal{S} \otimes \mathcal{E}).$$

Then the Atiyah-Singer index formula is the following.

THEOREM 2.1 ([**AS1**], [**AS2**]). *We have:*

$$(2.2) \qquad \operatorname{ind} \not{D}_{\mathcal{E}} = (2i\pi)^{-\frac{n}{2}} \int_M [\hat{A}(R_M) \wedge \operatorname{Ch}(F^{\mathcal{E}})]^{(n)},$$

where $\hat{A}(R^M) = \det^{\frac{1}{2}}\left(\frac{R^M/2}{\sinh(R^M/2)}\right)$ *is the $\hat{A}$-form of the Riemann curvature R^M and* $\operatorname{Ch}(F^{\mathcal{E}}) = \operatorname{Tr} \exp(-F^{\mathcal{E}})$ *is the Chern form of the curvature $F^{\mathcal{E}}$ of $\mathcal{E}$.*

Let $k_t(x, y)$ denote the heat kernel of $\not{D}_{\mathcal{E}}^2$. Then by the McKean-Singer formula we have

$$(2.3) \qquad \operatorname{ind} \not{D}_{\mathcal{E}} = \int \operatorname{Str} k_t(x, x) \qquad \text{for any } t > 0,$$

where $\operatorname{Str} = \operatorname{Tr}_{\mathcal{S}^+ \otimes \mathcal{E}} - \operatorname{Tr}_{\mathcal{S}^- \otimes \mathcal{E}}$ denotes the supertrace. Therefore, the index formula of Atiyah-Singer follows from:

THEOREM 2.2. *As $t \to 0^+$ we have*

$$(2.4) \qquad \operatorname{Str} k_t(x, x) = [\hat{A}(R_M) \wedge \operatorname{Ch}(F^{\mathcal{E}})]^{(n)} + \mathrm{O}(t),$$

with $\mathrm{O}(t)$ in $C^\infty(M, |\Lambda|(M))$.

This theorem, also called Local Index Theorem, was first proved by Patodi, Gilkey and Atiyah-Bott-Patodi ([**ABP**], [**Gi**]), and then in a purely analytic fashion by Getzler ([**Ge1**], [**Ge2**]) and Bismut [**Bi**] (see also [**BGV**], [**Ro**]). Moreover, as it is a purely local statement it holds *verbatim* for (geometric) Dirac operators acting on a Clifford bundle. Thus it allows us to recover, on the one hand, the Gauss-Bonnet, signature and Riemann-Roch theorems ([**ABP**], [**BGV**], [**LM**], [**Ro**]) and, on the other hand, the full index theorem of Atiyah-Singer ([**ABP**], [**LM**]).

PROOF OF THEOREM 2.2. First, it is enough to prove Theorem 2.2 at a point x_0. Furthermore, to this end we only need a Volterra parametrix Q for $\not{D}_{\mathcal{E}}^2 + \partial_t$ in local trivializing coordinates near x_0, for we have

$$(2.5) \qquad k_t(x_0, x_0) = K_Q(x_0, x_0, t) + \mathrm{O}(t^\infty).$$

In fact, we can use normal coordinates centered at x_0 and trivializations of $\mathcal{S}$ and $\mathcal{E}$ by means of parallel translation along the geodesics out of the origin. This allows us to replace $\not{D}_{\mathcal{E}}$ by a Dirac operator $\not{D}$ on $\mathbb{R}^n$ acting on the trivial bundle with fiber $\mathcal{S}_n \otimes \mathbb{C}^p$, where $\mathcal{S}_n$ is the space of spinors on $\mathbb{R}^n$.

Now, recall that as algebra $\operatorname{End} \mathcal{S}_n$ can be identified with the Clifford algebra $\operatorname{Cl}(n)$. The latter is isomorphic to the exterior algebra $\Lambda(n) = \oplus_{i=0}^n \Lambda^i(n)$,

$$(2.6) \qquad \Lambda(n) \xrightarrow{c} \operatorname{Cl}(n) \xrightarrow{\sigma=c^{-1}} \Lambda(n),$$

so that for $\xi, \eta \in \Lambda(n)$ we have

$$(2.7) \qquad \sigma[c(\xi^{(i)})c(\eta^{(j)})] = \xi^{(i)} \wedge \eta^{(j)} \text{ mod } \Lambda^{i+j-2},$$

$$(2.8) \qquad \zeta^{(l)} = \text{component of } \zeta \in \Lambda(n) \text{ in } \Lambda^l(n).$$

Here c is called the Clifford quantification map and σ the Clifford symbol map.

LEMMA 2.3 (Getzler [Ge1]). *For any a in* $\text{End} \$_n$ *we have:*

$$(2.9) \qquad \text{Str } a = (-2i)^{\frac{n}{2}} \sigma[a]^{(n)}.$$

From all this we see that as $t \to 0^+$ we have:

$$(2.10) \qquad \text{Str } k_t(0,0) = (-2i)^{\frac{n}{2}} \sigma[K_Q(0,0,t)]^{(n)} + O(t^\infty),$$

where Q is any Volterra ΨDO parametrix for $D\!\!\!\!/^2 + \partial_t$.

Now, recall that the Getzler rescaling [Ge2] assigns the following degrees:

$$(2.11) \qquad \deg \partial_j = \frac{1}{2} \deg \partial_t = \deg c(dx^j) = -\deg x^j = 1,$$

while $\deg B = 0$ for any $B \in M_p(\mathbb{C})$. It defines a filtration of Volterra ΨDO's with coefficients in $\text{End}(\$_n \otimes \mathbb{C}^p) \simeq \text{Cl}(\mathbb{R}^n) \otimes M_p(\mathbb{C})$ as follows.

Let $q \sim \sum_{l \leq m'} q_l$ be a Volterra symbol with values in $\text{End} \$_n \otimes M_p(\mathbb{C})$. Then taking components in each subspace $\Lambda^{(j)}(n)$ and using Taylor expansions near $x = 0$ allows us to get formal expansions,

$$(2.12) \qquad \sigma[q] \sim \sum_{l \leq m'} \sigma[q_l] \sim \sum_{l,j} \sigma[q_l]^{(j)} \sim \sum_{l,j,\alpha} \frac{x^\alpha}{\alpha!} \sigma[\partial_x^\alpha q_l(0,\xi,\tau)]^{(j)}.$$

Observe that each symbol $\frac{x^\alpha}{\alpha!} \sigma[\partial_x^\alpha q_l(0,\xi,\tau)]^{(j)}$ is Getzler homogeneous of degree $-|\alpha| + l + j$. Therefore, we have

$$(2.13) \qquad \sigma[q(x,\xi,\tau)] \sim \sum_{j \geq 0} q_{(m-j)}(x,\xi,\tau), \qquad q_{(m)} \neq 0,$$

where $q_{(m-j)}$ is a Getzler homogeneous symbol of degree $m - j$.

DEFINITION 2.4. *Using (2.13) we set-up the following definitions:*

- *The integer m is the Getzler order of Q,*
- *The symbol $q_{(m)}$ is the principal Getzler homogeneous symbol of Q,*
- *The operator $Q_{(m)} = q_{(m)}(x, D_x, D_t)$ is the model operator of Q.*

EXAMPLE 2.5. The operator $D\!\!\!\!/^2$ has Getzler order 2 and its model operator is

$$(2.14) \qquad D\!\!\!\!/^2_{(2)} = H_R + F^{\mathcal{E}}(0), \qquad H_R = -\sum_{i=1}^n (\partial_i - \frac{1}{4} R^M_{ij}(0)x^j)^2,$$

$$(2.15)$$

$$R^M_{ij}(0) = \frac{1}{2} \langle R^M(\partial_i, \partial_j)\partial_k, \partial_l \rangle(0)dx^k \wedge dx^l, \qquad F^{\mathcal{E}}(0) = \frac{1}{2} F^{\mathcal{E}}(\partial_k, \partial_l)(0)dx^k \wedge dx^l.$$

This uses the behavior near $x = 0$ in normal coordinates of the metric and of the coefficients of the Levi-Civita connection in the synchronous frame ([Ge2], [Po]).

The interest of the above definitions stems from the two results below.

LEMMA 2.6 ([**Po**]). *For $j = 1, 2$ let Q_j be a Volterra ΨDO of Getzler order m_j. Then*

(2.16) $$Q_1 Q_2 = c(Q_{(m_1)} Q_{(m_2)}) + O_G(m_2 + m_2 - 1),$$

where $O_G(m_2 + m_2 - 1)$ has Getzler order $\leq m_2 + m_2 - 1$.

LEMMA 2.7 ([**Po**]). *Let Q have Getzler order m and model operator $Q_{(m)}$.*
- *If $m - j$ is odd, then $\sigma[K_Q(0, 0, t)]^{(j)} = O(t^{\frac{j-m-n-1}{2}})$.*
- *If $m - j$ is even, then*

$$\sigma[K_Q(0, 0, t)]^{(j)} = t^{\frac{j-m-n}{2}-1} K_{Q_{(m)}}(0, 0, 1)^{(j)} + O(t^{\frac{j-m-n}{2}}).$$

In particular, for $m = -2$ and $j = n$ we get:

(2.17) $$\sigma[K_Q(0, 0, t)]^{(n)} = K_{Q_{(-2)}}(0, 0, 1)^{(n)} + O(t).$$

In particular, using Lemma 2.6 we can easily get:

LEMMA 2.8 ([**Po**]). *Let Q be a Volterra ΨDO parametrix for $\not{D}^2 + \partial_t$. Then:*

1) Q has Getzler order 2 and model operator $Q_{(-2)} = (H_R + F^{\mathcal{E}}(0) + \partial_t)^{-1}$.

2) $K_{Q_{(-2)}}(x, 0, t) = G_R(x, t) \wedge \exp(-t F^{\mathcal{E}}(0))$, where $G_R(x, t)$ is the fundamental solution of $H_R + \partial_t$, i.e. $(H_R + \partial_t) G_R(x, t) = \delta(x, t)$, where $\delta(x, t)$ is the Dirac distribution at the origin.

Noticing that H_R is a harmonic oscillator associated to the curvature at $x = 0$ we get:

LEMMA 2.9 (Melher's Formula). *We have:*

(2.18) $$G_R(x, t) = \chi(t)(4\pi t)^{-\frac{n}{2}} \hat{A}(t R^M(0)) \exp(-\frac{1}{4t} \langle \frac{t R^M(0)/2}{\tanh(t R^M(0)/2)} x, x \rangle),$$

where $\chi(t)$ is the characteristic function of the interval $(0, +\infty)$.

Now, from (2.10) and Lemma 2.7 we deduce
(2.19)
$$\text{Str } k_t(0, 0) = (-2i)^{\frac{n}{2}} \sigma[K_Q(0, 0, t)]^{(n)} + O(t^\infty) = (-2i)^{\frac{n}{2}} K_{Q_{(-2)}}(0, 0, 1)^{(n)} + O(t).$$

Thus using Lemma 2.8 and Lemma 2.9 we get:

(2.20) $$\text{Str } k_t(0, 0) = (-2i)^{\frac{n}{2}} [G_R(0, 1) \wedge \exp(-F^{\mathcal{E}}(0))]^{(n)} + O(t),$$

$$= (2i\pi)^{-\frac{n}{2}} [\hat{A}(R_M(0)) \wedge \text{Ch}(F^{\mathcal{E}}(0))]^{(n)} + O(t).$$

Comparing this with the heat kernel asymptotics (1.6) for $\not{D}^2$ we get Theorem 2.2, and so complete the proof of the Atiyah-Singer index formula.   □

The main new feature in the previous proof of the Atiyah-Singer index formula is the use of Lemma 2.7 which, by very elementary considerations on Getzler orders shows that the convergence of the supertrace of the heat kernel is a consequence of a general fact about Volterra ΨDO's. It incidentally gives a differentiable version of Theorem 2.2 as follows.

PROPOSITION 2.10 ([**Po**]). *Let $\not{P}$ be a differential operator of Getzler order m and let $h_t(x,y)$ denote the distribution kernel of $\not{P}e^{-t\not{D}}{}_M$.*

- *If m is odd, then* $\operatorname{Str} h_t(x,x) = O(t^{\frac{-m+1}{2}})$.

- *If m is even, then* $\operatorname{Str} h_t(x,x) = t^{\frac{-m}{2}} B_0(\not{D}_{\mathcal{E}}^2, \not{P})(x) + O(t^{\frac{-m}{2}+1})$, *where in synchronous normal coordinates centered at x_0 we have*

$$B_0(\not{D}_{\mathcal{E}}^2, \not{P})(0) = (-2i\pi)^{\frac{n}{2}} [(\not{P}_{(m)} G_R)(0,1) \wedge \operatorname{Ch}(F^{\mathcal{E}}(0))]^{(n)}$$

with $\not{P}_{(m)}$ denoting the model operator of P.

PROOF. By Proposition 1.10 we have

$$(2.21) \qquad h_t(x,x) = K_{\not{P}(\not{D}_{\mathcal{E}}^2 + \partial_t)^{-1}}(x,x,t) \sim_{t\to 0^+} \sum_{j\geq 0} t^{-(\frac{n}{2}+[\frac{m'}{2}])} b_j(P,\Delta)(x),$$

with asymptotics in $C^\infty(M, |\Lambda| \otimes \operatorname{End} \not{S} \otimes \operatorname{End} \mathcal{E})$. On the other hand, using Lemma 2.6 we see that $\not{P}(\not{D}_{\mathcal{E}}^2 + \partial_t)^{-1}$ has Getzler order $m - 2$. Combining all this with Lemma 2.7 gives the result. $\qquad\square$

3. The local index formula in noncommutative geometry

Rather than at topological or geometrical levels the local index formula holds in full generality in a purely operator theoretic setting ([**CM**]; see also [**Hi**]). This uses two main tools, namely, spectral triples [**CM**] and cyclic cohomology [**Co**].

A *spectral triple* is a triple $(\mathcal{A}, \mathcal{H}, D)$ where:

- $\mathcal{H}$ is a Hilbert space together with a $\mathbb{Z}_2$-grading $\gamma : \mathcal{H}_+ \oplus \mathcal{H}_- \to \mathcal{H}_- \oplus \mathcal{H}_+$;

- $\mathcal{A}$ is an involutive unital algebra represented in $\mathcal{H}$ and commuting with the $\mathbb{Z}_2$-grading γ;

- D is a selfadjoint unbounded operator on $\mathcal{H}$ s.t. $[D, a]$ is bounded $\forall a \in \mathcal{A}$ and of the form,

$$(3.1) \qquad D = \begin{pmatrix} 0 & D^- \\ D^+ & 0 \end{pmatrix}, \quad D_\pm : \mathcal{H}^\mp \to \mathcal{H}^\pm.$$

In addition we assume that $\mathcal{A}$ is *smooth* in the sense that $\mathcal{A}$ is contained in $\cap_{k\geq 0} \operatorname{dom} \delta^k$.

Recall that the datum of D above defines an index map $\operatorname{ind}_D : K_*(\mathcal{A}) \to \mathbb{Z}$ so that

$$(3.2) \qquad \operatorname{ind}_D[e] = \operatorname{ind} eD^+e,$$

for any selfadjoint idempotent $e \in M_q(\mathcal{A})$.

EXAMPLE 3.1 (Dirac spectral triple). A typical spectral triple is a triple

$$(C^\infty(M), L^2(M, \not{S}), \not{D}_M),$$

where M is a compact Riemannian spin manifold of even dimension, $\not{S}$ is the spin bundle of M and $\not{D}_M : C^\infty(M, \not{S}) \to C^\infty(M, \not{S})$ is the Dirac operator on M. Moreover, under the Serre-Swan isomorphism $K_0(C^\infty(M)) \simeq K^0(M)$ for any Hermitian bundle over M we have:

$$(3.3) \qquad \operatorname{ind}_{\not{D}_M}[\mathcal{E}] = \operatorname{ind} \not{D}_{\mathcal{E}}.$$

Next, the *cyclic cohomology groups* $HC^*(\mathcal{A})$ are obtained from the spaces,

(3.4) $C^k(\mathcal{A}) = \{(k+1)\text{-linear forms on } \mathcal{A}\}, \qquad k \in \mathbb{N},$

by restricting the Hochschild coboundary,

(3.5)
$$
\begin{aligned}
b\psi(a^0, \cdots, a^{k+1}) =\ & (-1)^{k+1}\psi(a^{k+1}a^0, \cdots, a^k) \\
& + \sum(-1)^j \psi(a^0, \cdots, a^j a^{j+1}, \cdots, a^{k+1}),
\end{aligned}
$$

to cyclic cochains, i.e. those such that

(3.6) $\psi(a^1, \cdots, a^k, a^0) = (-1)^k \psi(a^0, a^1, \cdots, a^k).$

Equivalently, $HC^*(\mathcal{A})$ can be obtained as the second filtration of the (b, B)-bicomplex of (arbitrary) cochains, where the horizontal differential B is given by

(3.7) $B = AB_0 : C^{m+1}(\mathcal{A}) \to C^m(\mathcal{A}),$

(3.8) $(A\phi)(a^0, \cdots, a^m) = \sum (-1)^{mj} \psi(a^j, \cdots, a^{j-1}),$

(3.9) $B_0\psi(a^0, \cdots, a^m) = \psi(1, a^0, \cdots, a^m).$

The *periodic cyclic cohomology* is the inductive limit of the groups $HC^k(\mathcal{A})$ with respect to the the cup product with the generator of $HC^2(\mathbb{C})$. In terms of the (b, B)-bicomplex this is the cohomology of the short complex,

(3.10) $C^{\mathrm{ev}}(\mathcal{A}) \overset{b+B}{\leftrightarrows} C^{\mathrm{odd}}(\mathcal{A}), \qquad C^{\mathrm{ev/odd}}(\mathcal{A}) = \bigoplus_{k \text{ even/odd}} C^k(\mathcal{A}),$

the cohomology groups of which are denoted $HC^{\mathrm{ev}}(\mathcal{A})$ and $HC^{\mathrm{odd}}(\mathcal{A})$.

We have a pairing between $HC^{\mathrm{ev}}(\mathcal{A})$ and $K_0(\mathcal{A})$ such that, for any cocycle $\varphi = (\varphi_{2k})$ in $C^{\mathrm{ev}}(\mathcal{A})$ and any selfadjoint idempotent e in $M_q(\mathcal{A})$, we have

(3.11) $\langle [\varphi], [e] \rangle = \sum_{k \geq 0} (-1)^k \frac{(2k)!}{k!} \varphi_{2k} \# \operatorname{Tr}(e, \cdots, e),$

where $\varphi_{2k} \# \operatorname{Tr}$ is the $(2k+1)$-linear map on $M_q(\mathcal{A}) = M_q(\mathbb{C}) \otimes \mathcal{A}$ given by

(3.12) $\varphi_{2k} \# \operatorname{Tr}(\mu^0 \otimes a^0, \cdots, \mu^{2k} \otimes a^{2k}) = \operatorname{Tr}(\mu^0 \dots \mu^{2k}) \varphi_{2k}(a^0, \cdots, a^{2k}),$

for $\mu^j \in M_q(\mathbb{C})$ and $a^j \in \mathcal{A}$.

EXAMPLE 3.2. Given a compact manifold M let $\mathcal{A}$ be the algebra $C^\infty(M)$ and let $\mathcal{D}_k(M)$ denote the space of k-dimensional de Rham currents. Then we have a morphism from $\mathcal{D}_{\mathrm{ev/odd}}(M)$ to $C^{\mathrm{ev/odd}}(\mathcal{A})$,
(3.13)
$$
C = (C_k) \to \varphi_C = (\tfrac{1}{k!}\psi_{C_k}), \qquad \psi_{C_k}(f^0, f^1, \dots, f^n) = \langle C_k, f^0 df^1 \wedge \dots \wedge df^k \rangle.
$$

such that $b\varphi_C = 0$ and $B\varphi_C = \varphi_{d^t C}$, where d^t denotes the de Rham boundary for currents. Therefore, we get a morphism from the even and odd de Rham's homology groups $H^{\mathrm{ev/odd}}(M)$ to $HC^{\mathrm{ev/odd}}(C^\infty(M))$. This even gives rise to an isomorphism if we restrict ourselves to continuous cyclic cochains.

Furthermore, we have:

(3.14) $\langle [\varphi_C], \mathcal{E} \rangle = \langle [C], \operatorname{Ch}^* \mathcal{E} \rangle \quad \forall \mathcal{E} \in K^0(M),$

where $\operatorname{Ch}^* : K^0(M) \to H^{\mathrm{ev}}(M)$ is the Chern character in cohomology.

Now, we can compute the index map by pairing the K-theory a cyclic cocycle as follows. First, we say that the spectral triple $(\mathcal{A}, \mathcal{H}, D)$ is p-*summable* when we have

$$(3.15) \qquad \mu_k(D^{-1}) = O(k^{-1/p}) \quad \text{as } k \to +\infty,$$

where $\mu_k(D^{-1})$ denotes the $(k+1)$'th eigenvalue of D^{-1}.

Next, let $\Psi^0_D(\mathcal{A})$ be the algebra generated by the $\mathbb{Z}_2$-grading γ and the $\delta^k(a)$'s, $a \in \mathcal{A}$, where δ is the derivation $\delta(T) = [|D|, T]$.

DEFINITION 3.3. *The* dimension spectrum *of* $(\mathcal{A}, \mathcal{H}, D)$ *is the union set of the singularities of all the zeta functions* $\zeta_b(z) = \operatorname{Tr} b|D|^{-z}$, $b \in \Psi^0_D(\mathcal{A})$.

Assuming p-summability and simple and discrete dimension spectrum we define a residual trace on $\Psi^0_D(\mathcal{A})$ by letting:

$$(3.16) \qquad \fint b = \operatorname{Res}_{z=0} \operatorname{Tr} b|D|^{-z} \qquad \text{for any } b \in \Psi^0_D(\mathcal{A}).$$

This trace is an algebraic analogue of the Wodzicki-Guillemin noncommutative residue trace ([**Wo**], [**Gu**]) for the algebra $\Psi^0_D(\mathcal{A})$. Moreover, it is local in the sense of noncommutative geometry since it vanishes on those elements of $\Psi^0_D(\mathcal{A})$ that are traceable.

THEOREM 3.4 ([**CM**]). *Suppose that* $(\mathcal{A}, \mathcal{H}, D)$ *is p-summable and has a discrete and simple dimension spectrum. Then:*

1) The following formulas define an even cocycle $\varphi_{\mathrm{CM}} = (\varphi_{2k})$ *in the* (b, B)-*complex of* $\mathcal{A}$.

- *For $k = 0$, we let*

$$(3.17) \qquad \varphi_0(a^0) = \text{finite part of } \operatorname{Tr} \gamma a^0 e^{-tD^2} \text{ as } t \to 0^+,$$

- *For $k \neq 0$, we let*

$$(3.18) \qquad \varphi_{2k}(a^0, \ldots, a^{2k}) = \sum_\alpha c_{k,\alpha} \fint \gamma P_{k,\alpha} |D|^{-2(|\alpha|+k)},$$

$$(3.19) \qquad P_{k,\alpha} = a^0 [D, a^1]^{[\alpha_1]} \ldots [D, a^{2k}]^{[\alpha_{2k}]},$$

where $\Gamma(|\alpha| + k)c_{k,\alpha}^{-1} = 2(-1)^{|\alpha|}\alpha!(\alpha_1 + 1) \cdots (\alpha_1 + \cdots + \alpha_{2k} + 2k)$ *and the symbol* $T^{[j]}$ *denotes the j'th iterated commutator with* D^2.

2) We have $\operatorname{ind}_D[\mathcal{E}] = \langle[\varphi_{\mathrm{CM}}], \mathcal{E}\rangle$ *for any* $\mathcal{E} \in K_0(\mathcal{A})$.

REMARK 3.5. Here we have assumed that the spectral triple $(\mathcal{A}, \mathcal{H}, D)$ was $\mathbb{Z}_2$-graded. In the terminology of [**CM**] such a spectral triple is *even* and an ungraded spectral triple is said to be *odd*. In the latter case we can similarly compute the Fredholm index with coefficients in $K_1(\mathcal{A})$ by pairing $K_1(\mathcal{A})$ with an *odd* cyclic cocycle, also called CM cocycle, and which is given by formulas of the same type as (3.18)–(3.19) (see [**CM**] for a more detailed account).

4. Computation of the CM cocycle for Dirac spectral triples.

Here M^n is a compact Riemannian spin manifold of even dimension and $\not{D}_M :$ $C^\infty(M, \not{S}) \to C^\infty(M, \not{S})$ is the Dirac operator on M. Then $(C^\infty(M), L^2(M, \not{S}), \not{D}_M)$ is an even n-summable spectral triple with dimension spectrum $\{k \in \mathbb{Z}; k \leq n\}$.

Therefore, the CM cocycle is well defined and allows us to compute the index map, so that for any hermitian bundle $\mathcal{E}$ equipped with a unitary connection we have

$$(4.1) \qquad \langle[\varphi_{CM}], [\mathcal{E}]\rangle = \mathrm{ind}_{\not{D}_M}[\mathcal{E}] = \mathrm{ind}\not{D}_{\mathcal{E}}.$$

PROPOSITION 4.1. *The components of the CM cyclic cocycle $\varphi_{CM} = (\varphi_{2k})$ associated to the spectral triple $(C^\infty(M), L^2(M, \$), \not{D}_M)$ are given by*

$$(4.2) \qquad \varphi_{2k}(f^0, \ldots, f^{2k}) = \frac{1}{(2k)!} \int_M f^0 df^1 \wedge \cdots \wedge df^{2k} \wedge \hat{A}(R_M)^{(n-2k)},$$

for $f^0, f^1, \ldots, f^n$ in $C^\infty(M)$.

PROOF. First, using the Mellin transform,

$$(4.3) \qquad |\not{D}_M|^{-2s} = \frac{1}{\Gamma(2s)} \int_0^\infty t^{s-1} e^{-t\not{D}_M^2} dt, \quad \Re s > 0,$$

we can rewrite $\varphi_{CM} = (\varphi_{2k})$ as
$$(4.4)$$
$$\varphi_{2k}(f^0, \ldots, f^{2k}) = \sum_\alpha \frac{2c_{k,\alpha}}{\Gamma(|\alpha| + k)} \cdot \{\text{coeff. of } t^{-(|\alpha|+k)} \text{ in } \mathrm{Str} \not{P}_{k,\alpha} e^{-t\not{D}_M^2} \text{ as } t \to 0^+\}.$$

Notice also that

$$\not{P}_{k,\alpha} = f^0[\not{D}_M, f^1]^{[\alpha_1]} \ldots [\not{D}_M, f^{2k}]^{[\alpha_{2k}]} = f^0 c(df^1)^{[\alpha_1]} \ldots c(df^{2k})^{[\alpha_{2k}]}.$$

Therefore, we can make use of Proposition 2.10 to compute the CM cocycle. Indeed, let $h_{\alpha,t}(x, y)$ denote the kernel of $\not{P}_{k,\alpha} e^{-t\not{D}_M^2}$. Then using Lemma 2.6 and Proposition 2.10 we obtain:

- If $\alpha = 0$, then $\not{P}_{k,0} = f^0 c(df^1) \ldots c(df^{2k})$ has Getzler order $2k$ and model operator $f^0 df^1 \wedge \ldots \wedge df^{2k}$, so that we have

$$(4.5) \qquad \mathrm{Str}\, h_{0,t}(x, x) = \frac{t^{-k}}{(2i\pi)^{\frac{n}{2}}} f^0 df^1 \wedge \ldots \wedge df^{2k} \wedge \hat{A}(R_M)^{(n-2k)} + \mathrm{O}(t^{-k+1}).$$

- If $\alpha \neq 0$, then $\not{P}_{k,\alpha}$ has Getzler order $\leq 2(|\alpha| + k) - 1$, and so

$$(4.6) \qquad \mathrm{Str}\, h_{\alpha,t}(x, x) = \mathrm{O}(t^{-(\alpha+k)+1}).$$

Combining this with (4.4) we get the formula (4.2)) for $\varphi_{2k}(f^0, \ldots, f^{2k})$. $\qquad \square$

Finally, let us explain how this enables us to recover the local index formula of Atiyah-Singer. Notice that Formula (4.2) means that φ_{CM} is the image under the map (3.13) of the current that is the Poincaré dual of $\hat{A}(R^M)$. Therefore, using (3.14) we see that for any Hermitian bundle $\mathcal{E}$ with curvature $F^{\mathcal{E}}$ we have:

$$(4.7) \qquad \langle[\varphi_{CM}], [\mathcal{E}]\rangle = \langle[\varphi_{\hat{A}(R^M)}], [\mathcal{E}]\rangle = (2i\pi)^{-\frac{n}{2}} \int_M [\hat{A}(R_M) \wedge \mathrm{Ch}(F^{\mathcal{E}})]^{(n)}.$$

On the other hand, we have

$$(4.8) \qquad \langle[\varphi_{CM}], [\mathcal{E}]\rangle = \mathrm{ind}_{\not{D}_M}[\mathcal{E}] = \mathrm{ind}\not{D}_{\mathcal{E}}.$$

Thus, we obtain

$$(4.9) \qquad \mathrm{ind}\not{D}_{\mathcal{E}} = (2i\pi)^{-\frac{n}{2}} \int_M [\hat{A}(R_M) \wedge \mathrm{Ch}(F^{\mathcal{E}})]^{(n)},$$

which is precisely the Atiyah-Singer index formula.

REMARK 4.2. When $\dim M$ is odd we can also compute the odd CM cocycle associated to the corresponding Dirac spectral triple ([**Po**]). In this setting we recover the spectral flow formula of Atiyah-Patodi-Singer [**APS**].

Acknowledgment. The work of the author was partially supported by the European RT Network *Geometric Analysis* HPCRN-CT-1999-00118 and the NSF grant DMS 0074062.

References

[ABP]   Atiyah, M., Bott, R., Patodi, V.: *On the heat equation and the index theorem*. Invent. Math. **19**, 279–330 (1973).

[APS]   Atiyah, M., Patodi, V., Singer, I.: *Spectral asymmetry and Riemannian geometry. III.* Math. Proc. Camb. Philos. Soc. **79**, 71–99 (1976).

[AS1]   Atiyah, M., Singer, I.: *The index of elliptic operators. I.* Ann. of Math. (2) **87**, 484–530 (1968).

[AS2]   Atiyah, M., Singer, I.: *The index of elliptic operators. III.* Ann. of Math. (2) **87**, 546–604 (1968).

[BGS]   Beals, R., Greiner, P., Stanton, N.: *The heat equation on a CR manifold*. J. Differential Geom. **20**, 343–387 (1984).

[BGV]   Berline, N., Getzler, E., Vergne, M.: *Heat kernels and Dirac operators*. Springer-Verlag, Berlin, 1992.

[Bi]   Bismut, J.-M.: *The Atiyah-Singer theorems: a probabilistic approach. I. The index theorem.* J. Funct. Anal. **57**, 56–99 (1984) .

[CH]   Chern, S., Hu, X.: *Equivariant Chern character for the invariant Dirac operator*. Michigan Math J. **44**, 451–473 (1997).

[Co]   Connes, A.: *Noncommutative geometry*. Academic Press, San Diego, 1994.

[CM]   Connes, A., Moscovici, H.: *The local index formula in noncommutative geometry*. GAFA **5**, 174–243 (1995).

[CFKS]   Cycon, H. L.; Froese, R. G.; Kirsch, W.; Simon, B.: *Schrödinger operators with application to quantum mechanics and global geometry*. Texts and Monographs in Physics. Springer-Verlag, Berlin, 1987.

[Ge1]   Getzler, E.: *Pseudodifferential operators on supermanifolds and the Atiyah-Singer index theorem*. Comm. Math. Phys. **92**, 163–178 (1983).

[Ge2]   Getzler, E.: *A short proof of the local Atiyah-Singer index theorem*. Topology **25**, 111–117 (1986).

[Gi]   Gilkey, P.: *Invariance theory, the heat equation, and the Atiyah-Singer index theorem*. Publish or Perish, 1984.

[Gr]   Greiner, P.: *An asymptotic expansion for the heat equation*. Arch. Rational Mech. Anal. **41**, 163–218 (1971).

[Gu]   Guillemin, V.: *A new proof of Weyl's formula on the asymptotic distribution of eigenvalues*. Adv. in Math. **55**, 131–160 (1985).

[Hi]   Higson, N.: *The local index formula in noncommutative geometry*. Lectures given at the CIME Summer School and Conference on algebraic K-theory and its applications, Trieste, 2002. Preprint, 2002.

[Le]   Lescure, J.-M.: *Triplets spectraux pour les variétés à singularité conique isolée*. Bull. Soc. Math. France **129**, 593–623 (2001).

[LM]   Lawson, B., Michelson, M.-L.: *Spin Geometry*. Princeton Univ. Press, Princeton, 1993.

[Me]   Melrose, R.: *The Atiyah-Patodi-Singer index theorem*. A.K. Peters, Boston, 1993.

[Pi]   Piriou, A.: *Une classe d'opérateurs pseudo-différentiels du type de Volterra*. Ann. Inst. Fourier **20**, 77–94 (1970).

[Po]   Ponge, R.: *A new short proof of the local index formula and some of its applications*. Comm. Math. Phys. **241** (2003) 215–234.

[Ro]   Roe, J.: *Elliptic operators, topology and asymptotic methods*. Pitman Research Notes in Mathematics Series 395, Longman, 1998.

[Ta]   Taylor, M. E.: *Partial differential equations. II. Qualitative studies of linear equations*. Applied Mathematical Sciences, 116. Springer-Verlag, New York, 1996.

[Wo]   Wodzicki, M.: *Noncommutative residue. I. Fundamentals. K-theory, arithmetic and geometry* (Moscow, 1984–1986), 320–399, Lecture Notes in Math., 1289, Springer, 1987.

DEPARTMENT OF MATHEMATICS, OHIO STATE UNIVERSITY, COLUMBUS, OH 43210, USA
E-mail address: ponge@math.ohio-state.edu